Acknowledgements

In the production of this book, many people have helped me.

I am especially grateful to those workshop proprietors who over the years have allowed me full access to their facilities for photography and the collection of information — including **Leon Vincenzi** (Awesome Automotive), **John Keen** (DAT Racing), **Craig** and **Adam Allan** (Allan Engineering), **Bill Keen** (Adelaide Turbo Services), **Bill Hanson** (Bill Hanson Engine Developments) and **David Keen** (Turbo Tune).

My thanks also to **Clive Humphris** (Ford Motor Company of Australia) who commented on the Aerodynamics chapter, and **Jim Gurief** (Whiteline Suspension) who read the chapter on Tyres, Suspension and Brakes. However, any errors present in these chapters are, of course, mine!

Robert Edgar and **Jamie Campbell** read each of the chapters and made detailed and insightful comments. **John Keen** read many chapters and also provided valuable information, advice and feedback. **Leo Simpson** gave permission for the use of artwork previously published in *Silicon Chip* magazine, as did **Dean Evans** for artwork from *Fast Fours & Rotaries* magazine. **David Bryant** and **Michael Knowling** freely contributed their photographs.

My heartfelt thanks to each of these people.

Author

Julian Edgar

Illustrations

David Heinrich

Photography

Julian Edgar, David Bryant, Michael Knowling, Greg Brindley, Georgina Cobbin

Design

David Heinrich

Production Editor

Kerry Boyne

Publisher

Martin White

21st Century PERFORMANCE

Published by **Clockwork Media** Pty Ltd, PO Box 5277 Erina Fair NSW 2250 Australia
Phone (02) 4367 3751, Fax (02) 4365 3191.
Printed by Times Offset (M) Sdn Bhd.
Distribution enquiries to Head Office 001161(02) 4367 3751.

ISBN 0 947216 90 1

ACN 090 383 225

Introduction

Cars now use electronics to control the addition of fuel to the engine, to set the ignition timing advance angle, to monitor idle speed and perhaps to control turbo/supercharger boost and camshaft timing. The engine is covered with sensors measuring coolant temperature, intake air temperature, crankshaft and camshaft position, throttle opening, the occurrence of detonation — and a host of other factors. The drive system may be to all four wheels, and all of those may be steered.

But does all this make modification of the car impossible? No! Does it make DIY budget changes impossible? Again, no! Does it mean you need to have a degree in electronic engineering before you can modify the car? No!

But you must know what to do.

It's a fact that many people make changes that are simply unnecessary. They have a new exhaust fitted without checking how well the standard system flows. They buy expensive intercoolers without ever measuring how well the factory intercooler is working. And since both checks cost very little, that's simply crazy! Or they spend money on a random this-is-sure-to-make-it-go-hard basis. Such as dropping a new aftermarket filter into the standard box. Or fitting a product a friend swears has made a huge improvement to his/her car — but who has no evidence of any performance gain.

Making the most of what you have — both in your pocket and under your bonnet — comes from finding out what **really** works before spending the money. Measuring the performance of a car is vital. You can easily test:

- On-road performance
- Engine power
- Intake system performance
- Exhaust system performance
- Aerodynamic flows

and a host of other factors cheaply and easily. And with that information available, discovering what changes are needed becomes a helluva lot easier!

In this book I have attempted to give you only information that is directly useful in letting you improve your car's performance. This book is aimed at the normal car owner who wants to make it perform better. It doesn't assume you are a fully qualified mechanic — just that you're interested and enthusiastic.

In these pages I have resisted the urge to throw numerous mathematical equations at you and to fill the book with theory that looks impressive when you first flip through it but which you'd never bother reading again. I have tried to be practical with the content; to cover approaches and techniques you can read about, reflect on and **then go out and do!**

For some people, this book will not be technical enough. If this is the case, I suggest you go to a good library and consult the many available books on automotive theory — the range of publications from Robert Bosch GmbH and the technical papers published by the Society of Automotive Engineers make excellent starting points. The SAE also has an excellent website. If, on the other hand, you believe this book is **too** technical, concentrate on the direct suggestions — things like making a good cold air intake for your engine, picking the right mufflers to give the best flow and avoiding spending money you can better keep in your pocket.

In this book I have not listed suppliers of parts — such lists invariably date rapidly. By far the best way of finding current suppliers is to use the World Wide Web, where all major sources of automotive performance parts have sites. The Web is also an excellent way of communicating with others who have modified the same sort of car as you have. However, always assess the credibility of information provided before following any advice!

Some readers will be annoyed by the mix of units I have used — kilowatts and horsepower for power, inches for exhaust pipe diameters, pounds per square inch for boost pressures. Where it has been easy to include, I have listed both metric and Imperial units in the text. Where it hasn't been straightforward and measurements are expressed in units you're not familiar with, you are directed towards the Appendix at the back of the book where you will find an extensive series of conversion tables.

It's been suggested that the car is the most complex mechanical/electronic device in mass production in the 21st century. And those who say this are probably right. But it can also be the most satisfying device to modify — not only giving you a feeling of achievement when you improve performance but also raising the hairs on the back of your neck when that newfound urge pushes you down the road at a rate that's simply awesome...

There's nothing like modifying your own car and feeling the difference. Enjoy!

Julian Edgar

Contents

Engine & Driveline

Alfa Romeo

THE ENGINE

In the distant past, the gaining of more power automatically required a complete engine disassembly. The blueprinting of an engine (that is, returning it to the manufacturer's **standard** specs!) was obligatory because engines were so badly assembled in the first place. Porting and polishing of the head(s), boring of the main bearing tunnels, prepping of the rods and crank, all consumed many hours and much money.

These days, most standard engines are superbly designed and built. In fact, those of the 'old school' are often astonished when they first see inside a high-performance 21st century engine. The quality of the components used, the engineering and the sheer gas flow of these engines are extremely impressive. Consequently, to improve substantially on what has been achieved by the manufacturer (in terms of both build quality and engineering) requires incredibly exacting work up to professional race standards — and the extensive development and testing of new components. Many rebuilt, modified late-model engines are less durable and often no more powerful than standard!

As an example, take the Honda S2000's engine. In standard form, the 2-litre, naturally aspirated four develops 180kW — and with ultra-clean emissions. To suggest you can pull the head off, disappear into the backyard shed and make substantial power improvements by employing your grinder on the ports is the stuff of fantasy. You may be able to achieve a gain of a few per cent or, more likely, you may well go backwards in power. And a similar story applies to other modern engines with high specific power outputs.

So does this mean you are stymied — that there's no point in looking under the bonnet for more power? Not at all! As you will see in subsequent chapters, there are a great many bolt-on mods that can be used to release power. But it **does** mean that if you want to venture inside the engine and perform major work, expect to pay a great deal of money to achieve good results. If these funds aren't available, in general restrict yourself to external bolt-ons, where good power gains can be made much more cheaply.

One modification that's not quite in the 'bolt-on' category — but also requires only a little engine disassembly — is the changing of the camshaft(s). A cam swap can yield good power increases, even in very well-designed engines. This is because an enthusiast like you is usually happy to trade off some idle quality and bottom-end response in exchange for more power.

CAMSHAFTS

To begin to have an understanding of how camshafts affect the development of power, you really need to go back to the four strokes of an Otto Cycle engine. In this description of the basic model of a four-stroke engine, I'll assume it's a two-valves-per-cylinder engine and the logic behind the valve events will initially be kept as simple as possible.

A camshaft consists of a shaft with lobes (the bumps) on it. As the camshaft rotates, the action of the lobes opens and closes the valves. The shape of the lobes and their orientation relative to each other have a dramatic effect on the power development of an engine.

First, when the piston is at Top Dead Centre (TDC) — that's as high up the bore as the piston will go — the inlet valve opens. With the inlet valve open, an air/fuel mixture is drawn into the cylinder as the piston then descends. This is called the **Intake Stroke**. When the piston reaches the bottom of the bore at Bottom Dead Centre (BDC), the intake valve closes, trapping the air/fuel mixture inside the cylinder. The piston then starts to move back up the bore, compressing the air/fuel mix as it does so. This is called the **Compression Stroke**. At or near to TDC (remember, that's Top Dead Centre), the spark plug fires. The resulting controlled explosion pushes the piston down with great force, producing the **Power Stroke**. When the piston reaches the bottom of its travel at BDC (Bottom Dead Centre), the exhaust valve opens. The piston then moves back up the bore, pushing the exhaust gas out through the open exhaust valve. This is called the **Exhaust Stroke**.

The cycles always follow in this sequence: intake > compression > power > exhaust (though, of course, it doesn't matter which one you start with).

In that series of four strokes, the crankshaft turned twice and the camshaft (driven thorough a reduction mechanism) turned once. The two rotations of the crank are expressed as 720 degrees (360 degrees x 2 = 720) of rotation. At a certain point in that 720 degrees of rotation, the inlet valve opened, the inlet valve shut, the exhaust valve opened, and the exhaust valve shut. This means the inlet valve was held open for a number of crankshaft degrees (that is, it was open for the time it took for the crank to turn this number of degrees of rotation) and the exhaust valve was also open for a number of degrees of crank rotation.

From the emphasis I am placing on the **timing** of the events (that is, at what number of crankshaft degrees each opening and closing occurred) you might guess that this is a very important aspect. And you would be right! In addition to the time at which the events occurred, another important measurement is the maximum distance each of the valves was lifted off its seat — how far they were opened, in other words.

Camshaft Terms

A camshaft consists of a shaft with specially shaped bumps on it called lobes. As the shaft turns, the lobes either bear on the ends of the valve stems through bucket tappets or operate a variety of mechanisms that transfer that movement to the valves. In cross-section, the characteristics of the various parts of the cam lobe can be identified as follows.

The base circle is the part of the lobe whose radius from the centre of the cam core is constant — it is the smallest diameter part of the cam lobe. The clearance ramps are

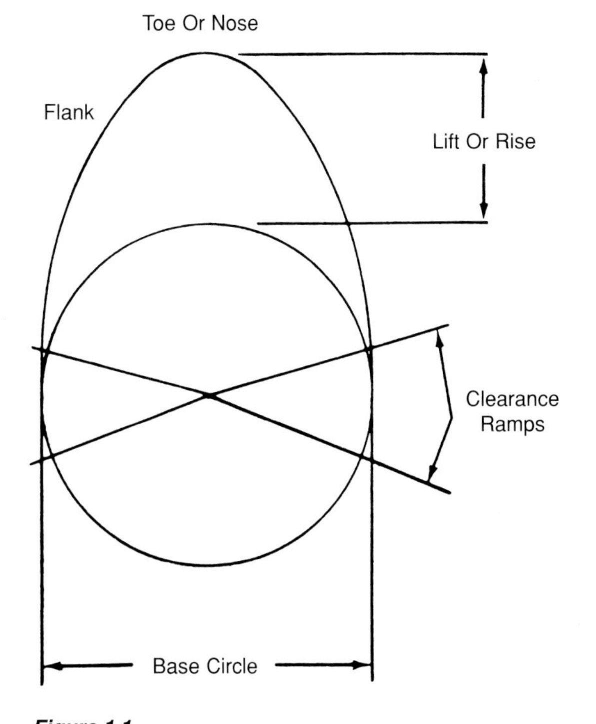

Figure 1.1
The different parts of a cam lobe. Increased duration (the valve is open for longer) is achieved by making the flanks steeper; increased lift (the valve is opened higher) by making the lobe taller; and changed valve timing (the valve opens at a different time relative to the movement of the piston) by grinding the lobe at a different orientation on the camshaft.

The cam on the right is standard while the one on the left is an aftermarket high-performance cam. Note the increased lift (the lobe on the left is slightly taller) and its greater duration (it's broader at the top).

the transitions from the base circle to the lobe proper, while the flank is the part of the lobe between the ramp and the nose. When the lifter (the part of the valve gear that rests on the cam lobe) is in contact with the base circle, the valve is shut. As the cam rotates and the lifter passes onto the clearance ramp, the valve starts to accelerate open. The opening acceleration continues as the lifter passes across the flank with the valve lifted further and further off its seat, the rate slowing as the lifter reaches the nose (toe) of the lobe. For the instant the lifter is sitting right on top of the nose, the valve is stationary, before it then begins to accelerate shut as the cam continues to turn. Figure 1.1 (previous page) shows these parts of the cam lobe.

Changing the height of the lobe will change the amount of valve lift that occurs, but how is the length of time the valve is held open changed? A cam needs to have a base circle of 280-320 crankshaft degrees. This gives time for the shut exhaust valve to transfer heat to the valve seat, and for the complete valvetrain to recover from the violent shock of opening and closing the valve. The ramp angle on a production cam will have 60-80 crankshaft degrees of duration, while the flank will be on average 120-140 degrees of crank duration. To increase the opening period of the valves, the cam needs to have decreased duration on the ramp, reduced to perhaps 40-60 degrees of crank duration. This means the valve is opened faster so it can be kept open longer.

Valve Timing

In engines not using variable valve timing, the times at which valves open and shut are dictated by the mechanical design of the camshaft. This (along with the amount of lift) is often called the camshaft 'grind'. The grind of the cam is extremely important because all is not as it first seems.

Remember how in an earlier paragraph I said the inlet valve closes at Bottom Dead Centre? I lied! In fact, the inlet valve usually closes quite a few crankshaft degrees after BDC, when the piston is already on its way up the bore. But why?

Inlet Valve Closing

If the inlet valve is still open and the piston has started heading upwards, you'd expect lots of the air/fuel mixture that had already been inhaled to be pushed out through the open inlet valve, wouldn't you? It's very important to understand why this **doesn't** happen. When the Intake Stroke is occurring, the air/fuel mix is rushing into the cylinder extremely quickly. In fact, in well-designed engines it's piling through the intake ports at about 800km/h. The reason it's rushing in is that the descending piston has created a partial vacuum in the cylinder and the air/fuel mix is flowing in to fill this vacant space.

This torrent of moving air has mass — and so momentum. This means that it keeps on flowing when the piston has stopped travelling downwards, **and even when it has started on its way up again!** However, as you'd expect, this situation doesn't go on forever — at some point, the rising piston will start pushing the air/fuel mix out through the intake valve. When this process occurs, it's called reversion. The best cylinder filling occurs when the intake valve closes sufficiently late that a touch of reversion is allowed to occur.

The point at which the intake valve closure occurs can be optimal for only one engine rpm. If the engine speed is high, the column of air entering the engine is travelling very fast and has more momentum. For this reason the intake valve closure can be delayed longer. However, at low engine speeds, this same closing point will allow lots of reversion — and so at low revs the engine will be down in power because it's not breathing its full quantity of air/fuel mix. **The time at which the intake valve closes is the single most important aspect of cam event timing.**

Exhaust Valve Opening

So if the intake valve doesn't close until **after** BDC, what about the exhaust valve which I said opens at BDC? If you guessed it doesn't, you're right. In fact, the exhaust valve opens **before** BDC is reached. The piston is descending on its power stroke, hot expanding gas pushing it down to develop power and push the car along. So why would you open this valve early and let some of that pressure on the piston escape?

To understand why the exhaust valve opens a little before BDC it helps to think of what the **next** stroke is. It's the exhaust stroke, where the piston is travelling upwards, pushing the exhaust gas out through the open exhaust valve. To push this exhaust gas out requires power, power that's taken from the crankshaft and so lost from the engine's output. Opening the exhaust valve when there is still some residual pressure in the cylinder allows some of these exhaust gases to escape before the piston starts its upwards stroke. This process is called blowdown. With some of the exhaust gases already gone, the pumping losses that occur as the piston rises (with less exhaust gas to push out of the cylinder) are reduced.

Another reason for an early opening of the exhaust valve is that when the first burst of exhaust gas passes into the exhaust, it creates a very strong pressure wave. If tuned length extractors (headers) are being used, the time that this first pressure wave can escape will determine when the exhaust cycle its reflection comes back to the exhaust valve. (That will also depend on the extractor design, of course.) If the opening of the exhaust valve is sufficiently early for a strong scavenging (sucking) wave to arrive back at the exhaust valve while it is still open, then even more exhaust gas will be encouraged to flow out of it.

The correct timing of the exhaust valve opening is a trade-off between losing some of the power-producing gas pressure of the power stroke and reducing some of the pumping load. Again, the timing which is optimal for this event varies, being dependent on engine speed.

Overlap

So the piston is on its way up, pushing out what remains of the exhaust gas after blowdown. When it gets to TDC, the exhaust valve will shut and the intake valve will open, right? Not quite! The exhaust gas is rushing out of the exhaust valve as the piston rises. Just like the intake air stream, this rush of gas has momentum. That means it keeps going, even after the piston has reached the top of its travel and started to descend. As a result, its closure can be delayed until after TDC.

But we want to make sure that as much fresh air/fuel is breathed in during the Intake Stroke as possible. After all, it is the burning of that mixture that's going to develop the pressure on the piston crown! To accomplish the best inhalation, the intake valve is opened **before** TDC. That's right — both the exhaust valve and the inlet valve are open at the same time! This is called the overlap period.

So why doesn't the rising piston push the exhaust gases straight out through the open intake valve? After all, it's pushing the exhaust gas out past the exhaust valve. One reason it doesn't is that a well-designed set of extractors (or, to a lesser extent, any good exhaust manifold) will have a low pressure, scavenging wave arrive during the overlap. This will help draw exhaust gases out through the exhaust valve faster than they would otherwise travel. The rapid flow of exhaust gases out through the exhaust valve will also keep going through its own momentum, creating a low pressure in the cylinder which is filled by the fresh air/fuel mix flowing in through the inlet valve.

However, at the wrong engine speeds, or if a scavenging is not effective in the exhaust system, exhaust gases will be driven out through the intake valve. This will cause reversion of the intake air/fuel mixture, resulting in over-rich mixtures on the next (successful) intake stroke. Emissions can suffer as a result. (However, this internal exhaust gas recirculation can also be used to reduce oxides of nitrogen emissions in engines with variable cam timing, as discussed below.) Engines with a long overlap are peaky in their power production, but can often develop high maximum power. Engines with shorter overlap periods have exhaust scavenging pressure waves that stay effective over a wider rpm band, but peak power is not as high.

So the exhaust valve shuts a little after the inlet valve opens, which in turn stays open a little longer than you'd initially expect it to ... and then you go back to the begining of the process again!

Cam Designs

While I have covered the basic opening and closing logic of the valves, what aspects of the mechanical design of the cam determine when these occur? Again, for the sake of simplicity, I'll use a single-cam engine as the example.

In a single-cam engine the following events are for the most part fixed by the mechanical design of the cam:

- Intake valve opening
- Intake valve closing
- Exhaust valve opening
- Exhaust valve closing
- Intake valve lift
- Exhaust valve lift

From these parameters the following cam specs can be derived:

- Intake duration
- Exhaust duration
- Lobe centre angle
- Valve overlap
- Intake centre angle (or centreline)
- Exhaust centre angle (or centreline)

It's very easy to become confused by all these terms, especially when they are abbreviated and/or have numbers associated with them, in quick conversation! The first six parameters I have already briefly explained. But what about the second six? After all, these are more often quoted in cam catalogs.

To envisage what each of these factors means, it helps to examine a cam timing diagram. On this type of diagram, crankshaft degrees are marked off on the horizontal axis, with 0 degrees (Top Dead Centre) placed in the middle. Valve lift is shown on the vertical axis, expressed most often in thousandths of an inch — Figure 1.2 (over the page) shows a typical example.

Start at 0 degrees crankshaft rotation — that's TDC. At this point both the exhaust and intake valves are open during the overlap period. In fact, the intake valve started opening 24 degrees before TDC (abbreviated to BTDC), while the exhaust valve has another 24 degrees of opening time left. In this case, the exhaust valve is said to close 24 degrees after TDC (ATDC).

Following across to the right, the intake valve opens more and more until, at 108 degrees after TDC, the intake valve reaches its maximum lift of 0.450 inches. It then starts to close, being fully shut 60 degrees after **Bottom** Dead Centre. So the intake valve started opening 24 degrees before TDC and closed 60 degrees after BDC. How long, then, was it open? This is worked out by: 24 degrees **before** TDC + 180 degrees from TDC to BDC + 60 degrees **after** BDC = 264 degrees inlet valve duration. Look closely at the diagram until that addition is clear in your head!

After the inlet valve has shut, the engine continues through the compression and power strokes (not fully

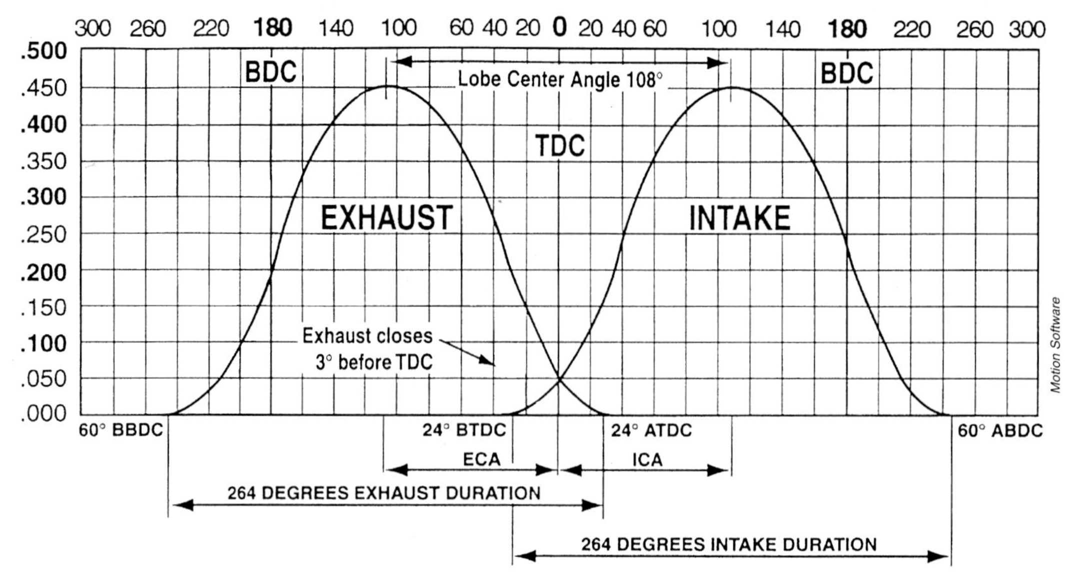

Figure 1.2
The grind of a camshaft can be summarised in a valve timing diagram like this. At 0 degrees crankshaft rotation (TDC), both the exhaust and intake valves are open. The intake valve started opening 24 degrees before TDC (BTDC), while the exhaust valve has another 24 degrees of opening time left. In this case the exhaust valve is said to close 24 degrees after TDC (ATDC). Following across to the right, the intake valve opens until, at 108 degrees after TDC, it reaches its maximum lift of 0.450 inches. It then starts to close, being fully shut 60 degrees after Bottom Dead Centre (ABDC). It was open for 264 crankshaft degrees — so it has 264 degrees duration. After the inlet valve has shut, the engine continues through the compression and power strokes (not fully shown in crankshaft degrees on the diagram) until the exhaust valve starts to open. This occurs with this particular cam 60 degrees before BDC (BBDC). It continues to open until it's fully open at 108 degrees before TDC, shutting 24 degrees after TDC (ATDC). In this case, both the intake and exhaust valves have the same duration.

shown in crankshaft degrees on the diagram) until the exhaust valve starts to open. This occurs with this particular cam 60 degrees **before** BDC. It continues to open until it is fully open at 108 degrees before TDC (and also has 0.450 inches lift), then going on to shut at 24 degrees **after** TDC. So how long was this valve open? 60 degrees BBDC + 180 degrees from BDC to TDC + 24 degrees ATDC = 264 degrees. In this case, therefore, both the intake and exhaust valves have the same duration.

This cam can be described as a 24 — 60 — 60 — 24, meaning the intake valve opens 24 degrees BTDC and closes 60 degrees ABDC, while the exhaust valve opens 60 degrees BBDC and closes 24 degrees ATDC. Where the intake and exhaust cams have the same specs, the description of this cam is sometimes abbreviated even more to '24/60'.

In this case, the intake valve was fully open at 108 degrees ATDC and the exhaust valve was fully open 108 degrees BTDC. Thus the crankshaft rotational difference between the exhaust valve being fully open and the intake valve being fully open is 216 crankshaft degrees.

However, this camshaft lobe centre angle is always expressed in camshaft degrees, which in this case is 108 degrees. The camshaft lobe centre angle will affect the amount of overlap the cam provides. As the lobe centre angle is decreased, the amount of overlap is increased.

So what about intake centre angle (centreline) and exhaust centre angle (centreline)? These measurements refer to how many crankshaft degrees BTDC the exhaust valve is fully open, and how many crankshaft degrees ATDC the inlet valve is fully open. On the symmetrical cam shown in Figure 1.2 the intake and exhaust centre angles are both 108 crankshaft degrees.

Cam Data

Unfortunately, in the aftermarket there is no universal method used to display camshaft specs. This makes the comparison of camshaft grinds difficult, and the mental picturing of how a particular cam will affect engine performance even more difficult! While cam timing specifications are often expressed for a certain lifter movement (eg 0.050 inches), seat-to-seat specifications are much more accurate. These figures show the point at which the valves actually start to move.

It is when cam events occur — rather than the span between the events — that's most important in determining the resulting engine power. For example, an intake duration figure does not tell you when the inlet valve is opened and closed — only how long it is open.

Thus, if the available cam specs list only duration and lift figures, estimation of the engine power and torque characteristics (either by computer software modelling or experience) is difficult — if not impossible — to undertake.

To be able to accurately predict the performance of a camshaft, the following seat-to-seat and lift specifications are needed:

- Intake valve opening
- Intake valve closing
- Exhaust valve opening
- Exhaust valve closing
- Valve lift

Variable Camshaft Timing

From the previous discussion it can be seen that the timing of the camshaft events is of great importance in engine operation. If the camshaft timing can be altered to provide optimal performance for different engine operating conditions, very substantial improvements can be made to power, emissions and economy. Variable camshaft timing designs most commonly change the inlet camshaft timing; however, there are systems available that alter the timing of both the inlet and exhaust cams. In either case, the change in timing can be infinitely variable, or can consist of a single step. A number of different systems are employed, with most using an ECU-controlled solenoid valve to direct oil pressure to an actuator which, through a special mechanism, varies timing.

On the 911, Porsche uses its VarioCam system to advance the intake camshaft timing by 25 crankshaft degrees when the engine reaches 1300rpm. This increased overlap results in better combustion chamber filling and scavenging, which provides improved torque. When engine speed reaches 5920rpm, the intake camshaft timing is retarded by 25 degrees (ie back to the original idle setting) for optimal high-speed operation. When engine oil temperature is high, the camshaft advance takes place at 1500rpm. On the Porsche Boxster, the timing of the intake cam can be altered by 15 degrees. At low speeds and at idle, retarded intake cam timing is used, which results in shorter valve overlap periods and reduced hydrocarbon emissions. At low loads and moderate engine speeds, the camshaft timing is advanced.

Jaguar's AJ-V8 uses a system able to advance the inlet cam by 30 degrees; in the advanced mode, the valve overlap is 35 degrees. The advanced setting is used at moderate throttle openings at low to medium engine rpm. The large valve overlap creates high levels of exhaust backflow, effectively giving internal exhaust gas recirculation and so reducing the emissions of oxides of nitrogen. In addition, this approach gives lower hydrocarbon emissions than would occur with an external EGR system. At wide-open throttle, the timing is advanced before 4500rpm and then retarded at higher rpm.

Slightly altering the rotational relationship of the cam to the crank (ie changing the cam timing) can allow the engine to develop more power. This is especially the case in DOHC engines, where vernier timing wheels like this are often used to change the timing of the cams to one another as well as to the crankshaft.

The variable valve timing system used on BMW engines is dubbed VANOS. Single VANOS adjusts the timing of only the intake camshaft (either in a single step or steplessly, depending on the model), while double VANOS steplessly adjusts both the inlet and exhaust camshafts. The most recent single VANOS systems allow up to a 40-degree alteration in the timing of the intake cam, with the required cam timing determined on the basis of a map with throttle position and rpm axes. Interestingly, in cars with the single VANOS system, BMW's 250km/h governed top speed is enacted by varying the cam timing. The mighty 294kW (400hp) double VANOS 5-litre V8 used in the M5 has no less than four VANOS actuators, one on each camshaft and working as two pairs.

BMW engines with double VANOS can alter **each** camshaft's timing by up to 60 degrees! However, in the M5 V8, the inlet cam is altered by 'only' 54 degrees and the exhaust cam by 39 degrees. As is shown later (Figure 1.10), this allows quite radical cam timing figures. Overlap, for example, can vary from a massive 80 degrees to **negative** 12 degrees — the latter where the exhaust cam closes 12 degrees before the inlet opens! However, while this degree of flexibility is possible, the timing changes made don't necessarily allow either of these extremes to ever actually occur.

Fitting a performance cam often requires that the valve springs and other parts of the valvetrain are also upgraded, especially if higher revs are going to be used.

Lexus VVTi engines use a system that's able to steplessly change inlet cam timing by 60 degrees. The camshaft timing that is selected is based on the inputs of engine rpm, throttle position, coolant temperature and intake airflow. Five different 'running' modes are used, with another two used to cater for engine starting and low temperature operation. Figure 1.3 shows these modes, with Figure 1.4 illustrating how modes 1-5 relate to engine load and speed.

manifold, the performance changes that are possible become very major indeed. However, as shown by the Lexus data, selecting optimal camshaft timing in a modified engine (especially one using a steplessly variable system) is likely to require extensive dyno mapping or computer simulation (covered below). The setting-up of a single-step variable camshaft timing system is easier; if full throttle performance is most critical, dyno runs can be carried out with the cam in each of its two positions. The rpm where the power curves cross becomes the changeover point.

Range	Conditions	Operation
1	Idle	The valve timing is set to 0 degrees (most retarded angle) and because of the lack of overlap, idle is stabilised.
2	Medium load range	The valve timing is advanced to increase the amount of valve overlap. Thus the internal exhaust gas recirculation rate is increased and the pumping loss is decreased, resulting in improved fuel economy.
3	Low load range	The valve timing is retarded to decrease the amount of valve overlap, thus ensuring the engine's stability.
4	High load, low-to-medium speed range	The valve timing is advanced to advance the timing of the intake valve. The volumetric efficiency is thus improved, resulting in improved low-to-medium speed range torque.
5	High load, high speed range	The valve timing is retarded to retard the timing of the closing of the intake valve, resulting in improved volumetric efficiency in the high speed range.
	Engine started and stopped	When the engine is started and stopped, the valve timing is set to its most retarded state.
	Low temperature operation	The valve timing is set to 0 degrees (most retarded angle) without any valve overlap. This prevents reversion of the air/fuel mixture and minimises the need for acceleration enrichment. Furthermore, because of the stable idle speed, fast idle rpm can be lowered, improving fuel economy and emissions during low-temperature engine operation.

Figure 1.3
The different operating modes for the infinitely variable intake camshaft timing on VVTi Lexus engines. Timing is able to be steplessly changed by a massive 60 degrees!

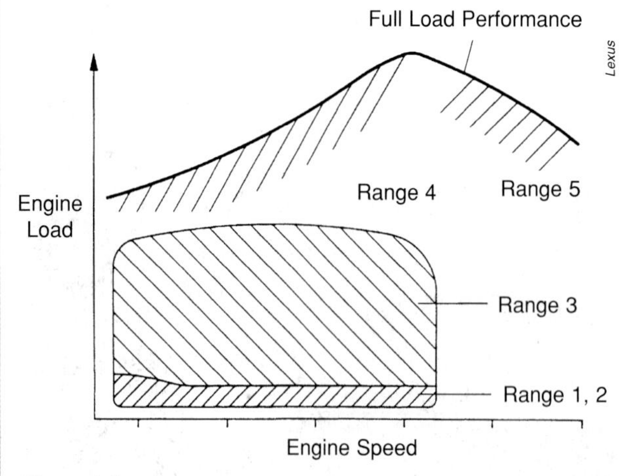

Figure 1.4
The relationship in the variable valve timed Lexus engines between the different camshaft timing modes shown in Figure 1.3 and engine speed and load.

Variable camshaft timing under the direction of a programmable aftermarket ECU has the potential to dramatically improve the driveability, fuel economy and emissions performance of a modified engine. If used in conjunction with an engine featuring a variable inlet

Honda's VTEC system is slightly different from systems using variable camshaft timing; in many ways it's a much cruder system. The acronym stands for Variable valve Timing and Electronic lift Control and allows valve lift, duration and camshaft timing to be altered in one step. Three different VTEC systems are available, but all use the same basic approach.

In addition to the normal number of camshaft lobes (two inlet and two exhaust per cylinder), the camshafts used in VTEC engines have two extra, 'hot' cam profile lobes and two extra rocker arms. The rocker arms for the high lift and duration cam lobes 'freewheel' at low rpm — their movement is not transferred to the valves. However, at a certain (usually high) rpm, an electronically controlled and hydraulically actuated pin locks all three rocker arms together, resulting in the hotter cam profile coming into play. Figure 1.5 shows this approach.

In its highest performance guise, both the two inlet and two exhaust valves are operated in this manner. The second VTEC approach opens one inlet valve by a

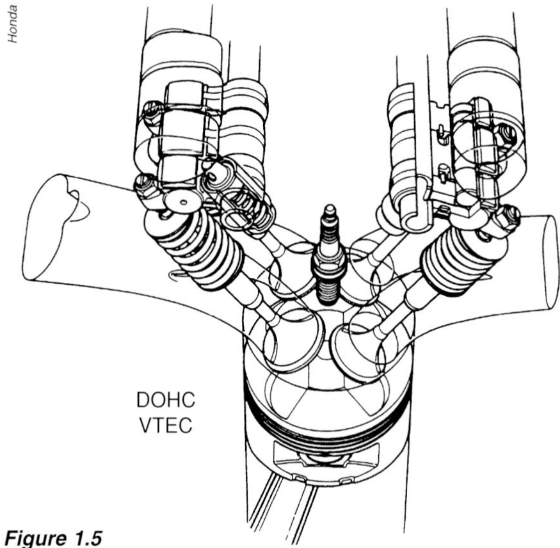

Figure 1.5
The camshafts used in VTEC engines have two extra, 'hot' cam profile lobes and two associated extra rocker arms. The special rocker arms 'freewheel' until an electronically controlled and hydraulically actuated pin locks all three rocker arms together, resulting in the hotter cam profile coming into play.

greater amount than the other at low engine speeds, promoting strong swirl in the combustion chamber. The third variant operates only the intake valves, with the exhaust valves being actuated conventionally.

While the 'performance' variant of VTEC has gained much attention, the 'emissions' variant (opening one inlet valve a greater amount than the other at low speeds) is significant; it has the potential to allow very clean emissions with high performance. Under cold start conditions, Honda engines equipped with this technology can run astonishingly lean 15.5-16:1 air/fuel ratios, where a conventional engine needs to be far richer at 13-13.5:1. Taking this approach results in a 45 per cent reduction in cold-start hydrocarbon emissions.

Camshaft Computer Simulation

Using a good computer engine simulation program is by far the most practical way of deciding on the best camshaft specs for a particular engine. Not only is it simple, cheap and effective, it also allows you to gain a 'feel' for the effect of making specific cam changes on the engine in which you are interested. Such modelling cannot, of course, guarantee that the results you see on the screen match the engine performance — but it's a lot better than picking cams at random!

The engine software package Dyno2000 from Motion Software was used to model the effect of a number of camshaft grinds in the Lexus 1UZ-FE 4-litre, 4-valves-per-cylinder V8. In all the following simulations, the compression ratio, intake and exhaust systems, valve lift and all other engine factors were unchanged — only valve opening duration and timing were altered.

The blue lines in Figure 1.6 (over the page) show the Dyno2000 simulation results for the standard engine. It should be noted that while peak power is very close to the manufacturer's data (a modelled 256hp at 5500rpm versus the manufacturer's 260hp at 5400rpm), the torque figure of the simulation is just over 10 per cent higher and occurs a little low in the rev range. The standard cams have the following timing: intake valves open at 6 degrees BTDC and close at 46 degrees ABDC, and

The BMW M5 uses a 5-litre 32-valve V8 with adjustable timing of both the intake and exhaust cams. The result? A tractable, refined engine with a smooth idle, good emissions — and 400hp!

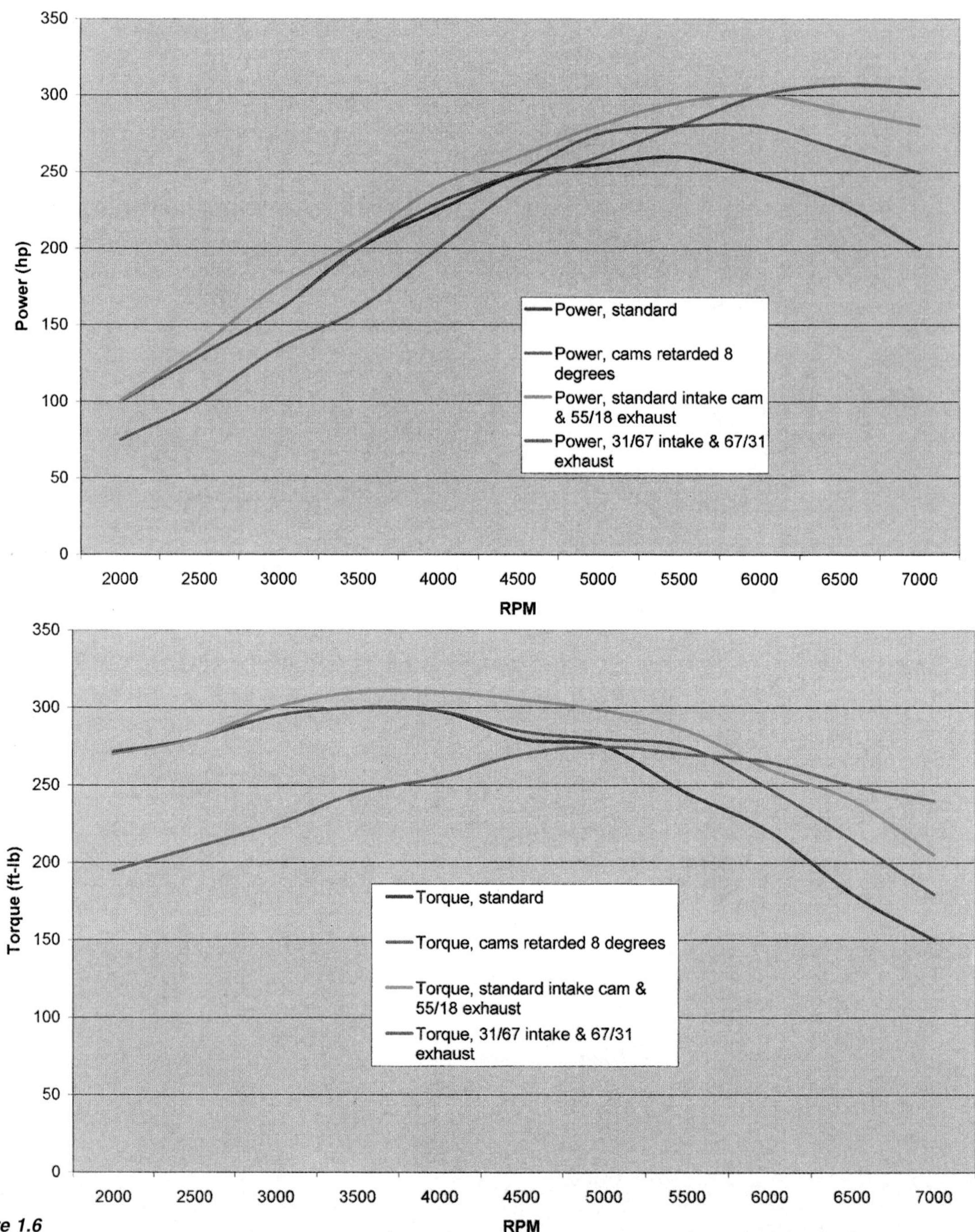

Figure 1.6
The power and torque curves resulting from the use of the Dyno2000 engine simulation software to model different camshaft configurations in the Lexus 1UZ-FE 4-litre V8. The blue lines show the power and torque curve with the standard cams — these match well with the manufacturer's figures. The red lines show the improved top-end power resulting from retarding both the intake and exhaust cams by 8 degrees; the green lines show the effect of retaining the standard intake cam but using an exhaust cam with a revised opening of 18 degrees BBDC and closing of 18 degrees ATDC; while the brown lines show the results of fitting new cams — intake opening 31 degrees BTDC and closing 67 degrees ABDC, and exhaust opening 67 degrees BBDC and closing 31 degrees ATDC. While such computer simulations will not always perfectly match reality, they represent an excellent way of assessing the likely results of cam changes.

exhaust valves open at 46 degrees BBDC and close at 3 degrees ATDC. This gives an intake duration of 232 degrees, an exhaust duration of 229 degrees and an overlap of 9 degrees. The power and torque curves show the strong bottom-end response favoured by the manufacturer of an auto-transmission, large luxury car.

Next, the intake and exhaust cams were retarded by 8 degrees, with no alteration made to the actual profiles. This type of change could be achieved with the fitting of adjustable ('vernier') cam wheels. The red lines in the graphs show these results. The torque and power curves below 4000rpm are unchanged, but power is now increased by 9 per cent at similar rpm. Power at 6500rpm is up by 18 per cent — so there is a significant gain if the engine is to be revved to 6500 before each gear change. Of course, after making cam timing changes in this way, the engine should always be checked for adequate piston/valve clearances.

The green lines on the graphs show the results of retaining the standard intake camshafts but changing the exhaust cams for those featuring an exhaust valve opening 55 degrees BBDC and closing 18 degrees ATDC. This gives a new exhaust duration of 253 degrees and an overlap of 24 degrees. In this form, bottom-end grunt until 3500rpm remains unchanged, but from there on, power improves considerably. Peak power has risen to 298hp at 6000rpm (a gain over standard of 16 per cent) and the whole top-end power curve is extremely strong.

Finally, the brown lines show the results of changing the intake and exhaust camshafts. With the intake valves opening 31 degrees BTDC and closing 67 degrees ABDC, and the exhaust valves opening 67 degrees BBDC and closing 31 degrees ATDC, there is a high 62 degrees of overlap and an equal exhaust and inlet duration of 278 degrees. Incidentally, the Dyno2000 package describes this as a 'High Performance Street' cam. Fitting these cams decreases bottom-end torque substantially; at 2000rpm, the simulation shows a drop of 25 per cent, with peak torque also moved from the standard engine's 3500rpm to 5000rpm. However, peak power is 28 per cent up over standard at 6500rpm, with the power curve showing that if the engine was revved past this figure, excellent power would still be available.

So which cams would be best in the Lexus V8? The answer can only be found by actually trying them in the engine, but on the basis of this simulation, simply fitting new exhaust cams to the cylinder heads — a relatively cheap step (just two cams needed, not four) — appears to give excellent results for a road car.

But what about engines with variable valve timing? Computer simulation of the camshafts used in the Jaguar AJ-V8 was carried out, again using the Dyno2000 package. As used in the Jaguar XK8, the AJ-V8 has exhaust valves that open 50 degrees BBDC and close 10 degrees ATDC (240 degrees duration). The intake cam Variable Valve Timing (VVT) is able to alter the cam timing by 30 degrees. With the VVT switched off, the intake valves open 5 degrees ATDC and close 65 degrees ABDC (also 240 degrees duration). With the VVT activated, the intake valves open 25 degrees BTDC and close 35 degrees ABDC (giving, of course, the same 240 degrees duration). Figure 1.7 shows the results of modelling this standard engine — again, these curves are very similar to the manufacturer's power and torque curves. With the VVT switched on (ie intake cam advanced), low rpm torque is improved, while retarding the cam at higher rpm improves power.

But how will the variation in cam timing that occurs with the VVT switched in and out affect the performance of modified cams? Dyno2000's iterative testing function was used to simulate more than 5000 different valve timing configurations. The cams that gave best power while observing a 7500rpm redline had an exhaust valve opening 66 degrees BBDC and closing 31 degrees ATDC, giving a duration of 277 degrees. The inlet cams opened 8 degrees BTDC and closed 69 degrees ABDC (257 degrees duration), giving an overlap of 39 degrees. The modelled power curve is shown in Figure 1.8 (over the page). However, unlike the results gained with the standard cams, advancing the inlet cams by 30 degrees (ie switching VVT on) gave **no** modelled improvement in torque in the lower rev range. To make full use of the variable cam timing feature would require further modelling of different camshafts!

I strongly suggest that any proposed cam change be modelled, using a good engine simulation program, before money is spent. While no program will be infallible, it will be a far better guide as to what to expect than any other approach that doesn't involve actually testing the cams in the engine.

Selecting a Cam

While using a computer simulation program or even looking at cam specs in a catalogue may lead you to believe

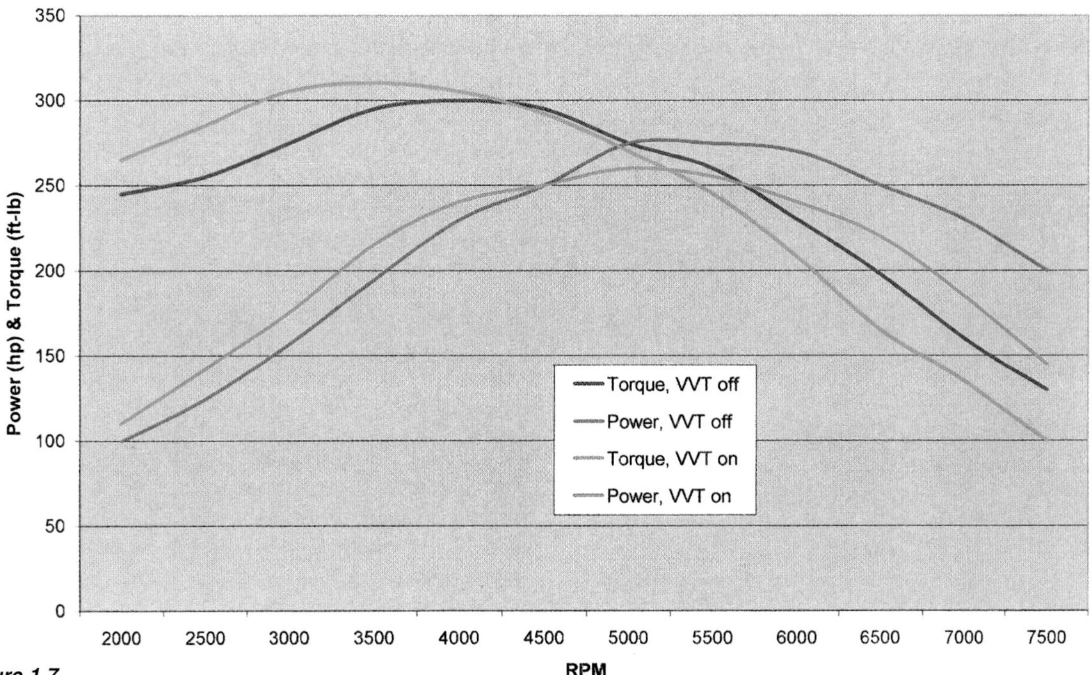

Figure 1.7
The results of computer-simulating the variable camshafts used in the Jaguar AJ-V8. The Variable Valve Timing (VVT) can alter the intake cam timing by 30 degrees. With the VVT switched off, the intake valves open 5 degrees ATDC and close 65 degrees ABDC. With the VVT activated, the intake valves open 25 degrees BTDC and close 35 degrees ABDC. With the intake cam advanced, low rpm torque is substantially improved, while retarding the cam at higher rpm improves power.

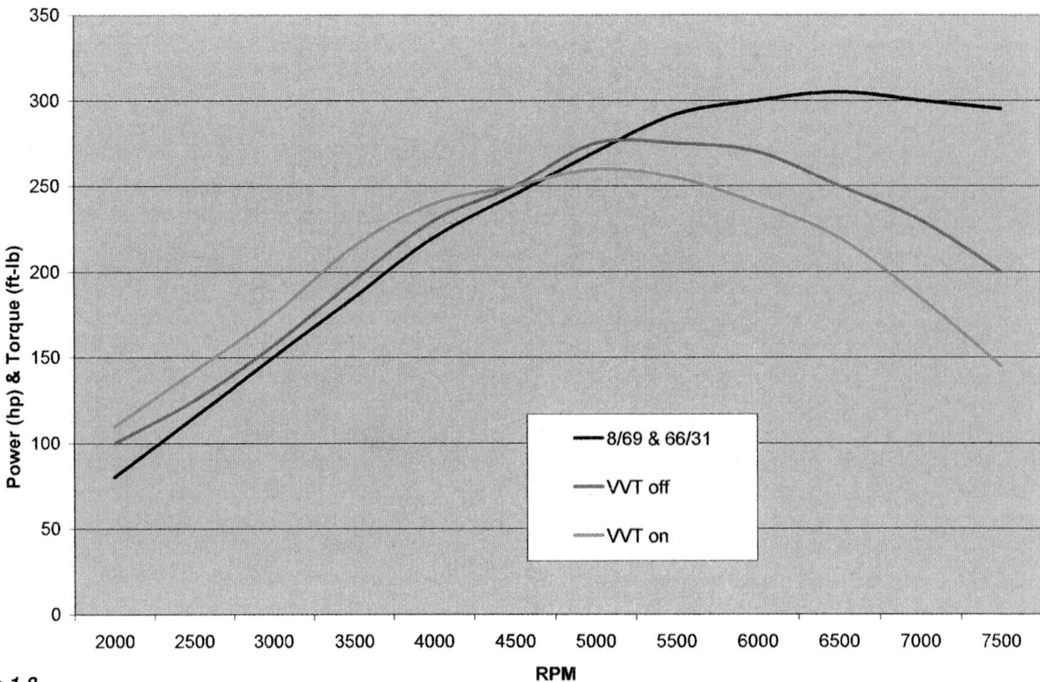

Figure 1.8
Dyno2000's iterative testing function was used to simulate more than 5000 different valve timing configurations to find cams that gave good peak power on the Jaguar V8. The black line shows the cams that gave best power. These cams used an inlet valve opening of 8 degrees BTDC and closing of 69 degrees ABDC, with the exhaust valves opening 66 degrees BBDC and closing 31 degrees ATDC. However, unlike the results gained with the standard cams (red and blue lines), advancing the modified inlet cams by 30 degrees (ie switching VVT on) gave no modelled improvement in torque in the lower rev range.

there's an enormous amount of power just waiting to be unleashed, the reality for a road car is that you should always lean to the 'mild' side when selecting a new cam. A cam that has too much lift will cause valvetrain problems; a cam that has too much overlap will give a bad idle and reduce low-load vacuum (so affecting the efficiency of the brake booster); and a cam that gives good power at only high rpm is simply no fun to drive in day-to-day traffic.

Matching a cam to its application is extremely difficult. Each of the following factors will have an impact on the best cam configuration:

- Required idle quality
- Required emissions performance
- Type of valve timing (fixed or variable)
- Type of intake manifold (runner length, fixed or variable geometry)
- Compression ratio
- Valve and head flows
- Exhaust system
- The use of forced induction
- Transmission type (auto or manual)
- Maximum permitted engine speed
- Engine management system

Unfortunately, most camshaft selection 'rules' do not hold true across a variety of engines. Further, many older references give camshaft recommendations that totally ignore emissions and (to a lesser extent) fuel economy. You should therefore be wary of those who make cam recommendations not based on test and development carried out on your specific type of engine.

However, one way to get a good 'feel' for the specs of camshafts that work well in late-model engines is to examine the valve timing used by the major manufacturers. Figure 1.9 (next page) shows the seat-to-seat specifications of a variety of standard camshafts. Each of these high-performance engines has had to satisfy emissions, fuel economy and idle quality criteria, in addition to achieving good power. Note the lower amount of overlap used in those forced induction engines running cams unique to that engine, and the relatively short opening durations of all of the cams.

Figure 1.10 (page 13) shows the camshaft timing specifications of some variable valve timed engines. The specific torque figure (ie Nm/litre) for each of these powerful engines is very high in comparison with naturally aspirated engines that do not have the luxury of altering their valve timing. Figure 1.11 (page 13) shows some typical aftermarket cam specs for a high-performance four-valves-per-cylinder engine, in this case the Suzuki Swift GTi DOHC 1.3-litre that develops 75kW (~100hp) in standard form.

As can be seen from the manufacturer's cam data shown in Figure 1.9, the use of multi-valve heads that flow extremely well (and the necessity to meet emissions legislation) has resulted in the use of relatively mild cam grinds.

Older two-valve engines — even in standard form — can tolerate higher amounts of overlap resulting from increased duration, as they are less prone to reversion. Head flows in these engines are much more restrictive than in modern three-, four- and five-valves-per-cylinder engines, and the cam timing of older cams reflects this.

	Nissan SR20 16 valve 2 litre four	**Subaru EJ-20** 16 valve 2 litre turbo four	**Subaru EJ-20** 16 valve 2 litre four	**Mazda F2** 12 valve 2.2 litre turbo four*	**Nissan RB30E** 12 valve 3 litre six	**Nissan RB30ET** 12 valve 3 litre turbo six	**Nissan RB26 DETT** 24 valve 2.6litre twin turbo six	**Lexus 1UZ-FE** 32 valve 4 litre V8
Power	104kW at 6400 rpm	147kW at 6000 rpm	85kW at 5600 rpm	108kW at 4300 rpm	114kW at 5200 rpm	150kW at 5600 rpm	209kW at 6800 rpm	190kW at 5600 rpm
kW/litre (# = turbo)	52	74#	43	49#	38	50#	80#	48
Torque	178Nm at 4800 rpm	260Nm at 3600 rpm	164Nm at 4400 rpm	258Nm at 3500 rpm	247Nm at 4000 rpm	296Nm at 3200 rpm	360Nm at 4400 rpm	360Nm at 4400 rpm
Nm/litre (# = turbo)	89	130#	82	117#	82	99#	138#	90
Intake Opens (degrees BTDC)	13	3	4	10	10	6	7	6
Intake Closes (degrees ABDC)	55	55	52	49	58	66	53	46
Intake Duration (degrees)	248	238	236	239	248	252	240	232
Exhaust Opens (degrees BBDC)	57	55	48	55	52	60	63	46
Exhaust Closes (degrees ATDC)	3	11	12	12	16	12	7	3
Exhaust Duration (degrees)	240	246	240	247	248	252	250	229
Overlap (degrees)	16	14	16	22	26	18	14	9

* the same camshaft is also used in the naturally aspirated version of this engine

Figure 1.9
The standard camshaft timing specifications for a variety of engines. These engines develop good power while retaining a smooth idle, good fuel economy and legal emissions.

For example, the old L-series 2-litre Datsun engine (designed in the late 1960s) has a standard cam with the following specs: intake opens 16 degrees BTDC and closes 52 degrees ABDC (248 degrees duration); exhaust opens 54 degrees BBDC and closes 14 degrees ATDC (also 248 degrees duration), giving an overlap of 26 degrees. This 2-litre engine developed just 70kW. The Nissan SR20DET with the same capacity (but, of course, with engine management and a 16-valve head) has milder cams (the same intake duration, 8 degrees less exhaust duration and 10 degrees less overlap), yet develops 104kW.

The lesson is that engines with heads that flow very well do not need the same 'amount' of cam to develop good power. This is another reason for leaning towards mild performance cam specs when dealing with modern engines. It also means that new performance camshafts should be chosen only after examining the specs of the factory-fitted cams, so you can at least ascertain the starting point!

However, even if you pick a mild performance camshaft, or use a computer to simulate the cam's effects (or compare the grind with other cams that have been known to produce good results in the same engine), you cannot be sure of the outcome until you actually run the engine with the new cams fitted. This is because the shape of the valve lift curve (ie how fast the valve reaches maximum lift and then closes) will determine to a major degree the camshaft performance. A cam that causes an intake valve to open 15 degrees BTDC may only slowly crack open the valve — or it may open it with a rush. Closing speeds may also vary; both characteristics influencing the area under the cam lift curve — the factor that really determines the flow of a given port and valve.

The number of variables that affect the performance of the chosen cam require that, when selecting a performance cam, you should buy from a reputable cam-grinding company that's experienced with your model of engine. You should be able to get meaningful answers when you pose questions like:

- Over what rev range will this cam work?
- At what revs will peak power and peak torque be developed?
- What will idle quality be like?
- What effect will the cam have on emissions?
- What effect will the cam have on cruise fuel economy?

The best cam companies will also have dyno curves available to show engine performance with their various cams fitted. Most cam companies stock cams in a number of different tunes — for example, specified from 'standard' to 'full race'. Where four or five cams are available for an engine, road cars are usually best served by the selection of the first or second performance cam grind listed.

	Toyota 4A-G 20 valve 1.6 litre four		**Jaguar AJ-V8** 32 valve 4 litre V8		**BMW S62 B50** 32 valve 5 litre V8
Power	118kW at 7400 rpm		216kW at 6100 rpm		294kW at 6600 rpm
kW/litre	74		54		59
Torque	162Nm at 5200 rpm		393Nm at 4250 rpm		500Nm at 3800 rpm
Nm/litre	101		98		100
Cam Timing Status	**VVT off**	**VVT on**	**VVT off**	**VVT on**	**Infinitely Variable**
Intake Opens (degrees BTDC)	0	30	5 **ATDC**	25	2 **ATDC** – 52
Intake Closes (degrees ABDC)	70	40	65	35	74 - 20
Intake Duration (degrees)	250	250	240	240	252
Exhaust Opens (degrees BBDC)	54		50		78 – 39
Exhaust Closes (degrees ATDC)	16		10		10 **BTDC** - 29
Exhaust Duration (degrees)	250		240		248
Overlap (degrees)	16	46	5	35	**Negative** 12 - 81

Figure 1.10
The camshaft timing specifications of three variable valve timed engines. As can be seen, the specific torque figures (ie Nm/litre) of these powerful engines are very high in comparison with the engines that do not have the luxury of variable valve timing. Looking at the variation possible in cam timing shows how the cams can vary from 'mild to wild' without change in durations.

	Standard Swift GTi	Standard Cultus	BD10	BD14	Factory Rally	Rally Race	Road-race Rally	Road-race only	4000-8200 rpm	4500-9000 rpm
Intake Opens (degrees BTDC)	8	8	10	11	28	30	48	42	39	39
Intake Closes (degrees ABDC)	36	36	62	63	56	62	72	73	79	71
Intake Duration (degrees)	224	224	252	254	264	272	300	295	298	290
Exhaust Opens (degrees BBDC)	42	42	59	60	54	64	73	73	79	75
Exhaust Closes (degrees ATDC)	10	10	13	14	22	28	41	42	39	35
Exhaust Duration (degrees)	232	232	252	254	256	272	294	295	298	290
Overlap (degrees)	18	18	23	25	50	58	89	84	78	74
Inlet Valve Lift (mm)	7.5	8.0	8.8	9.0	9.0	8.9	9.0	9.1	10.4	10.3
Exhaust Valve Lift (mm)	7.5	8.0	8.8	9.0	9.0	8.9	9.0	9.1	10.4	10.3

Ivan Tighe Engineering

Figure 1.11
A variety of camshafts suitable for the Suzuki Swift GTi engine, which in standard form develops 75kW (~100hp) at 6600rpm. Note that when valve lifts over 10.2mm are required in conjunction with engine speeds over 7500rpm, revised valve springs must be used. Lifts over 9.1mm require the resetting of lifter pre-load.

CYLINDER HEAD PORTING

The amount of power an engine can develop is dependent on the mass of air it can breathe per unit time. In fact, engine builders can often estimate quite closely the maximum power achievable from an engine simply by measuring on a flowbench the airflow that can pass through the cylinder head; some workshops use the following equation for naturally aspirated engines:

0.43 x maximum port flow (in cfm at 10 inches of water) x number of cylinders = peak hp

This equation explains why some workshops state that a particular head 'flowed X horsepower'. This is a little deceptive because the actual airflow efficiency (how much air is consumed for how much power is developed) varies from engine to engine. When they are tested on an engine dyno, good engines like the Nissan SR20DET consume about 1cfm for each horsepower developed. Also remember that peak power is not the most important aspect of a road car engine — average power across the working range is critical. In fact, one cylinder head porter said, "If we gain 10cfm at low valve lifts and lose it at high lifts — great!" This is because the valve spends more time at lower lifts than full lift, and so low lift flow is more important. However, gains at low valve lift will not increase the 'horsepower figure' that the head flows!

The porting of cylinder heads on modern engines will not give the same gains that were once easily seen on two valve, iron head engines. However, some improvements can still be made.

The restrictions to flow in the cylinder head comprise both the intake and exhaust ports and the intake and exhaust valves. Multi-valve engines have large curtain areas — the area found by multiplying the circumference of the valve by its lift. Thus, in many cases, the ports are the major restricting factor, rather than flow past the valves. Note that this large valve curtain area means reverse flow can easily occur when the piston moves from BDC towards the closing point of the intake valves, reducing torque at low engine speeds in engines with lots of valve overlap.

As mentioned, a flowbench can be used to evaluate the volume of air that can flow through the port and past the intake valve. The flow measurement is carried out at a

The flow of the intake valves is more important than that of the exhausts. This three-valves-per-cylinder engine uses two small intake valves and a single exhaust valve to give excellent flow, even with mild cams.

fixed pressure differential — normally very small (10, 20 or 28 inches of water). The flow is expressed in cubic feet per minute (cfm) and is measured at various valve lifts, invariably expressed in thousandths of an inch.

The larger the 'hole' (the cross-sectional area of the port and curtain area of the valves), the higher will be the airflow measured on the bench. From this it could be assumed that bigger holes — and so higher airflow readings — will always be better. However, this is definitely not the case. The combustion pressure — and so developed torque — is dependent on the burn characteristics of combustion in addition to the mass of fuel/air breathed. Manufacturers build swirl and tumble into their head designs, resulting in more complete combustion. Factors determining the swirl and tumble characteristics include the amount of 'turn' the port has, the angle from the horizontal at which the port approaches the short side radius, and the shape of the valves and the chamber. Simply making the port larger without taking into account these airflow characteristics will not always yield increased engine power.

Another important factor is that a larger port area will result in a decreased flow speed. This may result in decreased cylinder filling at low engine speeds, reducing bottom-end torque. Effectively, the larger the diameter of the port, the higher will peak torque be moved in the rev range. To indirectly indicate port area, port volume is

Engines with four or five valves per cylinder have excellent head flows. Coupled with turbocharging, even this tiny 660cc Daihatsu three-cylinder engine can develop a lot of power.

This cutaway view of a cylinder head port shows the important short radius (left) and the angle with which the port approaches the valve.

often measured. With the valve in place and closed, the port volume can be measured using a burette and kerosene. As a rule of thumb, an increase in 3cfm measured head flow should not involve the removal of more than 1cc of material (ie a 1ml increase in port volume).

Major porting involves the careful removal of metal from the passages, enlarging them and often also changing their shape and even position. I don't recommend that you undertake any major porting of a modern engine. While there are some professional workshops that successfully carry out this type of work, these workshops are always equipped with a flowbench that's used extensively. Further, most of these professional porters have developed techniques they will not share with mere mortals like you or me! Also, even professional workshops can spend many hours porting a head with only limited success.

An example of only moderate gains can be seen in the porting carried out on a four-valves-per-cylinder Honda CRX head. Before the porting, a spare Honda head was sectioned so the porter could see casting thicknesses — carving into a water jacket isn't unknown! The head was then treated to many hours of reshaping of the combustion chamber and intake and exhaust ports. So what

Lips and abrupt steps in the port walls will cause turbulence. Removing this type of obstruction will give a guaranteed increase in port flow.

were the results on the flowbench? On the intake side at all valve lifts over 0.275 inches there was a gain in flow, peaking with a 9 per cent improvement at 0.450 inches lift. On the exhaust side, flow was the same as stock **or lower** until a valve lift of 0.250 inches, with the biggest gain being 18 per cent at 0.450 inches lift. However, as already indicated, the valves are at their maximum lift position for only a very short time.

In contrast is the porting of a Suzuki Swift GTi cylinder head. Already a very efficient engine developing 58kW/litre (77hp/litre), the G13B four-valves-per-cylinder head responded well to extensive porting. The intake flow was substantially improved, with flow at 0.100 inches valve lift up by 28 per cent. This declined to a gain of 14 per cent at 0.250 inches lift, dropping to only a 1 per cent gain at 0.350 inches lift. On the exhaust side, flow was very substantially improved all the way from 0.100 to 0.350 inches lift, being up an average of 36 per cent! (These flow gains at 0.350 inches (8.9mm) lift can be considered in the context of the maximum valve lifts of the various GTi cams shown in Figure 1.11.)

This view of a port in an AlfaSud 1.5-litre head shows the smoothly contoured short radius achieved by simple porting.

Another good example of porting was that carried out on a Lexus 1UZ-FE DOHC V8 engine. The porting that was carried out is depicted in the photos, with Figure 1.12 showing the flow figures.

When compared with this type of professional (and expensive!) porting, how effective is some simple cleaning-up of the heads by a relatively unskilled amateur? With this approach, obvious casting marks, steps and jumps in the port walls are smoothed. The port shape is not changed and the port position remains standard. While not carried out on a four-valves-per-cylinder engine, 'before' and 'after' figures are available on an AlfaSud 1.5-litre flat four. In standard form, the 1.5-litre engine develops 75kW (100hp) from its 1490cc — that's 50kW (67hp) per litre. This is respectable for a two-valves-per-cylinder, unmodified engine.

Four different porting modifications were undertaken:

- The sharp lip on the short radius from the intake port to the valve was smoothed.
- The rough surfaces of the ports were smoothed.
- The sharp edges within the combustion chamber were radius'd.
- The intake manifold was port-matched.

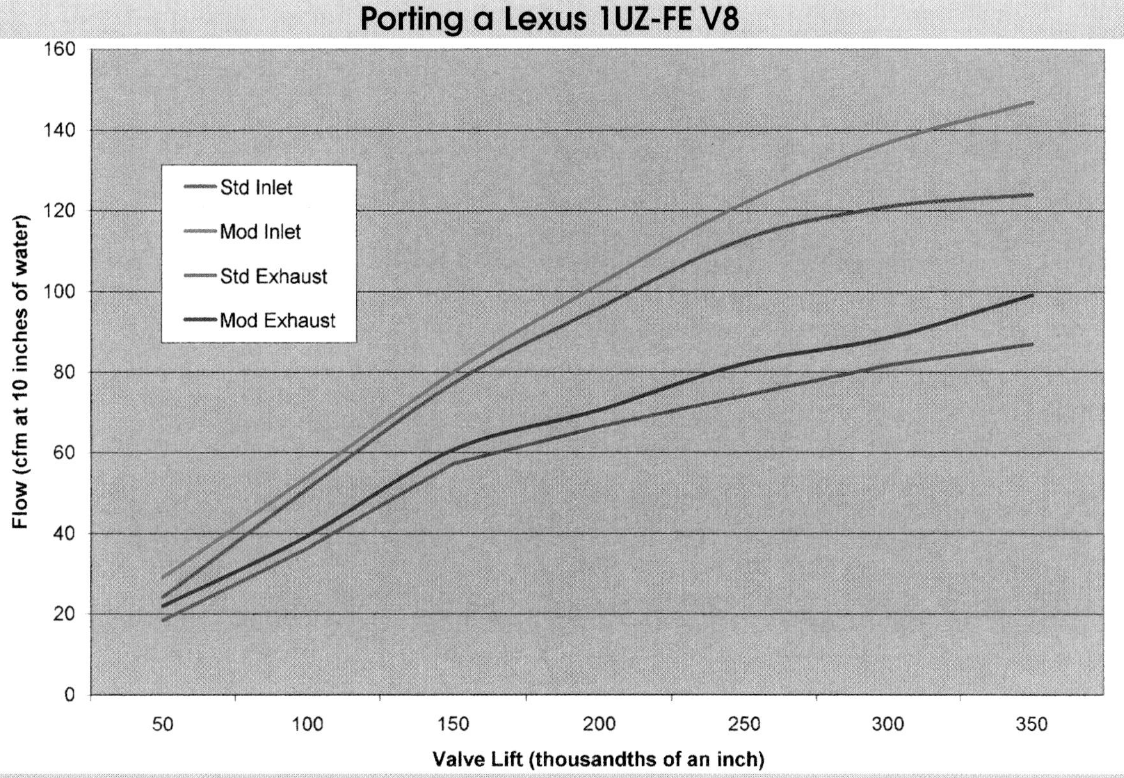

Figure 1.12
The gains Frank Intini achieved by porting the heads of a Lexus 1UZ-FE V8. At a valve lift of 0.350 inches (8.9mm), the intake flow was increased by 19 per cent and the exhaust flow by 14 per cent. Importantly, the porting improved both the inlet and exhaust flow through the whole valve lift range.

The view looking down into the valve bowl of the intake ports, taken from the combustion chamber side. Note the standard casting marks in the righthand port and the shape of the port upstream and downstream of the valve guide.

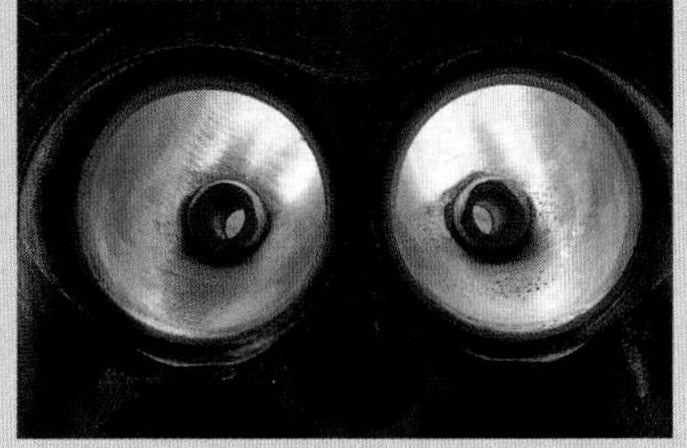

After porting of the head, the lump upstream of the valve guide is still present but its shape has been streamlined. Gains have been made especially by the removal of material on the outer sides of the guides in each port. The surface finish uses a 40- to 60-grit cartridge roll finish — a mirror polish finish is not needed.

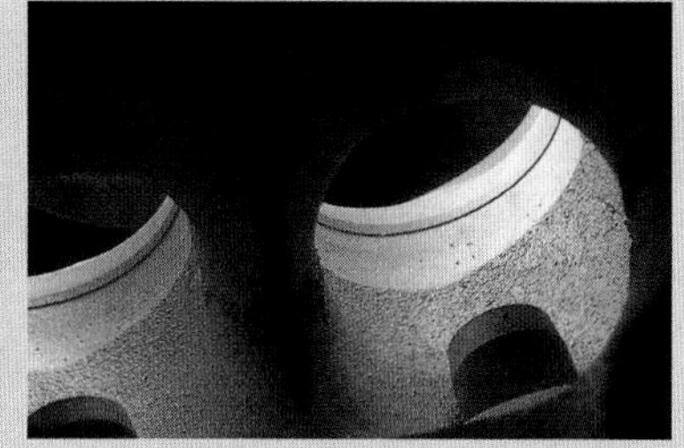

The view from inside the standard intake port, looking past the intake valve seat into the combustion chamber. Note the lips below the valve seats that have been left by the factory tooling, and the width of the dividing pillar between the ports.

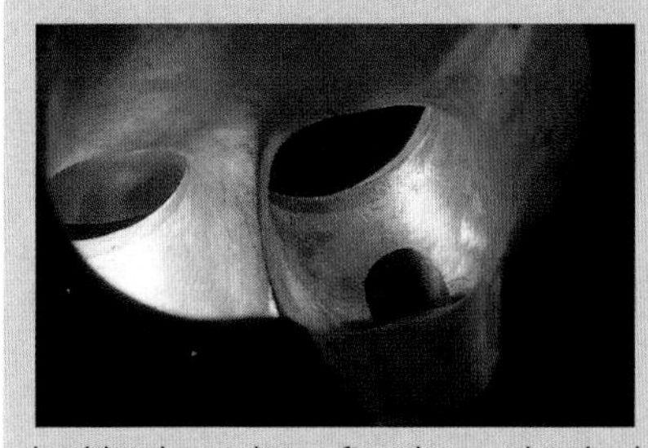

In this view taken after the porting had been carried out, it can be seen that the seat/wall intersections have been blended, resulting in an improvement in low valve lift flow. The cross-sectional areas past the valve guides have also been opened up (enlarging the bowl area), so decreasing the width of the central dividing pillar. The knife edge that results doesn't give a major gain in itself.

The view inside the standard intake port, with the head oriented as it is on the engine. The very important short-side radii can be seen at the base of the ports.

One of the porting approaches taken by Frank Intini is to make the intake floor of the port (at the bottom in this view) flat and wide, on the basis that low-speed air will travel around the short-side radius and that the wide, flat port floor will encourage this flow. The floors of adjoining ports for the one cylinder can also be angled to the horizontal to encourage swirl through each intake valve.

This is a view of the inlet valve seat in the standard combustion chamber. Note the lack of a smooth contour from the edge of the seat into the chamber.

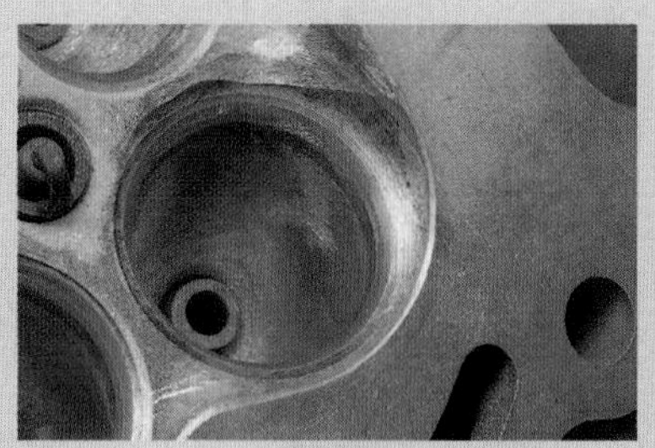
While not finished, this view shows how the intake valve has been unshrouded to promote mid-lift flow. The sharp edge remaining in the combustion chamber will be later smoothed, taking care that as little material as possible is removed, so maintaining a high compression ratio.

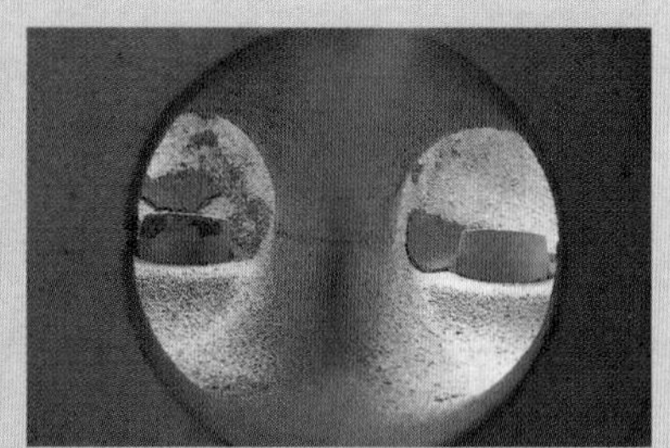
Looking in through the exhaust port shows a layout similar to the standard intake ports. The single branch of the Y-port normally has sufficient flow — porting improvements should concentrate on the valve bowl area.

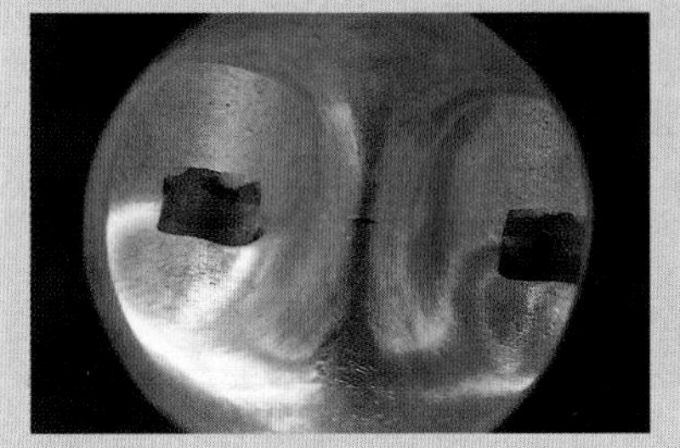
After the porting, the larger passage created past the valve guides can be seen. The surface finish of the exhaust ports should be a little finer than for the inlets.

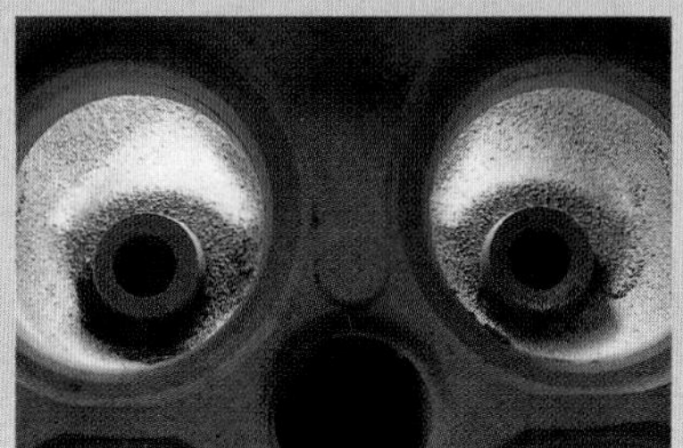
The standard exhaust ports show the same rough wall finish as the standard inlet ports. Note how the boss that supports the valve guide extends across the whole width of the port.

The modified exhaust ports are smoothed and have improved flow past each side of the valve guide. Note the shape that has been achieved on the downstream side of the valve guide support.

Each cylinder required about five hours work — so 20 hours for both heads.

The flowbench results showed the inlet ports improved in flow by an average of 9 per cent, with improvement recorded at all valve lifts from 0.100 to 0.500 inches. The peak improvement was 12 per cent at both 0.400 and 0.500 inches lift. On the exhaust side, the average gain was 3 per cent, with the peak improvement being 7 per cent at 0.500 inches lift. Again, at no valve lifts did exhaust flow fall.

The minor cleaning-up of the ports and combustion chambers is still a viable proposition on current engines, but very extensive (and expensive) porting can be carried out on sophisticated heads with no guarantee that the improvement will match the money spent. However, if the head(s) have been disassembled and more power is being sought, it costs little to have them tested on a flowbench so that at least informed decisions about head modification (and cam lift selection) can be made.

HEAD SWAPS

Where a manufacturer produces in the same family of engines both a large-capacity, two-valves-per-cylinder engine and also a smaller engine with a much better flowing head (DOHC, four valves per cylinder, for example), thought can be given to the transplant of the sophisticated head onto the larger-capacity bottom end. This is quite possible, although it's usually only worthwhile when the greatest possible amount of power is required.

Aspects to consider before embarking on such a project include:

- Stud patterns, bolt sizes and gallery matching
- The resulting compression ratio
- A source for the new cam belt or chain
- The requirement for a new timing belt (chain) cover
- The activation of variable camshaft mechanisms
- The plumbing of oil feed and return lines if the new head is equipped with a turbo (or even two turbos!)

Engine builder John Keen working on a Toyota 3S-GTE turbo four. Assembling a modern high-performance engine requires a very high degree of precision. Just buying the required measuring tools so you can do this work yourself is likely to be more expensive than paying an expert engine builder to perform the assembly...

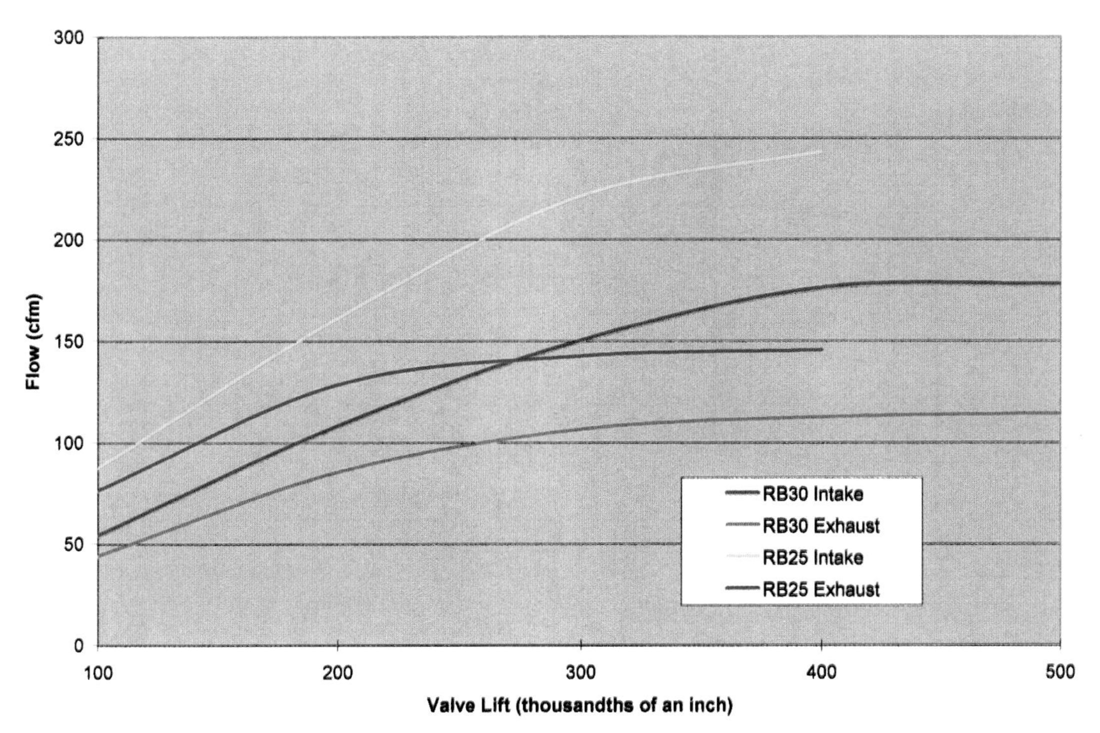

Figure 1.13
A comparison of the flow figures of two standard Nissan RB-series six-cylinder engine heads. The RB25 head uses four valves per cylinder while the RB30 head has only two. The advantage of the extra valves is obvious!

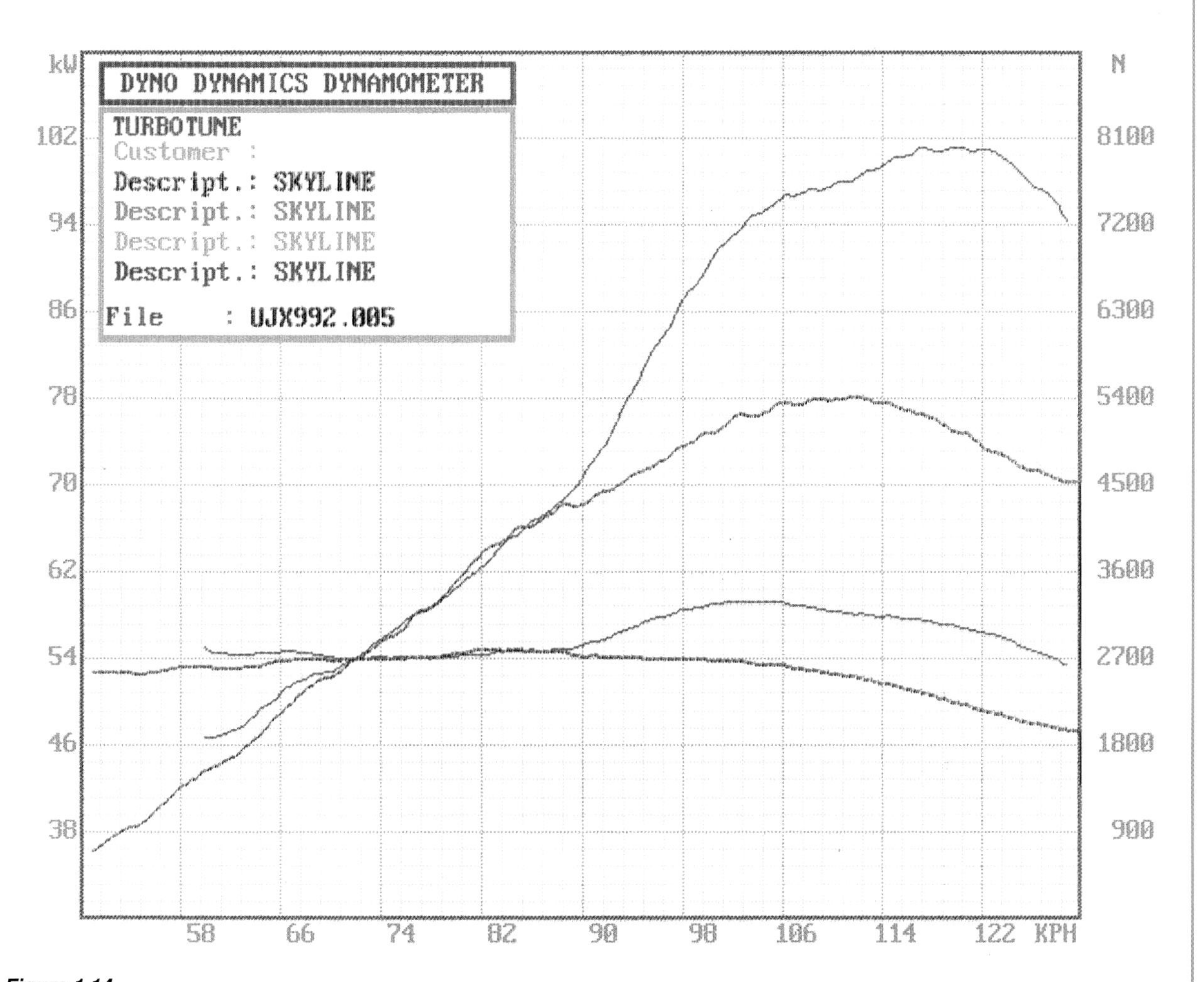

Figure 1.14
A comparison of the power and tractive effort developed on a chassis dyno by a Nissan two-valves-per-cylinder RB30 engine when equipped with the standard head (grey lines), and with the four-valves-per-cylinder RB25 head (red lines). Peak power improved by just under 30 per cent.

Ouch! An 8600rpm seizure in this BMW M3's engine at Mount Panorama in 1989 resulted in an interesting block modification... However, in road use, even major power upgrades can be made without block failure.

However, the power gain that's possible by taking this route makes this an attractive proposition. Figure 1.14 (previous page) compares the head flows of the 3-litre, SOHC, two-valves-per-cylinder Nissan RB30 in-line six with the head from the 2.5-litre, DOHC, four-valves-per-cylinder RB25. As can be seen, at 0.400 inches lift, the smaller engine's head has an inlet flow approximately 70 per cent greater and an exhaust flow 45 per cent higher! The power and tractive effort gains resulting from placing the RB25 head on the RB30 block are shown in Figure 1.15 (previous page). Even though there was a decrease in compression ratio to about 8.5:1, the power gain (nearly 30 per cent over the standard RB30 engine) was still impressive, with no low rpm power decrease.

THE BOTTOM END

As a rule of thumb, most well-designed modern engines will cope with power increases of up to about 80 per cent **without** changes needing to be made to the pistons, rods or crankshaft. However, whether this is actually the case will depend on the strength of these components in standard form and the way in which the extra power is being gained.

Some engines are simply extremely strong. For example, the Nissan SR20DET, RB30ET and the older L-series engines are immensely durable. These engines can be used in configurations that yield double the standard power output with no reliability problems — even in race applications! In normal road car use, where peak power is used for only a tiny percentage of the time, and if appropriate mixtures are held and detonation does not occur, there is simply no need to change bottom-end components. However, other engines are nowhere near as strong. Relatively flimsy conrods, small conrod bolts and pistons that are known to give problems can all be present. In these engines, doubling the power output is likely to lead to a broken engine very quickly!

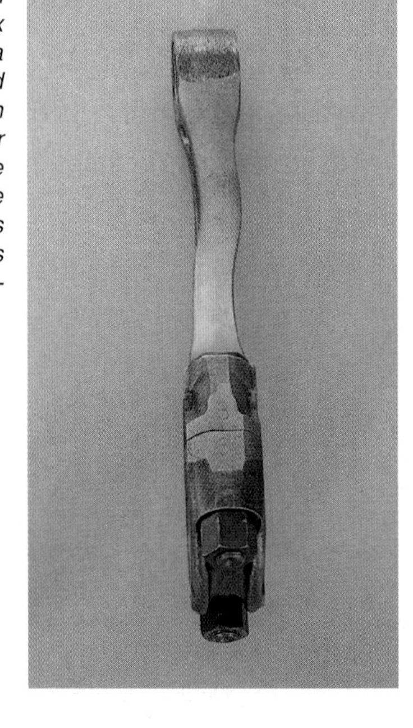

While this Nissan RB30ET 3-litre turbo six conrod has bent like a banana, this engine had to develop over 650Nm (more than a 100 per cent increase in torque over standard) before the conrod failed in this way. Many engines have very strong standard conrods.

This Lexus 1UZ-FE V8 bottom end has as standard six-bolt mains, generous bearing areas, large oil pump pick-up, alloy block.... As with many modern engines, leave this bottom end alone unless you intend more than doubling the engine's power.

So how do you know which class your car's engine falls into? There are three approaches that can be taken:

- Increase the power and see what happens.
- Have the engine pulled down and an expert inspection made before modification.
- Talk to those who have gained a great deal more power from the engine and question them about durability problems.

When increasing power substantially, or in highly modified forced-induction engines, the standard cast pistons should be replaced with forged units. Note the five valve recesses on this Toyota 4A-E 20 valve piston.

In addition to the basic engine's strength, the other point to consider is how the extra power is being gained. In modified forced-aspiration engines, the maximum speed to which the engine needs to be revved is unlikely to be much higher than standard — the extra power comes from the increased torque output rather than from higher rpm. If the engine's standard redline is not being exceeded, the likelihood of broken rods is much reduced. This is because the stretching loads as the pistons pass through Top Dead Centre are usually much higher than the compressive loads caused by combustion. However, if high torque is being maintained at greater than standard revs (so giving the power increase), the chance of engine destruction is much higher. For example, an engine with a 6000rpm redline, modified to develop peak power at 9000rpm, is much more likely to break than if it is modified to produce more power at 6000rpm. This is a strong argument in favour of forced induction — engine speeds can be kept relatively low, aiding engine durability.

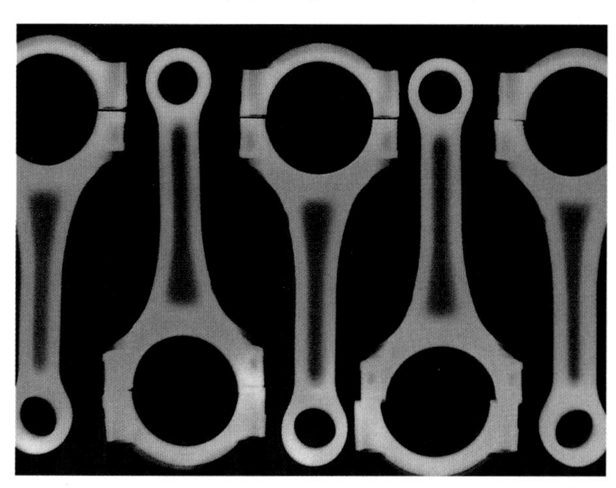

Standard conrod strength can be improved by using metal testing techniques (x-raying and magnetic crack-testing) to select only the best standard rods, followed by linishing and shotpeening.

H-beam, high-performance aftermarket rods are available for many engines and are a worthwhile investment if very high rpm and/or torque loadings are going to be applied.

If the maximum engine speeds are being kept near standard but higher combustion pressures (developing more torque) are being used to generate the increased power, the components most likely to need replacement are the pistons. Standard engines use cast pistons which, while cheap, are not as strong as forged pistons. Forged pistons should be used where the standard pistons are known to be weak or where peak torque is going to be increased by about 50 per cent or so. A naturally aspirated engine modified by the fitting of a blower or turbo should be fitted with forged pistons as a matter of course.

Anti-friction coatings on piston skirts (piston on left) and ceramic coatings on piston crowns are necessary only on very highly stressed engines.

If maximum engine speeds are to rise substantially, forged pistons and new conrods need to be used. H-beam rods and new high-performance rod bolts are available for most engines. Billet crankshafts, main bearing stud kits and the like are also available for some engines. In a maximum-performance engine — power increases of 100 per cent and upwards — all these goodies should be considered. However, enormous power gains of this magnitude are, frankly, quite uncommon in naturally aspirated modified road cars. As a result, running a modern engine full of trick internals is the exception rather than the rule.

In this chapter I have assumed you will not be carrying out an engine rebuild yourself. As touched on earlier, rebuilding a modern engine requires race-standard precision — just buying the measuring tools to allow this level of assembly will cost far more than paying an expert engine builder to assemble the engine. Thus, if you are not assembling engines on a frequent basis, it's not

Kits that increase capacity are available for some engines. This A'PEXi kit takes the Nissan SR20 engine from 1998cc to 2164cc via a 1mm increase in bore and a 5mm increase in stroke. However, this is a very expensive way of gaining extra power and torque.

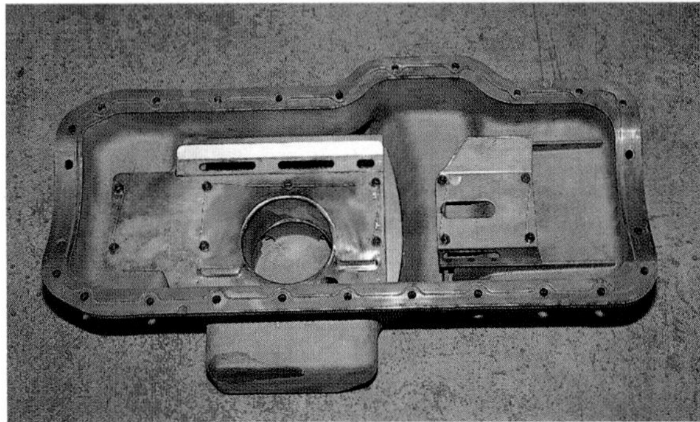

An engine retaining the standard sump and mounted in a car that can accelerate, brake and corner very hard is highly likely to experience oil starvation, especially if the oil level is a little low. This TRD sump for a Toyota 7M-GTE six features effective baffling and a crank scraper to reduce windage and return more oil to the pick-up.

worthwhile to put an engine together yourself — if appropriate accuracy is to be maintained, anyway! If changes to internal engine components are being considered, the advice of such a professional engine builder should be sought.

However, it should be noted that many people underestimate the strength of the standard internals of engines. For example, they may blame weak pistons or rods for failure that, in fact, was caused by detonation, lean mixtures or over-revving. The quality of the engine management is a major ingredient in keeping together an engine that's developing a lot more power than standard — engine management is covered in later chapters.

ENGINE SWAPS

Replacing an engine with the more powerful one from another car or model is a time-honoured way of achieving greater performance. It can also be very effective — giving you factory high performance, reliability, good emissions and better fuel economy in one fell swoop. If you select another engine from the same family, it's also quite likely that the new engine will simply bolt into place. If this is the case, such a swap can be very straightforward in RWD, FWD or 4WD cars.

The easiest way of carrying out a swap that involves changing engine families is to install both the engine and gearbox as one unit — in that way, problems of incompatible gearboxes, insufficient gearbox strength, sourcing a unique clutch and so on are avoided. Most engine/gearbox swaps are carried out on RWD cars, although if you are prepared to have new driveshafts made, FWD engine/gearbox swaps can also be done.

When extreme power is required, extreme engine components will be needed. This A'PEXi crankshaft to suit the Nissan RB26DETT twin turbo six is good for engines developing 900kW (~1200hp)!

An engine transplant can provide a cost-effective improvement in power, economy and emissions. This Lexus V8 is being tested for fitting in a VR Commodore.

Before doing any DIY development work, always make sure there's not an adaptor kit for the proposed swap already available off the shelf.

The first thing to do when considering an engine swap is to check that the engine will actually fit in the car. If you're prepared to perform surgery to the firewall, inner guards and so on, pretty well anything can be fitted into anything! But such a major engineering revision to the car's basic bodywork will require extensive work and also create legal problems. When considering the installation of a new engine, there are three vital dimensions to measure: the distance between the chassis rails, the height of the engine and its length.

If the engine is wider (longitudinal engines) or the engine and gearbox combination is longer (transverse engines) than the space between the chassis rails, it makes the swap very difficult without bodywork changes. If the engine is too high, a bonnet scoop or bulge can often be used to cover the injection plenum. If the sump hits the crossmember or suspension components, investigate what other sumps are available for the engine. If there are no alternative sumps, the standard unit may need to be modified. If the engine is longitudinally mounted and is too long by only 5-10cm, the radiator can be moved forward and thin electric fans used.

Because there are so many dimensions that need to be correct, it's best to do a trial-fitting (after you have concluded from some basic measurements that the engine is likely to fit), temporarily installing both the new engine and transmission as one unit.

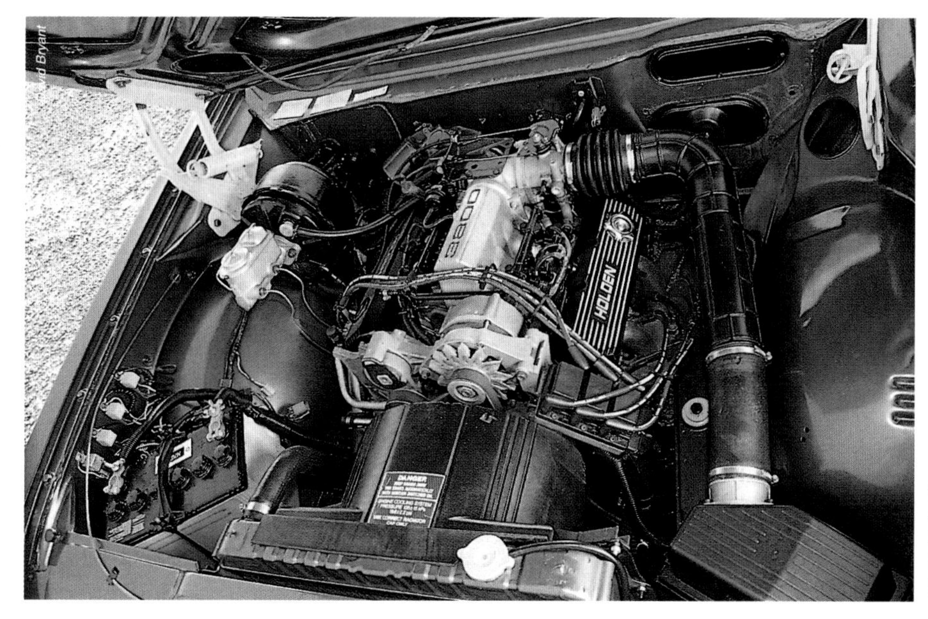

A Holden V6 transplanted into a Holden HG ute. The battery has been moved to the other side of the engine bay to make room for the factory airbox.

This old Datsun C210 Skyline uses an L-series 2.4-litre in-line six. Nissan followed this series of engines with the RB-series, with the new and old engines having very similar dimensions. As a result...

The actual location of the engine/transmission within the engine bay will depend on the most important fitting criterion that needs to be satisfied. In a transverse FWD car, the driveshaft angularity will be important; in a longitudinal engine car, the engine is normally placed as far rearwards as possible to aid weight distribution. Alternatively, it may be decided that it's important that the tailshaft length be kept stock, or that the old and new engine mounts line up with one another. In most cases, a compromise between these often contradictory requirements is necessary. However, make sure that at least something lines up, or you'll have to adapt everything!

If the block of the new engine is of different design, it's likely that new engine and gearbox mounts will need to be fabricated — and this is an area where some people do horrible things! First, inspect both the new and old engine mounts. Can the old mounts be used with the new engine? Can the rubber block from the new engine's mounts be used with the old engine's metal brackets? Check all the possible mixing and matching options using factory-produced mounts first. But if they won't fit, you will need to make or adapt your own. Engine mounts should have a straight load-bearing path between the engine and the chassis. Some gussets (triangular reinforcing pieces) welded into the corners will give the bracket greater strength. If fabricating from scratch, use at least the same gauge material as in the **new** engine's mounts. At the gearbox end, either carefully adapt the factory mount or make a new one from heavy gauge steel.

The new engine will probably not match the original radiator's water hose connections. If the radiator has the

...an RB20DET 2-litre in-line turbo six can be easily transplanted into a C210 Datsun Skyline, replacing the L2.4. This swap gives an instant 60 per cent power upgrade, with much better economy and emissions as well!

outlet (or inlet) on the wrong side, a radiator workshop can alter its position — this is much neater than running long crossover hoses. When sourcing appropriate new hoses, use welding wire to bend up a guide to the shape of the required hose and then browse through a selection of hoses at a big auto parts shop. A moulded hose that fits accurately is much better looking (and also flows better) than a 'universal' convoluted hose, or multiple sections of hose joined to steel tube with hose clamps. If the engine power is being increased very considerably, you will need to use a larger radiator, which can be selected to have hose locations to suit the new engine.

Other points to keep in mind when considering a swap are the compatibility of the:

- Throttle cable
- Clutch actuation mechanism
- Heater hoses
- Starter, oil pressure, coolant temperature and alternator wiring
- Power steering hoses and pump
- Air-conditioning compressor and hoses
- Speedo cable
- Exhaust pipe
- Throttle body location
- Airbox location

This list excludes the wiring and fuel supply changes associated with installing an engine using a new management system — a complete topic in itself!

CLUTCHES & FLYWHEELS

In any standard car boasting a power increase, it's likely the clutch will need upgrading. This is especially the case with rear-wheel- and four-wheel-drive cars; front-wheel-drive cars not fitted with slicks normally just spin their wheels when given the stick off the line! Even those four-wheel-drive cars without a lot of power are likely to have clutch problems when driven hard. When the clutch is dumped at high revs, the huge amount of traction from all four wheels means something has to give — and it's usually the clutch. Note that many manufacturers intend this to be the case — better to have the clutch slip than a driveshaft or gearbox component break. (And, of course, that means that if the clutch is upgraded so it **doesn't** slip, expect to break other driveline components!)

An upgraded clutch can be revised in several key areas: friction material, clutch plate design, pressure plate clamping force and pressure plate design. The friction material on the clutch plate can be of a number of compositions — organic, Kevlar or ceramic. Organic linings are used in most standard applications where they provide good durability, smooth engagement and low price. However, with a coefficient of friction of about 0.32, organic linings require high clamping forces to work effectively in performance applications. Kevlar linings have a similar coefficient of friction (0.30-0.35) but can

The burnt and blackened flywheel from a constant four-wheel-drive Subaru Impreza WRX. Four-wheel-drive cars place an immense loading on their clutches.

withstand much higher temperatures than organic linings. Kevlar clutch plates are smooth in their operation. Sintered bronze/ceramic has the highest coefficient of friction of the three lining types (0.48-0.55) but can cause excessive wear of the flywheel and pressure plate because of this. Shudder on engagement can also be a problem with this type of material.

Another factor that affects clutch grip, engagement properties and life is the design of the clutch plate. Most clutch plates use springs mounted so the engagement of the plate is cushioned. Plates equipped like this are said to have 'sprung centres' as opposed to 'solid centre' plates. The clutch plates fitted to most cars are of the sprung-centre, full-face design — the latter meaning the friction material is complete round the whole of the

Once a conventional clutch plate has worn down to the level of the rivets it will start to slip. This is a worn, full-face, organic-material clutch plate of the type used as standard in nearly all cars.

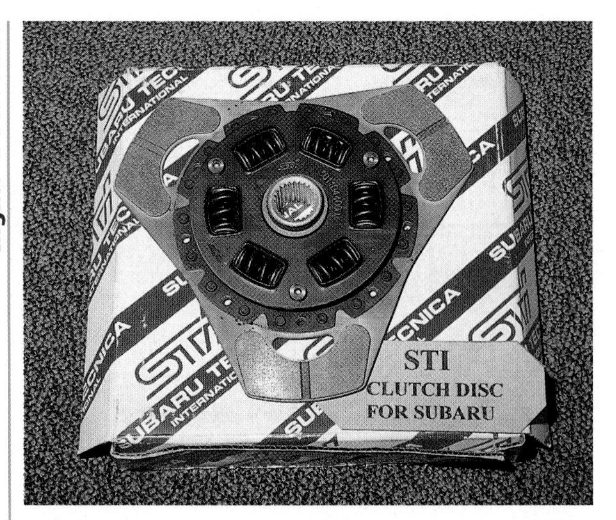

A Subaru Tecnica International three-puck paddle clutch. This competition-type clutch plate will handle high torque loads but be almost undriveable on the road, especially if manoeuvres like reverse parking need to be carried out.

clutch plate. This type of plate gives the smoothest engagement but also requires a high clamping force. If the material on the clutch plate is divided into discrete sections, the resulting plate is termed a paddle, puck or button clutch. With this approach, the clutch plate itself is often shaped into a series of separate paddles, or the clutch plate can be conventional in appearance but use discrete blocks of friction material. Without changing either the friction material or the pressure plate, fitting a clutch plate of this type will give a greater torque capacity before slippage occurs.

However, paddle/puck/button clutch plates can have very significant trade-offs in a road car — much more sudden engagement and an increased rate of lining wear. The 'grabbiness' of these types of clutch plates can sometimes be so bad that, for example, reverse parking is impossible. Furthermore, an engine with poor bottom-end torque (eg a small-capacity turbo engine) will be very easy to stall when moving away from a standstill. All road car clutch plates should use a sprung centre so some of the shock loadings are absorbed but, even with such a centre, paddle clutches can be bad news in a road car unless you are prepared to make significant trade-offs in driveability.

As the name suggests, twin-plate clutches use two friction plates. These have a significantly increased torque capacity over single-plate clutches, but again do not have quite the driveability of a standard, full-face, organic-lining clutch. However, with some heel-and-toeing of down-changes and some care in manoeuvring, twin-plate clutches can be quite acceptable on the road. A twin-plate clutch consists of a completely new clutch assembly and flywheel, so this type of clutch is much more expensive than the alternatives.

A five-puck clutch will give more gentle engagement than one with only three pucks. However, how user-friendly the clutch actually is will depend on the torque characteristics of the engine and the type of friction material used.

A modification that can be made to the standard clutch is to increase the pressure plate clamping force. One way this can be achieved is by the use of a heavier diaphragm spring. Since most current performance cars (especially Japanese) in standard form have quite a light clutch pedal effort, increasing the clamping force (while retaining a clutch plate and material design that still give good driveability) is an excellent option for an upgraded clutch. This is especially the case for cars that use hydraulically operated clutches, with or without vacuum assistance.

When refacing is necessary, the friction surface of a flywheel should be ground rather than machined.

Another way of achieving a higher clamping force is to move the fulcrum (the 'knife edge') against which the diaphragm spring within the pressure plate works. Once this has been done, the clutch release bearing (and so the pedal) will have to travel further to disengage the clutch — but the clamping force will be higher. By using one or both of these techniques, a major increase in clamping force can be obtained with some pressure plates, the individual design determining what upgrades are possible.

Figure 1.15 (next page) shows some of the advantages and disadvantages of taking different approaches to clutch modification. In addition to the points in the table, keep in mind the type of engine idle speed control system that's being used. Some programmable management systems use quite crude idle speed control — a clutch that engages suddenly will more easily cause the engine to stall in these cars than in cars using better programmable management or the factory engine management system. Finally, note that when a clutch is being replaced, it's good practice to also replace the engine's rear main seal and the seal on the gearbox input shaft; oil leaks are responsible for many a slipping clutch.

Before you buy a new high-performance clutch, it's **vital** that you drive a car equipped with the type of clutch you are considering purchasing. Some people's "it's as

		Advantages	Disadvantages
Friction Material	**Riveted Organic**	• Smooth engagement • Cheap • Commonly available	• Limited torque capacity • Slippage results in early failure
	Bonded Organic	• Cheap • Smooth engagement	• Possibility of shudder • Harsher engagement
	Steel-backed organic	• Durable	• Poor availability
	Ceramic	• Very durable	• Not as smooth in engagement as other materials
	Kevlar	• Very durable • Fairly smooth engagement	• High price • Requires higher rated pressure plate
Friction Plate Design	**Full face**	• Smooth engagement • Common • Cheap	• Limited torque capacity • Slippage results in early failure
	Button/Puck/Paddle (large number of buttons eg 6)	• High torque capacity • Can take slippage without disintegrating • Less affected by oil contamination than full face	• More sudden engagement than full-face • Can be prone to shudder, squeal • With slipping, may wear flywheel and/or pressure plate
	Button/Puck/Paddle (small number of buttons eg 3)	• Very high torque capacity • Less affected by oil contamination than full face	• Very sudden engagement • May cause drivetrain breakages eg gearbox
Pressure Plate Modifications	**Stiffer diaphragm spring**	• Increased torque capacity • Simple • Cheap	• Modification cannot be done to all pressure plates • Higher clutch pedal effort • Cable or hydraulic clutch actuation can be overloaded
	Moved diaphragm pivot point	• Increased torque capacity without increased pedal effort	• Expensive • Longer pedal travel and/or less internal clutch clearance
Twin plate clutch		• Smooth, near standard feel • Excellent torque capacity	• Very expensive • Not widely available • Bulkier assembly • More complex

Figure 1.15
The advantages and disadvantages of different high-performance clutches.

smooth as stock — you wouldn't even know it's a high-performance clutch" can be other people's kangaroo-hopping, stalling, jerking monstrosity. It's **extremely** easy to spend a lot of money on a new clutch and be unhappy with the result — drive before you buy!

Reducing the mass of the flywheel is a simple and relatively cheap step to take while the clutch is being replaced. The lightened flywheel will take less power to speed it up when the engine is accelerating — as a result, more power will be available to accelerate the car linearly. Flywheels are lightened by being machined in a lathe, with this process able to be undertaken by workshops specialising in motorsport machining. The reduction in mass will be most beneficial if the material is removed from the periphery of the flywheel rather than towards the centre. Once machined, the flywheel must be balanced. Alternatively, some cars have available aftermarket flywheels that are lighter than stock and are already balanced.

Contrary to popular belief, a flywheel lightened by a moderate amount (eg 30 per cent) will not cause detectable changes in idle quality or make hillstarts any more difficult. Instead, better acceleration will occur — especially in lower gears — and the engine will feel more

A twin-plate clutch for a Subaru Impreza WRX. While expensive, a twin-plate clutch has a very high torque capacity and can be quite driveable on the street.

responsive to throttle movements. On a Nissan Skyline GTR, reducing the mass of the flywheel by 29 per cent resulted in a measured improvement in the rolling 60-90km/h time from 1.8 to 1.7 seconds. Having a flywheel machined, balanced and ground flat does not cost a great deal of money — about a fifth the price of a high-performance clutch — so is a worthwhile process to undertake while the gearbox is out of the car.

GEARBOXES

Manual Gearboxes

Most gearboxes are able to handle the increased torque resulting from normal engine modifications. Even in those gearboxes that are being stretched in capacity, using good oil and top-quality bearings is generally sufficient to cope with a 50-100 per cent power increase, especially if the power increase is achieved at higher rpm so that the increase in peak torque is reduced. However, one way gearboxes can be broken is when an engine swap has been undertaken but the original gearbox has been left in place. For example, if a factory turbo engine has been installed with the original gearbox and the new engine is then modified for greatly increased power, gearbox problems are probable. In this situation it's wise to install the gearbox that was originally sold with the more powerful engine.

If your type of car is involved in professional motorsport, it's likely there are gearbox upgrades available. However, before investing in new internals for the standard box (or a completely new gearbox) make sure the competition gearbox is suitable for road use. Most competition gearboxes lack synchro rings and are also very noisy.

A better alternative for rear-wheel-drive cars is to look at the availability of kits allowing the installation of a gearbox from another car. Taking this route can be a cost-effective and relatively simple way of upgrading gearbox strength, and often the number of ratios as well. When assessing the suitability of a new gearbox, look at the peak torque of the engine behind which the gearbox originally sat, and also check the difference in gear ratios between the new box and the standard one fitted to your car.

The quality of the gearshift can make a substantial difference to the speed of shifts and the general feel of the gearbox. Lapping-in the synchro rings will improve gearshift quality and is especially effective if the box crunches on fast shifts. Fitting a short-shift gear lever is a cheap and effective way of improving shift speed and feel, as is using synthetic high-quality oil. However, note that if the gearbox has started crunching, it's usually too late to fix the problem by using good oil!

Automatic Gearboxes

Automatic gearboxes normally require substantial upgrading when placed behind engines boasting a major increase in torque output. There are three areas where changes are made:

- The torque converter
- The frictional components
- The control system

Most modifications to engines increase top-end power at the expense of bottom-end torque. Fitting a larger than standard turbo, using short-length intake runners or fitting warmer cams will cause a reduction in bottom-end torque. So that an automatic car continues to launch off the line as hard as possible, the stall point of the torque converter is usually revised upwards.

Changing the stall point of the converter will mean that when the car is stationary in gear with the brakes on and some throttle applied, the engine will be able to rev to a higher speed. When the brakes are released, the car will accelerate more quickly than it would with a lower rpm stall converter. This is the case for two reasons: the engine will be developing more torque at the higher rpm and the change in speed of the engine will be less in the first few seconds after launch, so decreasing the effect of rotational inertia. In a turbo or centrifugally supercharged car, boost will also be higher. (Incidentally, turbo engines and automatic transmission cars work very well together.) Note that even when fitted with a high-stall torque converter, the car will still be able to be driven at lower rpm than the stall — part throttle drive will still occur. Changing the torque converter stall speed requires that the converter be removed from the car, cut open and internally modified.

When the wet clutches and bands inside a transmission are subjected to much higher loads than normal they are very likely to slip. The friction material on these components can be replaced with higher specification material, or the transmission can be modified so that more clutches or wider bands are used. Note that the overheating of transmission fluid will result in a substantially decreased transmission life and a much higher chance of slippage. In hot climates, the transmission oil cooling circuit through the radiator should be disconnected and a substantial oil cooler used to replace it. A copper-cored condenser from a used industrial or domestic air-conditioner provides a very cheap and effective oil cooler. Where the transmission is working very hard, an oil temperature gauge should also be fitted.

The traditional modification of transmissions equipped with pure hydraulic control has involved the installation of a 'shift kit', which makes changes to the hydraulic control system, normally referred to as the valve body. Shift kits consist of a new separator plate — a thin metal plate

Lightening of the standard flywheel will allow the engine (and car!) to accelerate faster. This is the rear view of a standard flywheel that has a mass of 9.5kg...

... and this is what it looks like when it has been machined. In this form, its mass has dropped to just under 7kg.

containing openings that direct the passage of the fluid through the valve body. Sometimes a drill bit has also been provided to allow the enlargement of a particular hole, and stiffer spool valve springs could also be fitted. These modifications increased clamping pressure of the friction components, increased the rpm at which upshifts and downshifts occurred and quickened the shift time. In transmissions that are still largely hydraulic in nature, this approach can still be carried out. For example, a transmission that uses electronic control only of the overdrive function, the torque converter lock-up clutch and the like can still be modified in this manner.

However, most auto transmissions today are fully electronically controlled. While hydraulic actuators are still used to apply clamping forces, the flow of fluid through a valve body is not used to direct the operation of these components. Instead, a dedicated Electronic Control Unit (ECU) — or one integrated with the engine management ECU — decides when and with what force the actuators are applied.

Making electronic control changes alone is not sufficient to substantially improve the torque-handling capacity of this type of transmission. Instead, extra clutches, better friction material and so on should be installed. Once these mechanical changes have been carried out, electronic changes can then be made to increase line pressure and quicken shifts. As is the case with engine management modifications (outlined in subsequent chapters), these electronic changes can be carried out by:

- Intercepting signals travelling to (or from) the electronic control unit and making changes to these signals
- Having the main transmission control PROM chip re-written
- Using a dedicated software package that allows the complete reprogramming of the factory software (eg: Kalmaker for GM-Delco systems)

The electronic changes can also be made so an otherwise unmodified transmission behaves in a more sporty manner — holding gears to higher rpm or having a more distinct difference between 'power' and 'economy' modes, for example. This type of change can be very effective; despite this, chip companies seem to be reluctant to offer dedicated transmission control chips.

DIFFERENTIALS

All two-wheel-drive cars are equipped with a differential, a device that allows the left and right driven wheels to turn at different speeds when cornering, while still transferring torque through each wheel. Most two-wheel-drive

The front face of the standard flywheel shows the blackened surface — evidence of too many hard launches!

The flywheel after it has been machined (note the three areas where lightening has occurred on this side of the flywheel) and the ground face.

cars use an 'open' diff. This means that when a wheel begins to spin, all engine torque is (unfortunately) directed to the spinning wheel. This gives poor traction in slippery conditions, when exiting corners and/or when the car is powerful. Constant four-wheel-drive cars have three differentials — one between the front wheels, one between the rear wheels and a 'centre' diff that allows variation in the front/rear torque transfer. (Note that the centre diff may be located near the front of the car!) Without the centre diff, the transmission of a four-wheel-drive car would wind up as the front and rear axles turned at different speeds while cornering. To prevent single wheelspin occurring, many RWD performance cars are equipped with limited-slip differentials (LSD). As the name suggests, an LSD limits the speed difference that can occur between the wheels. Constant four-wheel-drive cars often use LSD rear and centre diffs, while just a handful of powerful FWD cars use an LSD.

LSDs are available in a number of types:

1. Eaton Locker

During normal driving conditions, this type of diff operates as a conventional open design. However, as soon as wheel slip occurs, a flyweight governor spins rapidly, catching a latching bracket and initiating lock-up. During lock-up, a self-energising clutch system causes a cam plate to ramp against a side gear. This ramping compresses clutch packs that are inserted between the side gears and the case, limiting the difference in speed. This type of LSD is most frequently used in RWD cars.

2. Frictional

This type of LSD uses either friction plates or friction cones to limit the amount of slippage between the wheels. When both wheels have the same amount of torque applied to them, the friction surfaces are not engaged and the differential acts conventionally. However, when an increased amount of torque is fed to just one wheel, springs and pinion/side gear reaction forces cause the friction surfaces to become engaged. The amount of allowable slip can be altered by changing the strength of the internal springs, the number and/or size of the clutch discs, or by altering the properties of the differential oil. This type of LSD is most frequently used in RWD cars.

After lightening, the flywheel must be balanced, followed by the balancing of the combined flywheel/pressure plate assembly.

Lapping-in the synchro rings (indicated) can provide improved gearshift speed. This process can be carried out only with the gearbox in pieces.

3. Viscous

This uses a design that splines multiple plates to each driveshaft. The plates are held in close proximity to each other and are immersed in silicone fluid. When the shafts rotate at the same speed, there is no limited-slip action. However, when one shaft rotates more rapidly than the other, the silicone fluid is heated and quickly becomes much more viscous, limiting the difference in the speed of the shafts. Unlike a frictional LSD, the limited-slip action occurs in reaction to the difference in speed rather than the amount of torque being transmitted. The viscous LSD is frequently used as the centre diff in constant four-wheel-drive cars and also sometimes in those rare FWD cars that use LSDs. The viscous LSD unit is sealed — the fluid cannot be easily changed.

A Nissan RWD viscous LSD. If the manufacturer makes a diff in a variety of versions (open, viscous LSD or friction LSD) a swap is normally straightforward.

4. Torsen

This type of design uses a variation on worm and ring gears. Forces between the diff housing and the output shaft (ie engine torque being transmitted to the wheels) are directly coupled, whereas forces between the two output shafts (the difference in wheel speeds during cornering) cause internal gears to rotate, permitting differentiation. Torsen diffs can be used in all automotive diffs — FWD, RWD and 4WD. Later-model Torsen diffs use a different system called parallel-axis gearing, which relies on internal friction to transfer torque.

The centre viscous LSD of a Subaru Impreza WRX. This diff apportions torque on the basis of relative shaft speeds, feeding an increased amount of torque to the shaft (and so wheels) that are spinning more slowly.

5. Quaife

This type of diff is similar in basic principle to the Torsen design except that internal springs are used and the angle at which internal gear teeth are cut helps influence the behaviour of the limited-slip action. As with Torsen designs, the limited-slip action is gentle and progressive. Quaife aftermarket diffs are available for a very wide range of FWD, RWD and constant four-wheel-drive cars.

6. Active

An active differential uses electronic control to alter the torque split. This type of diff most often comprises a wet multiplate clutch, which is engaged by hydraulic pressure. Infinite variations in torque transfer are able to occur, either fore/aft (when used as a centre diff) or from side to side (when a rear diff). Some FWD cars use systems described as Electronic Differential Locks and similar, but as these simply brake spinning wheels, they are more akin to traction control than an LSD.

The front diff and gearbox of a Subaru Impreza WRX. As with most constant four-wheel-drive cars, this is an open diff.

If you have a powerful RWD car with an open diff, the on-road performance will be severely limited. Most RWD cars have factory LSDs available for them and it's simple to fit one of these off-the-shelf units. The more sophisticated the LSD used (locker, frictional, viscous, or Torsen/Quaife — in that order!) the better the end result. However, such a wide choice is usually unavailable and any LSD is better than none. Note that power oversteer will **increase** in most powerful RWD cars when an LSD is fitted, because instead of spinning away the torque through one wheel when cornering, both wheels will spin when a lot of torque is supplied — resulting in lateral movement. However, corner exit speed and straight-line traction will both be substantially improved. Frictional-type LSDs can be modified to provide less slippage, but the higher the level of lock-up, the potentially greater will be understeer as the rear wheels attempt to push the car straight on. (Chapter 12 looks in more detail at vehicle handling.) Welded-up diffs and mini-spools should be avoided in road cars.

A powerful FWD car is absolutely transformed with the fitting of an LSD. Both in straightline traction and cornering performance, the difference is **very** substantial. Locker and frictional LSDs are too harsh in their action for front-wheel drives, so if an LSD is to be fitted, viscous, Torsen or Quaife designs need to be used. Unfortunately, the availability of FWD LSDs is limited. Some manufacturers have produced FWD cars factory-fitted with LSDs, and sometimes these units can be adapted to similar gearboxes used in other models. Quaife produces aftermarket LSDs for a number of FWD cars, as does Cusco.

The LSDs used in constant four-wheel-drive cars seldom need to be modified — traction is not usually a problem for these cars! However, changing the slip characteristics of the centre diff will result in altered handling. Some four-wheel-drive road cars also used in rallying have special motorsport 'stiffer' centre viscous couplings that transfer torque to the slipping end more quickly, while others have a driver-definable degree of lock-up. However, the latter is often designed for only loose or slippery surfaces and causes transmission wind-up when activated on dry bitumen.

A two-wheel-drive car without an LSD can become a one-wheel-drive car very easily when there is a surplus of power available...

Electronic Engine Management

Understanding how electronic engine management works is very important if you are to successfully modify the system. Retaining the driveability, fuel economy and emissions performance of the standard car are all possible, but not if the modifications you make are carried out in complete ignorance of how the management system operates!

However, this chapter takes only a general overview of the components and concepts common to most management systems — it doesn't attempt to provide car-specific information. If you intend modifying your car in the pursuit of power and wish to retain elements of the standard management system, I strongly advise that you buy the official manufacturer's workshop manual for your car and study it closely. Many enthusiasts never even think to do this, instead preferring to buy a supermarket-level guide on how to change the plugs and the oil! But the very best of the manufacturers' workshop manuals are quite brilliant — almost textbooks on the design and technology of the cars they cover. Even if the manuals cost you the equivalent of a couple of high-performance tyres, they're usually worth the purchase price in the long run.

SPARK AND FUEL

All engines require that the combustion air and fuel are mixed in the right proportions and that each cylinder's spark plug is fired at the right time. These two points are very important to remember because it's easy to lose sight of these basics when making modifications or considering complex engine management systems. The amount of air mixed with the fuel is described as the air/fuel ratio. A 'rich' mixture (eg 12:1) uses lots of fuel and is good for power, while a 'lean' mixture (eg 17:1) uses less fuel so is more economical. With normal fuel, an air/fuel ratio of 14.7:1 is termed stoichiometric and is the chemically correct ratio for complete combustion. The ignition timing refers to the point in the crankshaft's rotation at which the spark plug fires. This is normally expressed in crankshaft degrees. If the spark is fired when the crankshaft is 15 degrees Before Top Dead Centre, the spark timing is referred to as having '15 degrees of advance'. The ignition advance needs to vary because at high revs there is less time available for combustion to occur, so the spark needs to start the burn earlier. Changes in timing are also needed at different engine loads, different temperatures and with different octane fuels. Note that even when the timing is retarded, the spark plugs are almost always still fired Before Top Dead Centre.

THE INPUTS

In order that the engine's Electronic Control Unit (ECU) makes the right decisions about how much fuel to inject and when to fire the spark, it needs to know precisely the operating conditions of the engine. Is the car travelling at full throttle up a long hill at 3000rpm in very hot conditions? Or is the cold engine idling after it has been started for the first time that day? It's the input sensors that provide this type of information to the ECU.

1. Temperature Sensors

The Coolant Temperature Sensor is used by the ECU to determine engine temperature. This sensor is normally mounted on the thermostat housing and comprises a device whose resistance changes with temperature. Usually, as the sensor gets hotter, its resistance decreases. The sensor is fed a regulated voltage by the ECU and the ECU then measures the changing voltage coming back from the sensor. From this, the ECU can work out the temperature of the coolant. The temperature of the intake air is also sensed, with this device located on the airbox or on an intake runner. Like the coolant temperature sensor, the Intake Air Temp Sensor is normally a temperature-dependent resistor. Some cars use other temperature sensors to measure fuel and cylinder head temps.

The coolant temperature sensor is often located on the thermostat housing. Here the engine management temp sensor has a yellow plug, while directly behind it, the sender for the temperature gauge uses a black plug.

2. Airflow Meters

Many cars use airflow meters to measure engine load. Airflow is relevant because the amount of power being developed is proportional to the mass of air the engine is breathing at that rpm. If the engine is drawing in a lot of air (because the throttle is fully open at 6000 rpm with the car travelling quickly up a steep hill, for example), lots of fuel will need to be injected to keep the air/fuel ratio correct. The high load that the engine is undergoing will also influence the ignition timing that the ECU selects.

There's a number of different types of airflow meters available. In a hot-wire airflow meter the intake air flows past a very thin, heated platinum wire. This wire is formed into a triangular shape and suspended in the intake air path. The platinum wire forms one arm of a Wheatstone bridge electrical circuit and is maintained at a constant temperature by the electricity flowing through it. If the mass of air passing the wire increases, the wire is cooled and its resistance drops. The heating current is then increased by external circuitry to maintain the bridge balance. The value of the heating current (as measured by the voltage drop across a series resistor) is therefore related to the mass intake airflow.

Hot-wire airflow meters have very quick response times and are internally temperature compensated. To make sure the platinum sensing wire remains clean, it's heated red-hot for a short time after the engine is switched off, burning off any dirt or other contamination. (Note that in some airflow meters the hot wire is replaced with a hot-film resistor.) Large hot-wire airflow meters pose very little restriction to intake air — the wire mesh screens used at each end of the meter, in fact, cause most of the disruption to the intake flow. Hot wire airflow meters can have either a 0-5 volts or variable frequency output signal. In the US, a management system using this approach is sometimes called MAF — mass airflow.

One of the oldest automotive airflow meters is the vane airflow meter. This employs a pivoting flap which is placed across the inlet air path. Once the engine starts, low air pressure is experienced on the engine side of the vane, causing the flap to open a small distance. As engine load increases, the flap is deflected to a greater and greater extent. To prevent the flap overshooting its 'true' position (and also erratic flap movements being caused by intake pressure pulsing) another flap is connected at right angles to the sensing vane. This secondary flap works against a closed chamber of air, thus damping the motion of the primary vane. Mechanically connected to the pivoting assembly is a potentiometer (a 'pot') usually made up of a number of carbon resistor segments and a metal wiper. As the vane opens in response to the airflow, the wiper arm of the pot moves across the segments, changing the electrical resistance of the assembly. A regulated voltage is fed to the airflow meter, so that as the vane moves in response to airflow variations, the output voltage from the meter changes.

A spiral spring with an adjustable pre-load is used to relate the angle of the flap to the airflow and to ensure the flap closes when there is no airflow. An air bypass is constructed around the flap, with an adjustment screw positioned in this bypass, allowing control over idle mixtures. Because they measures only air volume (not mass), vane airflow meters require built-in temperature sensors. With both volume and temperature inputs, the ECU can work out the mass of air being breathed.

The intake air temperature sensor is often located on the plenum chamber or on one of the intake runners. Here the temp sensor is located at the beginning of the plenum chamber, in the foreground.

The vane airflow meter always causes a restriction to the inlet airflow. At low loads, the vane partially blocks the intake air path, and at all loads the relatively small internal cross-sectional area of most vane meters causes a pressure drop. Backfires can cause distortion of the alloy vane — when operating properly, the flap in a vane airflow meter should be able to be moved through its full travel with just light finger pressure.

A fuel pump switch is built into many vane airflow meters. This is designed to shut off the fuel pump if there is no air passing through the meter and so disables the pump when the vane is in its fully closed position. Because of the need to measure air volume and air temperature,

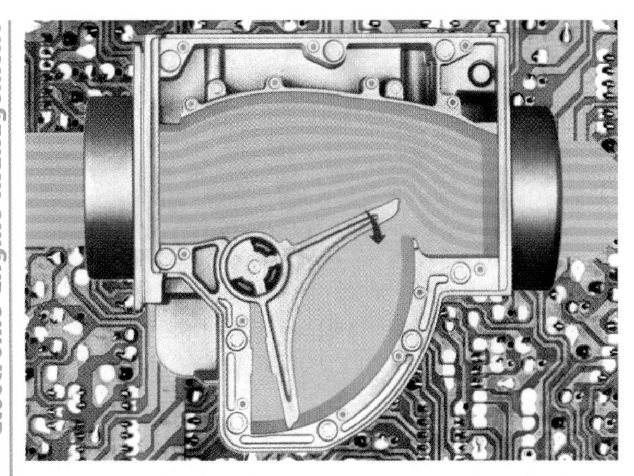

A vane airflow meter has a flap that's deflected open by the passage of the intake air, as is being done manually here. This flap is connected to a second flap that works inside a closed chamber, helping to damp pulsations of the primary vane.

and to operate the fuel pump, vane airflow meters can have up to seven connections within their plug. Note also that the pin-outs (the electrical function of each connection) don't always match from manufacturer to manufacturer! Vane airflow meters typically have 0-5 volt signal outputs, but some use a 0-12 volt output range.

Karman Vortex airflow meters generate vortices whose frequencies are measured by an ultrasonic transducer and receiver. They use a flow-straightening grid plate at the inlet to the meter and can be quite restrictive. This type of meter has a variable frequency output.

3. MAP Sensors

Some cars don't use any form of airflow meter, instead relying on the input of three sensors to work out how much air is being breathed by the engine. One is the inlet air temp sensor, the other is an rpm sensor (we'll get to that in a moment) and the third is a manifold vacuum sensor. This sensor is called a MAP (Manifold Absolute Pressure) sensor. A MAP sensor continuously measures the pressure in the intake manifold.

The pressure in the intake manifold is dependent on rpm, throttle opening and load. In a naturally aspirated car, the measured pressure will be below atmospheric in all conditions except full throttle (where it will be the same as atmospheric pressure). In a car with forced aspiration,

Beneath the black cover on this vane airflow meter is an adjustment mechanism that allows the air/fuel ratio to be changed.

the pressure will be above atmospheric when the engine is on boost and below atmospheric when it's not. In a naturally aspirated car, when the MAP sensor registers a low manifold vacuum and the rpm sensor indicates that the engine is at high rpm, a high load situation is signalled to the ECU. High manifold vacuum at high **or** low revs means the throttle is closed — indicative of a low load situation. In the US, the MAP sensor approach is somtimes called 'speed density'.

A MAP sensor is fed manifold pressure via a hose connected to a port situated after the throttle butterfly. Because intake air does not have to flow through the MAP sensor, the sensor does not cause any intake restriction at all. The same MAP sensor can be used on a very wide variety of engines, so it's the type of load sensor used with most aftermarket programmable management systems.

A hotwire airflow meter senses mass airflow through the cooling of a heated platinum wire or film resistor. Hot-wire airflow meters usually pose little restriction to airflow.

MAP sensing has some limitations when engine modifications are being made. While to a greater or lesser extent a modified car with airflow metering will measure the extra air needed and add the correct amount of extra fuel, a modified MAP-sensed car may run lean enough to actually decrease power. This is because the ECU cannot measure the extra airflow. All it knows is there's a certain manifold vacuum at a certain rpm with a certain inlet air temperature.

The other point to be aware of with a car running a MAP sensor is that if you fit a hot cam, the car will run poorly at light loads. This is because the light load vacuum signal is changed by the cam, confusing the ECU. This makes ECU modification vital when a new cam is fitted to a MAP-sensed engine.

MAP sensors are often referred to as being 1-Bar, 2-Bar or 3-Bar devices. This refers to their measuring range, with 1-Bar sensors being used on naturally aspirated engines, 2-Bar sensors being used on forced induction engines running up to 1-Bar (14.5psi) boost, and 3-Bar sensors on engines using up to 2-Bar (29psi boost).

Programmable management systems often mount the MAP sensor within the ECU (with a hose connecting it to the plenum chamber), while cars using MAP sensors as original equipment normally have the sensor mounted on the firewall.

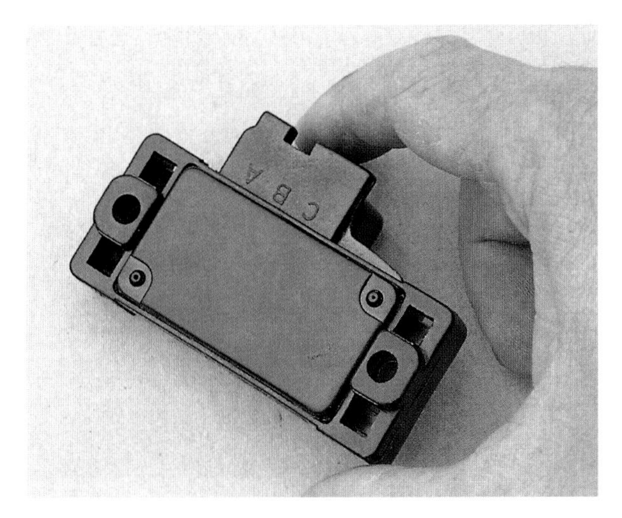

A Manifold Absolute Pressure (MAP) sensor can be used by the ECU (in conjunction with rpm and air temperature inputs) to allow the calculation of mass airflow. MAP sensors are connected to the plenum chamber by a hose.

4. Position Sensors

The ECU needs to know how fast the engine crankshaft is rotating and where it is in its rotation. It uses one or more position sensors to collect this information. Some cars use both camshaft and crankshaft position sensors, allowing the ECU to sequentially fire the spark and injectors. There's a number of different types of position sensors.

Optical position sensors use a circular plate with slots cut into it. The plate is attached to the end of the camshaft and is spun past an LED. A sensor on the other side of the disc registers when it sees the light shining through one of the slots, with the ECU then counting the light pulses. Some optical sensors use 360 slots in the disc, allowing very fine resolution of engine speed. A Hall Effect position sensor uses a set of ferrous metal blades that pass between a permanent magnet and a sensing device. Every time the metal vane comes between a magnet and the Hall sensor, the Hall sensor switches off. This gives a signal whose frequency is proportional to engine speed. An inductive position sensor reads from a toothed cog. It consists of a magnet and a coil of wire, and as a tooth of the cog passes, an output voltage pulse is produced in the coil. Engines with sequential fuel or spark often have a second position sensor or use two crankshaft sensors mounted in the one package.

This is an optical position sensor. It uses a spinning slotted disc that interrupts the light being emitted from a LED. Many optical sensors have very fine resolution.

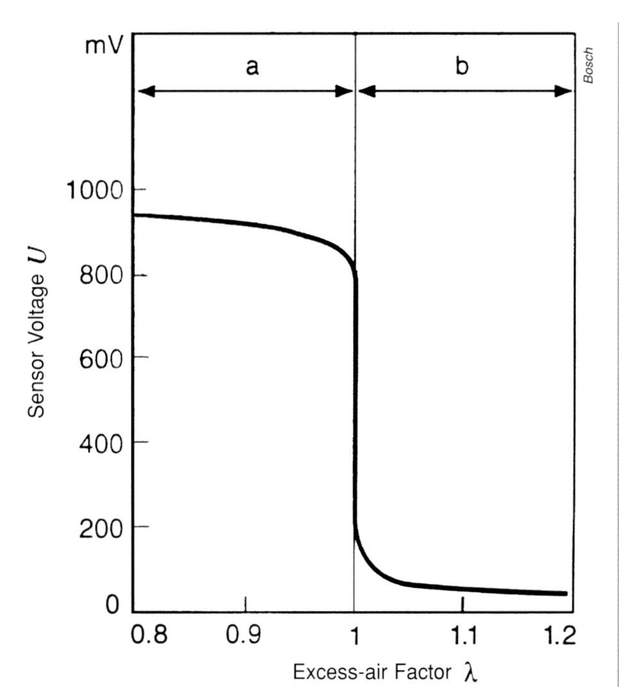

Figure 2.1
The output voltage curve of a typical oxygen sensor shows how the voltage suddenly switches at a precise mixture strength. On the diagram 'a' is a rich mixture and 'b' a lean mixture. (For fuels with a 14.7:1 stoichiometric ratio, Air/fuel ratio = excess air factor x 14.7:1.)

5. Oxygen Sensor

The Oxygen Sensor (sometimes called the Lambda Probe) is located in the exhaust close to the engine. It measures how much oxygen there is in the exhaust compared with the atmosphere and by doing so indicates to the ECU whether the car is running rich or lean. The sensor generates its own voltage output, just like a battery. When the air/fuel ratio is lean, the sensor emits a very low-voltage output, eg 0.2 volts. When the mixture is rich, the voltage output is higher, eg 0.8 volts.

The ECU uses the output of this sensor to keep mixtures around 14.7:1 in cruise and idle conditions. To facilitate this, the voltage output of the sensor switches quickly from high to low (or low to high) as the mixtures move through the 14.7:1 stoichiometric ratio. Note that this means the raw voltage output of the oxygen sensor is **not** directly proportional to the air/fuel ratio. Figure 2.1 shows the output of a typical oxygen sensor.

Older oxygen sensors use just a single wire (and engine earth). These sensors are unheated, while later three-wire sensors have a 12-volt heating element within them. More recent designs include four- and even five-wire sensors. Generally, the greater the number of wires, the better the quality of the sensor! Oxygen sensors can be contaminated by the use of inappropriate silicone sealants and leaded petrol. They can also become carboned-up over time, slowing their response.

Some cars use multiple oxygen sensors, positioned before and after the catalytic converter(s). Cars equipped with sensors arranged in this manner have ECUs that can assess the effectiveness of the cat converter operation. Twin turbo and V engines usually have at least two oxygen sensors.

The oxygen sensor screws into the exhaust. It detects the ratio of oxygen in the exhaust gases compared with that in the atmosphere, and so indicates the air/fuel ratio.

Other Input Sensors

A number of other sensors are also common to most engine management systems. The throttle position sensor indicates to the ECU how far the throttle is open. This information is used, for example, during acceleration enrichment. Most throttle sensors use a variable pot, but some comprise just two switches — one for idle and the other for full throttle.

The vehicle speed sensor lets the ECU know how fast the car is travelling. The sensor can be mounted on the gearbox or in the speedometer.

A throttle position sensor detects the amount of throttle opening. A potentiometer-type sensor such as this can also be used by the ECU to detect how quickly the throttle is opening or closing.

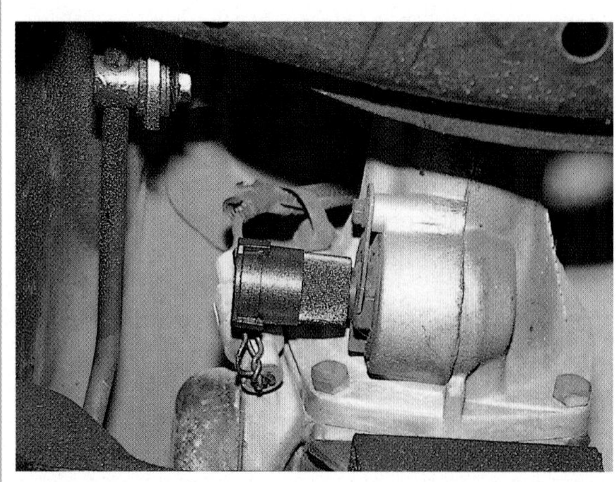

A road speed sensor can be mounted in either the gearbox (as here) or in the speedo. Optical, inductive and Hall Effect designs can all be used.

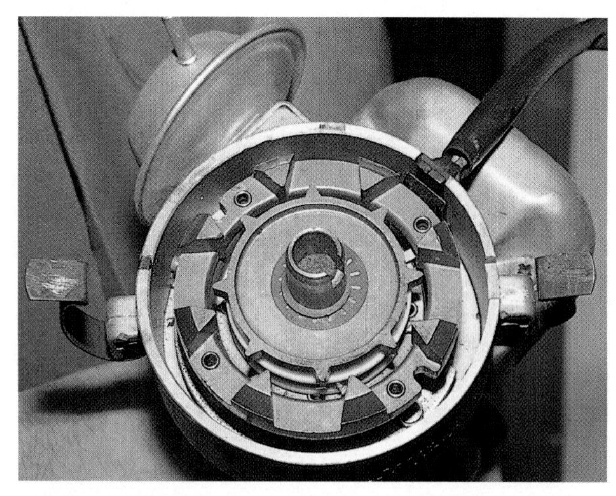

This is an inductive position sensor. It consists of a toothed cog, a magnet and a coil of wire. As a tooth of the cog passes each pole of the magnet, an output voltage pulse is produced in the coil.

The knock sensor is like a microphone listening for the sounds of knocking (detonation). It's screwed into the block and works in conjunction with filtering and processing circuitry in the ECU to sense when knocking is occurring. Many engines with engine management run ignition timing very close to knocking, so the information provided by this sensor is vital if engine damage isn't to occur on a bad batch of fuel or on a very hot day. This is more important in forced-induction cars, so most OE turbo and supercharged cars use knock sensing. Some cars (especially those with V engines) run two knock sensors. Knock sensors in older engine-managed cars are notoriously inaccurate, often false-alarming.

Other inputs used in engine management systems can include:

- Atmospheric pressure sensor
- Water temperature, vacuum, brake and clutch switches
- EGR position sensor
- A neutral position switch

This is a Hall Effect position sensor. It uses a set of metal blades that pass between a permanent magnet and a sensing device.

The ignition coil switching signal is one of the most important ECU outputs. Engine-managed cars can use a single coil as here...

...or multiple coils as shown here. These coils mount directly on the spark plugs.

THE OUTPUTS

The most important components the ECU controls are the injectors. The amount of fuel that flows through the injectors is determined by the length of time the ECU switches them on. This opening time — called the pulse width — is measured in milliseconds. When an injector is open, fuel squirts from it in a fine spray. The pulsing on and off of the injectors is the clicking noise you can hear if you listen closely to an idling injected engine.

While it varies from car to car, the injectors in many cars squirt once for each two rotations of the crankshaft. Injectors are often fired in banks of three of four at a time, while some ECUs trigger them sequentially. Sequential injection means the injectors are operated individually, being fired just before the opening of that cylinder's intake valve(s). The percentage of time the injectors are open is called the duty cycle. An injector open for half the time has a 50 per cent duty cycle, while if it is open for three-quarters of the available time it has a 75 per cent duty cycle. At maximum power in a standard engine, the injectors might have an 85 per cent duty cycle. That means the injectors are flowing at 85 per cent of their full capacity.

In addition to controlling the injectors, the ECU in engine-managed cars also controls the ignition timing. However, in most cars, the ECU doesn't directly control the ignition coil. Instead, an ignition module is used to switch the power to the coil(s) on and off. The ECU tells the ignition module exactly when it needs to switch to provide the correct ignition timing.

Older engine management systems are equipped with a single coil and a distributor. The distributor uses a rotor spinning around inside the cap to send high-voltage electricity to the spark plug at the right moment. In this type of system, the crankshaft position sensor is usually located inside the dizzy body. However, most cars nowadays use a multi-coil arrangement that can involve either a single coil for each spark plug or double-ended coils. These direct-fire ignition systems do not use a distributor because the coils are fired individually. With double-ended coils, two of the engine's spark plugs are fired at the same time — one on a cylinder that's on the compression stroke (which does something) and one on the cylinder on an exhaust stroke (which doesn't do much). On these engines, the crankshaft position sensor is usually mounted on the crank itself.

Idle speed control is carried out by changing the amount of air that can bypass the nearly closed throttle butterfly. Some cars used a pulsed valve (a little like an injector in the way it switches on and off) to regulate the amount of air that can bypass the throttle body. If the idle speed needs to be lifted, the duty cycle of the valve is increased and more air passes through. Stepper motors are also used in some systems, especially on those cars that have electronic throttle control. Other cars use an

Operating the injectors is one of the most important output functions of the ECU. Injectors are pulsed on, spraying fuel when they are open.

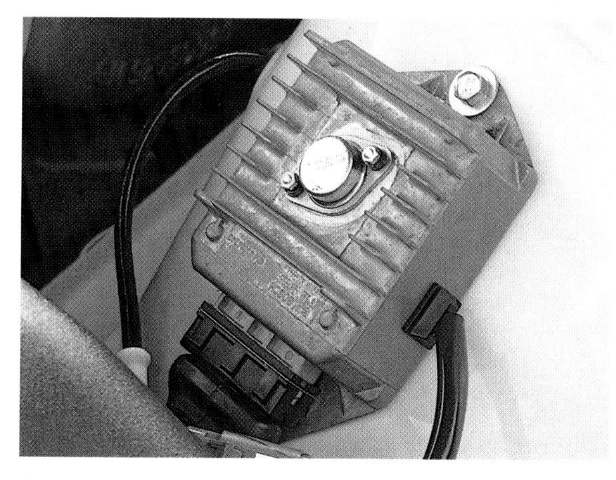

In most engine management systems the ECU controls an ignition module that switches the coil(s) on and off. The ignition module is usually heavily heat-sinked.

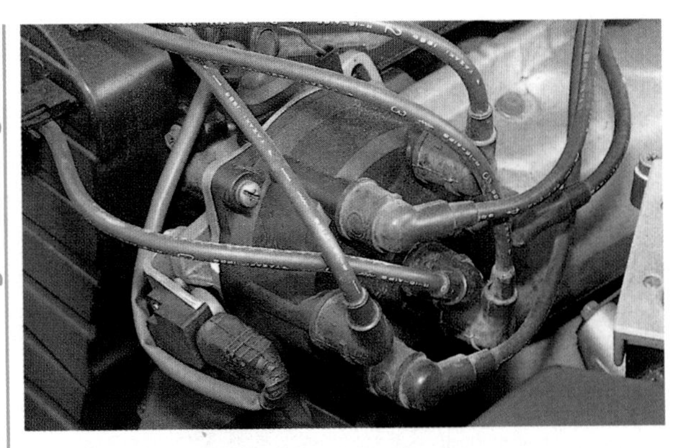

In this system a single power transistor (foreground) is used to switch the coil on and off. The distributor directs the high voltage to the correct spark plug, while inside this dizzy is an optical crankshaft position sensor.

additional valve to control idle speed when the engine is not yet up to operating temperature. Typically, the valve has engine coolant passing through its body and is also equipped with an electric heating element. As it warms, an open passage through the valve gradually closes, reducing the amount of air bypassing the throttle. This allows more air to be breathed by the engine when it is

Another important output of the ECU is idle speed control. In the system shown here, the idle speed control motor is located on top of the throttle body.

cold, keeping the idle speed high. Other engines use 'idle-up' solenoids that allow extra intake air to bypass the throttle when loads such as the air-conditioner are switched on.

There's a host of other things that can be operated by the ECU. The radiator fan is a common item switched on and off by the ECU. In turbo cars, the boost level is often

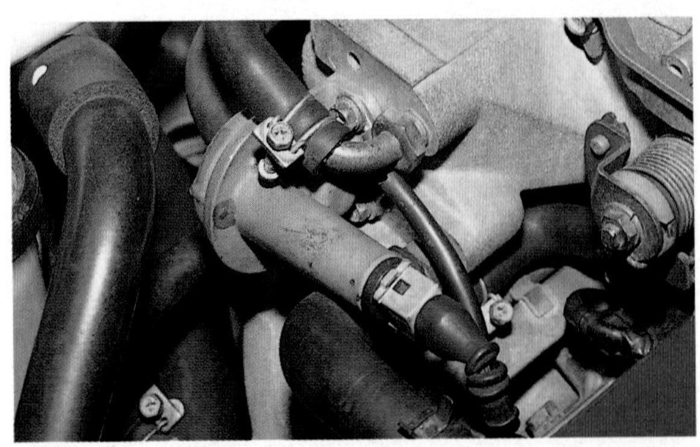

Some cars use a secondary valve to control idle speed during warm-up. This valve is heated by both the passage of coolant and an electric coil. When the valve is hot, it stops air bypassing the throttle body.

ECU-controlled. The auto transmission control is frequently integrated into the engine management system, and in four-wheel-drive cars, the engine management ECU often communicates with the four-wheel-drive system's controller. All engine-managed cars control the fuel pump through the ECU and a separate relay. Finally, the Check Engine Light is an important ECU output. Not only does it indicate when there is an engine-management problem, it can also be used to communicate what the problem actually is.

THE ELECTRONIC CONTROL UNIT

The view inside an ECU can be quite overwhelming — your first thoughts will probably be who (other than an electronics engineer) will be able to modify this? But rest easily — you **don't** need to know the function of every tiny electrical component to understand something about how the ECU works. Instead, it's easiest to simply think of it as an electronic decision-making box that has inputs and outputs. I've already briefly covered the input sensors and the outputs, so what about the decision-making part?

Most of the components inside an ECU are concerned with switching on and off the electrical loads like the injectors and converting the analog output of the coolant and other sensors into a digital form that can be understood by other parts of the ECU. The actual program that makes the engine management decisions is normally located on only one chip. GM makes this easy to see because in some of their ECUs they put all of the program (and a few other things) onto a detachable plastic module called the Mem-Cal, short for Memory Calibration.

So what goes on in the software that actually runs the engine? Rather than try to decipher everything, I will look at some of the general strategies adopted by the ECUs in most cars.

1. Closed Loop

When you are driving along steadily at city road speeds, there's very little throttle being used. In these conditions, the ECU is programmed to keep the air/fuel ratio close to 14.7:1 — the air/fuel ratio at which the catalytic converter works best at cleaning up the exhaust. The oxygen sensor (that's the one sniffing the exhaust gas) sends a voltage signal back to the ECU, indicating to the ECU whether the car is running rich or lean. If the engine is running a little rich, the ECU will lean it out. If it's a bit lean, the ECU will enrich the mixtures. The oxygen sensor then checks on the effect of the change. Figure 2.2 (next page) shows this process.

When the ECU is working in this way it's called closed-loop running. Closed-loop running on most cars occurs when:

- The engine is up to operating temperature
- The throttle opening is not very large or very small
- The throttle opening is fairly constant
- In certain speed ranges

Engines with variable valve timing usually use ECU control of the cam timing to allow the development of maximum torque at all rpm.

In normal driving, this means that the ECU operates in closed loop mode for a great deal of the time.

However, the ECU can switch out of closed loop running in an instant. You can be driving along gently in closed loop and then mash the throttle to the floor. The ECU instantly switches out of closed loop, ignoring the output of the oxygen sensor and substantially enriching mixtures. None of this can be felt by the driver, although a sensitive listener can sometimes detect at idle the very slight changes in engine note as the closed-loop cycle occurs. Closed-loop running relies on having the oxygen sensor in good condition — if it is defective, the car won't go into closed loop and fuel economy and emissions will suffer.

2. Self Learning

In addition to closed-loop running, the oxygen sensor is also used as part of the ECU's self-learning system. Imagine for a moment that the fuel filter in your car is a little blocked, causing the mixtures to be always a little lean. The oxygen sensor measures this and in response the ECU enriches the air/fuel ratio. But it's a pretty inefficient system if every day the ECU has to respond in the same way! Instead, what happens is this. The ECU knows the mixtures are always a bit lean, so it **permanently** enriches the mixtures. It has 'learned' that richer mixtures are needed, so always runs this correction. If you change the filter, the mixtures will be a little rich until the ECU gradually re-learns the new requirements. This self-learning process occurs in most ECUs and is totally dependent on the health of the oxy sensor. However, a typical oxygen sensor used in mass-produced cars can measure the

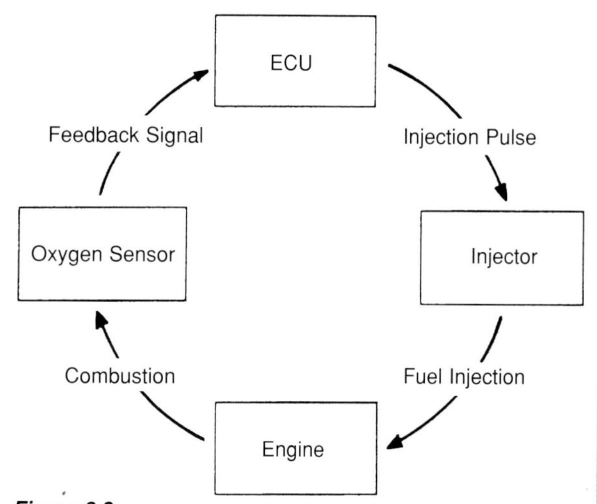

Figure 2.2
In closed-loop running, the feedback of the oxygen sensor is used to adjust the air/fuel ratio.

The radiator cooling fan is controlled by the ECU working through a relay. If the ECU is in limp-home mode, the fan is usually switched on permanently.

In sequential twin turbo cars, the ECU determines when each turbo operates. One of the turbo actuators of a Toyota Supra is shown here.

The software program containing the maps used to control the engine is usually on just one chip. The rest of the ECU is devoted to switching loads and conditioning the signals received from the sensors.

air/fuel ratio only over a quite limited range. As stated previously, the sensor is designed to be very sensitive around 14.7:1 air/fuel ratios, because that's where its output is most important.

3. Lean Cruise

We already know that rich mixtures are needed for power and leaner mixtures for cruise. But what about a long, gentle drive? Even leaner mixtures can then be used, improving economy further. And that's just what a 'lean cruise' ECU does. It takes note of how long the car has been maintaining a steady speed, how much throttle is being used and whether the engine coolant is up to operating temperature. If all these factors are OK, the ECU will start to lean out the mixtures. Second by second the air/fuel ratio will gradually become leaner, until the engine is running at an air/fuel ratio as lean as 18:1 or 19:1! If you put your foot down, the ECU instantly forgets all about lean cruise — until the right conditions are again met. Not all ECUs have a lean cruise function but most cars of the past few years are so equipped. Lean cruise is a good example of where the standard engine management system is in most cases more sophisticated than an aftermarket programmable system.

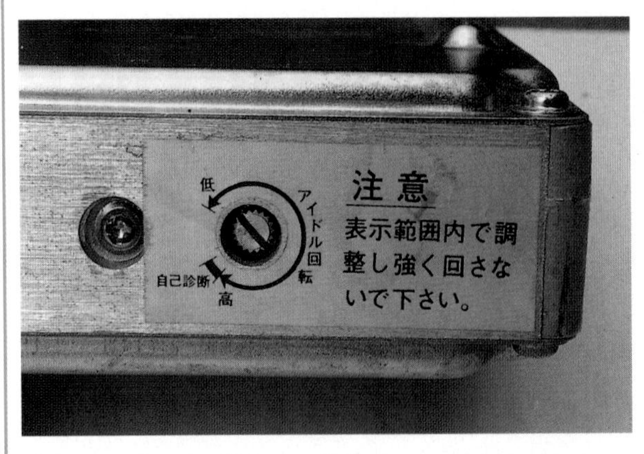

ECUs can be placed in a self-diagnosis mode by turning a control on the ECU or (more commonly) by bridging two terminals in a connector.

4. Open Loop

At full throttle, the oxygen sensor is always ignored. This is called open-loop running. In this situation, the ECU bases its fuelling decisions totally on the information that has been programmed into it. If the ECU senses a high load, it will open the injectors for a long time and spray in large amounts of fuel. The ECU uses a table of information (called a map) that tells it how to long to open the injectors at all the different engine loads. This characteristic means self-learning cannot be relied on to cater for the increased full throttle fuel supply required for engine mods that increase power. However, self learning often does help in the changed requirements occurring in part throttle conditions. Most ECUs also use their pre-programmed maps of information to select the right ignition timing, based on information derived from the input sensors. However, a few cars use the feedback of the knock sensor in a self-learning approach similar to that which occurs with the oxygen sensor on the fuel side of things.

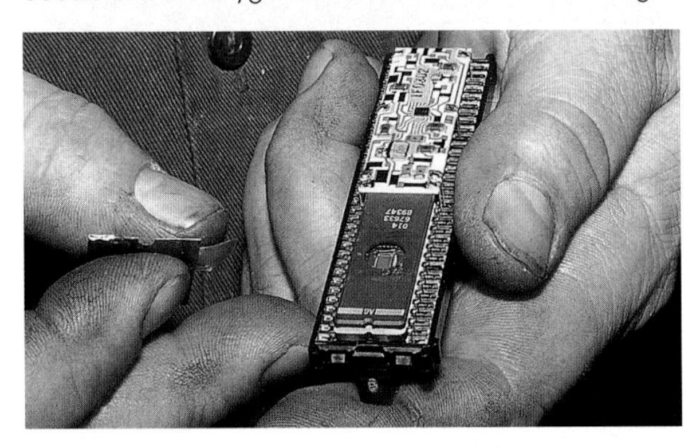

Here the Mem-Cal (Memory Calibration) module from a GM-Delco system can be seen. The EPROM (Erasable Programmable Read Only Memory) chip in the foreground contains the software that runs the engine.

5. Rev Limiting and Injector Cuts

All engine management systems use a rev limiter. Some limiters cut off fuel completely at the prescribed engine speed, withholding it until you're 500rpm below the limit. Hitting this 'bed of nails' limiter makes you think you've just broken the crankshaft! Other rev limiters cut off the spark or injectors of individual cylinders one after the other so you can barely feel you have reached the maximum allowable rpm. These soft limiters mean the car can be used right to the rev limit without a worry.

When zero throttle is used at higher speeds, the ECU switches off the injectors. For example, this situation occurs when you're approaching a red traffic light on a main road and have lifted the throttle completely. The injectors only resume flow when engine revs drop to around 500rpm above idle. If you watch the tacho closely, you can sometimes see the needle flick slightly when the injectors resume their flow. This injector shut-off benefits both fuel economy and emissions and is one of the ECU outcomes reliant on the input of the speed sensor.

6. Limp Home and Self Diagnosis

When engine-managed cars were first released, it was widely suggested they would be very unreliable, but this has proved not to be the case. One reason is the ECU has

An important ECU output is the Check Engine light. In addition to alerting the driver to faults, the light is often used during self diagnosis to flash fault codes.

internal back-up values and strategies to adopt if a sensor fails. For example, if the coolant temp sensor becomes defective (or the wire to the sensor is damaged), the ECU is programmed to ignore the incorrect input. Instead of measuring coolant temp, the ECU might rely only on the intake air temp sensor. Alternatively, the ECU might replace the coolant temperature sensor's input with a pre-programmed value. When an ECU does this it's said to be in limp-home mode. Some ECUs can do without all but a couple of input sensors and still keep the engine running. Incidentally, changing ECU inputs to achieve high-performance outcomes (discussed in the next chapter) needs to be carried out with care if the ECU isn't to go into limp-home mode, believing a sensor has become defective.

All OE engine management systems of the past 15 years or so have what's called a 'self diagnosis' ability. That means you can ask and they'll tell you what's wrong with them! For example, suppose the intake air temp sensor wire is broken. If you put the ECU into self-diagnosis mode (usually by just linking two terminals in a connector or by turning an adjustment on the ECU), the ECU will indicate it's the air temp sensor input that has the problem. Most ECUs communicate this information by flashing codes on the dash-mounted Check Engine light. The workshop manual will show you how to put the ECU into self-diagnostic mode and the list of error codes. More complex information can also be gained from the ECU. With appropriate hardware and software, the output of all the ECU sensors can be displayed, as well as information such as the ignition timing advance, rpm and injector opening time. All this can be done in real time as the engine is being run. Some systems can even save this information, allowing the playing back of all these factors in slow motion.

FAULT FINDING

Modification of a car should never be attempted if there is a fault present in the engine management system. Doing this is a recipe for failure — the chance of engine damage occurring is obviously much higher if (for example) the knock sensor or oxygen sensor is defective! Whole books have been written on fault-finding engine management (and the best will be the manufacturer's workshop manual for your car!), so here I will cover the subject only very briefly. It's important to realise that the engine management system is commonly blamed for faults that are actually present elsewhere in the vehicle. Poor fuel economy may well be occurring because of dirty injectors or a faulty oxygen sensor — but it may also be the result of lower tyre pressures or a dragging brake! Mechanical factors external to the engine and engine management system should always be assessed before attention is turned to the ECU.

The following preliminary inspection should be undertaken when a problem is present.

1. Check the quantity and quality of the engine oil (dirtiness, viscosity, etc).
2. Check the quantity and quality of the coolant (dirtiness, presence of anti-freeze).
3. Check the battery and battery terminals (quantity of electrolyte, voltage and terminal condition).
4. Check the intake system (dirtiness of filter, security of ducting).
5. Check the drive belts (looseness, cracking).
6. Check the spark plugs (clean, inspect the gap and deposits).
7. If it is fitted, check and adjust the distributor (tip gap, cap condition, check output of signal generator).
8. If it can be set, check and adjust ignition timing.
9. Check vacuum and hose connections.
10. Check all connectors and looms (looseness, broken wires).
11. Check fuel hoses (clamp tightness, leaks).

If these checks do not reveal anything untoward, the ECU's self-diagnosis function can be triggered. The procedures to do this and also the steps to be taken when the self diagnosis reveals specific problems can be found in the car's workshop manual.

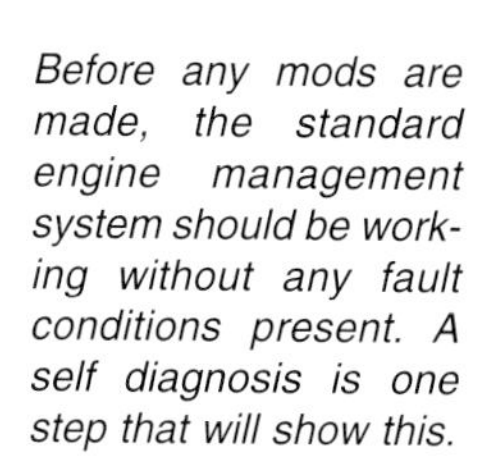

Before any mods are made, the standard engine management system should be working without any fault conditions present. A self diagnosis is one step that will show this.

Modifying Engine Management

If you wish to modify an engine-managed car there are many ways of going about it. But it's very important to realise that the best path to take depends on what you want to achieve and, most importantly, the available budget.

If you buy a brand-new, hi-tech, expensive car and decide you want to double the power output, **head straight towards programmable engine management and full engine dyno tuning**. That way, you'll get excellent driveability, superb power and very good reliability. (Programmable management is covered in the next chapter.) But if, on the other hand, you've just scraped together enough money to buy a 15-year-old turbocharged EFI car, to suggest you fit programmable management is laughable — the management system alone would cost you as much as (if not more than) the car!

However, if you **are** in this budget situation, there's a great many ways in which you can modify the engine management system. Some of the techniques won't always be the absolute best for reliability, power or driveability, but at a cost perhaps 1/100th the price of doing it the 'proper' way, they still have a strong place in the overall scheme of things. That's because, as with all car modification over the past hundred-odd years, ingenuity still counts for a lot in the end result. And that applies more than ever to cars equipped with electronic control of fuel and ignition.

As an example, the car pictured above opening this chapter — a three-cylinder 660cc turbo Daihatsu Mira — developed over 87kW (116hp) at the wheels (that's well over 200hp per litre at the flywheel!) while still using the factory ECU. The car featured (among other modifications) larger injectors and a disabled over-boost fuel cut, both implemented using techniques covered in this chapter. For the money, it's probably the best fun car I have ever owned!

WHEN ACTION IS NEEDED

It's important to realise that when you modify an engine to gain more power, you may not need to do anything at all to the engine management system. The use of airflow mass metering (the ECU knows how much air the engine is breathing), detonation sensing (the ECU knows there's a combustion problem) and exhaust oxygen sensing (the ECU knows roughly what the part throttle mixtures are) means power-ups of 10 or 15 per cent can often be coped with. In a turbo car, the power increase possible without problems is more likely to be 15-20 per cent. So before you start shelling out money on a chip or the like, see how well the modified car runs with no changes made to the engine management at all.

But you need to be conservative and careful when you take this approach. That means that grabbing a 15-year-old turbo car, screwing up the boost, fitting a big exhaust and driving the ring out of it will result in just one ending — the death of the engine. If the engine doesn't run lean on one cylinder because an injector is blocked, the piston slap and low oil pressure will destroy it, anyway!

Some engines can develop more power than standard with only small engine management modifications (or even none at all!). This factory supercharged Toyota 4A-GZE engine develops substantially more power than standard through exhaust and intake mods, and required only a small change to the vane airflow meter spring adjustment to achieve correct mixtures.

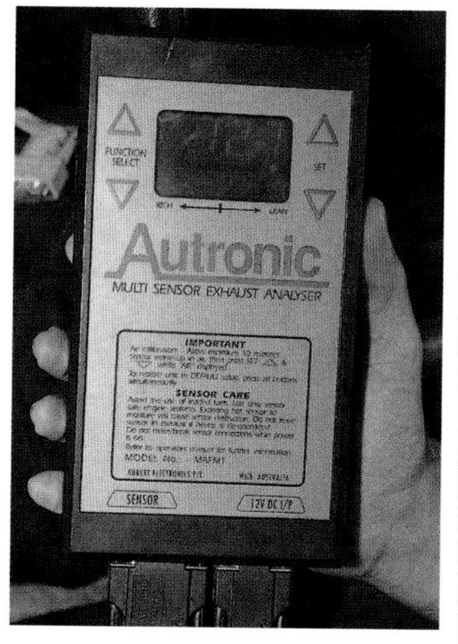

If the air/fuel ratio is correct through the rev and load range and the engine does not detonate, generally no damage will occur to the engine, even if it's modified to develop much more power than standard.

Always make sure the car is good mechanical condition before starting any modifications. By 'good mechanical condition', I mean the car should be able to run acceleration figures fairly close to those achieved when the car was new. In other words, if the car did the 0-100km/h sprint in nine seconds flat in its new car tests, it shouldn't be slower than about 9.5-10.0 seconds 10 or 15 years later. If it runs 11 or 12 seconds, fix the problems **before** you start to modify the car!

Another point to consider is that the standard engine management is generally very good indeed. When engine durability, fuel consumption, exhaust emissions and power are all taken into consideration, the standard system was among the best then available. To think you can blithely improve on it assumes quite a lot of ability on your part! In fact, changes made to just the ECU usually result in very little power improvement.

Of course, if you have made modifications that have very substantially improved the **potential** power output, the software program the original manufacturer envisaged as the ideal ECU mapping will no longer be correct.

Many workshops will lend (or hire out) their air/fuel ratio meters, allowing this sort of full load testing to be done easily and fairly cheaply at a drag strip!

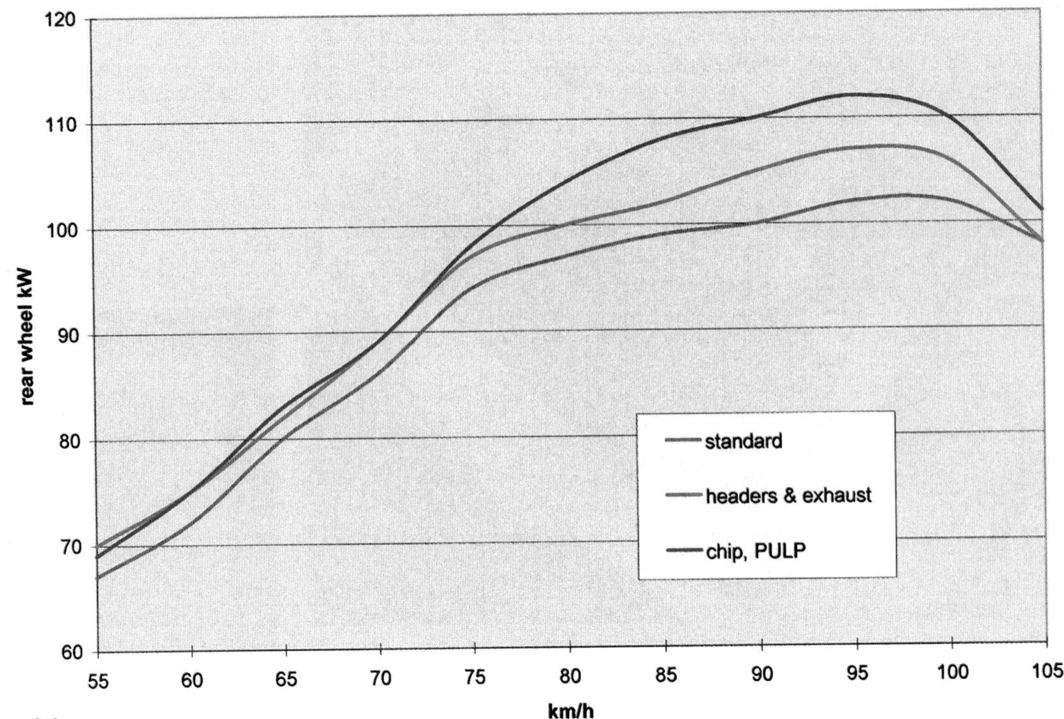

Figure 3.1
The dyno testing carried out on an SOHC six. Fitting headers and an exhaust gave the type of improvement expected, while installing a new chip gave a good upper rpm increase in power. However, when the weather changed, the engine started performing poorly. Installing the standard chip instantly solved the problem.

CHIPS

I want to be really blunt about this — in many cases a new chip does nothing at all to improve performance. Over the years of performance testing I have done, off-the-shelf performance chips have fallen into much the same category as miracle anti-friction additives and super mufflers that are supposed to create a vacuum in the exhaust. (And both of those I tested with lousy results!) The very **worst** result for a so-called 'performance chip' comes when it's used in a near-standard, naturally aspirated car. Simply, putting in more (or less) fuel and dialling in more (or less) timing does not usually make a great deal of difference to power unless the engine is already running with the wrong fuel and timing! And very few near-standard engines have the wrong fuel and timing... When it's expressed in this way, it becomes clear that you can't expect much of a power gain from a chip used on a near-standard, naturally aspirated car.

In many cases, a new chip does nothing at all to improve performance. The effectiveness of a chip change depends entirely on how well the new engine management maps have been developed and how well they match the engine modifications.

At the time of writing this book I have been saying exactly that about chips for seven years. I am also one of the **very** few journalists writing in popular magazines who has said it. I have maintained that view since the first on-road test I did of a 'hot' chip car proved it to have only standard performance, and I have been repeating the view after seeing the results of every dyno and road test I have since done on a chip. In that time, two major chip companies have threatened legal action, at many car workshops I have been declared *persona non grata*, and lots of other people with vested interests have expressed strong displeasure with my views.

Over the past seven years I have seen some chip companies make what must be hundreds of thousands of dollars selling a product I knew in many cases didn't make any difference. Just last year, the head honcho of a major chip company said to me that — at best — his chips had a 70 per cent success rate. Or, to put it another way, around one-third didn't work... A few years ago, the research and development premises of another chip company comprised just an office and two desktop computers. No dyno, no gas analyser, no mechanical expertise. This company even produced and sold chips for cars **they had never even seen**, let alone rigorously mapped on a dyno! The typical scenario in many chip companies is to whack in a bit of timing and some more fuel. And that's it.

Further, when chip companies attempt to decode the hex code the standard engine management software is expressed in, they are very often making stabs in the dark. This means that sometimes there are quite unexpected outcomes from chip modification. Figure 3.1 shows the dyno testing carried out on a car using an SOHC six-cylinder engine. Fitting headers and an exhaust gave the type of improvement expected, while installing a new chip gave a good upper rpm increase in power.

To crack the code on a chip, an emulator is used. The emulator is plugged into both the ECU and a PC and is loaded with the standard chip's software. With all this plugged back into the running car, the operator can see on the PC's screen which parts of the program are being accessed by the ECU. In this way, the location of the engine management maps can be found.

However, when summer came, the engine started to stall at traffic lights and then the idle speed increased to 1800rpm! Installing the standard chip instantly solved both of these problems.

Another example of random chipping can be seen in Figure 3.2. This dyno test of a 3.8-litre V6 was carried out with the car standard, fitted with a 2.5-inch (64mm) exhaust, and then with the exhaust and chip. As can be seen, the chip did give more power, but it also lost power at the low and high ends of the rev range and didn't gain any at 4000rpm. Properly mapped, the new power curve should be smooth (without dips) and not lose power anywhere!

The normal approach taken by chip companies to decoding a chip is to temporarily replace the main memory chip of the ECU with an emulator. As the name suggests, an emulator appears to the ECU to be just a normal memory chip — it 'emulates' the standard memory. Within the emulator are two programs that can be swapped over. Changes can be made to these programs so that when the swap takes place, a modified program is running the engine.

Most emulators used for this process can also identify which part of the code on the chip the ECU is accessing at any one time. That is, with the ECU hooked up to both an emulator and the car on a dyno, the actual bytes of information being accessed can be seen. Changes are then made to these bytes and the results assessed. For example, increasing a number may enrich or enlean the air/fuel ratio. In this way, a picture of which parts of the code control which parts of the engine management system can be developed. But all is not always as it seems. For example, when it's found that changing a

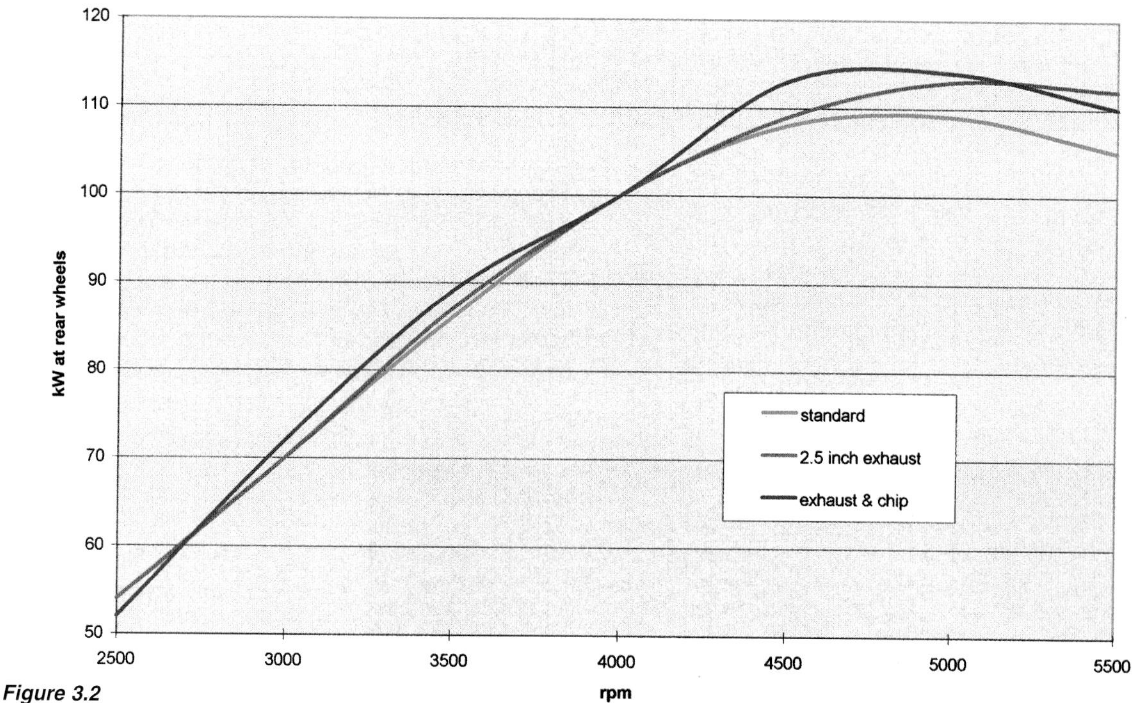

Figure 3.2
This dyno test of a 3.8-litre V6 was carried out with the car standard, fitted with a 2.5-inch (64mm) exhaust and then with the exhaust and a chip. The chip gave more peak power, but it also lost power at the low and high ends of the rev range and didn't gain any at 4000rpm.

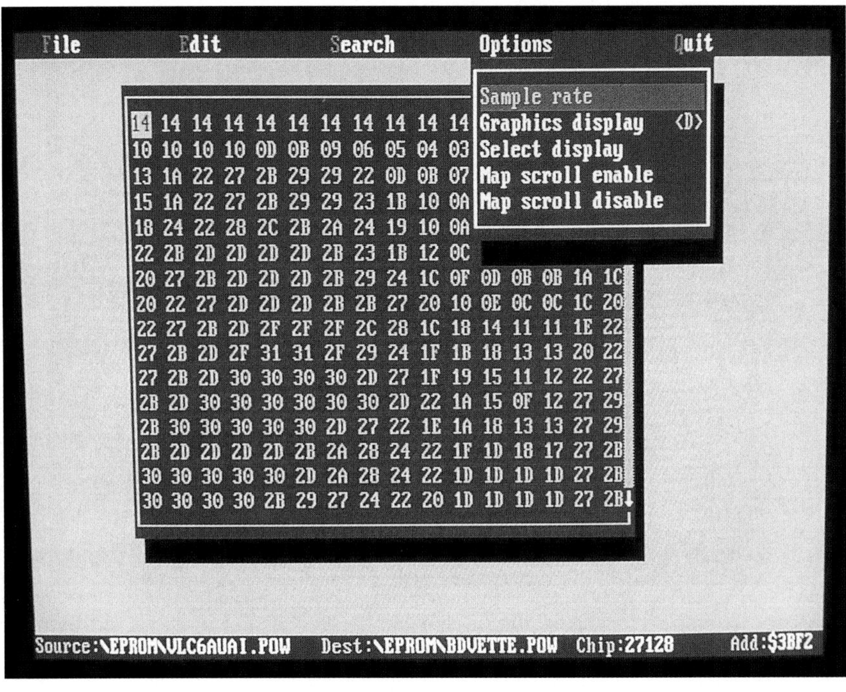

The data on a standard engine management system's main memory chip is in hex code, as shown here. This information controls the engine, but which byte controls which characteristic of the management system? Manufacturers of modified chips need to find out before they can make successful changes.

number in the hex code adds more fuel, it cannot be assumed that the number is part of the main air/fuel ratio table. After all, it might also be part of a correction table to compensate for cool intake air temperatures, or even be a number that indicates the injector size! Properly deciphering the code on a chip takes hundreds — if not thousands — of hours, so many hot chip companies understand only a fraction of the engine management maps for any one car.

But — and it's a **very** big but — a remapped standard engine management system on a modified car can work absolute wonders for both power and response. Here's the huge difference: **using a complete software package, the chip needs to be re-mapped in real time on the dyno with your car on the rollers**. That way, individual fuel and ignition sites can be identified and then mapped to give the very best results. If your engine needs a little more fuel at a manifold vacuum of X and an engine rpm of Y, the programmer can give it what it needs. If the compression ratio is a little higher in your engine (either because you have made it that way or it was always a good 'un from the factory), the engine management programmer can pull back the timing over that which he/she would normally use in this engine. This sort of custom, blueprinted chip designed expressly for your car is **utterly** different from an over-the-counter one-size-suits-all item.

The test is simple. If you make a phone call and the company says that — sight unseen — they can provide you with a chip that increases your naturally aspirated engine's power by X per cent, you know it's a total hit-and-miss affair. (And 'X' can vary in magnitude from 5 to a whopping 15 per cent!) If they question you extensively about the mods made to your engine and then cautiously suggest they can provide you with a chip that will **probably** be something close to the mark, OK. But if they state that you simply must come in to have your car programmed on the dyno in real time — full marks. It isn't the cheapest way, it isn't the easiest way — but it works.

If you are buying the chip as part of a tightly specified package of performance mods designed for your car's engine, the chip will probably work OK. It still won't be as good as specific mapping for your car, but a performance package comprising, for example, a camshaft, extractors, exhaust and chip should work without too many problems. The same can apply for a complete package of an upgraded turbo, lifted boost pressure and a new exhaust.

In fact, the situation regarding chips for turbo cars is quite different from that for naturally aspirated cars. Because most turbo cars have ECU-controlled boost, an off-the-shelf chip for a turbo engine can give a guaranteed power increase. The reason for this is simple: the boost is turned up. More boost usually equals more gas flow, which in turn equals more power. The programmer will also normally increase the fuelling to cope and may advance or retard the timing depending on how brave he/she feels! This means if you own a turbo car and want

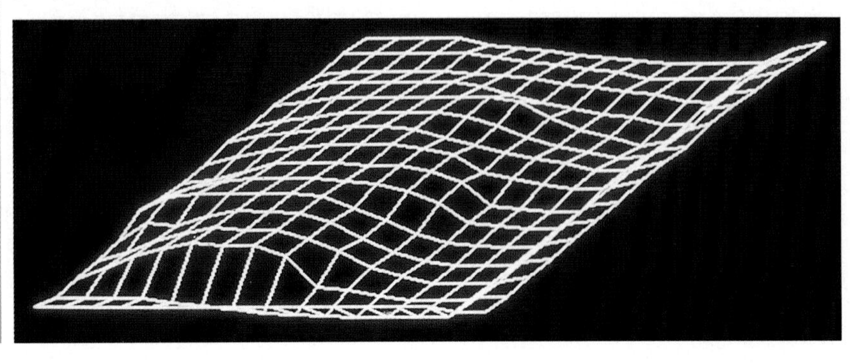

Graphing the data from the standard chip is also used to help find the management maps. This sort of graph is indicative of a management map, whereas a random graph would appear when non-map parts of the software are displayed in this manner.

a single plug-in power upgrade, doing it with a hot chip is certainly one approach you can take. However, if you want to tinker, making your own boost controller is likely to be much cheaper and for the sort of power upgrades we're talking of, the vast majority of turbo engine management systems will cope without problems.

A good chip can give you the very best in results, while a poor chip can at best cost you the money spent on it and at worst cause engine damage. The best way to modify the factory engine management system is with a purpose-mapped chip done especially for your car. There are literally no downsides to taking this approach, other than expense. But if you fork out the money, make sure it's being done in the proper way. But if having read all this you still decide to buy an over-the-shelf chip for your car, at least get a money-back guarantee and perform some before-and-after testing of the car's performance.

ALTERING THE TIMING

In many engine-managed cars the ignition timing can be changed by adjustment of the crank angle sensor. This is analogous to altering timing by adjusting the position of the distributor in a car that uses a traditional ignition system. Changing the position of the crank angle sensor (or moving the dizzy, for that matter) changes the whole ignition map by the same amount. For example, if the standard idle ignition timing is 10 degrees BTDC, adjusting it to 5 degrees BTDC retards the whole timing map by 5 degrees. In a boosted turbo car prone to detonation, retarding the whole map can be effective at stopping detonation when on boost — but it has the downside of giving poorer low rpm performance as well! In some cars, advancing the whole timing map (and/or using a higher-octane fuel) can give some minor power gains. However, as discussed above, this is the exception rather than the rule.

Figure 3.3 shows the result of varying the ignition timing on a Mazda MX5 (Miata) 1.6. The standard timing is 12 degrees BTDC, with the power resulting from this setting shown by the red line. As can be seen, power peaked at just under 65kW at the wheels. Advancing that by 3 degrees to 15 degrees BTDC resulted in the power curve shown by the blue line. There was a drop in peak power of about 1kW, with a 2kW decrease in the mid-range. However, at very low rpm, the power improved slightly. Finally, an ignition timing setting of 8 degrees BTDC was tried. This decreased power by a few kilowatts at low and high rpm. So in this case, altering the ignition timing map by 7 degrees made very little difference (just 2-3 per cent) to the power that the engine developed. This can be contrasted to a Nissan RB20DET turbocharged 2-litre six that responded quite strongly to changed ignition timing. While no figures are available, the car varied in power noticeably with changed timing settings. If advancing the timing in this way, note that for safety, the fuel octane value should be increased to avoid the occurrence of detonation, and testing should be conducted very carefully.

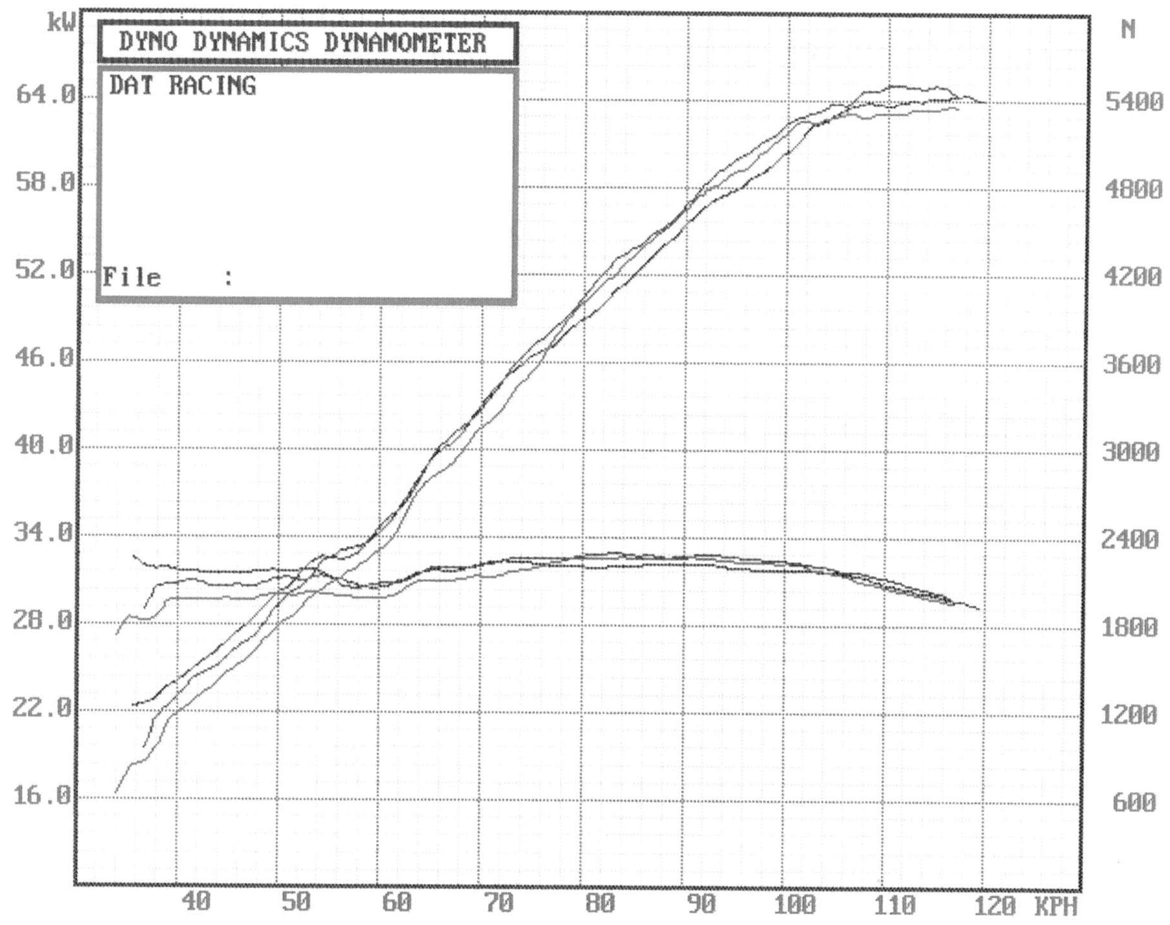

Figure 3.3
*Altering the ignition timing on a Mazda MX5 (Miata) 1.6 map by 7 degrees around the standard advance angle made very little difference (just 2-3 per cent) to the power the engine developed. However, some cars **are** responsive to ignition timing changes.*

A chip change in a turbo car such as this Subaru Impreza WRX is guaranteed to improve power because the new software increases the ECU-controlled boost level. Depending on the cost of the chip, this can be a very good way of getting increased grunt. However, in some cases it can be vastly cheaper and as effective to simply increase the boost by other means.

ALTERING THE FUEL PRESSURE

If an adjustable fuel pressure regulator is installed, it can be used to change the fuel rail pressure and so the amount of fuel that flows through the injectors for a given pulse width. On a naturally aspirated car, increasing the fuel pressure in this way will probably enrich the mixture at all loads and revs. Depending on how well the oxygen sensor feedback loop works, the ECU might then be able to lean out closed-loop running, giving richer mixtures only in open loop running (eg at full throttle). However, unless the car was running lean previously, this will not make a great deal of difference to power! On an engine with forced aspiration, a rising rate fuel pressure regulator can be used to squeeze more fuel through the injectors at high loads. This approach is often taken on turbo and blower kits because it's cheap and can be quite effective if the power increase is not great.

The ignition timing in many engine-managed cars can be adjusted by changing the position of the crank angle sensor. Here John Keen alters the timing on a Mazda MX5 (Miata) 1.6. The results of dyno testing the changed ignition advances are shown in Figure 3.3 (previous page).

An adjustable fuel pressure regulator is a cheap way of flowing more fuel through the standard system. It works well on engines that have a modest increase in power courtesy of an added turbo or supercharger.

CHANGING ECU INPUTS & OUTPUTS

Interceptor Modules

Modules that plug in between the standard wiring loom and the standard ECU are called interceptors. They are termed this because they take the signals from sensors or from the ECU and modify them. For example, an interceptor designed to overcome a speed cut function modifies the output signal of the speed sensor so it never reaches the level at which the ECU drops engine power. Some interceptors alter both an ECU output (for example, the ignition module signal to alter timing) and also an ECU input (for example, the airflow meter signal to change fuelling). The best interceptors have two inputs (for example, load and rpm) so the changes can be properly mapped in two dimensions (ie 2D mapping).

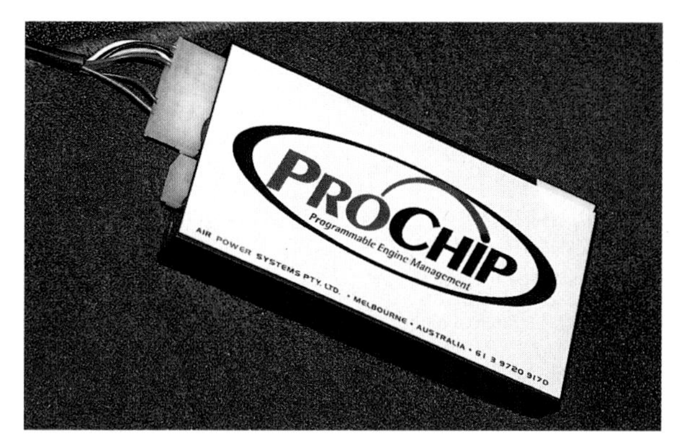

An interceptor module changes the ECU outputs or (more commonly) the input signals. Some interceptors, such as the one shown here, do both. The ECU then provides the altered fuelling or ignition timing that's required.

Any add-on module that retains the standard ECU and alters the engine management functioning is an interceptor of one type or another.

Interceptors have advantages and disadvantages. They are at their best where there is a single signal that requires a relatively simple change if the ECU is to perform as required. The speed sensor example mentioned above is a good use for an interceptor module. However, when it's required that the ECU's output signals be changed in a major way, there can be some problems. Running multiple interceptors designed to (for example) defeat an over-boost cut, change a speed sensor signal, alter ignition timing and change fuelling means there can be a nightmare of interrelating, altered signals. In this sort of case, the outputs of the ECU seldom reflect what would be chosen if the programming of each output variable could be done separately. It is **always** better to fit a proper programmable management system rather than run multiple interceptor boxes — and the cost is much the same.

A major disadvantage of using an interceptor is that the ECU software doesn't know anything about the new product. As a result, part-throttle mixture changes, for

The Safeguard interceptor uses the input of an added knock sensor to retard the timing on a cylinder-specific basis.

example, cannot be easily made to cars equipped with an oxygen sensor. This is because, during closed loop operation, the ECU immediately corrects for changes to the air/fuel ratio. In these conditions the ECU actively maintains an air/fuel ratio of about 14.7:1, so changing the output of the load sensor to alter mixtures is doomed to failure. To actually enact (for example) a lean cruise mode using an interceptor, the ECU input needs to be changed to the extent that the self-learning behaviour of the ECU is pushed past its limits. In other words, the mixtures have to be made so lean that the ECU runs out of correction capability. This can be done, but if the battery is disconnected and the ECU loses it learned data, the car will run very badly indeed until the maximum correction values are again reached.

The ignition retard of the Safeguard interceptor is achieved by delaying the signal reaching the coil. Here it has been wired into place for testing purposes.

Another example of where an interceptor approach has severe limitations is when an engine is converted to forced aspiration and an ignition timing interceptor used to pull back timing. Taking this approach will not change the way in which the timing retard occurs if the standard knock sensor detects detonation. The management software in a factory forced-induction engine (or one fitted with programmable management featuring a knock sensor input) will be programmed to pull off timing very quickly when detonation is detected, re-advancing the timing only slowly. This cannot be achieved with an interceptor.

Before you can use an interceptor, you must be absolutely sure of what you want to achieve and how the signal you are modifying actually works. This is seldom done, leading to different outcomes from those expected! An example of this can be seen in the fitting of a Fueltronics AMFC Pro airflow meter/MAP sensor interface module to a Daihatsu Charade. A single input interceptor, the AMFC Pro allows the PC programming of 20 load point voltages. Each load voltage can be trimmed by plus/minus 100 per cent. It was decided to fit the interceptor because the MAP-sensed Daihatsu was running very rich (~10:1 air/fuel ratio) at full throttle.

Dyno testing using an air/fuel ratio meter showed the Daihatsu's overly rich mixtures occurred only at high rpm, full load. At full throttle at lower engine speeds, the mixtures were fine. However, the interceptor module working on the MAP sensor output was unable to make the

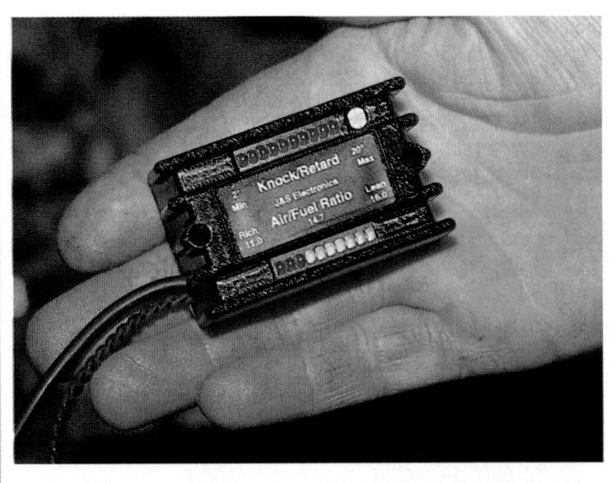

The Safeguard interceptor uses a dash-mounted display that shows the amount of timing retard and also (from the output of the standard oxygen sensor) the approximate air/fuel ratio.

desired changes to just the high load mixtures. As soon as full throttle was used, the manifold vacuum dropped to zero. That was the case whether the engine revs were 1500rpm or 6000rpm — or any engine speed in between! This meant that changing the output voltage of the MAP sensor when zero vacuum was present altered the air/fuel ratio **all the way** through the rpm range by the same amount. So with the engine running rich at the top end of the full throttle rev range (but having the correct air/fuel ratio at the bottom end), nothing could be done with this single input interceptor module to correct the mixtures throughout the full throttle rev range.

Another problem is that some sensors do not perform as they're expected to. The same Fueltronics AMFC Pro was fitted to the vane airflow meter of a Toyota Sprinter running the 4A-GZE supercharged 1.6-litre engine. However, it was discovered that in this application the output of the vane airflow meter is **not** proportional to all variations in load. In fact, the airflow meter snaps fully open when the throttle is floored at **any** rpm. The factory ECU obviously uses the airflow meter signal only in cruise and idle conditions, calculating at full load the injector pulse widths from rpm and perhaps also throttle position. With the vane airflow meter output voltage remaining constant at

The manner in which an interceptor module splices into the ECU loom can be clearly seen here. The top plug connects to the standard loom while the bottom plugs connect to the ECU. This module overcomes boost and speed limit cuts.

any full throttle engine power output from 15kW to 130kW, altering the vane airflow meter output voltage had little effect on full-throttle mixtures.

Single input interceptors designed to alter the output of the load sensor are most suitable when airflow meter or injector swaps are being undertaken. They can then be used to shift the entire load signal, with some minor trim changes able to be made to the offset used.

Interceptor modules are available to:

- Change the output of the airflow meter
- Replace an airflow meter with a MAP sensor
- Change the output of a speed sensor
- Prevent a turbo boost cut being enacted
- Change ignition timing
- Retard ignition timing on the input of a new knock sensor

Interceptors should be used sparingly, despite their widespread use in multiple numbers in some examples of domestic Japanese high-performance cars! Make sure you differentiate between an interceptor-style box and full programmable management when selecting products.

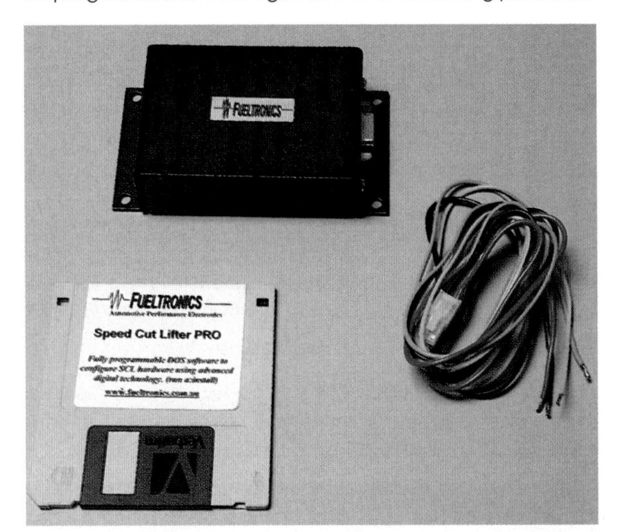

This Fueltronics interceptor removes vehicle speed cuts by altering the output signal of the speed sensor. Single-purpose, relatively simple interceptors like this generally give the desired results.

Simple Changes to Input Sensors

The same cautions and limitations that apply to the use of single input interceptor modules also apply to the simple changes to input sensors covered in the following sections. However, there is one major difference — the following modifications cost almost nothing and in some cases can achieve the same outcome as a much more expensive interceptor module! However, again it should be noted that the best results from altering inputs to the ECU come when it's known how the ECU uses the input signal and the exact characteristic of the sensor whose output is being modified.

Another point to consider when making changes to input sensors is that it can be easy to trigger the ECU's back-up or limp-home mode. When this occurs, a Check Engine

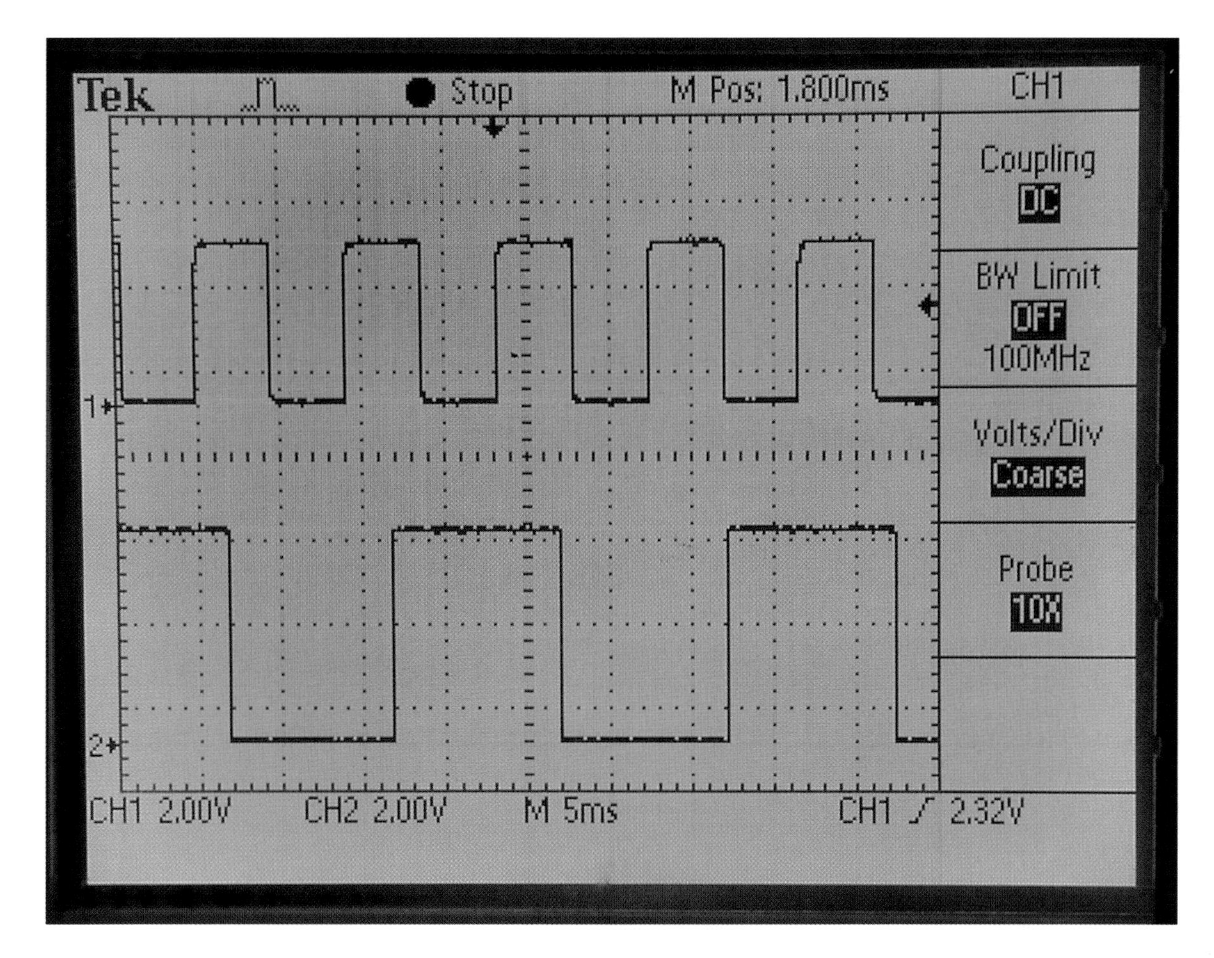

This digital oscilloscope shows the input and output speed signals to the Fueltronics Speed Cut Lifter. The input signal (top trace) shows the high frequency indicative of the car travelling quickly, while the output of the interceptor (bottom trace) is the much lower frequency normally occurring when the car is travelling more slowly. The ECU sees only the bottom trace!

(or similar) light may be illuminated and the engine's performance limited. Entering this mode can occur if the ECU sees a sensor output well out of the range it has been programmed to consider 'normal'. This can occur if the sensor does not vary in output as would be expected, or if its output is unexpectedly high or low. Note that the ECU may make this judgement on the basis of other input sensors — for example, if the inputs of other sensors indicate the car is travelling at 60km/h at a large throttle opening, but the airflow meter signal still reflects idle conditions, the ECU may determine that one of these sensors is defective. Modifications should always be carried out so the Check Engine light is not continually illuminated, or important (and real!) malfunctions will not be able to be seen when they occur.

1. Temperature Sensors

The coolant temperature sensor controls almost exclusively the amount of fuel enrichment during times of cold running. This means that changing the output of the coolant temp sensor can cause more fuel to be injected. However, note that the extra enrichment may not be made in the same proportion throughout the load and rev range — the manufacturer doesn't expect that cold enrichment will be required at full load, 5000rpm! Another temp sensor that can be usefully modified is the intake air temperature sensor. The output of this sensor is frequently used by the ECU to determine the final ignition timing.

Before you modify the output of the temperature sensors, you should know the sensor operating parameters. Most temperature sensors decrease in resistance as the temp goes up. The table below shows the relationship between coolant temp and resistance in one sensor:

Temperature (degrees C)	Resistance (ohms)
0	6000
20	2500
30	1800
40	1200
70	450
90	250
100	190
110	110

This sort of information is available in workshop manuals, or you can test a sensor using the sensor, a thermometer, a multimeter and a saucepan of hot water on the stove. To carry out this test, simply connect the multimeter to the sensor and place it in the water. Measure the temperature of the water and note the sensor resistance. Then heat the water, measuring the changed resistance of the sensor every 10 degrees.

In the example shown above, the sensor has a resistance of 250 ohms at a normal coolant operating temperature

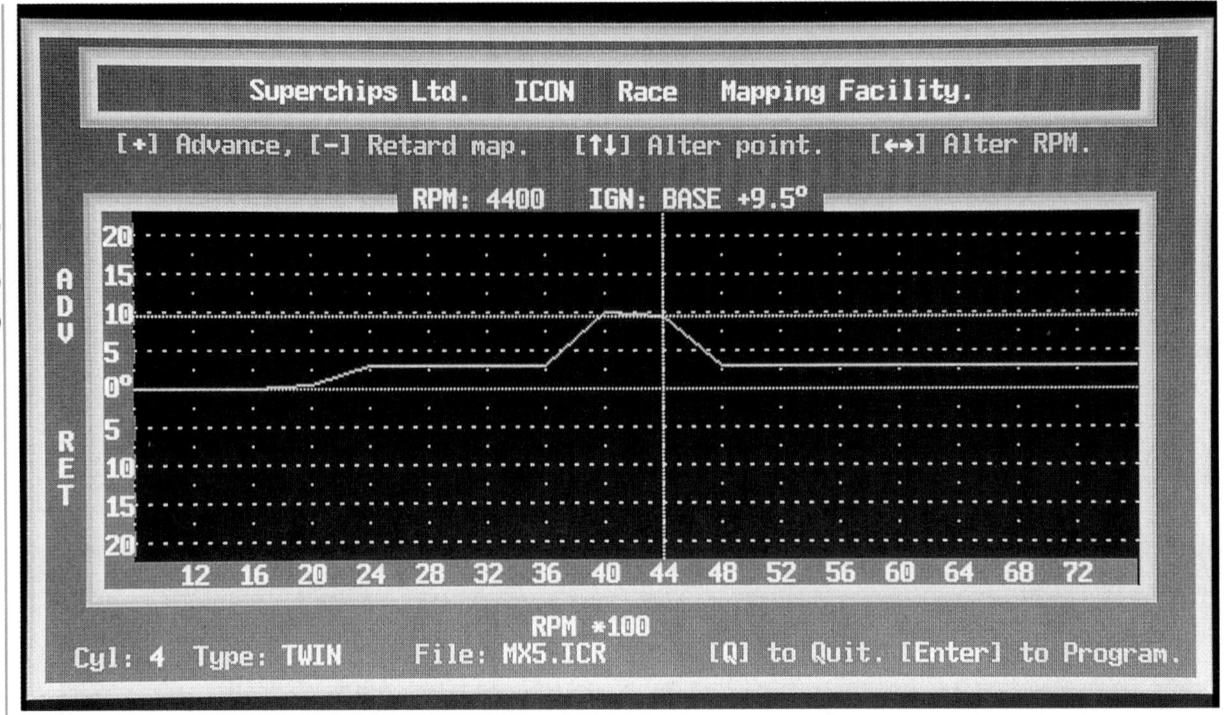

The Icon Race interceptor uses a laptop PC to display and set the ignition timing changes made. Here it can be seen that the timing has been advanced by 10 degrees at 4000rpm and 4400rpm.

of 90 degrees. If the resistance of the sensor is artificially changed so it has a new resistance of 6000 ohms, the ECU will 'think' the coolant is at 0 degrees, even when it's still 90 degrees! To achieve this, we need to increase the resistance of the sensor by 5750 ohms. A 10 kilo-ohm potentiometer can be used to give an adjustable range of extra resistance around this figure. Such a variable resistor is available from electronics shops for less than the price of a candy bar. Figure 3.4 shows how you wire it up.

If the opposite effect is needed — you want the ECU to believe the temp is higher than it really is — you need to wire the resistor in parallel with the sensor. This type of wiring makes it a little more difficult to work out the value of resistor you need to use, but it still isn't very hard. For example, assume the air temp sensor has a resistance of 2000 ohms at 20 degrees C and 400 ohms at 80 degrees C. To make the ECU think the intake air temp is 80 degrees when it's really only 20 degrees, we need to remove 1600 ohms of resistance. But how do you work out the value of the new resistor?

The resistance of resistors in parallel is worked out by:

$$\frac{1}{\text{total}} = \frac{1}{R1} + \frac{1}{R2}$$

In this example we want a total resistance of 400 ohms but the sensor resistance is currently 2000 ohms. The equation can be re-arranged:

$$\frac{1}{R1} = \frac{1}{400} \text{ minus } \frac{1}{2000}$$

which gives a value of R1 (the new resistor) of 500 ohms. Doing these sums is a lot easier if the '1/x' button on a scientific calculator is used during the process.

With the installation of a 500-ohm resistor in parallel with the air temp sensor, the ECU will think the inlet air temp is 80 degrees C when the intake air temp is really 20

Load sensor interceptors (MAP or airflow meter) cannot alter the part-throttle air/fuel ratios because the ECU corrects the injector opening times to retain a stoichiometric (14.7:1) mixture. This is one limitation of this type of device.

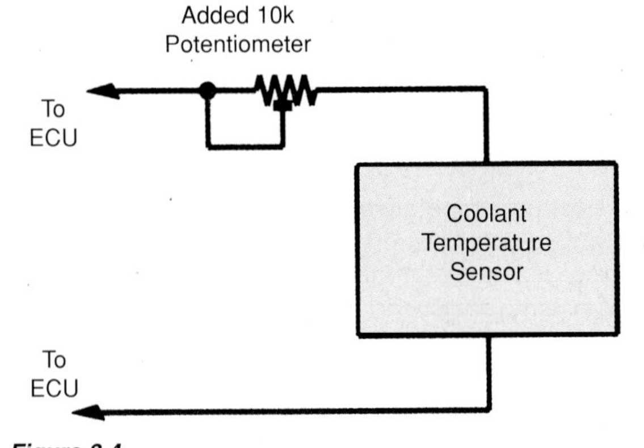

Figure 3.4
Adding a pot in series with the coolant temp sensor can be used to enrich mixtures a little. This is especially effective at low loads and rpm, so can be used to stabilise idle when running a hot cam.

degrees. This is very likely to retard the timing, but you won't know the exact effect until you've tried it. Figure 3.5 shows how to wire in this modification. Note that a 10K pot has been used here also, but the actual value of the pot should be close to the resistance that you calculate is required.

As indicated earlier, changing the output of temperature sensors is most frequently carried out with coolant temp sensors (to change the air/fuel ratio) and with intake air temp sensors (to change the timing). However, any sensor that uses a variable resistance can have its output changed in this way. Other sensors can include throttle position, fuel temperature and transmission temperature. The changed sensor output can be easily switched in and out with a throttle-triggered microswitch or a manifold pressure switch if the modification is wanted in only some situations.

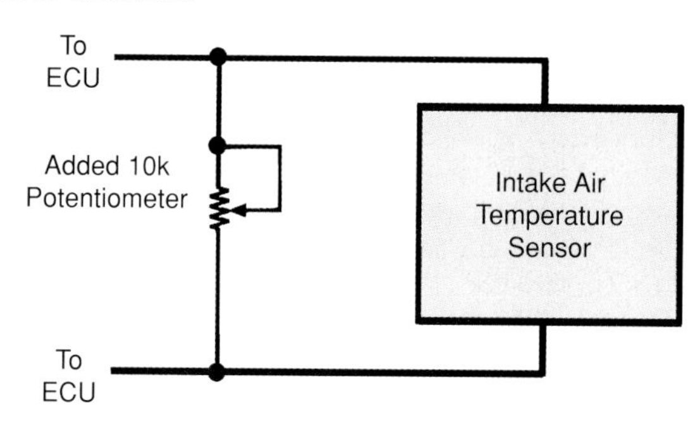

Figure 3.5
In many cars, placing a pot in parallel with the inlet air temp sensor will retard the ignition timing. This can be switched in and out as needed, by the action of a manifold pressure switch, for example.

2. Increasing Injector Size

Changing the output of the coolant temp sensor is good for relatively small changes — but what if you want to dramatically increase fuel flow? In this case the injectors will need to be upgraded for larger items. Swapping in larger injectors can revitalise the life of the standard ECU. It's most often needed when the injector duty cycle of the standard units has reached 100 per cent.

When making injector swaps there are a few aspects to keep in mind. First, the new injectors need to have the same coil resistance as the standard injectors. This resistance value can be measured with a multimeter. If the readings are, for example, 13 ohms versus 14 ohms, this small variation will be tolerated by the ECU without any problems. However, replacing a 16-ohm injector with a 4-ohm unit wouldn't be on. Second, it helps if the new injectors are physically very similar to the original units. Machining intake manifolds to accept new injectors can be difficult and expensive, while the mounting system at the other end of the injector can also vary enough to cause problems. This means that if you can source larger injectors from another engine made by the same manufacturer, you're more likely to end up with injectors that can be simply slotted into place.

Finally, the increase in flow should be appropriate for the application. Chapter 10 covers the selection of injectors and includes a table that lists the specifications for nearly 100 injectors, including their flow and resistance. Don't

Using two potentiometers to shift the output signal of a MAP sensor or airflow meter can be an extremely powerful tuning tool when larger injectors have been fitted or the airflow meter changed. Here the settings have been finalised and the knobs glued into position.

go up massively in flow capability or the idle and low speed driving may suffer. This is because a large injector loses flow accuracy at the small pulse widths that are required of it at low loads. Some factory cars actually change their injector pulsing to occur only once per four-stroke cycle at low loads, while squirting twice per cycle at high loads. In this way, very small pulse widths aren't required at low loads. If you pick a massively oversized injector there might be low-power problems.

Once fitted, you need to make the new injectors flow properly to match the air/fuel ratio needs of the engine. It's likely the air/fuel ratio will be very rich with the large injectors fitted — everywhere except possibly at max power in a modified engine! This makes sense since at the same pulse widths (injector opening times) you'd expect the bigger injectors to flow more fuel. There's a number of different ways around this dilemma.

First, the ECU software chip can be rewritten so the injector pulse widths match the new injectors' flow. If it's done in real time on the dyno, using a good air/fuel ratio meter and working with a computer programmer who understands cars as well as hex code, the results will be very good. Second — believe it or not — no changes at all may need to be made, especially if the new injectors are only a little bigger than standard. With the oxygen sensor telling the ECU what mixtures are being run in most conditions and with many ECUs capable of self-learning, a long drive may be all it takes for the ECU to compensate for the new injectors' flow rate. Unless the car won't even

It's not a good idea to watch an air/fuel meter, make tuning changes and drive in city traffic all at the same time. This is what happened when I tried that particular trick...

While modifications to the engine management system are necessary when lots of extra power is being gained, many standard systems will cope with a 10-20% power lift without problems.

start, try driving the vehicle for a while first. Over-rich mixtures run for a short time won't do any damage and the ease and cost-efficiency of this approach (if it works) leaves everything else behind.

However, the most likely requirement is that the output of the major load sensor will need to be changed. Depending on the car, this sensor will be either a MAP sensor or airflow meter. If the ECU thinks there's less load than there really is, it will reduce the pulse width (opening time) of the injectors. This reduced pulse width, coupled with the bigger injectors, means the right amount of fuel can then flow. In other words, if the ECU thinks only enough fuel for 50kW is needed while a demand for 70kW is actually required, it will pulse the injectors to provide 50kW of fuel. However, with the bigger injectors in place, enough fuel for 70kW will actually be injected! So, making the computer sense a lower engine load than is actually occurring will straighten up the mixtures again.

This computer misinformation can be provided either mechanically or electrically. In an airflow meter car, it's possible to place an air bypass around the meter. With less induction air actually flowing through it, the meter will signal to the ECU that a shorter pulse width is needed, so mixtures will be normalised. Changing the diameter of the bypass will adjust the mixtures. Doing it electrically is far easier and more cost-effective, though. But before discussing the electrical approach, a few words of warning are required.

- Blowing up an engine (especially a turbo one) developing a lot of power can result from one power run with lean mixtures.
- Mixtures should be monitored by an air/fuel ratio meter. Doing this with the car on a dyno gives the best control over what is happening.
- Start off rich and then lean things out — don't go the other way.

The reason for these warnings is that — with the following approaches — changing the mixtures from full rich to full lean anywhere in the rev range is very easy.

If the car runs a vane airflow meter, the signal output of the meter can be changed by adjustment of the preload spring mounted within the meter's base. If this spring is tightened, the vane will open a smaller amount, so the injectors will be operated with shorter pulse widths. This means the airflow meter can be recalibrated so the larger injectors provide the right amount of fuel. Note that the new airflow demands of the engine will mean you will probably need to use a larger airflow meter than standard.

Another approach also electrically fools the ECU into giving the right mixtures. First, you will need to examine the pin-outs of the airflow meter or MAP sensor. A power feed to the sensor (usually five volts), an earth and a signal wire should be able to be located. In vane-type airflow meters, an air temp sensor is also located inside the meter, but you can ignore that. Whether the sensor's output voltage rises or falls with increasing engine load doesn't matter when using this circuit. However, this approach **cannot** be used with airflow meters or MAP sensors that have a variable frequency output.

Two potentiometers (abbreviated to 'pots') are used. Both are 10 kilo-ohm linear designs that can be bought very cheaply from electronics stores. A pot works as a variable voltage divider, allowing the central wiper contact to be skewed to one voltage or another. The circuit is shown in Figure 3.6. One of the pots is connected between the five-volt supply wire and earth. With this pot's wiper at the five-volt end, that's the voltage available on the wiper terminal. With the wiper at the earth end, there will be no voltage available. Pot 2 connects between the wiper of Pot 1 and the signal output wire of the sensor. Moving the wiper arm of this pot (which goes to the ECU) will cause the signal to be closer in level to either the sensor output or the voltage provided by Pot 1.

The wiring is simple and the use of the pots even simpler. To start the car, set both pots to their central positions. Disconnect the oxygen sensor from the ECU so it can't go into closed loop mode and then use Pot 1 as a coarse

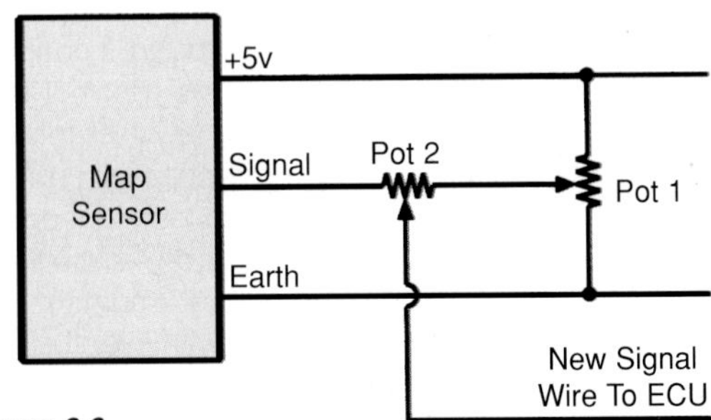

Figure 3.6
Two pots can be used to shift the entire load map up or down, allowing the fitting of larger injectors or, when used with an airflow meter, allowing the electrical calibration of a larger airflow meter.

Turbo over-boost fuel or ignition cuts can normally be circumvented easily and cheaply. Here a one-way valve and bleed has been placed in the pressure hose leading to a manifold pressure sensor, preventing it detecting full manifold boost while still allowing it to detect normal vacuum signals.

mixture control and Pot 2 as fine adjustment. Start off by setting light loads. When the light load adjustment is about right, lift the power level required and re-do the fine-tuning using Pot 2. Obviously, it's more important to get the high loads right than the light loads, but when the mixtures are set perfectly for high load conditions, they may not be right at light loads because of non-linearity in the engine's fuel needs when compared with its unmodified state. If the light loads look a bit too rich, don't worry — when the oxygen sensor is reconnected they'll lean out as the ECU learns its new mixtures.

When the mixtures are right, it's easiest to simply glue the adjustment knobs into place so they can't be changed further. Alternatively, fixed-value resistors of the same resistance as being provided by the pots can be substituted. Taking this potentiometer approach gives enormous power over the mixtures so should be used with extreme care. The results can be spectacularly successful, though.

3. Disabling Turbo Boost Fuel Cuts

Many turbo cars have an in-built safety mechanism that cuts fuel or ignition (or both) if a preset boost figure is exceeded. But when you are deliberately increasing boost, such a cut can be a real pain! The cuts are implemented in three quite different ways. In cars that use a MAP sensing system, the manifold pressure is constantly monitored. If it rises above the boost cut level, fuel and/or ignition is shut down. Other turbo cars use a secondary boost pressure sensor whose job in life is to activate the fuel cut. This boost pressure sensor sometimes also runs a dashboard boost gauge. Finally, some turbo cars simply monitor the airflow meter signal, and if it exceeds the preset level, the ECU assumes it must be because of excessive boost and shuts the engine down.

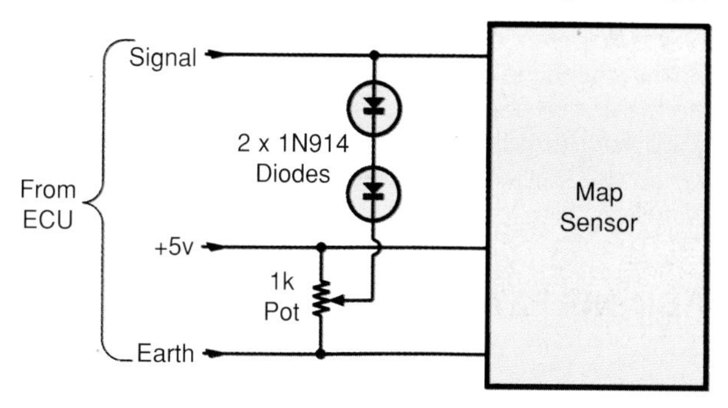

Figure 3.7
If an over-boost fuel cut is made on the basis of the output signal of a MAP sensor, this clipping circuit can be used to disable it.

If the cut is made on the basis of a sensor voltage that rises above the preset level, a very cheap and simple clipping circuit can be used. Figure 3.7 shows the circuit, which costs almost nothing. Its sensitivity can be adjusted by rotating the pot, which should be set so the circuit clips at as high a manifold pressure as possible. As with any of these techniques, the power over the car's electronics is very great, so extreme care should be taken that the engine doesn't suddenly run lean during the setting-up procedures.

Some turbo cars that use vane airflow metering will shut down the engine if the vane opens too far, indicating an excessive airflow. A mechanical stop that physically prevents the vane opening to this extent can easily be made. If a screw is used as the stop and it's inserted through a drilled and tapped hole, adjustment of the position of the stop is easily made. A locknut on the screw can be tightened when the stop is adjusted correctly. While on non-electronic techniques, if a manifold pressure sensor needs to be limited in its peak reading, a bleed can be teed into the hose leading to it. This will decrease the peak boost and vacuum signals that are read. If a one-way valve is inserted in the bleed, the sensor will still detect manifold vacuum as normal. Figure 3.8 (next page) shows how to implement this technique. Note the diameter of the bleed hose will determine how much boost the sensor sees; if it's required that some boost still be sensed, a restriction should be placed in the bleed.

If an airflow meter signal output reaches its ceiling or it's desired that its signal be reduced across all loads, a bypass can be installed. Here an extra filter allows airflow into the turbo intake that has not passed through the airflow meter.

CHANGING AIRFLOW METERS

If pressure drop testing (see Chapter 5) proves the airflow meter to be restrictive, or the meter is reaching its maximum output value well before peak power arrives, you may want to swap the standard airflow meter for a larger unit. Generally, it's simplest if you retain the same design of meter — vane or hotwire, for example — and

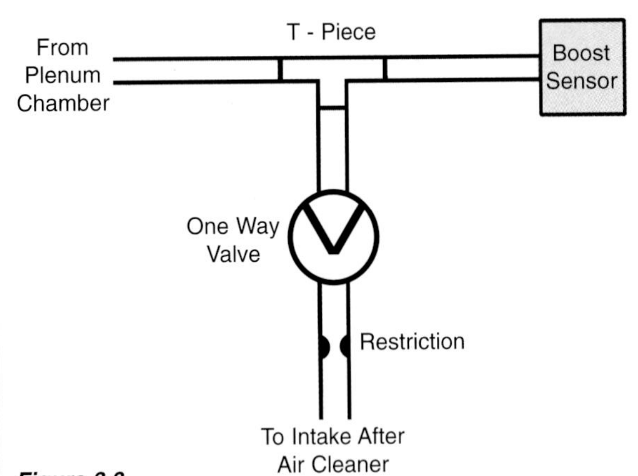

Figure 3.8
If it's desired that a pressure sensor receives a boost pressure signal lower than that which is actually occurring (to stop an over-boost fuel cut occurring, for example), a simple bleed can be used. The one-way valve will still allow the sensor to detect normal levels of manifold vacuum.

simply increase its size. Taking this approach means the electrical characteristics of the meter should be very similar — except that the whole output of the meter needs to be shifted within the airflow range.

Because over the years so many cars have been sold equipped with vane airflow meters, this design exists in sizes all the way from about 50kW (75hp) through to about 190kW (~250hp)! Hot-wire airflow meters are also available in a very wide range of engine power outputs. It's important to note with both types that the wiring pin-outs vary from car to car, so even if the plugs are interchangeable, the wiring should be checked for pin-by-pin compatibility before the swap is made. If the wiring is different, wiring changes can be most easily made within the loom plug.

Before changing the airflow meter, you should measure the operating voltage output of the meter you already have. To do this, a multimeter can be mounted inside the cabin so the output voltage is easily read off. Place an assistant armed with a pen and paper in the car to write down readings and then drive the car at a variety of specific loads and engine speeds. The amount of engine load is easily seen if you fit a manifold pressure gauge.

First, note the meter output voltage when free-revving the car at 1000rpm intervals to the red line. Next, at a constant throttle setting that corresponds to (say) -20kPa

Removing the mesh screens from this hot-wire airflow meter will improve flow, although making the meter more vulnerable to damage.

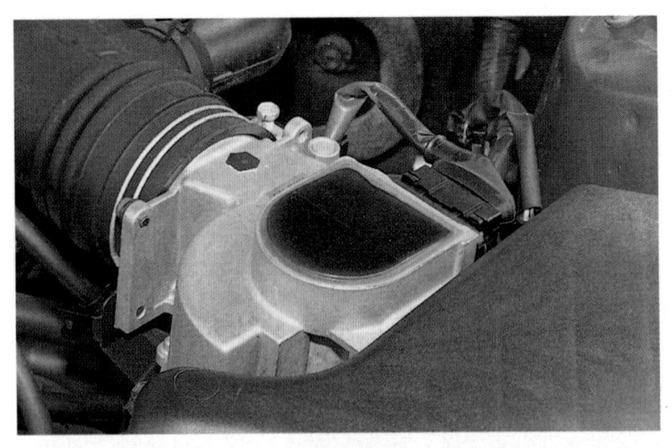

Under the black panel on this vane airflow meter is a spring adjustment mechanism. The internal spring adjustment mechanism can be used to re-set mixtures after a vane airflow meter swap has been performed.

manifold pressure, note the voltage output at 1000rpm intervals to the red line. Finally, make full-load measurements to the red line. Performing these measurements allows you to develop a simple map showing what input voltages the ECU is expecting at the different combinations of loads and rpm. To calibrate a new airflow meter to run the car in this way, you need only adjust it so it's providing the same voltage outputs as the standard meter.

Of course, if because of your mods the standard meter was wrong everywhere, taking that approach won't be any good! But it's much more likely that the standard meter's calibration became incorrect only at just the highest loads, or the pressure drop across it limited power only above full throttle at 4000rpm, for example. If your new meter can be calibrated to provide the right outputs at lower loads, it should be somewhere in the ballpark at higher loads. But if you don't take any measurements at all of the standard airflow meter outputs, you won't have any idea where to begin in its calibration.

As indicated above, vane airflow meters can be recalibrated by adjustment of the pre-load spring. This mechanism is normally hidden beneath the plastic on the top or bottom of the meter. Although it's glued on, the cover can be removed with some careful use of a knife. Before changing the spring adjustment, it's wise to place a daub of white paint on the wheel to indicate the 'starting' position. That way, you can always find your way back to that spot! Other types of airflow meters with a voltage output can use the 'two pots' approach shown in Figure 3.6 (on page 53).

Incidentally, note that if manometer testing (see Chapter 5) indicates there's a major pressure drop across a hot-wire airflow meter, the mesh screens placed at each end of the meter can be removed. This will improve flow, though of course make the meter more vulnerable to damage should the airfilter fail or objects be dropped in during handling.

ADDING EXTRA INJECTORS

If you need to provide the engine with more fuel but you are reluctant to upgrade all the injectors (perhaps because of cost), you can add extra injectors. These should be located prior to the plenum chamber so the

If extra injectors are to be added, they should be placed prior to the plenum chamber to permit as much air/fuel mixing as possible. Such an injector should work only at high loads where the chance of the fuel becoming separated from the air is lessened.

fuel is well mixed with the intake air and evenly distributed to each cylinder. Well, that's what's supposed to happen! One of the drawbacks of taking this approach is that the plenum and intake manifold are forced to adopt a new role in life — flowing a fuel/air mix, when the system was designed for only air. Another negative is that because the pulsing of the new injector(s) is often tied to an existing injector, fuel is added to the intake air in a series of discrete bursts. The timing of these pulses of fuel can mean one cylinder misses out because, by the time the fuel reaches the intake manifold, that cylinder's intake valve is always shut! However, adding an extra injector or two is cheap, easy and often quite effective. Just don't think it's the very best way of doing it...

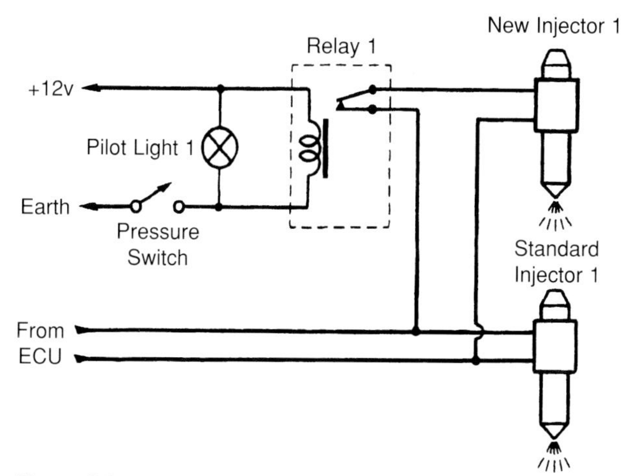

Figure 3.9
A simple way of adding more fuel is to wire a new injector in parallel with an existing injector, with the new injector pulsed by the standard ECU. The extra injector can be brought onstream via the action of a manifold pressure switch.

Extra Injector Controllers

There's a variety of add-on controllers that can operate an extra injector. Inputs are often rpm and manifold pressure, with the rate of gain and the injector switch-on point both able to be set. The capability of these controllers is generally very limited for the money outlaid. As a result, I advise that you look at either upgrading all the injectors and then modifying the load sensor output (using the two pots or an interceptor module), or saving the extra money for fully programmable management.

Extra injectors can be fed fuel from a T-piece inserted in the fuel line ***prior to*** *the fuel pressure regulator.*

Certainly if you can have fully programmable management for perhaps only twice the cost of an extra injector controller, you're well advised to take the programmable path.

Switched Extra Injectors

An extra injector can be added very cheaply if it's fired by the standard ECU. The injector is mounted in front of the plenum chamber and (through a switch and relay) is wired in parallel with an existing injector. When the switch is closed, the extra injector is pulsed. The new injector can be brought onstream by the action of a manifold or throttle position switch, for example. Figure 3.9 shows the wiring, with a dashboard pilot light used to indicate the operation of the extra injector. The extra injector(s) can be fed fuel from a T-piece placed in the fuel line before the pressure regulator.

Because the extra injector is pulsed by the standard ECU, the injector driver output transistor in the ECU has an increased load placed on it. This means the new injector should have the same (or higher) coil resistance as the standard injectors, and it also means there is the possibility that the ECU injector driver could be damaged. I have never heard of this happening (and many ECU injector drivers are protected against this type of failure, anyway) but the possibility remains.

Another disadvantage of this approach is that the mixture is very suddenly enriched — one moment there's no extra injector, a moment later the fuel being injected into the intake airstream has been increased by 25 per cent or more (depending on the number of injectors used by the engine and the flow of the extra injector). If the

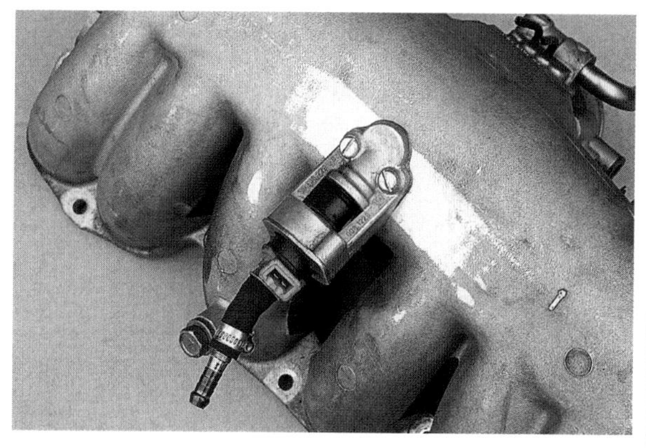

The location of this extra injector may cause uneven fuel distribution from cylinder to cylinder.

Injector Configuration	60-90 km/h time	Detonation
Single injector on at 12psi boost	4.1	Severe
Two injectors on at 12psi boost	4.3	Moderate
Single injector on at 10psi boost	4.1	Minor
Two injectors on at 10psi boost	4.3	Very minor
Injectors staged at 7psi then 10psi boost	3.9	Not audible

Figure 3.10
The results of a test session where two extra injectors were being set up on a 3-litre turbo car. The primary function of the injectors was to provide over-fuelling so detonation could be controlled. The tuning of the injector switch-on points made a major difference to the driveability of the car.

engine needs a lot more fuel (perhaps in a turbo car with the boost wound well up), using two extra injectors that are switched on in a staged manner can reduce the size of this sudden jump in fuelling. To do this, the circuit shown in Figure 3.9 is duplicated, with the second pressure switch set at a higher manifold pressure than the first.

Tuning of the operation of the extra injectors will help determine how well the system works. If two injectors are being used, altering their switch-on points can make a major difference to the driveability of the car. Figure 3.10 shows the results of a test session where two extra injectors were being set up on a 3-litre turbo six. The primary function of the injectors was to provide over-fuelling so detonation could be controlled. The car was using normal unleaded fuel, an intercooler and 15psi (~1 Bar) boost.

Extra injector controllers can be used to fire the injector, with the additional fuel proportional to load. However, make sure the controller doesn't cost a good proportion of a full engine-management ECU!

CHANGING THE IDLE SPEED

The idle speed is normally controlled by the ECU and is adjustable only within a very fine range. However, if modifications take the required idle speed out of the normal adjustment range, other techniques need to be adopted. If the idle speed is too low, adjusting the opening of the throttle butterfly can be used to increase it. If the speed is too high, closing the throttle butterfly right down (it's often cracked open in the standard position) or inserting an adjustable valve in the throttle body bypass hose can be used to reduce the speed. Cold idle-up can be achieved by using a solenoid-switched valve that opens to bypass the throttle body. A temperature switch on the radiator can control this valve, for example. Note these techniques should only be used when adjustment of the factory-installed valves and devices can't achieve the required idle speed!

OTHER ADD-ON MODULES

While the following are not engine management modifications as such, they do interface with the management system and are very useful in a modified car.

The first module is called the SmartShift. It's sold by Dali Racing and is designed to operate a shift light. A shift light is simply a lamp that illuminates to warn you the engine revs have risen to the extent that a gearchange is needed. It saves you having to watch the tacho — when your eyes are best on the road — and so is a good safety item as well as a performance part.

The SmartShift is housed in a neat, low-cost package, just a little bigger than a matchbox. The module requires that only four wires be connected to the car's wiring. The connections are:

- A 12-volt feed from an ignition-switched source
- Earth
- The tachometer output wire at the ECU
- The shift light or buzzer

The Dali Racing SmartShift module triggers a shift light at a preset rpm. It uses the tacho output signal of the ECU to sense rpm so is compatible with all types of ignition systems.

Instead of the traditional rpm-sensing approach of taking the engine speed signal from the ignition coil, the SmartShift simply monitors the tacho feed signal coming from the ECU. For this reason it will work on cars with single coils, double-ended multi coils, remote-mounted direct-fire coils and spark-plug mounted coils!

The SmartShift is calibrated by the use of an internal DIP switch that has eight separate on/off switches. Switches 1-5 are used to set the trigger rpm (200rpm increments from 4000rpm to 10,200rpm), switches 6 and 7 select the number of cylinders (4, 6, 8 or 10); and switch 8 sets the output mode (flashing or steady). One of the most elegant ways of installing the shift light is to use it to trigger a single high-intensity LED, aimed at the driver's eyes. Suitably modified with a relay and a diode to protect the switching transistor against back-EMF spikes, the module can also be used to switch any device at a predetermined engine rpm.

Another very effective add-on module is the Haligator from Haltech. Primarily designed as a fuzzy logic boost control, in addition the Haligator has functions for extra fuel injector control (two low- or four high-impedance injectors), a three-stage shift light and two further auxiliary outputs (eg water injection and intake manifold control). The inputs to the module comprise MAP, throttle position

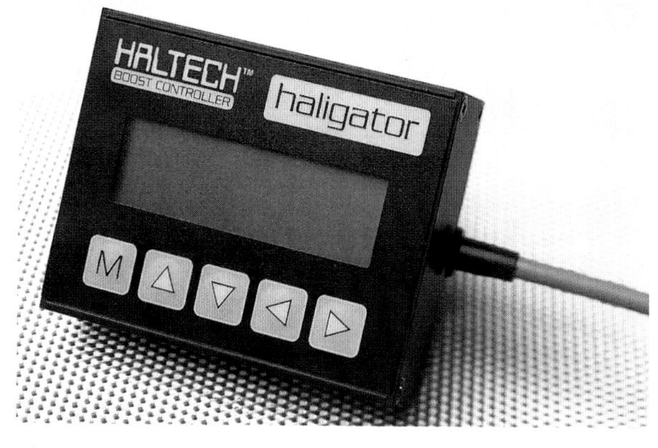

Another very effective add-on module is the Haligator from Haltech. The device is primarily a fuzzy logic boost control, but in addition the Haligator has functions for extra fuel injector control (two low- or four high-impedance injectors), a three-stage shift light and two further auxiliary outputs (eg water injection and intake manifold control).

and engine rpm signal. Programming of the unit is by a dedicated LCD hand-held pendant or PC. In a complex engine where fitting programmable management can be a very expensive process, an add-on module such as the Haligator can be a cost-effective way of introducing additional functions to the engine control system.

MONITORING AIR/FUEL RATIOS CHEAPLY

One of the most difficult aspects of modifying an EFI or engine-managed car on a small budget is knowing what air/fuel ratios are occurring at different combinations of loads and revs. The ignition timing maps won't usually need much tweaking for a budget car to run well, but more power needs more fuel and knowing how much extra fuel to add can be quite tricky. However, there is an easy and cheap way of getting a rough idea of the air/fuel ratios being used. It is certainly **not** a technique accurate enough to guide you when mapping programmable management, or to use when modifying a mega-dollar engine, but it is **far** better than tuning by 'feel' alone. It's also much cheaper than having professional exhaust gas analysis carried out on a chassis dyno every time you make a tuning change.

So how do you do it?

All unleaded petrol cars use an exhaust gas oxygen sensor. As the name suggests, this sensor is mounted in the exhaust flow — usually in the exhaust manifold — and sniffs the composition of the exhaust gas. Specifically, it measures the ratio of oxygen in the exhaust gas with that in the atmosphere. It does this to determine whether the air/fuel ratio is rich, stoichiometric or lean. The ECU uses this information as part of its self-learning technique, and in some cars a dash warning light is also illuminated if the mixture stays rich or lean for too long a period.

The most commonly used sensor generates its own voltage output, which varies between 0-1 volt. In round terms, if the sensor output is about 0.2 volts or less, the mixture is lean, and if the output voltage is over 0.8 volts, it is rich. However, the precise value of the output voltage is less important than its relative value — whether it is 'richer' or 'leaner' than the mid-point voltage.

Monitoring the sensor output can be done with a digital multimeter, but in practice the response time of many multimeters is too slow to keep up with mixture fluctuations. This is because when the ECU is in closed-loop mode, the air/fuel ratio fluctuates in a rapid rich-lean-rich-lean sequence. This occurs because, if the oxygen sensor output indicates the mixture is rich, the ECU leans it out a little. When the mixture is a little lean, the ECU enriches it. This constant process causes a cycling of the mixtures around the stoichiometric point. Depending on the particular engine management system (and the health of the EGO sensor), this can occur up to 10 times a second. That's what makes it hard to read on a digital multimeter! (Incidentally, don't use an analog meter — it loads the circuit down too much.)

The speed of change of the EGO sensor output means it's easiest read on a bar-graph display, with an LED bar graph fast in response and cheap. Ten coloured LEDs can be used, with two red LEDs indicating lean mixtures (remember 'red for danger'!), six green LEDs showing mixtures in a 'normal' range, and two yellow LEDs showing rich mixtures. If you wish, you can reduce the number of green LEDs and substitute more red and yellow ones. It's important that coloured LEDs be used (as opposed to an all-red bar-graph display, for example), because it's far easier to see at a glance the mixture strength by simply looking at the LED colour rather than its position in the display.

Figure 3.11(over page) shows a circuit diagram of the Mixture Meter. If you're unfamiliar with electronics, take it to a radio or TV repair technician or anyone with electronics experience. They should be able to make it in a few hours. In Australia, several electronics companies sell kit versions of the meter with its own easy-to-solder printed circuit board. Jaycar Electronics (www.jaycar.com.au/) is one, with the kit having the catalog number of KC5195.

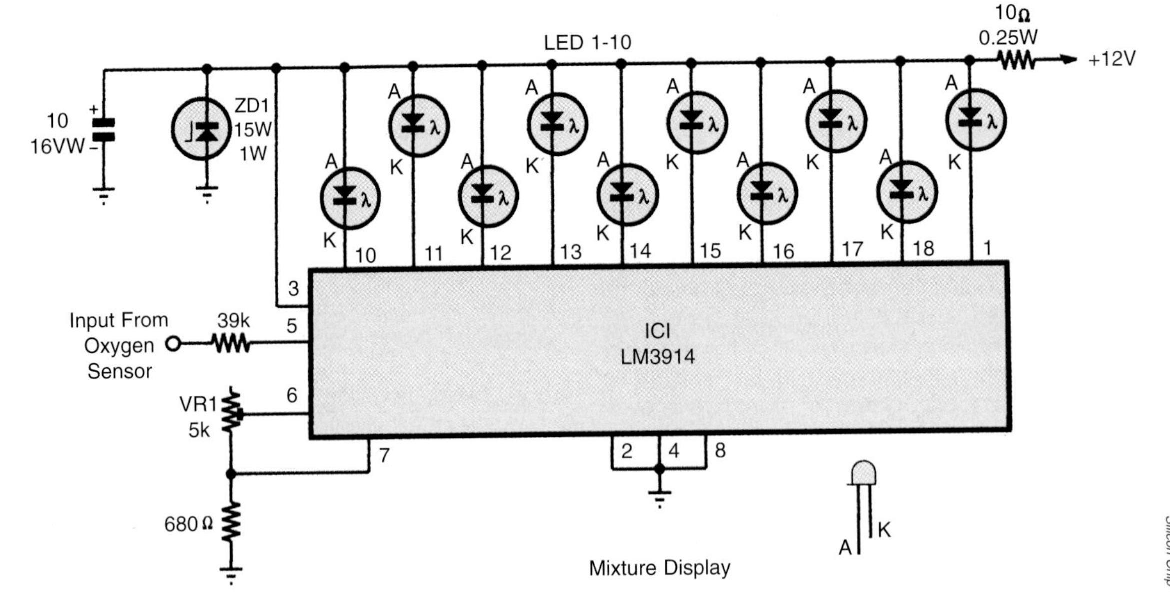

Figure 3.11
A simple-to-build LED Mixture Meter that can be used to monitor the air/fuel ratio. Commercial kits are available, allowing the easy construction of this design, or an electronics technician can put it together in a few hours.

The breakout box shows you how to construct the meter, step by step. Note the circuit shown here has a few additional components over the kit version; if the chip keeps blowing, add the extra components. These better protect the meter from voltage spikes and the like, but are not needed in most applications. The following assumes you have built a kit version of the meter, but if you have assembled the meter from scratch, the approach will be much the same.

The meter is powered by an ignition-switched volt source, so you will need to access such a supply at either the fuse box or another switched device (like the radio). Connect positive 12 volts to the '+12V' pin on the meter and connect the 'GND' pin to earth (the chassis or negative lead of the battery). Make sure these wires are connected the right way around — the meter is not protected against reverse power supply polarity! The final meter connection is to the signal output of the oxygen sensor.

Oxygen sensors are commonly available in single or three-wire configurations. If your car is fitted with a single-wire sensor, simply connect the signal lead from the meter to this wire. Don't disconnect the oxygen sensor output from the vehicle ECU; instead wire the meter in parallel. The Mixture Meter (unlike some commercially available air/fuel gauges) does not draw enough current to affect the reading the ECU is making of the sensor. The easiest way of making the connection is to access the EGO sensor wiring near to the sensor itself. Bare a short length of the wire by using a razor blade to carefully slice away the insulation covering, and then solder the mixture meter signal lead to it. Wrap the joint with good-quality insulation tape or seal the connection with heat-shrink tubing. Alternatively, you can access the oxygen sensor signal wire at the ECU.

If your car's sensor is the three-wire (or four-wire) type, you will need to locate the signal output wire. The extra wires found in this type of sensor are to power an internal heater, which brings the sensor up to temperature more quickly than occurs solely by heat transfer from the exhaust gas. With the car running and up to operating temp, one wire will have +12 volts on it, another 0 volts (the earth) and the final wire 0.4-0.6 volts. It's the latter that is the EGO sensor output, which must be connected to the '02INPUT' on the Mixture Meter.

With the car running, the connected meter should light some of its LEDs. If the EGO sensor is still cold, the 'lean' red LED may be the only one to light, but as the sensor comes up to temp, other LEDs will also glow. With the sensor up to temperature, a blip on the throttle should cause the lit LED to run up and down the scale. If all the LEDs light up at once — and there is a burning smell coming from the meter — switch off the car immediately and check the orientation of the IC. If no LEDs light, check the polarity of the power supply wiring and if you find you've connected the meter wrongly, buy another IC and try again!

There are two ways of calibrating the meter — on the road or on a chassis dyno. On the road is easier, although note this won't be appropriate in a car that has already been highly modified.

With an assistant installed in the passenger seat and the engine up to operating temp, drive at a constant speed — say 60km/h — and with a steady throttle opening. The lit LED should start oscillating up and down the display as the ECU makes the mixtures alternately rich and lean in closed-loop operation. Adjust the trim-pot so the oscillations in either direction are symmetrical around the middle LED. Now, use full throttle and watch what happens to the mixture readout. The display should instantly show a rich mixture (either of the two yellow LEDs lit) and this mixture should be constantly held. Lift the throttle abruptly and the display should blank, as the injectors reduce their flow on the overrun — and so the mixture goes full lean. At idle, the meter should again show the closed-loop oscillations.

If you're installing the meter on a highly modified engine, in-car calibration can still be done — but with the proviso that the mixtures may be all wrong to start with! With this

Building An LED Mixture Meter

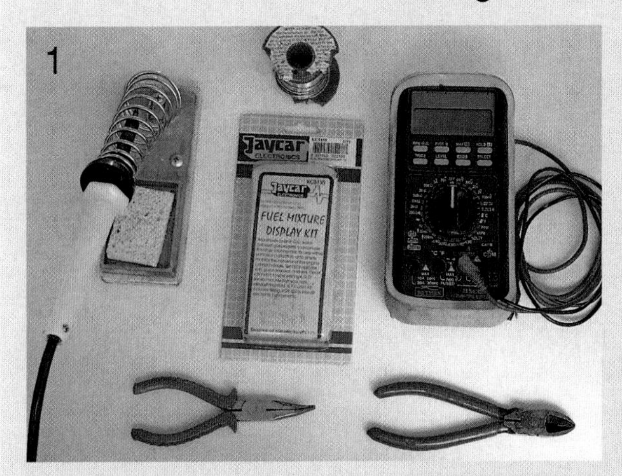

Even if you have no electronics experience, you will still be able to assemble the LED meter. You'll need a fine-tipped soldering iron, diagonal cutters and the kit (see text). It will be useful to also have a multimeter, some needle-nosed pliers and some more solder.

Make sure you know what each of the components looks like before you begin assembly of the kit. From top then clockwise: printed circuit board (PCB), hook-up wire, LEDs, potentiometer (pot), resistor, pins, capacitor, solder and (in the middle) the integrated circuit (IC).

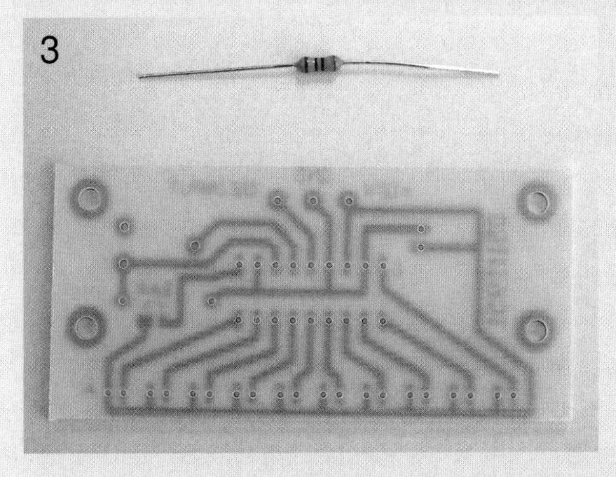

Each of these components needs to be soldered to the PCB with the wires going into exactly the right holes. The first component to be placed on the board is the resistor. This can be inserted either way around.

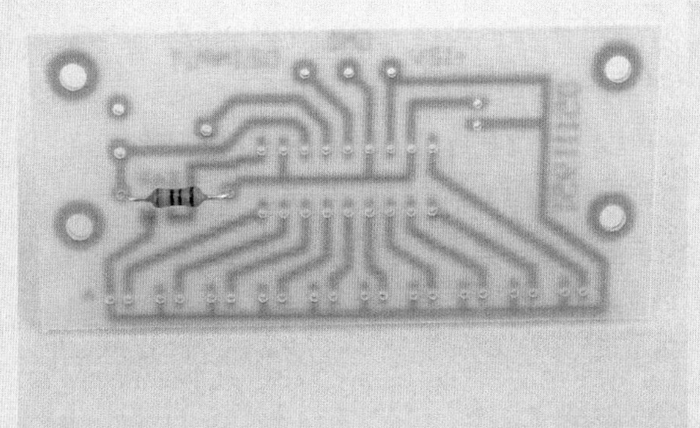

Using the needle-nosed pliers, bend the resistor wires at 90 degrees and insert the wires into the holes shown here. When working on the PCB, you should always orientate the board so the multiple row of holes at the base of the board is closest to you, as shown here.

Turn the PCB over and you will see the long resistor wires protruding through the holes in the silver PCB tracks. You now need to solder these wires to the track.

Touch the tip of the soldering iron to the wire and PCB track, applying solder at the same time. When the junction is heated sufficiently, the solder will flow around the join, forming a shiny solder blob and electrically joining the two. This should be a quick operation — don't leave the soldering iron applied for more than a few seconds. If the solder doesn't 'take' properly, apply the iron again.

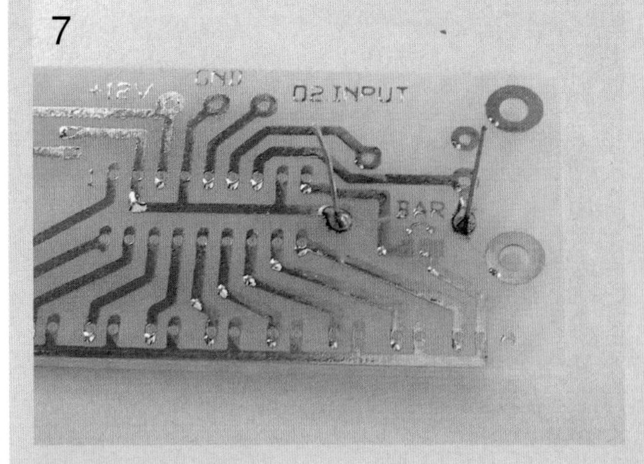

*After you have soldered the resistor into place, the board should look like this. Note that huge blobs of solder are **not** wanted — there should be only enough to surround the wire and cover the track on the PCB. Make quite sure you haven't formed a bridge to another PCB track. If you have, apply the iron and remove this surplus solder.*

Solder the other resistor wire and, once you're happy that the soldered joins are good, use the diagonal cutters to nip off the wires close to the PCB. Don't chop into the solder join itself; cut the wires off just proud of the solder. The soldering technique used for the resistor is used to attach each electronic component to the board.

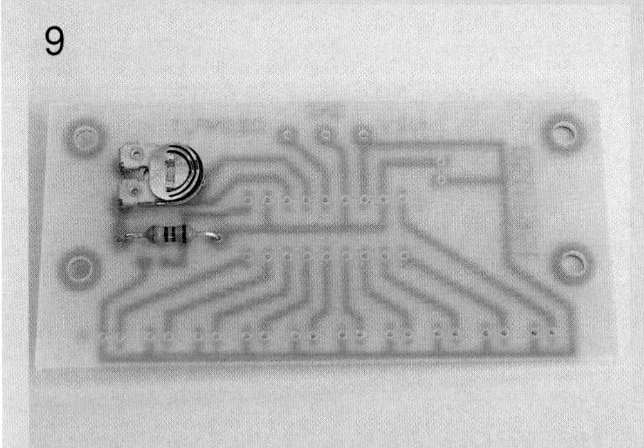

Next, insert the pot. This uses three wires and so fits onto the PCB only the one way around. The wires are already short, so they don't need to be cut off after you have soldered this component into place.

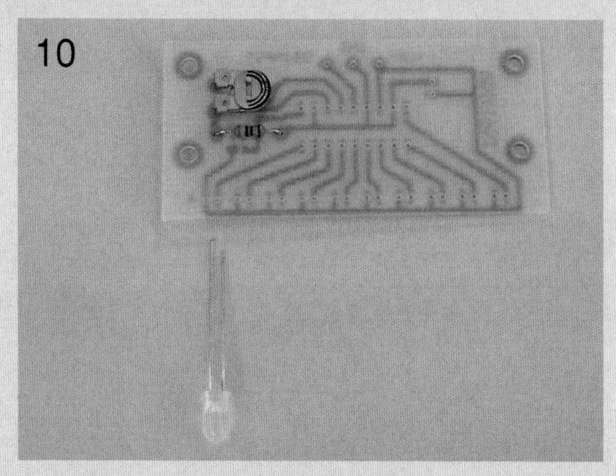

The next step is to solder the LEDs into place. These must be oriented the right way around or they won't work. Place one of the yellow LEDs next to the board as shown here. Make sure the longer lead (the anode, in technical speak) is at the lefthand side, nearest to the 'A' marked on the back of the PCB.

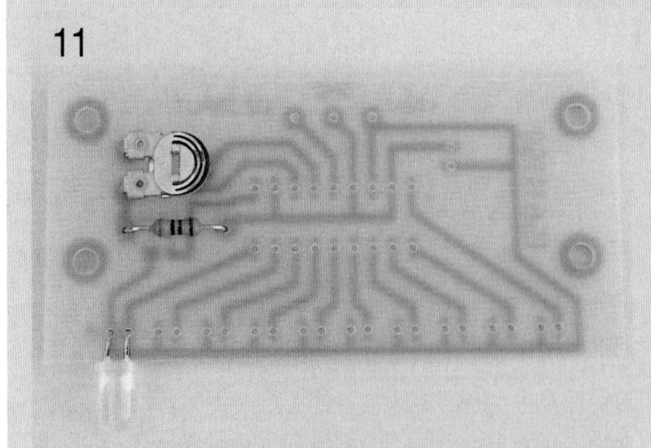

Bend the wires at 90 degrees and insert the yellow LED, pushing its body snugly up against the edge of the board. Doing this means all the LEDs will line up neatly. Solder the two wires into place and nip off the surplus wire.

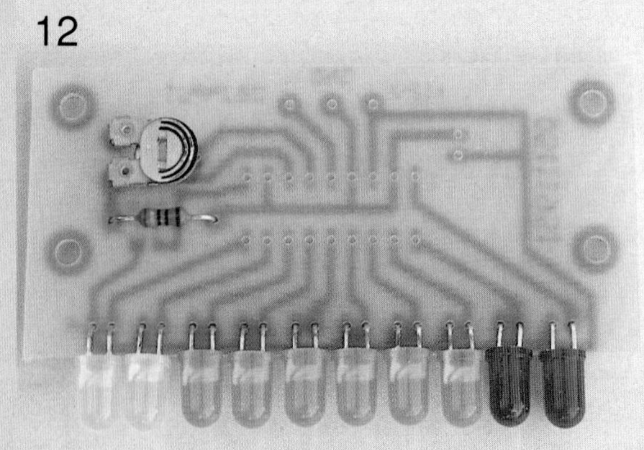

Insert the rest of the LEDs, soldering their leads into place as you go. Make sure you get all the LEDs oriented the right way around. One way to check this is to hold the board up to the light and check that the internal shapes of the LEDs are all oriented the same way.

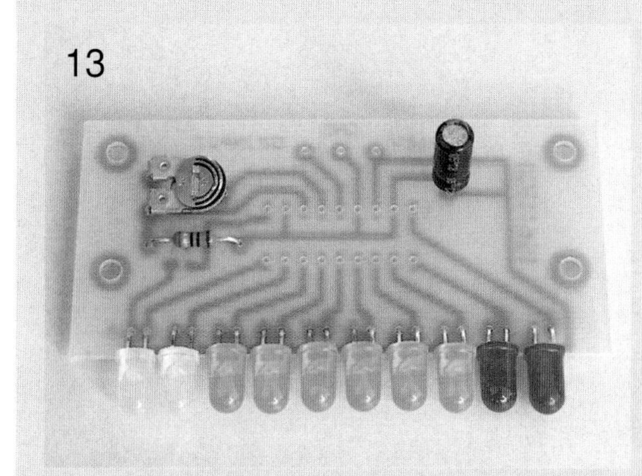

Next, the capacitor can be soldered into place. The capacitor must go onto the board the correct way around. If you look at the capacitor, you'll see that one lead is marked with a (-) on the body. This is the negative wire. The other wire is therefore the positive (although it's not marked) and the positive lead goes closest to the LEDs.

Nearly done! The IC socket is soldered on next. The IC socket has a cut-out at one end, and the socket is soldered into place so (with the board oriented as shown here) the cut-out goes to the right. The wires from the socket are quite short and won't need to be cut off after they have been soldered. Be very careful that solder bridges are not formed between the pins — they are spaced together tightly.

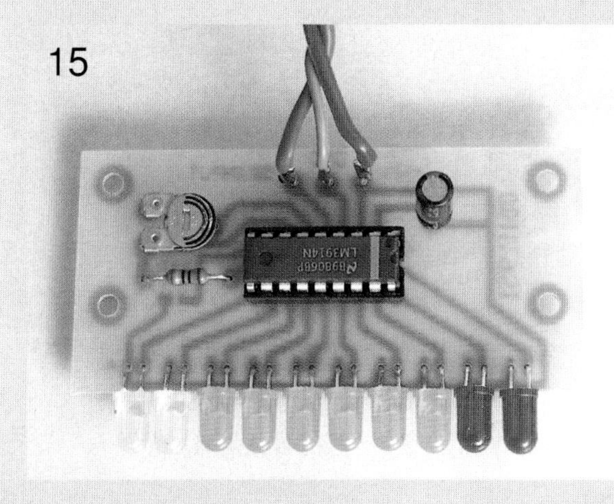

Solder the gold-coloured pins into place (they just make it easier to solder wires to the board) and then solder a different-coloured length of hook-up wire to each pin. Carefully insert the IC into its socket, making sure all the pins actually go in without being bent, and then the Mixture Meter is finished!

type of car it's safest to use a chassis dyno and an exhaust gas analyser so the mixture meter can be calibrated according to the gas analyser's readout.

It's important that you realise there are a couple of substantial limitations to the Mixture Meter. First, it's incapable of separating air/fuel ratios of (say) 11:1 and 12:1. The meter will just show full rich! Second, mixtures will always appear lean when the sensor is not up to temperature. The latter means you should be consistent in your testing procedure, so when you make adjustments to the air/fuel ratio, changes that appear on the meter are changes you have caused, not just the meter warming up!

If you have a leaded car that lacks an exhaust gas oxygen sensor, you can source a sensor from a wrecker and install it in the exhaust yourself. However, running leaded petrol will soon poison the sensor, so this approach should be used only for tuning purposes, with the sensor removed for everyday use.

Once you have installed an LED Mixture Meter on your dashboard, it's very hard to go back to having no mixture indication at all. You simply feel blind without it.

The engine in turbo cars is very easily damaged if a lean air/fuel ratio occurs under full power. A LED mixture meter will instantly show if this is happening, and if it is, you can get off the accelerator fast. Without the meter, you may never have known that the engine had been running lean.

Programmable Management

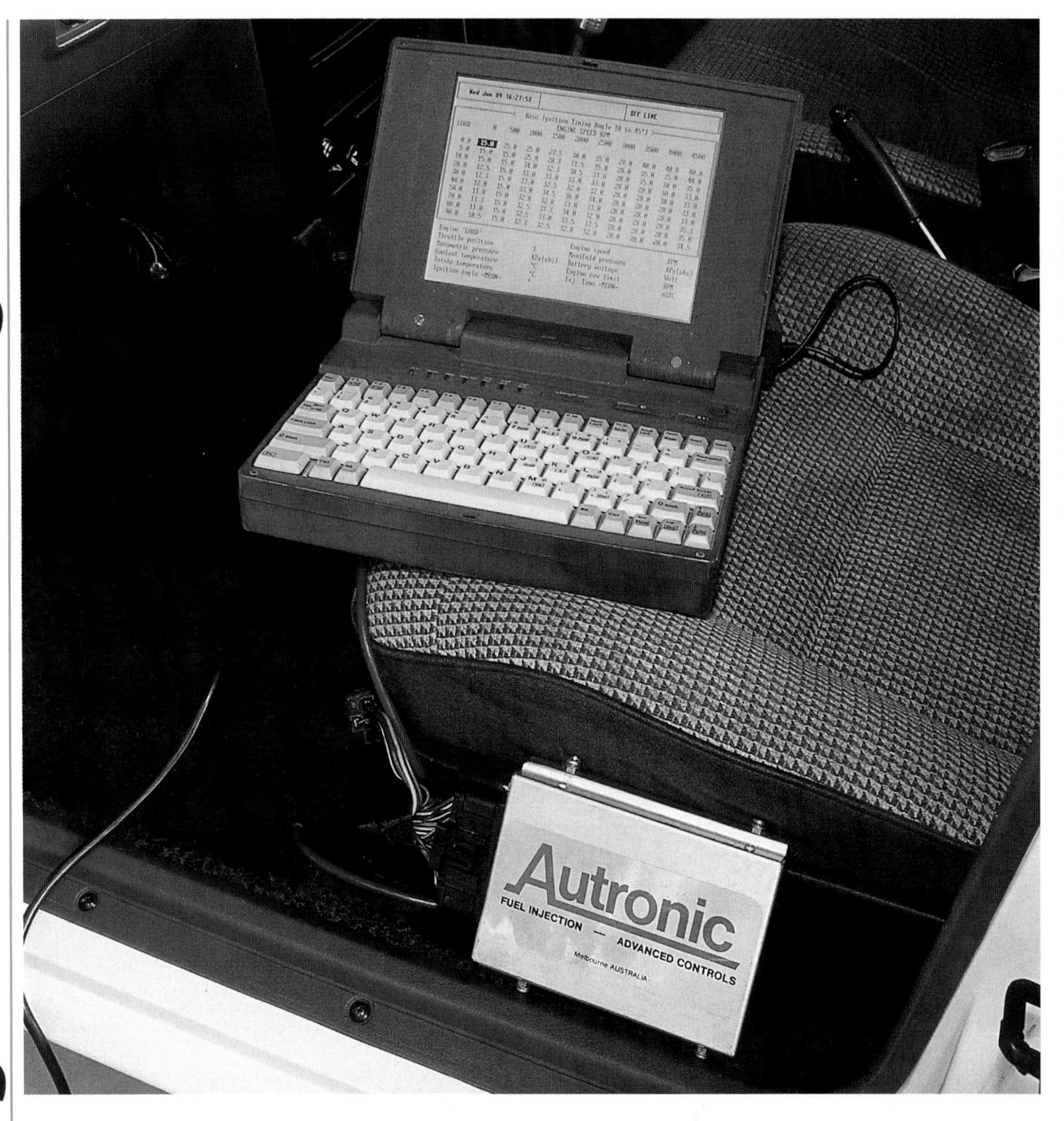

A programmable engine management system is one that allows the calibration of the engine management maps. These maps are the arrangements of data that determine the fuel injector pulse widths, the ignition timing and many other variables.

Programmable management systems typically have inputs from sensors measuring coolant temperature, inlet air temperature, manifold absolute pressure (ie MAP), crankshaft position and throttle position. In addition to fuel injector pulse width and ignition timing, other outputs that may be able to be programmed include idle speed control, radiator fan control, fuel-pump speed control, variable intake manifold control, turbo boost control, traction control, launch control and shift light control. Figure 4.1 (next page) shows the inputs and outputs for the MoTeC M8, a sophisticated programmable engine management system.

Due to their infinite flexibility, programmable management systems can be used effectively on almost any engine, whatever the power-producing modifications. This makes programmable management the best approach for any seriously modified road car.

TYPES OF SYSTEMS

Most programmable engine management systems evolved from racing. There it became obvious that to gain the best results in terms of power, reliability, responsiveness and fuel economy (vital in endurance racing), it was required that each facet of the engine's operation be precisely controlled. The first ECUs available at a less than stratospheric price were simple in design and provided only very basic adjustments. Typically, they were characterised by having screwdriver adjustment facilities, which obviously limited the resolution of inputs and the number of parameters that could be altered.

One problem with those early ECUs (that still remains today in some cheaper programmable systems) was inconsistent behaviour in different weather conditions. An

MoTeC

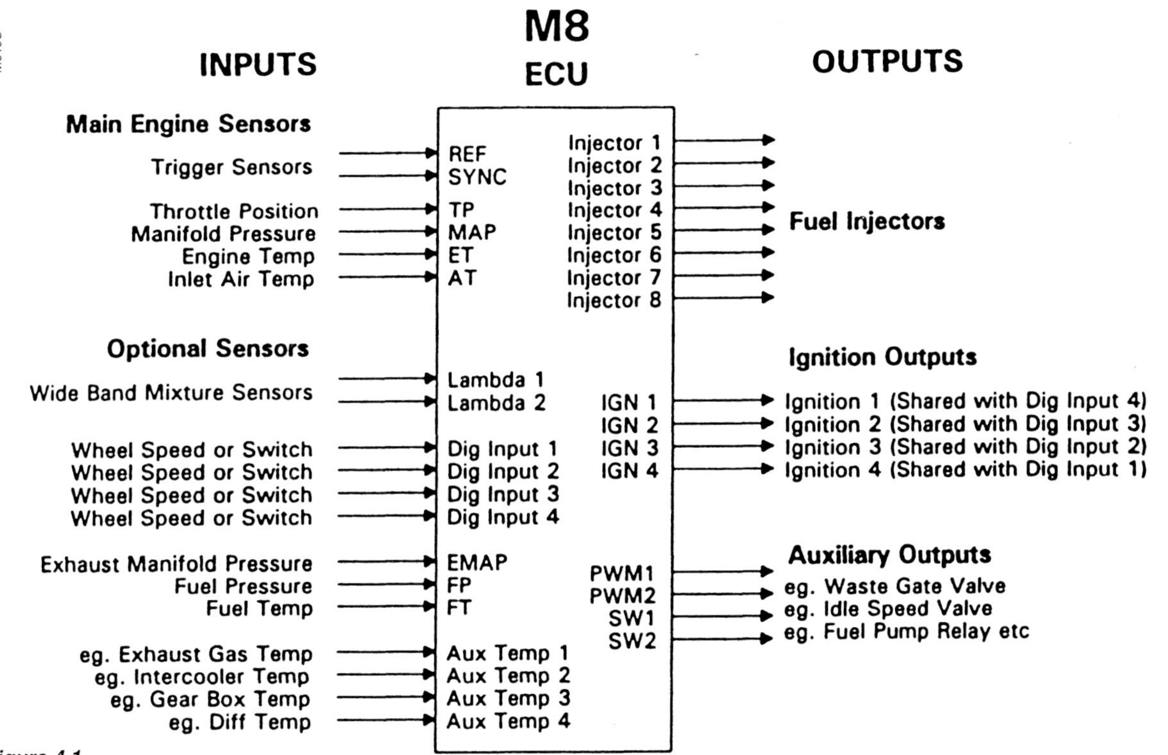

Figure 4.1
The inputs and outputs of the MoTeC M8 programmable ECU. This sophisticated ECU features traction control, sequential operation of the injectors and configurable outputs such as turbo wastegate control.

engine may have run smoothly and well on one day, but the next day, when the weather had changed, it wouldn't even start, let alone run! This was generally the result of insufficient control being available over coolant and intake air temperature correction parameters, factors that vary from engine to engine in the way they need to be calibrated.

ECUs with digital multi-function hand controllers were next on the scene. These used a two- or four-line LCD display on which various parameters could be selected for display. Much more precise calibration was available when compared with the screwdriver boxes, and the electronics and depth and breadth of engine control maps were increased substantially.

ECUs were then developed to interface with laptop PCs, allowing the display, graphing, logging and controlling of very detailed engine-management maps, sensor parameters and engine operating conditions. Throughout this development process, the price of programmable systems has continued to fall, at the same time as their complexity and capability has improved in quantum leaps!

A number of different types of programmable management systems is available. These are shown in Figure 4.2 (overleaf).

Each of the programmable ECUs discussed so far uses a replacement ECU box and a new wiring loom. Often the input sensors are also new, so that in effect, a completely new engine management system (that happens to be programmable) is installed. This has the major disadvantages that installation time can be lengthy and the engine is non-standard in management, so spare parts cannot be sourced from the car's manufacturer.

A'PEXi

An A'PEXi Power FC plug-in compatible ECU that's available for a number of Japanese high-performance cars. This ECU interfaces with all the car's standard sensors and output actuators, making installation a two-minute job. It's programmed via the handheld controller shown.

	Control Capability	Interface Type	Installation Process	Example System
Basic aftermarket ECU; fuel-only	Injectors; some extra outputs	Hand Controller	Re-wire some of engine, install mostly new sensors	Injec EM1
Basic aftermarket ECU; ignition-only	Ignition modules; some extra outputs	PC or hand controller	Re-wire some of engine, install mostly new sensors	Haltech IG5
Basic aftermarket ECU; fuel & ignition	Injectors and ignition module, shift light control, etc	PC or hand controller	Re-wire engine, install mostly new sensors	Microtech MT
Mid level aftermarket ECU; fuel & ignition	Injectors and ignition module, turbo boost control, cam timing control, etc	PC	Re-wire engine, install mostly new sensors	Autronic SMC
Advanced aftermarket ECU; fuel & ignition	Sequential firing of injectors, direct fire ignition coils, traction control, boost control, data logging, etc	PC	Re-wire engine, install mostly new sensors	MoTeC M8 pro

Figure 4.2
The different types of programmable management systems which require that the engine be re-wired.

An alternative approach when fitting programmable management is to retain elements of the standard engine-management system. As in the case of those programmable ECUs that plug straight into the standard loom and make use of the standard ECU sensors, sometimes with the addition only of a MAP sensor. This type of ECU is becoming increasingly common, but it is generally available only for cars that are popularly modified or used in motorsport.

Another approach is to modify the hardware and software of the standard ECU, allowing it to be programmed. This is different from simply installing a new chip in that the modified standard ECU behaves just like a fully programmable ECU — using a plain English display of the management maps and being able to be programmed in real time. At the time of writing, the only fully fledged system like this is Kalmaker, available for Holden (and some Chevrolet) GM-Delco ECUs.

These types of programmable management are shown in Figure 4.3.

FITTING PROGRAMMABLE ENGINE MANAGEMENT

The physical process of fitting a programmable management system to a car can take literally two or three minutes — or solid days of work. It very much depends on the system being used and the engine to which it is being fitted. As you would expect, the installation cost also varies very substantially!

The shortest procedure occurs when a pre-programmed ECU that plugs straight into the original wiring loom is being installed. The reduction in installation time associated with this type of programmable ECU can allow a **very** major cost saving. On the other hand, the longest

	Control Capability	Interface Type	Installation Process	Example System
Plug-in advanced aftermarket ECU; fuel & ignition	Full control of all standard functions; may not interface with other on-board computers	PC or hand controller	Plugs into standard wiring loom, may require a new MAP sensor	A'PEXi for Subaru Impreza WRX
Factory fuel & ignition with new software	Full control of all standard functions; normal interfacing with other on-board computers	PC	Standard wiring, standard sensors	Kalmaker for GM-Delco

Figure 4.3
The different types of programmable management systems which do not require that the engine be re-wired.

The MoTeC M8 is a sophisticated programmable ECU. In addition to controlling fuel and spark, it has facility for turbo boost control, traction control, inlet manifold control and many other functions.

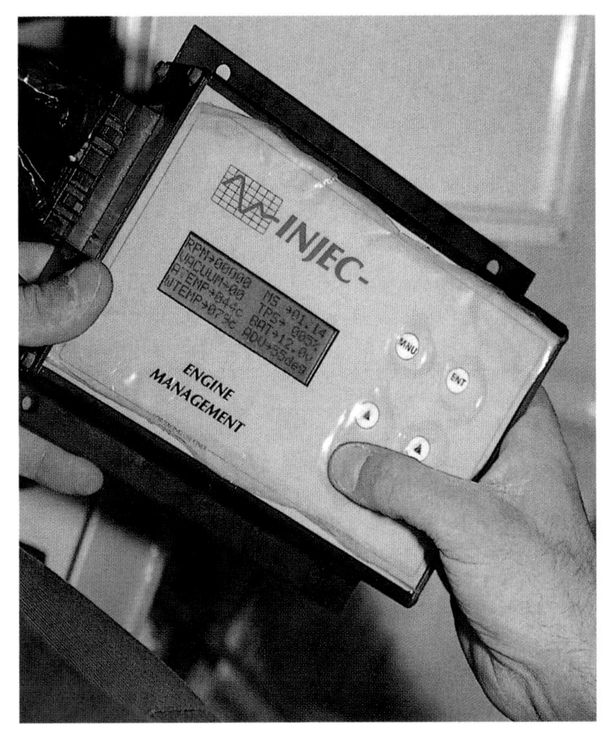

This Injec ECU integrates a two-line LCD display and menu buttons into the ECU itself. Displayed on the screen are engine speed, manifold vacuum, air and water temps, injector pulse width, throttle position, battery voltage and ignition advance.

The Haltech F9 is a fuel-only ECU — it can control the pulse width of the injectors but not the ignition timing. This ECU can be used in conjunction with an electronic ignition system.

The Haltech E6A is another high-specification programmable ECU. Like the pictured MoTeC unit, it's programmed using a laptop PC.

An early MoTeC hand controller that uses two LCD displays and individual LED indicators.

The Haltech IG5 is an ignition-only ECU — it can set the ignition timing through the control of ignition module(s) but cannot operate fuel injectors. This ECU can be used in conjunction with an LPG or carby fuel system, or with a separate fuel-only ECU.

This simple Holley ProJection fuel-only ECU uses coarse screwdriver-adjustment control. Note the simple control strategies, and how the owner has placed clear adhesive tape over the pots to prevent inadvertent changes to the settings.

procedure occurs when an engine sold originally with mechanical injection or carburettors is being converted to full aftermarket engine management.

1. The Air Intake System

If the engine to which the programmable management system is being fitted originally used carburettors or mechanical injection, changes will need to be made to the air intake system. By far the best way of sourcing an Electronic Fuel Injection (EFI) intake is to find an EFI version of the engine! Some engines were converted to EFI by their manufacturer late in the engine's life, so the intake manifold and plenum chamber can be simply bolted to the earlier version. However, before buying a later-model intake system, make sure any new intake **is** simple to fit. If nothing is available ex-factory, you can fabricate your own intake system. Chapter 5 has more on this topic.

2. Throttle Body

If it's required that a new throttle body be fitted, its size should be appropriate for the amount of engine power being developed. In addition, the type of throttle position sensor normally fitted to the throttle body should be assessed for compatibility with the new management system. It may be cheaper in the long run to buy a more expensive throttle body that comes with a suitable sensor rather than later have to replace an inappropriate sensor. In some systems, an idle air speed control solenoid is integrated into the throttle body. Again, the throttle body chosen should have compatibility of idle air system with the new engine management ECU.

The ProJection ECU controls two large single-point injectors that squirt past two throttle butterflies — shown is the fuel flow at full load. Unlike this unit, almost all programmable ECUs are able to control multipoint injection (ie one injector per cylinder).

Throttle position sensors used with programmable management systems must be of the potentiometer type; some original equipment manufacturers use switches that can detect only idle and full throttle positions. A check should therefore be made of the type of throttle position sensor fitted before the throttle body is purchased — especially if it's secondhand. Engine vibration can cause the throttle position sensor to wear and so, in high vibration environments, the sensor should be checked or replaced at relatively frequent intervals. High-pressure washing of the throttle position sensor should be avoided and care should be taken that the sensor does not act as the throttle stop!

3. Temperature Sensors

Programmable engine management systems require inputs from coolant and intake air temp sensors. Typical two-wire sensors used in this application include those from Bosch and Delco. Most management systems will

A Link programmable ECU that is plug-in compatible with a Nissan Skyline's standard wiring harness, requiring only the addition of a MAP sensor.

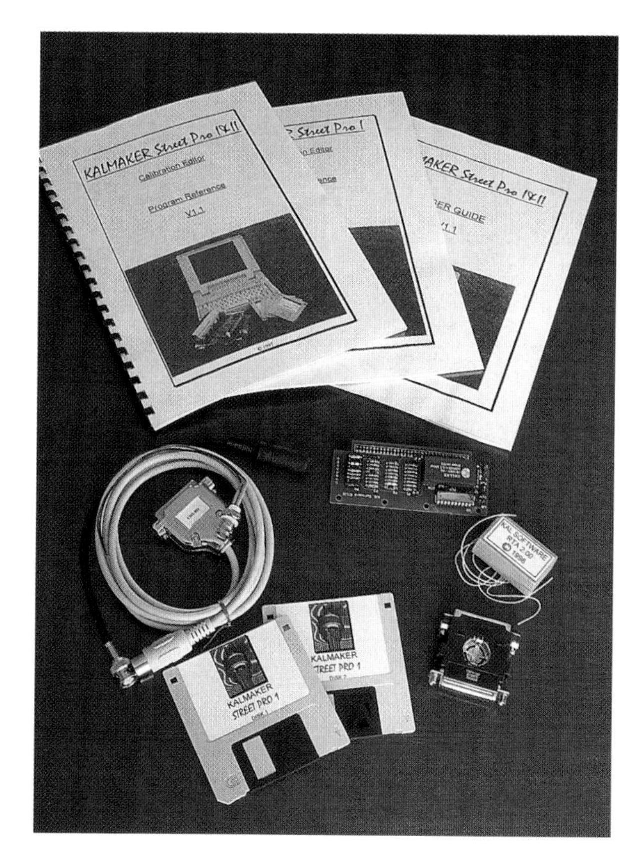

Kalmaker software and hardware allows a factory GM-Delco ECU to be converted to full programmable management, retaining all the original factory management maps and engine hardware.

work with either of these sensors, but note that some original-equipment temperature sensors will not be compatible with the new ECU. If this is the case, replacement generally involves welding up the original sensor's hole, drilling and re-tapping. The same applies for the coolant temp sensor, which is normally located prior to the thermostat on the thermostat housing. In air-cooled engines, the temperature sensor can be embedded in the engine block or can be used to sense oil temperature.

Turbocharged cars that are not intercooled (or have small intercoolers) should have a quick-response intake air temp sensor fitted. All intake air temp sensors should be mounted such that they will not be affected by fuel 'stand-off' (reverse flow of atomised fuel in the intake system), which could cause the sensor to be cooled.

4. Load Sensor

The majority of programmable management systems use a three-wire Manifold Absolute Pressure (MAP) sensor to detect manifold vacuum (or vacuum and boost). The sensor is connected to the intake manifold after the throttle body by a small-bore rubber hose. The GM-Delco sensor is used in many programmable management installations. This sensor is available in three versions: 1 Bar (to suit naturally aspirated engines), 2 Bar (up to 14.5psi boost) and 3 Bar (up to 29psi boost).

The MAP sensor can be adversely affected by vibration so should be mounted remotely to the engine. The manifold pressure take-off point should be at a position that best represents average manifold pressure and minimises pressure pulsing. In cars with a small number of cylinders and a small plenum chamber, the pulsing can be excessive. In these cases, the pressure signal can be damped by the use of a small restrictor in the hose; some manufacturers (eg Daihatsu) use special damping fittings as standard equipment on their MAP-sensed cars. The MAP sensor should be mounted so the pressure port faces downwards and the hose that connects the sensor to the intake manifold runs downhill to the engine, so that any condensation drains away from the sensor.

Some programmable ECUs mount the MAP sensor within the body of the ECU, with the sensor connected to the intake manifold by a long, small-bore rubber hose.

5. Position (Trigger) Sensor

It is the interfacing with the position sensor on the crankshaft and/or camshaft that causes the greatest difficulties in the installation of programmable management systems. While some programmable management systems are claimed to work with inductive pickups, best results come from the use of either optical or Hall Effect sensors. Conversions of the original equipment position sensors to either a different type (eg inductive to Hall Effect) or a different trigger mechanism (eg replacement of the optical chopper disc with another) are both very common.

Plug-in compatible ECUs (those that use the standard loom and sensors) require that no changes be made to the position sensors. This is a major advantage of these units.

A GM-Delco factory ECU that has been modified using Kalmaker hardware. Note the addition of the new printed circuit board (top).

Manufacturer	Position Sensor
Toyota	24 tooth Magnetic
Nissan	360 slot Optical
Honda	16 or 24 tooth Magnetic
Mitsubishi	4 slot optical or Hall Effect
Mazda (rotary)	24 tooth Magnetic
Mazda (piston)	4 slot Optical or Hall Effect
GM	Hall Effect, 360 slot Optical, Magnetic or Hall Effect with Direct Fire Ignition
Ford	Hall Effect in distributor, magnetic on crankshaft
Porsche	Magnetic on crankshaft with Magnetic or Hall Effect on cam
BMW	Magnetic on crankshaft with Magnetic or Hall Effect on cam

Figure 4.4
Manufacturers use a variety of position sensors on their engines. Interfacing the programmable ECU with the position sensor is vital if the system is to work successfully.

All engine-management systems need to be able to sense engine speed, referred to by some manufacturers as a 'ref' (reference) signal. In addition, management systems using multi-coil ignition and/or sequential injection require a further input of a single pulse per engine cycle so piston position can be ascertained. This is termed the 'sync' (synchronisation) signal.

All modern Original Equipment engine management systems have position sensors already fitted to the engine. Magnetic sensors are located in the distributor (16-24 teeth) or on the crankshaft (60 teeth). Optical sensors are usually found in the distributor or mounted at the end of a camshaft and have up to 360 slots. Hall Effect sensors are mounted on either the crankshaft or in the distributor and have 4-8 vanes. Where both sync and ref signals are required, these position systems usually make use of two sensors, normally built into the same housing.

If programmable management is being installed on an engine previously equipped with either mechanical injection or a carburettor, the air intake system will need to be altered. This is a Holley-pattern quad throttle body that contains four injectors and which can be bolted to a V8 inlet manifold.

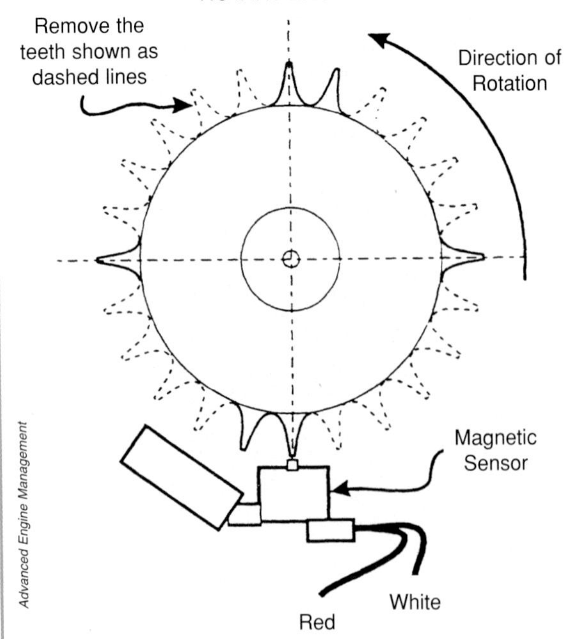

Figure 4.5
The modification required of the Mazda rotary engine 24-tooth magnetic sensor so it interfaces with a Wolf engine management system.

The position sensors commonly used by the various manufacturers are shown in Figure 4.4.

Some programmable management systems interface directly with the original-equipment position sensors, requiring only that the correct wiring connections be made to the sensor. Other programmable management manufacturers specify how the original sensor should be modified so it becomes compatible with the new ECU. Figure 4.5, for example, shows the modification required of the Mazda rotary engine 24-tooth magnetic sensor so it interfaces with a Wolf engine management system.

Where conversion to a distributor-located Hall Effect system is required, the dizzy is gutted of its original pickup or ignition timing components. These are then replaced

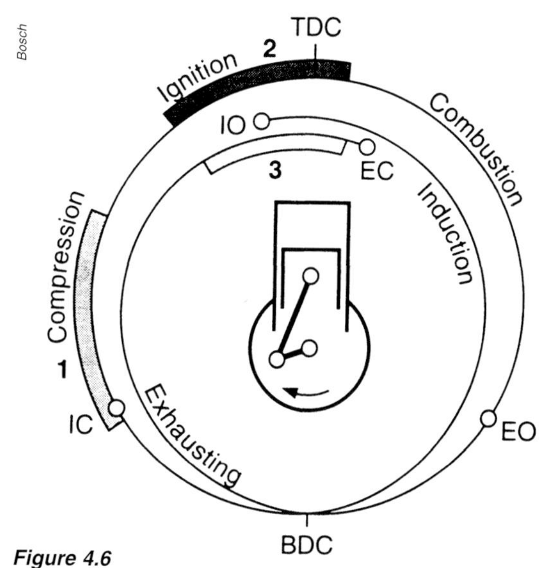

Figure 4.6
Waste-spark fires the plugs on two companion cylinders — those with the same TDC but on different strokes. This means that in each four-stroke cycle, the plug on each cylinder fires twice. (1) Switch-on range of primary current, (2) ignition range of the first ignition spark, (3) ignition range of the second ignition spark, (TDC) top dead centre, (BDC) bottom dead centre, (IO) intake valve opens, (IC) intake valve closes, (EO) exhaust valve opens, (EC) exhaust valve closes.

with a Hall Effect sensor, triggered by the passing of a ferrous metal vane. Suitable ref signal vanes for four-, six- and eight-cylinder engines can be found as standard components in some cars and so can be bought new as spare parts. In optical systems, the standard slotted disc is usually replaced with a disc containing fewer slots.

If a distributor is being converted to crankshaft sensing duties, care needs to be taken that the distributor rotor phasing remains correct. Many ECUs require the reference signal falls within a certain window — for example, the Haltech E6A ECU needs a signal within 60-100 degrees BTDC. If the sensor is installed so the trigger signal is received outside this window (and then the whole dizzy is rotated to correct for this problem), it's likely the rotor button will have passed the plug lead terminal, leading to arcing in the distributor cap. To prevent this occurring, the plate the sensor is mounted on (or the chopper disc) needs to be rotated relative to the rotor button.

Because interfacing the crankshaft position sensor with the new engine-management system is typically the hardest part of fitting a new programmable engine-management system, I strongly suggest that any question marks over position sensing be fully resolved before a management system is purchased.

A potentiometer throttle position sensor has a constantly variable output from zero to 100 per cent throttle. All programmable management systems use this type of sensor.

A Holden throttle body adapted to a Mercedes V8 intake manifold. On the throttle body are mounted the throttle position sensor (left) and the idle air control valve (right). This car was fitted with GM-Delco management under the control of Kalmaker software.

6. Ignition

There's a number of different ways in which the ignition system can be configured. If the engine was originally fitted with a distributor, this can be retained in conjunction with a single coil. If an aftermarket capacitor discharge ignition (CDI) unit is used with a single high-energy coil, good results can be obtained even on high-power forced-induction engines.

The use of multi-coil direct fire ignition is, however, now more common. If there are sufficient ignition outputs from the ECU, the coils can be configured with a separate coil and ignition module for each spark plug. If there is not a sufficient number of ignition outputs, a waste-spark system can be employed. Waste-spark pairs companion cylinders (those with the same TDC but on different strokes). Each coil fires sparks on a pair of cylinders, one on the compression stroke (the spark that achieves results) and the other on the exhaust stroke (the waste-spark).

Figure 4.6 shows how the process works. Figure 4.7 (overleaf) shows a waste-spark approach using multi-ended coils, while Figure 4.8 (page 72) shows a single-coil-per-cylinder waste-spark approach. Neither multi-coil system uses a distributor.

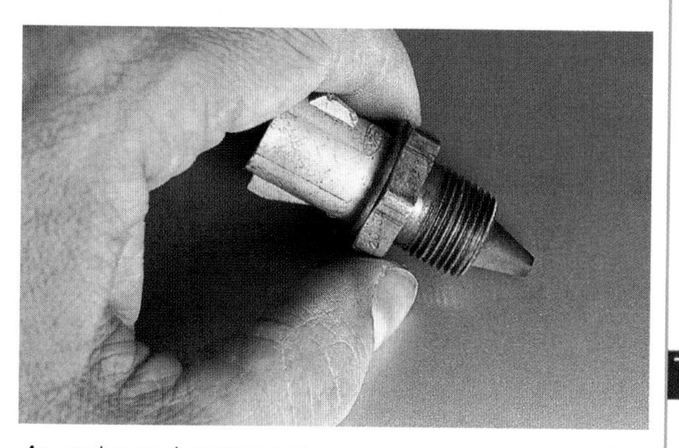

An engine coolant temperature sensor. In many cases, when programmable management is fitted, a new coolant temp sensor will also need to be installed.

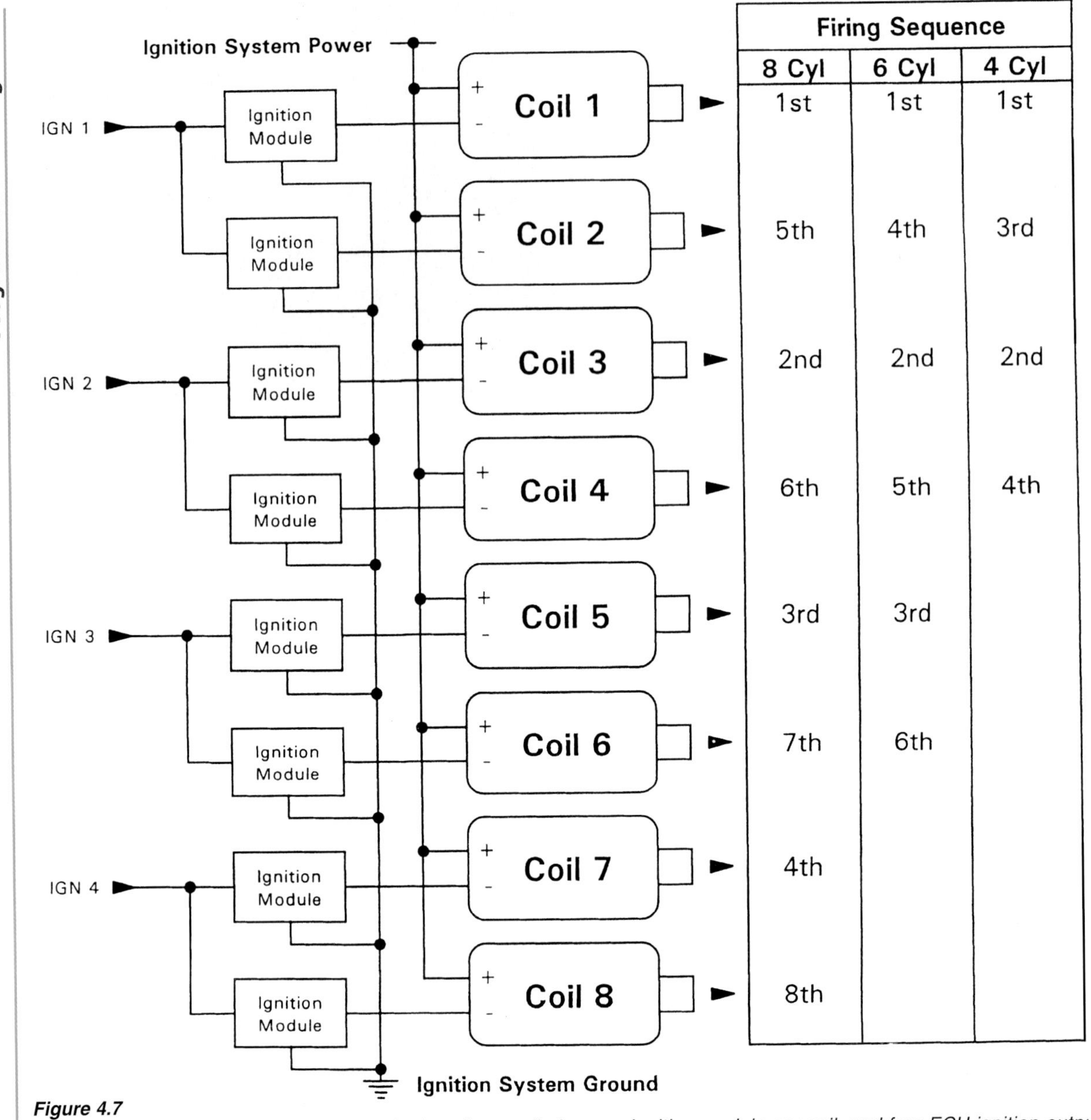

MoTeC

Figure 4.7
The ignition system shown here uses a single coil per cylinder, one ignition module per coil, and four ECU ignition outputs. The cylinders are paired in a waste-spark approach.

Installing programmable management is relatively simple because clear wiring diagrams are provided with systems. The most difficult task is interfacing the new ECU with the original crankshaft/camshaft position sensors.

An intake air temperature sensor mounted on a foam ram-pod intake trumpet. Intake air sensors should be located so they only sense the temperature of the intake air — not that of air that's passed through the radiator, for example!

This GM-Delco inlet air temp sensor has been installed on a Toyota engine. Swapping sensors in this way will often require the welding-up, drilling and re-tapping of original sensor mounting holes.

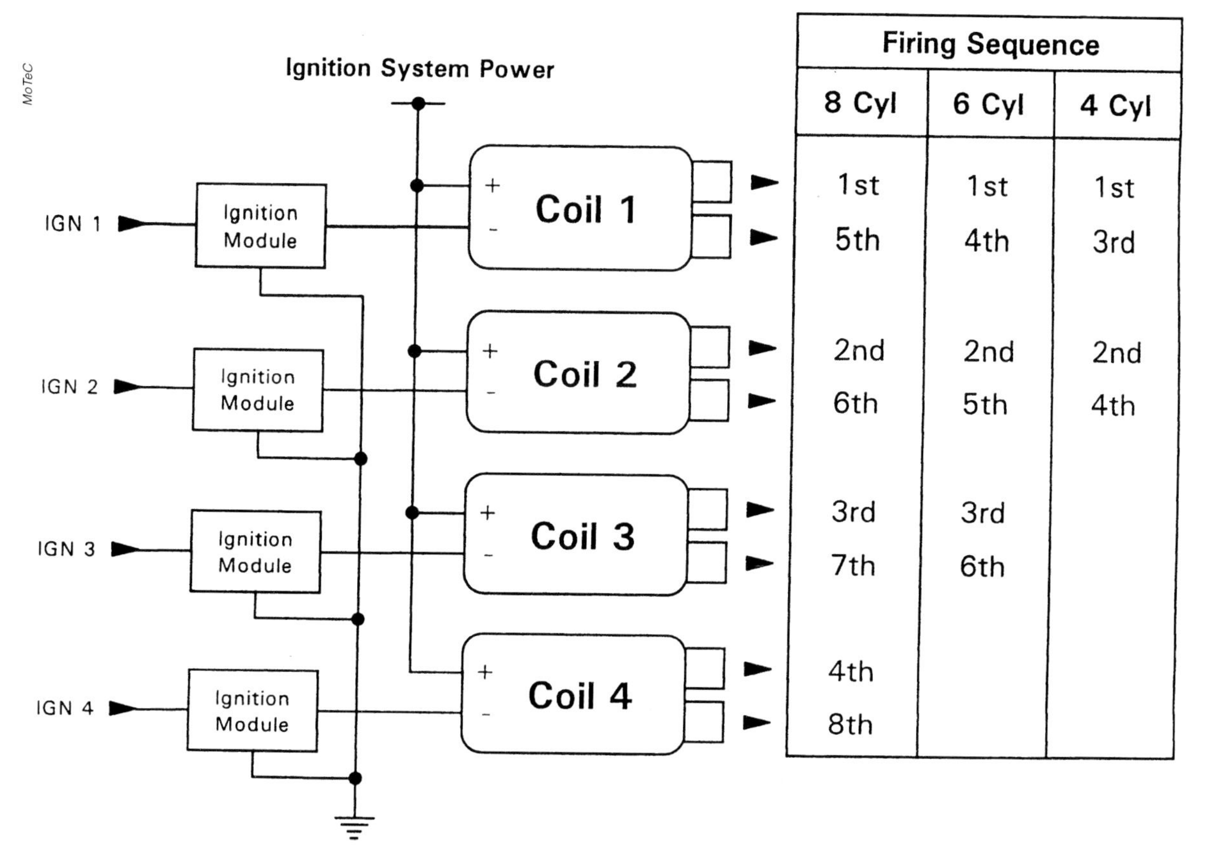

Firing Sequence		
8 Cyl	6 Cyl	4 Cyl
1st	1st	1st
5th	4th	3rd
2nd	2nd	2nd
6th	5th	4th
3rd	3rd	
7th	6th	
4th		
8th		

Figure 4.8
Here double-ended coils have been used in a waste-spark system, reducing the number of ignition switching modules needed over Figure 4.7's approach.

Interfacing the crankshaft (and/or camshaft) position sensor with the new programmable ECU is often quite difficult. This Nissan engine runs a camshaft position sensor that looks standard, but inside it some new parts have been used.

The ignition module(s) must be mounted away from hot underbonnet areas, such as near turbos or exhaust manifolds. They should also be mounted on a heatsink, with appropriate heatsink compound used between the module(s) and the heatsink. Heatsinks are available from electronics stores, while a short section of heavy-duty aluminium extrusion can be bought for nearly nothing from a non-ferrous scrap metal dealer.

7. Oxygen Sensor

If the new engine management uses a closed-loop feedback signal, an oxygen sensor will need to be fitted to the exhaust. So as it reaches 400 degrees C operating temperature, the sensor should be mounted close to the engine, but at the same time it should be positioned so it senses the gas flow from as many cylinders as possible. This means a location at the end of the exhaust manifold (or where header pipes come together) is usually appropriate. If the oxygen sensor is located more than

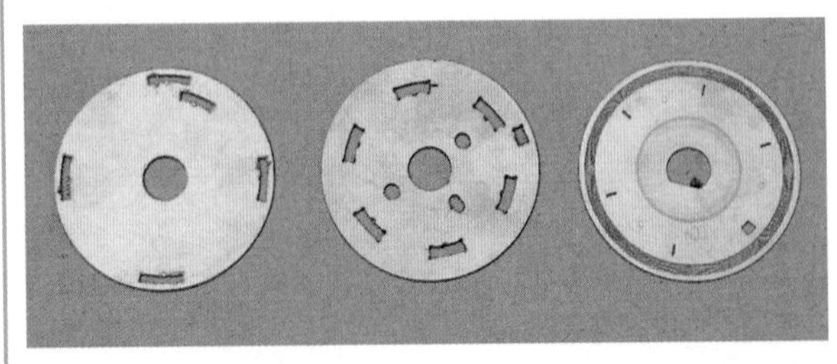

These are discs used in Nissan optical camshaft position sensors. On the left are two discs used to interface the standard sensor to programmable management, while on the right is the standard Nissan disc with its 360 + 6 slots.

A Nissan position sensor with the new chopper disc in place. An optical device, it works by shining an LED at a sensor. The disc spins at camshaft speed, interrupting the light beam.

30cm or so from the exhaust valves, a heated type should be used. Exhaust workshops have available weld-on spuds already drilled and tapped to take common oxygen sensor threads, so installing an oxygen sensor is usually straightforward. In cars with very short exhaust systems, the oxygen sensor should be located at least one metre from the end of the exhaust.

A GM-Delco MAP sensor of the type used with many programmable management systems. The '2' on the sensor refers to the pressure range for which this particular sensor is suitable — 2 Bar absolute, ie up to 1 Bar (14.5psi) boost.

8. Fuel System

An EFI car's fuel delivery system differs from a carby fuel system in that flow and pressure are both higher, so a high-pressure EFI fuel pump needs to be fitted if a carby car is being converted to engine management. The feed to the pump must be installed so it will never starve, even during cornering or acceleration. This requires that either the tank be well baffled (so fuel can't slosh away from

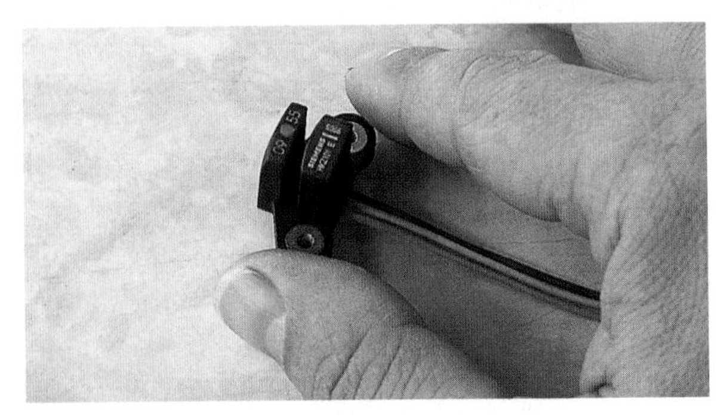

A three-wire Hall Effect sensor is used in conjunction with the metal vane. The passing of the vane through the slot between a magnet and the sensor switches off the sensor output.

A Hall Effect Vane mounted in a distributor. Adapting an engine to Hall Effect sensing can be done relatively easily using these components.

the pump pickup) or an external or internal swirl pot needs to be added. If programmable management is being fitted to an engine-managed car experiencing a major power upgrade, the fuel system will also need to be modified appropriately. Chapter 10 covers fuel supply systems in more detail.

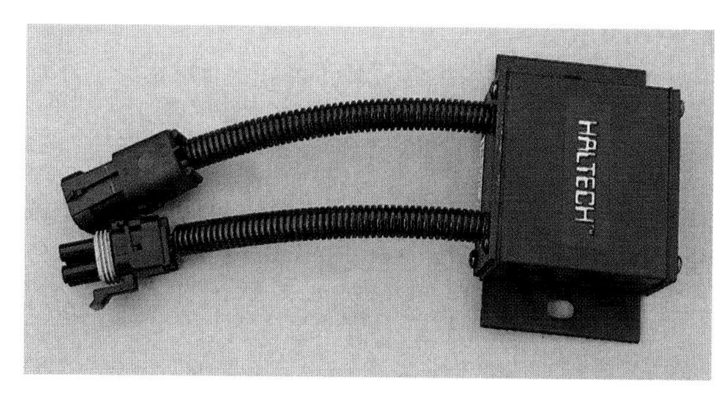

Haltech markets an adaptor module that converts the output of an inductive sensor into a form suitable for use with Haltech ECUs. However, best results still come from using Hall Effect or optical position sensing.

9. Mounting the ECU

The ECU needs to be mounted inside the cabin so it's located away from dust, heat and water. A number of locations can be used, depending on the size of the ECU and the room available in the car. Common mounting locations include under the dash, inside the passenger-side kick-panel or under a seat. The ECU should be mounted so vibration is minimised and air can circulate around it.

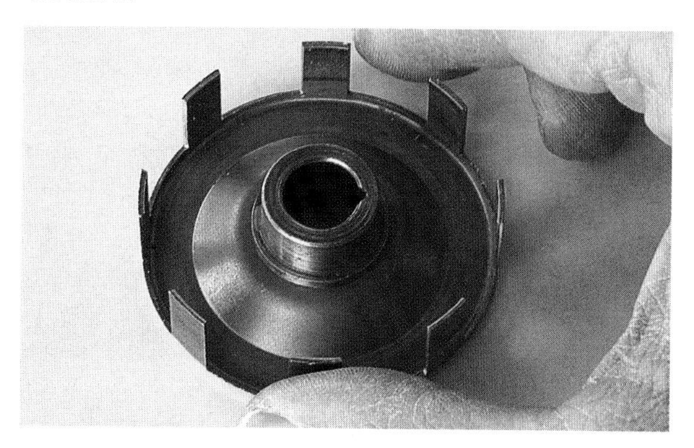

A ferrous vane used in a Hall Effect crankshaft position sensor. This vane is for an eight-cylinder engine; vanes are also available to suit four- and six-cylinder engines.

Engines requiring emissions compliance need to use a programmable management system that caters for oxygen sensor feedback. A heated sensor should be used if the sensor is to be installed more than about 30cm from the exhaust valves. Note that this cable is just a little tight...

When installing an ECU in the cabin, it can be difficult to pass the wiring loom through the firewall. Drilling a hole in the firewall large enough to push the engine connectors through and then feeding the loom through from the cabin side is the best technique. However, sandwich firewalls, sound deadening, all-curved firewalls and those with many small fittings often make this easier said than done! The loom hole needs to be adequately sealed after the wiring has been passed through, otherwise noise will be transmitted into the cabin. Using a rubber grommet with a long 'tail' that extends along the loom is one approach that works well.

Multiple coils wired in a direct-fire configuration give excellent spark. Here four Bosch coils (and four ignition modules which are not shown) are being used on a turbo Nissan CA18DET four-cylinder engine.

Single-coil ignition systems in high-powered cars benefit from the addition of a Capacitor Discharge Ignition unit like this Crane HI-6. These CDI systems are compatible with programmable management.

The engine-management system wiring in the engine bay should be installed neatly with multiple tie-downs. It should be kept away from hot areas and should never have long runs parallel and close to high-voltage ignition leads. Relays should be mounted tab-upwards and away from areas that could be splashed with water.

TUNING PROGRAMMABLE MANAGEMENT

The most important aspect of fitting programmable management is to get it tuned correctly. The very best system can run poorly in almost every aspect — driveability, starting, fuel economy, emissions and power — if it is not tuned well. And unlike 'hot chip' re-programmers who usually alter only a tiny proportion of the information already on a chip, many programmable engine management tuners start with a nearly blank slate! Filling in that blank slate with the optimal map data is not at all easy.

Here are two points you should consider when reflecting on management mapping. The first is that car company development engineers spend many hundreds of hours on engine dynamometers carrying out the mapping of stock management systems. This is then followed by the driving of test cars hundreds of thousands of kilometres in all sorts of conditions. The interesting fact is that the engineers are able to make changes to the mapping during the drive program — and often do.

If a programmable management system is being used that retains most or all of the standard engine components, the factory ignition module will be used. For example, this V8 unit is retained when Kalmaker programmable management is used with the GM-Delco ECU.

Before a programmable management system is bought and installed, consideration should be given to the fuel supply system. Increasing engine power dramatically will also require far greater flows of fuel, often at a higher pressure if a turbo engine is being boosted.

They might decide, for example, that the hot-start capability of the engine could do with a little improvement after the car has been driven hard and then parked for 15 minutes. The same problem may not occur if the car is parked for 30 minutes, though. Or they might decide the car equipped with a manual transmission is just a little too easy to stall when it's towing a trailer and needs to perform a hillstart. This level of detail is simply impossible for any normal workshop to cost-effectively perform when mapping a management system.

The other example of the difficulties of comprehensive mapping can be seen in a long-term test of an A'PEXi programmable ECU, carried out in my own car. The ECU came pre-programmed for my car's engine type and it was installed by being plugged straight into the car's standard loom — it used all the standard input sensors and output actuators. However, it was programmed for a higher-octane fuel than is available where I live, so some areas of the map controlling the main ignition timing advance needed to be retarded to prevent the engine detonating. Note that even with the ignition timing over-advanced, detonation was never actually audible, the ECU flashing the Check Engine light whenever it detected it and then automatically retarding the timing. Using the hand controller, the ignition timing was re-programmed for different loads and conditions until all seemed perfect.

The interesting aspect was that the test took place over nearly 12 months. For the first two months there were almost no problems, the exception being that just once or twice the Check Engine light flashed to show the ECU

Each ignition coil requires an ignition module that contains the power switching circuitry to turn the coil on and off. Ignition modules should always be mounted on an effective heatsink and located away from hot areas within the engine bay.

John Sidney watching Autronic management self-mapping on his engine dyno. Self-mapping of the air/fuel ratio is now available on some programmable ECUs; expect ignition timing self-mapping to be available soon.

had detected detonation at about 3000rpm on medium throttle openings. However, when the same driving situation was deliberately re-created, no detonation could be made to occur. The test commenced in winter and, by the time the hot summer arrived, I was confident that the ECU was programmed almost perfectly. The car had no flat spots, started just as it did on the standard management and generally behaved so well that mechanics who drove the car did not believe it had a programmable ECU in it!

The punch line is that at the height of summer, the flashing 'detonation detected' light went mad, **but only ever when climbing one totally unexceptional hill!** In all other conditions, the ignition timing was correct, but climbing this particular hill caused just the right combination of throttle, load and engine rpm to trigger the ignition timing that was a tad too far advanced. And the hot weather tipped it over the edge so that it became a noticeable problem. In a car without a knock sensing ignition retard facility, the engine would have detonated audibly, probably causing damage — even though it was apparently perfect in **all** other situations!

From these examples you can see that, if you want original-manufacturer driveability, there is much more to setting up programmable management than spending just a few hours on a dyno! It requires you have a good relationship with the workshop that programs the management system for you, because almost certainly you will return several times for minor mapping work. This extra work should be included in the original price you pay to have the management system mapped — and this point should be clearly understood by both parties!

Leon Vincenzi mapping a management system on a chassis dyno. The full mapping of a system can take from a few hours to days, depending on the system being used and the extent of the mapping required.

A chassis dyno is cheaper and easier to access for management mapping than an engine dyno. Good air cooling is a prerequisite, being provided here by the fan located in front of the car.

PICKING A SYSTEM

Selecting the right programmable management system for your car is an important decision. The system you pick will have a large bearing on how well the car runs — a bad system will remind you of its lack of ability in **all** driving conditions. Poor starting, poor driveability, poor economy and poor power are factors you will notice every day!

1. Legality

If legality is relevant, investigate this area very carefully. Some jurisdictions prohibit programmable ECUs in road cars, simply because, once one has been fitted, there is no control over the emissions of the car. That is, it's easy to change emissions away from those that may have been legal when the vehicle was tested to those that are totally illegal! Other authorities allow the use of programmable ECUs, but only if the car undergoes a full emissions compliance test procedure. (Obviously, the ECU maps are not supposed to be then changed.) As discussed in Chapter 11, this emissions test might be quite simple — or one that requires a fully accredited testing laboratory, of the sort usually operated by car manufacturers. In any case, if low emissions are a requirement, the programmable ECU selected will need to be able to work in closed loop, using the input of an oxygen sensor to keep mixtures close to stoichiometric during cruise and idle conditions. To pass emissions testing, the ECU will also need to be quite sophisticated in both its software and mapping.

2. Ease of Installation

Is there software available that makes the standard engine-management system in your car fully programmable? If so, the ease of installation and the sophistication of the finished system make this a very attractive proposition. At the time of writing, only Kalmaker for Holden (and some Chev) GM-Delco ECUs exists; however, by the time you read this, there may be systems available for other standard engine management systems.

If no system exists to make the factory ECU programmable, is there a plug-in compatible system that uses the original loom, sensors and output actuators? Such a system has huge advantages in installation costs, and the new ECU can be easily removed and replaced with the standard ECU if a return to standard is required. Furthermore, the reduction in installation costs will allow more money to be spent on programming.

If neither of these approaches is viable, is the selected ECU compatible with at least the standard engine position (crankshaft and/or camshaft) sensor? Interfacing with this sensor causes most of the sensor problems when installing a programmable ECU, so it's preferable that no modification of the standard position sensor be required.

3. Required Features

What do you actually require of the programmable ECU? At a minimum, full control of the fuel and ignition is usually needed. In a road car, good idle speed control (one that can cope with varying loads from the power steering and air-conditioning) is another requirement. How many extra outputs are needed? For example, if you wish to switch on a shift light, actuate a variable-length intake manifold and control turbo boost, you will require extra ECU outputs. These outputs fall into two types: those that are either on or off (eg suitable for the control of a dual-length intake manifold) or those that can be modulated (for example, a boost-control solenoid). On a sophisticated engine, all the following control outputs may be required:

- Turbo boost
- Sequential turbo valves
- Blow-off valve
- Intake manifold
- Camshaft timing
- Radiator fan
- Shift light
- Intercooler water spray
- Water injection
- Active aerodynamics

While this list may seem long, using the ECU to control variable functions in a car has a number of significant advantages. First, the sensors and associated wiring are already in place to allow the accurate determination of car operating conditions. Second, those functions that need to be controlled within strictly defined parameters can be accurately operated. For example, an intercooler fan can be switched on only when the inlet air temperature is above 45 degrees C and the vehicle speed is below 15km/h, or camshaft timing can be altered on the basis of engine speed and MAP pressure. Automatic transmission control is another important ECU function. However, at the time of writing, all programmable ECU manufacturers apparently overlook this area. But ECUs able to control the auto trans in addition to the engine are probably not far away.

Do you require data logging? In a serious road car this can be a useful adjunct, but the required logging period will be only quite short, so the ability to record all channels for more than a few minutes probably won't be needed. Is traction control needed? Is turbo anti-lag a requirement? These latter features are found in only a few programmable ECUs. Will the ECU you have selected work with the injectors you wish to use? If you are using high-flow injectors, they will probably be of low resistance, which may severely limit the number that can be driven by the ECU.

Can the ECU control the ignition system in the way you wish to configure it — ie single-coil or direct-fire ignition? In engines fitted with factory direct-fire ignition, the ECU must be capable of controlling this type of ignition system, while a distributor engine can either be converted to direct-fire or retain a single-coil, dizzy approach. Do you want perfect drivability in all conditions? If you do, the ECU you pick should have a very large number of maps, including intake air and coolant temp correction maps for the main running ignition and fuel charts — not just for cold-start conditions.

Does the ECU have self-diagnostics? Especially helpful during setting up, self-diagnostics can save many hours when faults need to be located. Is the software upgradeable or do ECU upgrades require the soldering-in of new chips? Some programmable management companies offer upgrade software free over the Web as it is released, allowing your ECU to continually improve in its capability without physical changes being necessary.

4. Programming Approach

How is the ECU programmed? Invariably, those systems that use just pendant hand controllers with small LCD displays are not as sophisticated as ECUs using laptop PC interfaces. This is the case for two reasons. First, ECUs using small LCD displays are normally being built down to a price — one so low that potential buyers may not have or cannot access a laptop PC. Second, showing a complete management map on a small display is very difficult. In fact, it's hard enough to see everything on a full-sized monitor, let alone a two-line LCD display! Also,

Whether an engine or chassis dyno is used to perform the majority of the mapping, light loads and acceleration/deceleration adjustments need to be made on the road.

Engine-management mapping is most accurately carried out using an engine dyno. This dyno room at Nizpro features two air/fuel ratio meters (gold and blue boxes); thermostatically controlled intercooler, coolant and oil temperatures (digital displays at right); and two computer monitors — one displaying engine data and the other, management maps.

data logging and the graphing of data cannot be carried out on LCD display, with a few notable exceptions.

Of course, if you are in the market for a very cheap programmable ECU, the fact that it has an LCD display is not necessarily a major concern. However, be aware you will be limited in programming and the information that can be displayed. Note that some LCD pendant ECUs also have full PC software available at additional cost.

5. Mapping Density

How many fuel and ignition points can be mapped? While all engine-management systems use interpolation (filling in the gaps between the mapped points), the more mapped points that there are, the better. For example, for the ignition timing, one mid-level system uses sixteen 500rpm intervals from 500-8000rpm in 16 stages of manifold pressure. In other words, the timing matrix comprises 256 points. Another more expensive system can accommodate up to 32rpm and 16 load sites (512 points) and, importantly, they may be chosen at random intervals. This means that if the engine is running poorly at a particular combination of load and rpm, the mapping density can be increased around this area. ECUs with sparse mapping points should be avoided — irrespective of how good the interpolation is said to be!

6. Service and Back-Up

As in the case of superchargers (covered in Chapter 7), some companies producing programmable management systems are very small, so investigate the background to the system you intend to buy. You may not be happy to invest your engine's life in a company comprising one young bloke building ECUs on the kitchen table...

One way to assess the back-up is to ask for the manual and (if relevant) the software, **before** you buy the system. In PC-programmed systems, the software is an especially good way of quickly seeing how sophisticated the system really is. Most major programmable ECU companies will give free downloads of their software, often via the Web. Manuals should be professionally produced — if the company can't even afford to put together a well-written manual, how good will their system be? Good-quality programmable management systems will also be widely used. This is often promoted by aftermarket management companies that advertise their dragstrip records; however, an ECU with good dragstrip performance does not necessarily make for a good road system! On the strip, driveability isn't very important, but on the road it's nearly everything.

In addition to being happy with the product, you must also be happy with the workshop that will map the system. If the workshop is a distributor for the particular brand of management system, the personnel will probably be more experienced with the system. This is important, because learning the programming complexities of a good engine management system can take literally years.

6. Driving a Car

Once you've narrowed your choice, approach the company and see if it can nominate a local car similar to yours that has been previously fitted with the programmable management system. Approach the car's owner and question him or her about the system. Ask particularly about driveability and starting behaviour in different weather — any programmable management system can pour in the right amount of full-load fuel and ignition timing; it's how it responds in light loads, different weather conditions and during starting that's more indicative of the quality of the system. If possible, drive the car with the system fitted. It should behave in every regard like a car equipped with factory engine management — no ifs or buts — as well as a factory management system in every single aspect. If it doesn't, take note of the problems and quiz the company about them. However, remember it may be the fault of the programming, not of the ECU itself.

The minimum piece of equipment needed when mapping a management system is an accurate, wide-range air/fuel ratio meter. This can sometimes be hired or borrowed from workshops and can be powered from the car's electrical system.

MAPPING EQUIPMENT

So that various engine loads and speeds can be simulated, a dyno is used while mapping programmable management. The dyno can be of two types: a chassis dyno or engine dyno. As discussed in Chapter 11, the use of an engine dyno requires that the engine be removed from the car, making engine dyno tuning more expensive. If a chassis dyno is used, it must be able to hold a constant load, not be of the inertial type.

A dyno is needed because the engine is held at one load and speed while that point in either the ignition chart or fuel chart is being mapped. So if the injector pulse width is being mapped at 30kPa (absolute) manifold pressure and 3000rpm, the engine needs to be held in a loaded state so these conditions are met. This makes full management mapping on the road extremely difficult. It's really only possible to get good results on the road if the ECU being used has at least some capability for self-tuning or if the ECU is sold pre-programmed and requires only minor trimming. But, on the other hand, it's also impossible to get every aspect of mapping correct using just a dyno! It's therefore normal to follow dyno mapping with some fine-tuning on the road of factors such as acceleration enrichment, deceleration enleanment and injector cut-off.

If a chassis dyno is used, great care needs to be taken that engine and transmission temperatures do not rise to excessive levels. It's easy to hold an engine under substantial loads for quite long times, during which intake air, engine oil and transmission oil temperatures all skyrocket!

From left to right: an A'PEXi plug-in programmable management system, Dyno Dynamics/Autronic air/fuel ratio meter, and the A'PEXi hand-held pendant controller. The A'PEXi ECU comes pre-programmed, requiring only the trimming of the maps. This can be done on the road.

A turbo engine with a small intercooler is especially likely to have temperature-related problems if mapped without the benefit of long cool-down spells. Using an engine dyno to tune engines of this type is safer because the cooling and engine monitoring equipment is usually more sophisticated than on a chassis dyno.

The more data available from the engine being mapped, the better. The minimum required equipment is an accurate air/fuel ratio meter using an exhaust gas oxygen probe. Note that the simple LED meter covered in Chapter 3 is not sufficiently wide in its measuring range to accurately set up fully programmable management. Instead, a meter capable of reading from at least 10:1 to 18:1 air/fuel ratios should be used. If emissions compliance is required, five gas exhaust analyses (carbon monoxide, hydrocarbons, carbon dioxide, oxides of nitrogen and oxygen) should also be carried out.

Use of an engine-mounted microphone, an amplifier/filter and headphones allows detonation to be heard very early. Taking this apparently simple approach is often better than relying on universal electronic knock-sensing systems.

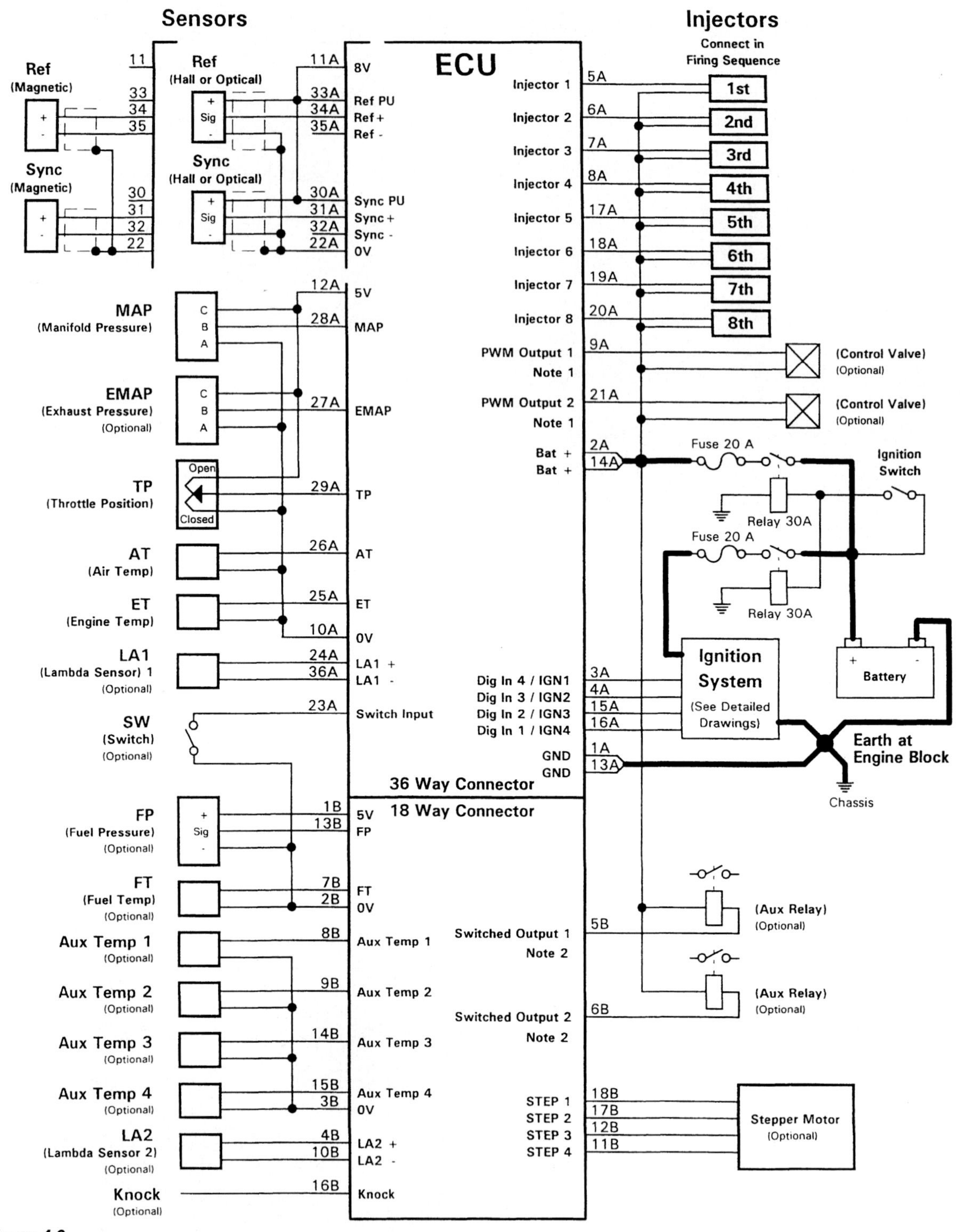

Figure 4.9
The wiring diagram for the MoTeC M8 programmable ECU shows a number of important features. The separate injector outputs allow sequential injection for up to eight cylinders, with the four ignition outputs allowing the use of waste-spark direct-fire ignition, also for a maximum of eight cylinders. Sophisticated inputs include exhaust back-pressure, fuel pressure and temperature, and a knock sensor.

The measurement of exhaust gas temperature (EGT) can also be made, preferably with individual thermocouples for each cylinder. An exhaust gas temperature higher on one cylinder indicates the air/fuel ratio is leaner on that cylinder. If individual injector trimming is available, this function can be used to equalise the actual air/fuel ratio of each cylinder. EGT can also be used to assess the maximum enleanment that can be safely used in lean cruise conditions.

Many workshops use knock detection equipment, and if this is reliable and accurate, it is an important input. However, knock detectors — especially those able to be used on a variety of engines — are often inaccurate. In the absence of such a detector, your ears can be used, aided by an engine-mounted microphone, amplifier/filter and headphones. These electronic stethoscopes are available from a number of suppliers or can be fairly easily built.

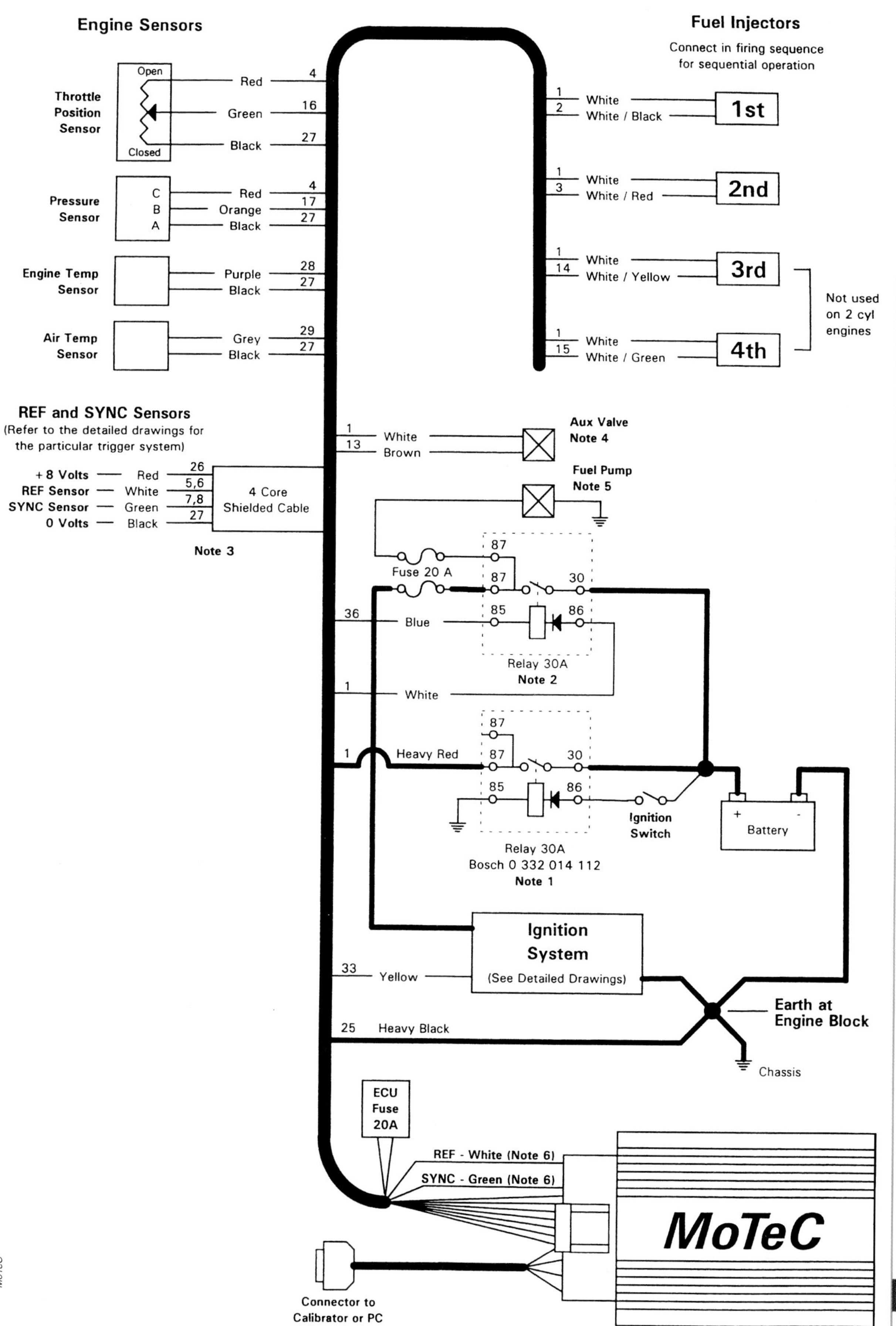

Figure 4.10
The wiring diagram for the MoTeC M4 fuel-only ECU is much simpler than for the M8! However, again sequential operation of the injectors is possible with major input sensors including ref and sync crankshaft position signals, MAP, throttle position, and coolant and intake air temps.

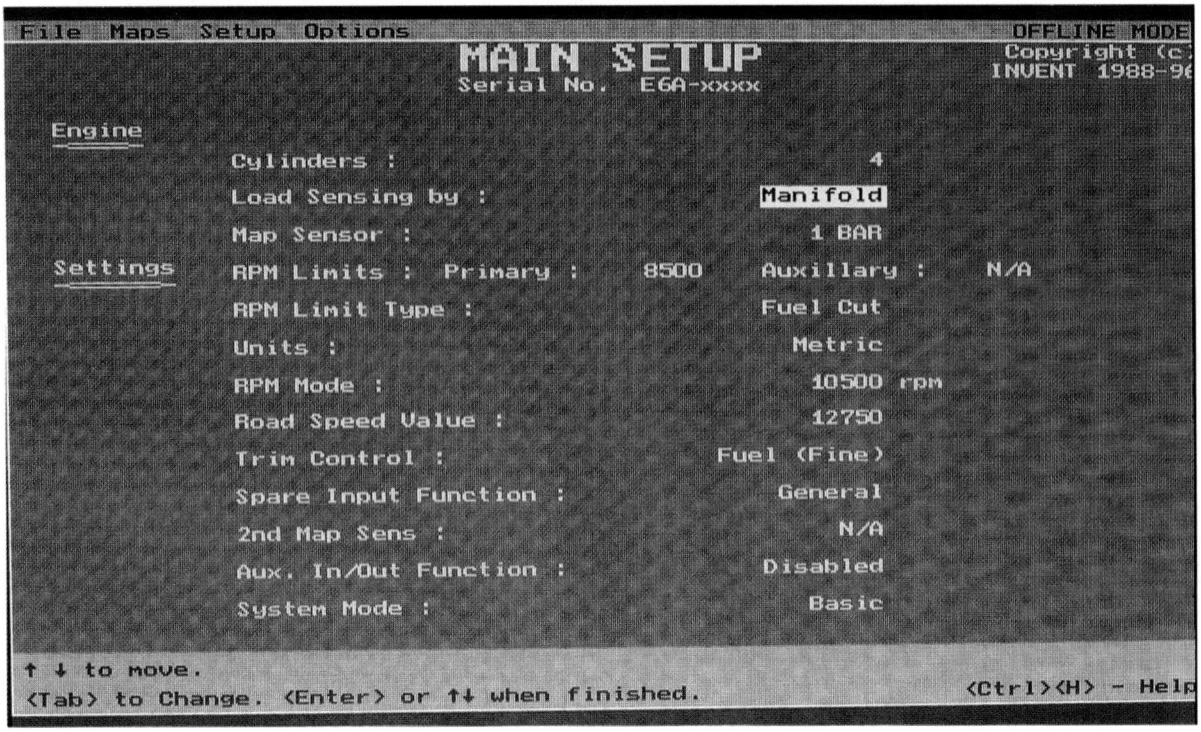

File Maps Setup Options OFFLINE MODE

MAIN SETUP

Serial No. E6A-xxxx

Copyright (c) INVENT 1988-96

Engine

Cylinders : 4

Load Sensing by : Manifold

Map Sensor : 1 BAR

Settings

RPM Limits : Primary : 8500 Auxillary : N/A

RPM Limit Type : Fuel Cut

Units : Metric

RPM Mode : 10500 rpm

Road Speed Value : 12750

Trim Control : Fuel (Fine)

Spare Input Function : General

2nd Map Sens : N/A

Aux. In/Out Function : Disabled

System Mode : Basic

↑ ↓ to move.

<Tab> to Change. <Enter> or ↑↓ when finished. <Ctrl><H> - Help

Programmable management systems have a 'set up' screen where basic data is input. This is for a Haltech E6A fuel and ignition management system.

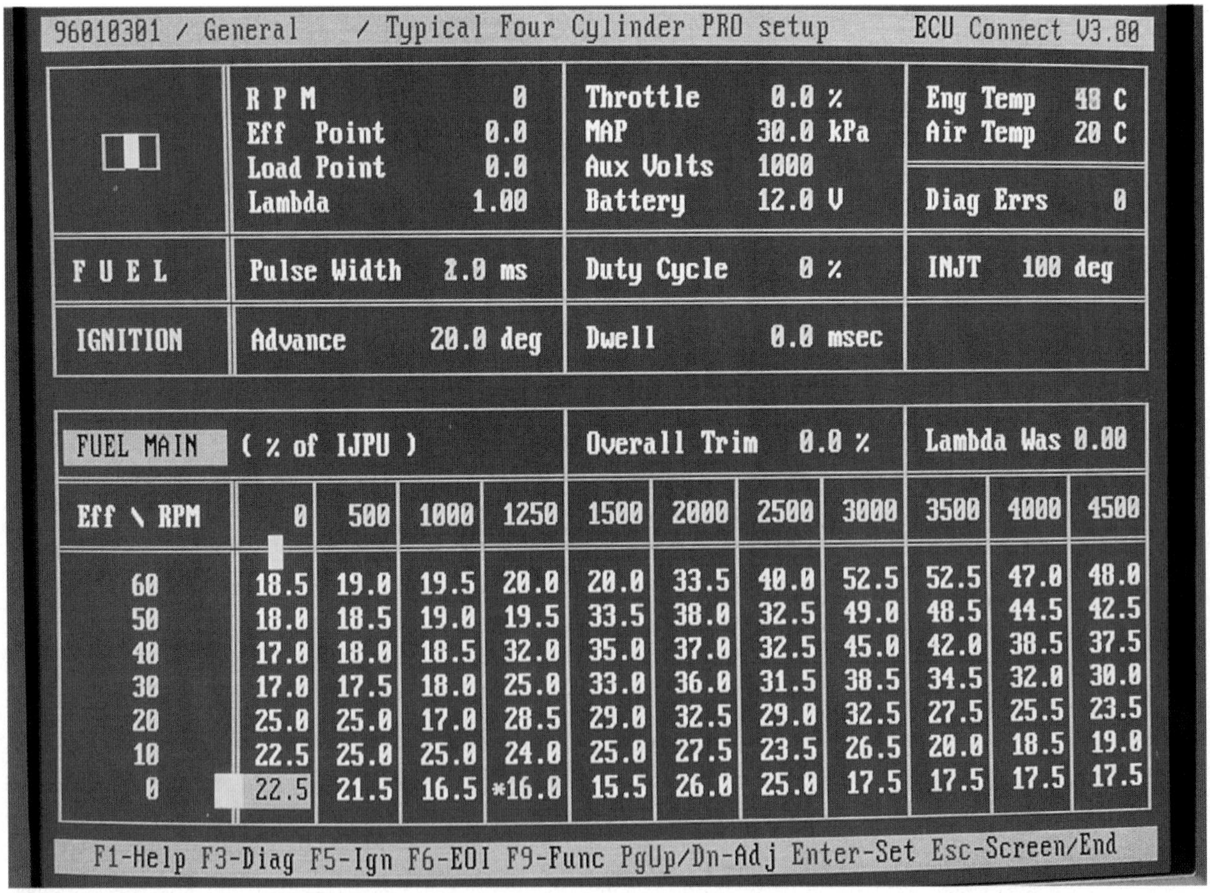

96010301 / General / Typical Four Cylinder PRO setup ECU Connect V3.00

RPM 0 | Eff Point 0.0 | Load Point 0.0 | Lambda 1.00

Throttle 0.0 % | MAP 30.0 kPa | Aux Volts 1000 | Battery 12.0 V

Eng Temp 48 C | Air Temp 20 C | Diag Errs 0

FUEL: Pulse Width 2.0 ms | Duty Cycle 0 % | INJT 100 deg

IGNITION: Advance 20.0 deg | Dwell 0.0 msec

FUEL MAIN (% of IJPU) | Overall Trim 0.0 % | Lambda Was 0.00

Eff \ RPM	0	500	1000	1250	1500	2000	2500	3000	3500	4000	4500
60	18.5	19.0	19.5	20.0	20.0	33.5	40.0	52.5	52.5	47.0	48.0
50	18.0	18.5	19.0	19.5	33.5	38.0	32.5	49.0	48.5	44.5	42.5
40	17.0	18.0	18.5	32.0	35.0	37.0	32.5	45.0	42.0	38.5	37.5
30	17.0	17.5	18.0	25.0	33.0	36.0	31.5	38.5	34.5	32.0	30.0
20	25.0	25.0	17.0	28.5	29.0	32.5	29.0	32.5	27.5	25.5	23.5
10	22.5	25.0	25.0	24.0	25.0	27.5	23.5	26.5	20.0	18.5	19.0
0	22.5	21.5	16.5	*16.0	15.5	26.0	25.0	17.5	17.5	17.5	17.5

F1-Help F3-Diag F5-Ign F6-EOI F9-Func PgUp/Dn-Adj Enter-Set Esc-Screen/End

Part of a MoTeC fuel chart. Displayed in the upper part of the screen are the engine operating parameters, while in the fuel chart, volumetric efficiency (how well the engine breathes) is listed on the vertical axis and engine rpm on the horizontal axis. The numbers in the table determine the opening period of the injectors.

96010301 / General / Typical Four Cylinder PRO setup ECU Connect V3.80

	R P M 1200 Eff Point 2.0 Load Point 2.0 Lambda 1.00	Throttle 2.0 % MAP 40.0 kPa Aux Volts 1000 Battery 13.2 V	Eng Temp 70 C Air Temp 20 C Diag Errs 0
F U E L	Pulse Width 1.9 ms	Duty Cycle 1 %	INJT 142 deg
IGNITION	Advance 32.8 deg	Dwell 0.0 msec	

FUEL Indiv (% Trim)

Cyl No	1	3	4	2					
	0	0	0	0					

F1-Help F3-Diag F5-Ign F9-Function PgUp/Dn-Adj Enter-Set Esc-Screen/End

In a sequential injection system, the fuel flow of injectors can be individually trimmed, as shown in this MoTeC screen display.

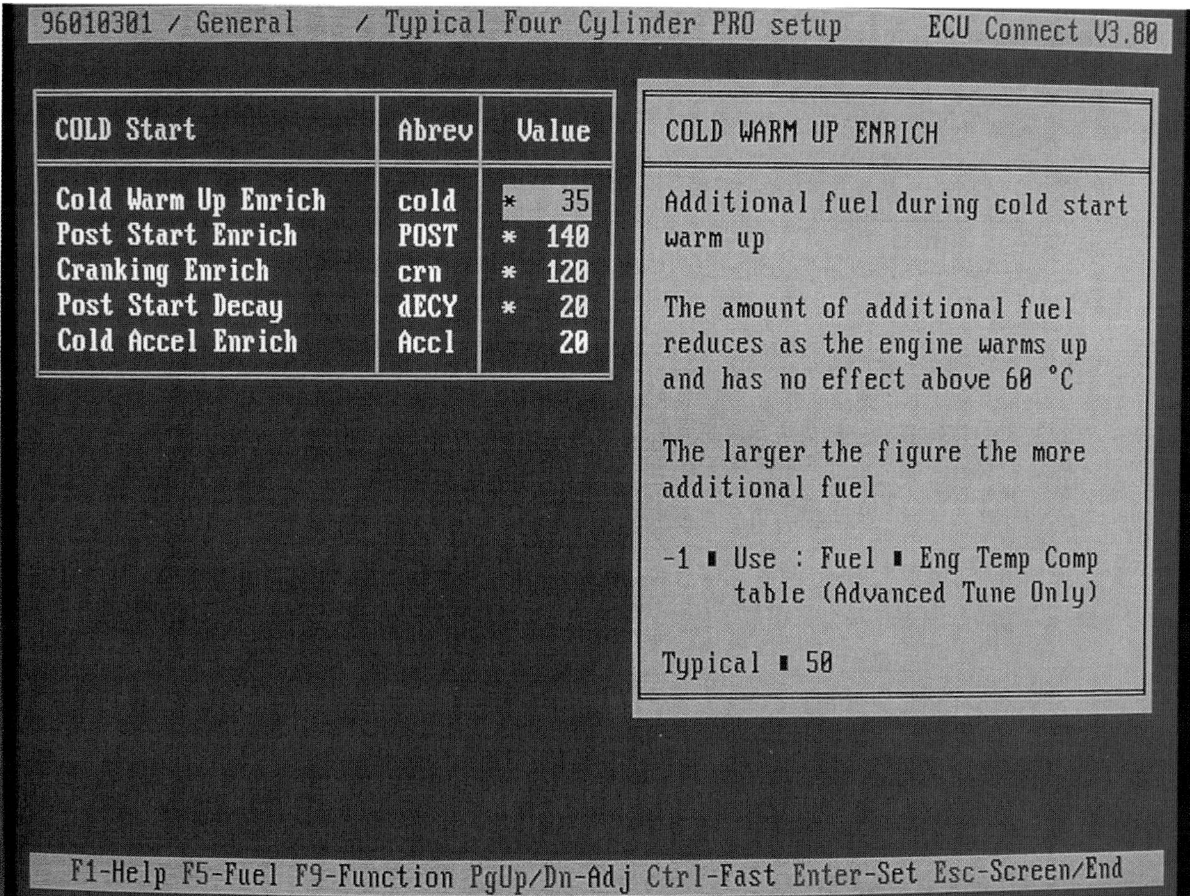

Having sophisticated engine warm-up maps is extremely important if emissions legislation must be met.

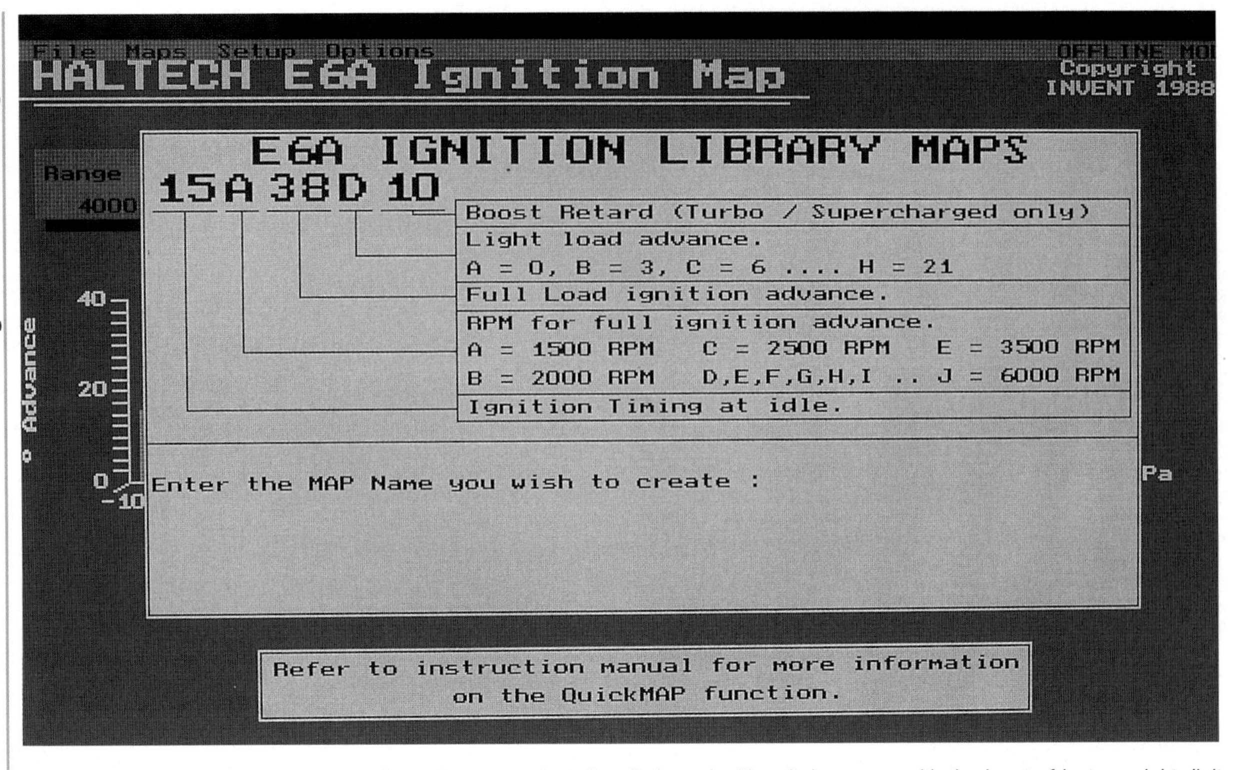

Haltech has a QuickMap facility that allows the generation of preliminary ignition timing maps with the input of just an eight-digit instruction.

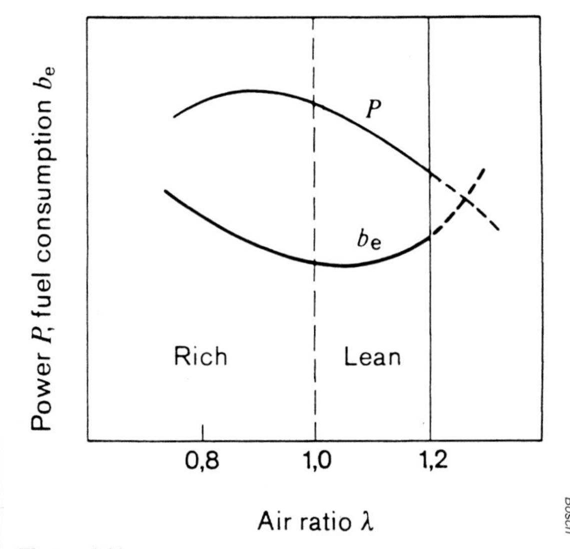

Figure 4.11
The relationship between the excess air ratio (14.7 x excess air ratio = air/fuel ratio) and power and fuel consumption. Power (P) peaks with richer mixtures, while as expected, economy (b_e) peaks with leaner mixtures. This means that at light loads the air/fuel ratio is usually tuned for economy, while at heavier loads the air/fuel ratio is much richer.

AIR/FUEL RATIO REQUIREMENTS

A well-tuned engine used in normal road conditions has an air/fuel ratio that's constantly varying. At light loads, lean air/fuel ratios are used, while when the engine is required to develop substantial power, richer (lower number) air/fuel ratios are used.

Bosch states that most spark ignition engines develop their maximum power at air/fuel ratios of 12.5:1-14:1, maximum fuel economy at 16.2:1-17.6:1, and good load transitions from about 11:1-12.5:1. Figure 4.11 shows the relationship between the excess air ratio (14.7 x excess air ratio = air/fuel ratio) and power and fuel consumption. (Note that the symbol lambda denotes a stoichiometric air/fuel ratio.) However, engine air/fuel ratios at maximum power are often richer than the quoted 12.5:1, especially in forced-induction engines where the excess fuel is used to cool combustion and so prevent detonation.

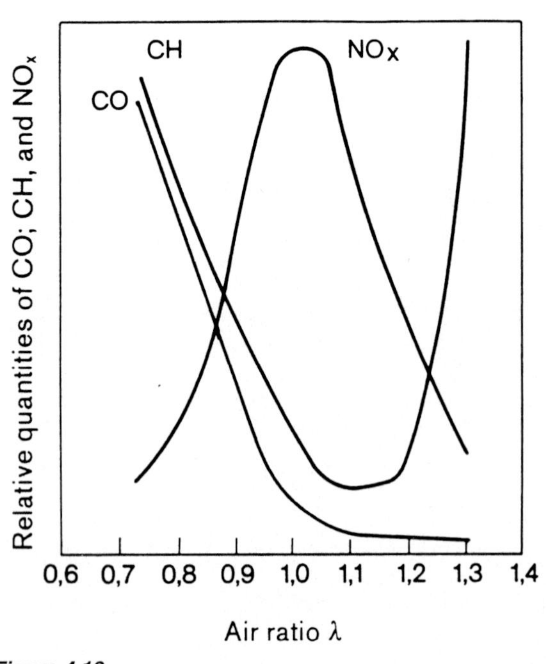

Figure 4.12
The relationship between the emission of hydrocarbons (CH), carbon monoxide (CO) and oxides of nitrogen (NO_X) and the excess air ratio (14.7 x excess air ratio = air/fuel ratio). At lambda = 1 (air/fuel ratio of 14.7:1) oxides of nitrogen peak, while hydrocarbons and carbon monoxide increase substantially as the air/fuel ratio richens.

The air/fuel ratio that is ideal varies, depending on the type of engine and the conditions under which it is operating.

There is no one air/fuel ratio where all emissions are minimised. Figure 4.12 (previous page) shows the relationship between the excess air ratio and carbon monoxide, hydrocarbons and oxides of nitrogen, measured prior to the catalytic converter. As can be seen, at lambda = 1 (ie air/fuel ratio of 14.7:1) oxides of nitrogen peak, while hydrocarbons and carbon monoxide (CO) increase substantially as the air/fuel ratio richens. Note the almost linear relationship between CO and rich air/fuel ratios, making the measurement of CO a very good indicator of mixture strength.

Cranking and Idle

The amount of fuel that needs to be added during cranking can best be determined by experimentation. This enrichment may be configured by just a one-dimensional variable based on engine coolant temperature, or it may be able to be controlled in a more sophisticated manner. Examples of the latter include post-start enrichment and enrichment decay time. Cold start is one of the dirtiest times in regard to emissions, so if emissions requirements are to be met, a sophisticated ECU with multiple-starting enrichment and decay maps should be used. Reducing the cold-start enrichment but increasing cold-acceleration enrichment will reduce the total amount of emissions. Some factory systems open the idle air bypass during cold deceleration, presumably to act as a form of exhaust air injection.

The air/fuel ratio required for a smooth idle will depend on the engine's combustion efficiency and the camshaft(s) used. Some engines with hot cams will require an air/fuel ratio as rich as 12-12.5:1 for a smooth idle, while others will run happily at 13-13.5:1. Engines with hot cams that are fitted with sequential injection management systems can run leaner idle mixtures than systems using bank or group fire. Those engines that can be configured to run in closed loop at idle will use an air/fuel ratio of about 14.7:1 when fully warmed, although they will still usually idle better at a slightly richer air/fuel ratio. However, keeping the engine air/fuel ratio as close to stoichiometric as possible will benefit emissions because the cat converter works most efficiently at this ratio.

Cruise

Light-load cruise conditions permit the use of lean air/fuel ratios. Ratios of 15-16:1 can be used in engines with standard cams, while engines with hot cams will require a richer 14:1 air/fuel. If a specific lean cruise function is available, air/fuel ratios of 17:1 or 17.5:1 can be used, normally at the standard light-load ignition advance. Running too lean a cruise mixture will cause the cat converter to overheat. If a dyno and exhaust gas temperature probe is available, the cruise air/fuel ratio can be leaned out until exhaust gas temperature becomes excessive for these load conditions (eg 600+ degrees C), or torque starts to significantly decrease. Remember, an engine in a road car will spend more time at light-load cruise than in any other operating condition. The air/fuel ratio used in these conditions will therefore determine to a significant degree the average fuel economy gained, especially on the open road.

High Load

A naturally aspirated engine should run an air/fuel ratio of around 12-13:1 at peak torque. The exact air/fuel ratio

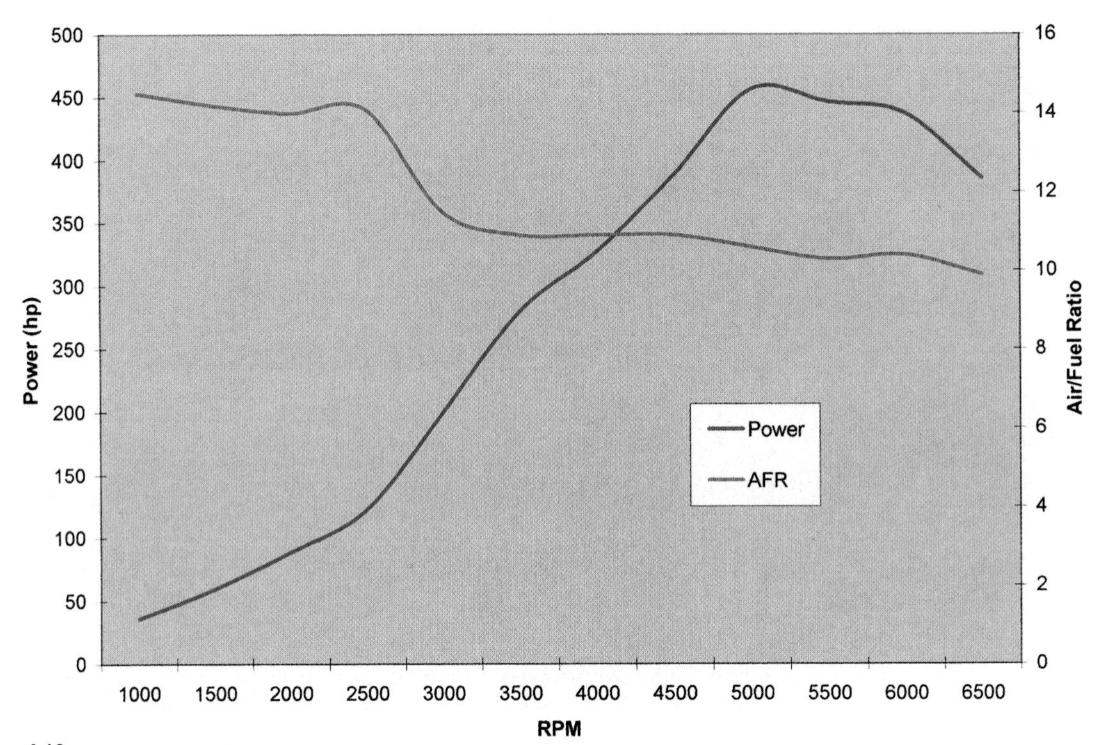

Figure 4.13
The relationship between power and the air/fuel ratio on a powerful turbo 3-litre six. It can be seen that the air/fuel ratio becomes richer (lower in number) as the power increases.

can be determined by dyno testing, with the ratio selected on the basis of the one that gives best torque. Rich air/fuel ratios can be used to control detonation, and this is a strategy normally employed in forced induction engines. Thus, on a forced induction engine, the mixture should be substantially richer — 11.6-12.3:1 on a boosted turbo car and as rich as 11:1 on an engine converted to forced aspiration without being decompressed. As is also the case for ignition timing, the air/fuel ratio should vary with torque rather than with power. Figures 4.13 and 4.14 show these relationships for a turbocharged 3-litre six-cylinder engine, mapped with programmable management.

Most factory forced-induction cars run very rich full-load mixtures, with 10:1 being common. This is done for safety reasons — in case an injector becomes slightly blocked, or the intake air temperature rises to very high levels.

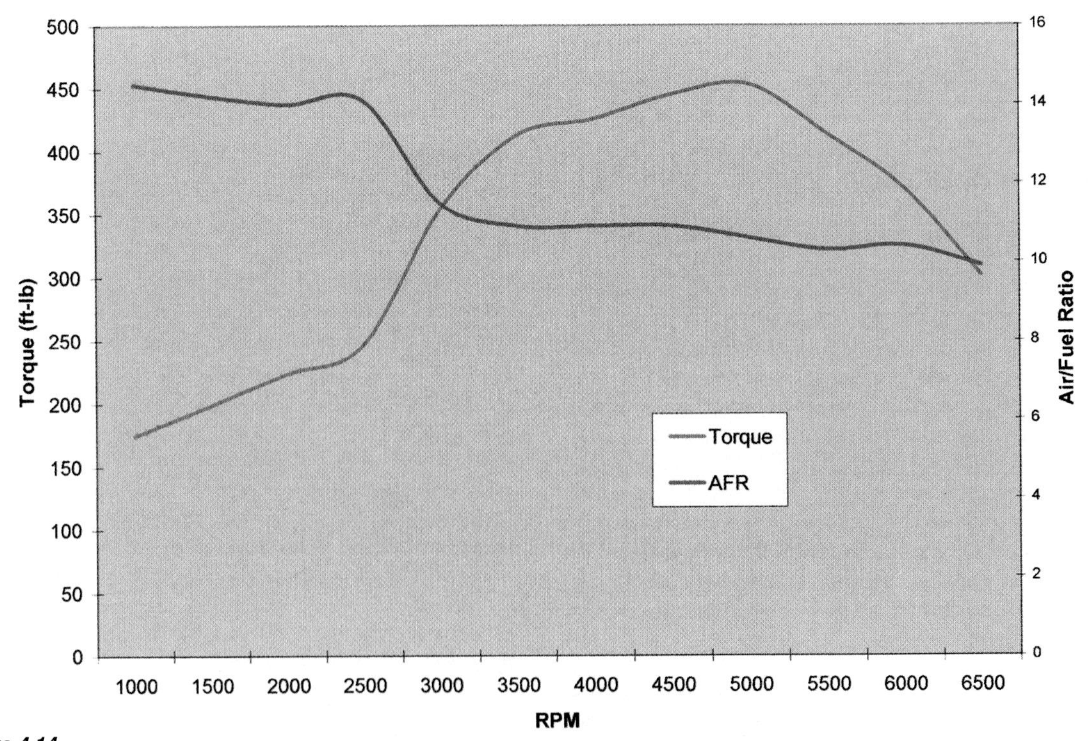

Figure 4.14
This graph is for the same engine as Figure 4.13, but this time the air/fuel ratio has been graphed against torque. The inverse relationship can be clearly seen — as torque increases, the air/fuel ratio is richened. The retention of a rich air/fuel ratio as torque declines high in the rev range is to cool combustion, so reducing the chance of detonation.

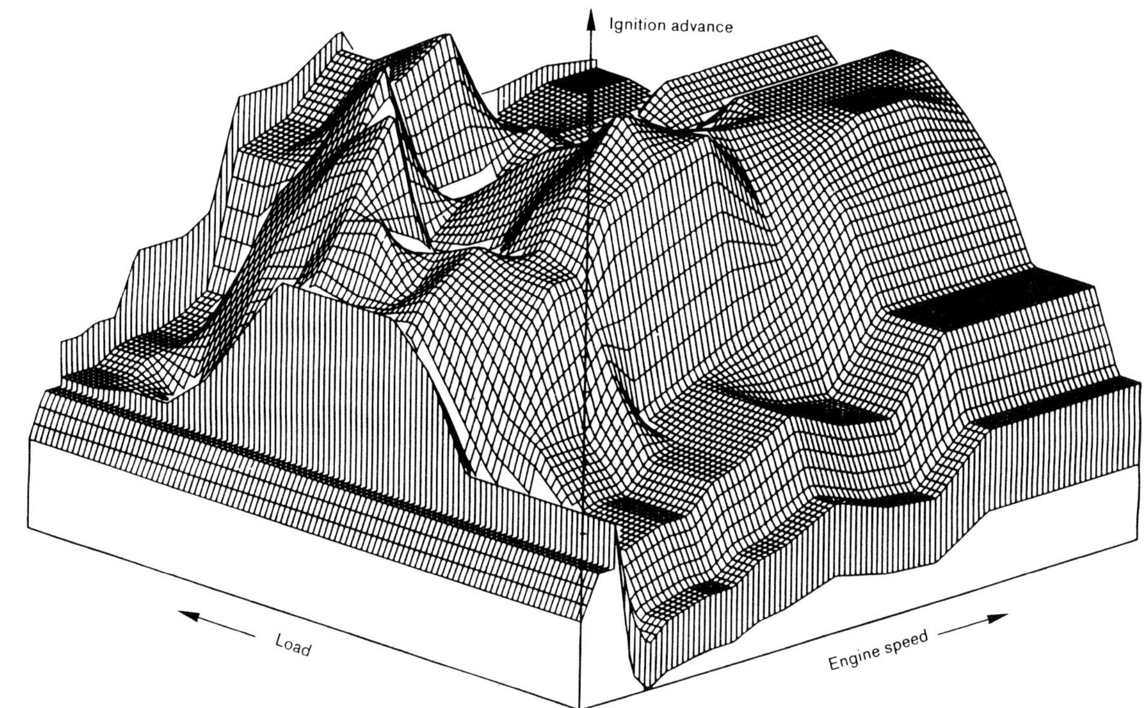

Figure 4.15
When the ignition timing angle is controlled by a management system, very complex patterns of ignition advance can be used.

These cars will normally develop more power if leaned out slightly. Note that emissions testing does not normally take place at full throttle, so full load emissions can be high without legal problems.

In the engine operating range from peak torque to peak power, a naturally aspirated engine should be slightly leaner at about 13:1, with the forced-induction factory engine about 12:1 and an aftermarket supercharged engine staying at about 11:1.

Acceleration

During acceleration, the engine requires a richer mixture than during steady-state running, with the extra fuel provided by acceleration enrichment. Under strong acceleration, the air/fuel ratio will typically drop 1-1.5 ratios from its static level. The amount of acceleration enrichment required is normally found by trial and error, and this is best done on the road. The acceleration enrichment should be leaned out until a flat spot occurs, then just enough fuel to get rid of the flat spot should be added. This approach usually gives the sharpest response. Note that both over-rich and over-lean acceleration enrichment will result in flat spots, and that a greater amount of acceleration enrichment is needed at lower engine speeds than higher speeds.

Overrun

In roadgoing vehicles, deceleration enleanment is used to reduce emissions and improve fuel economy. This normally takes the form of injector shut-off, with the shut-off often occurring at mid-rpm (such as 3000-4000rpm) and the injector operation re-starting at 1200-1800rpm. High rpm injector shut-off can, in some cases, have the potential to cause a momentary lean condition.

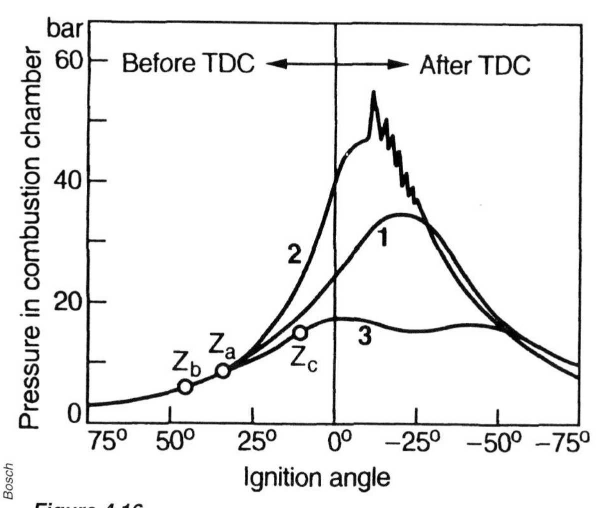

Figure 4.16
The pressure curve in the combustion chamber with various ignition advances. Curve 1 shows ignition (Z_a) at the correct time, Curve 2 shows ignition (Z_b) too early with detonation then occurring, and Curve 3 shows ignition (Z_c) occurring too late with a resulting lack of combustion pressure.

IGNITION TIMING REQUIREMENTS

Firing the spark at the right moment is critical in gaining good power, emissions and economy. The period between the spark firing and the complete combustion of the fuel/air mix is very short — on average only about two milliseconds. Ignition of the fuel/air mix must take place sufficiently early for the peak pressure caused by the combustion to occur just as the piston has passed Top Dead Centre (TDC), so is on its way down the cylinder bore. If the ignition occurs a little too early, the piston will be slowed in its upward movement, and if it occurs too

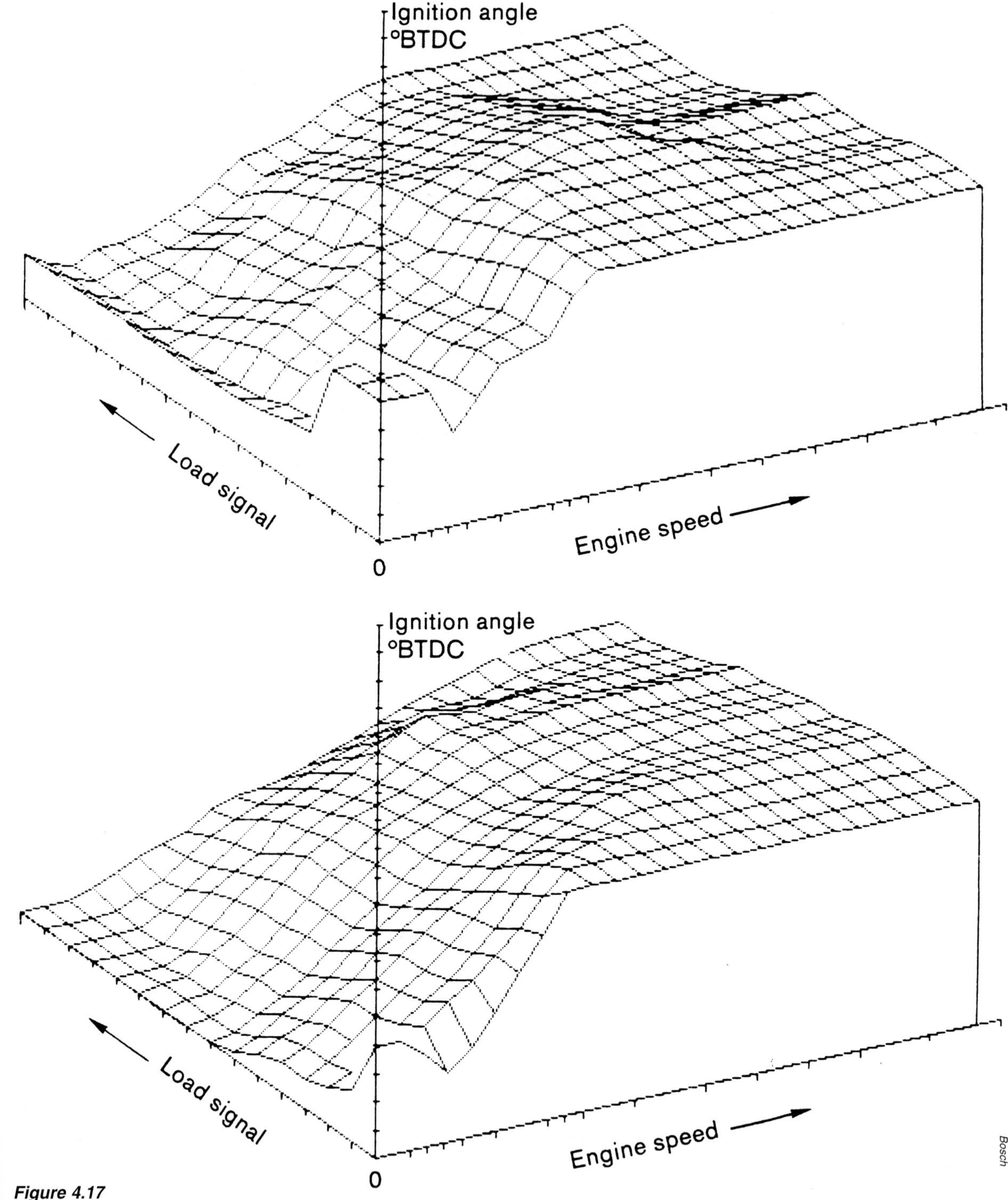

Figure 4.17
The lower ignition advance map is for higher-octane fuel and the upper map for standard-octane fuel. The two have very similar advance characteristics in the low load range but the lower map uses increased advance at higher loads because of the reduction in the tendency of the engine to detonate.

late, the piston will already be moving downwards, so reducing the work done on it. If the spark occurs **much** too early, the ignition pressure wave can ignite the mixture in various parts of the combustion chamber, causing detonation. Figure 4.16 (previous page) shows the behaviour of the combustion pressure with different ignition timing.

If the composition of the mixture were constant (and it isn't!), the elapsed time between ignition and full combustion would remain about the same at all rpm. So, if the ignition advance angle were set at a fixed angle before Top Dead Centre, then, as the engine speed increased, combustion would be shifted further and further into the stroke. This is because the faster-moving piston would be further down the bore by the time combustion actually occurred. To prevent this, the ignition advance must increase as engine speed rises.

The other major factor affecting the amount of advance required is the engine load. At light loads when lean mixtures are used, the speed of combustion is slowed, so more ignition advance is needed. But if only it were that simple! Not only does engine speed and load determine the best timing for the combustion of the mixture, but the following factors are also relevant:

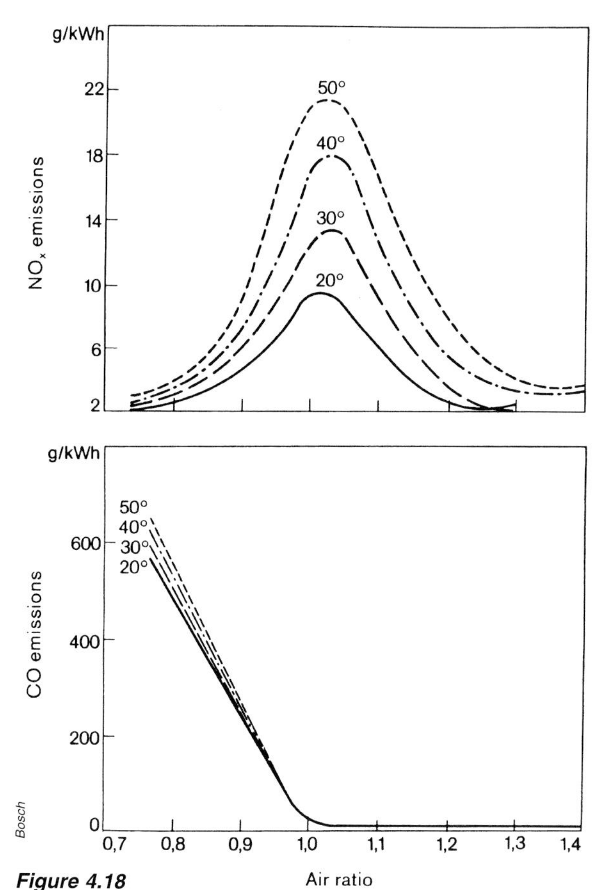

Figure 4.18
The relationship between the ignition advance and the production of oxides of nitrogen and carbon monoxide. Both these and the graphs shown in Figure 4.19 are for an engine in a light-load cruise condition. Oxides of nitrogen (NO_x) increase as ignition timing is advanced. However, the emission of carbon monoxide (CO) is affected very little by ignition timing, being much more heavily influenced by the air/fuel ratio.

- The design and size of the combustion chamber
- The position of the ignition spark(s) in the chamber
- The fuel characteristics
- The emissions levels required
- Engine coolant and intake air temperature
- The safety margin required before detonation occurs

These factors mean that not only does the ignition advance have to vary with load and rpm, but if optimal advance is to be used, the ECU must also be able to sense detonation that might occur on lower-octane fuel or in very hot conditions. Figure 4.17 (previous page) shows two ignition timing maps. The lower map is for higher-octane fuel and the upper map for standard-octane fuel. The higher-octane map has very similar ignition advance characteristics to the standard-octane map in the low load range, but uses increased advance at higher loads because of the reduction in the ten-dency of the engine to detonate. Thus the ignition timing maps on engines that use knock sensing are not fixed in value — the timing can alter depending on the input of knock sensor(s). The input of the knock sensor is an ignition timing correction; additional corrections of the main ignition timing map should be available from an intake air temperature sensor and other maps.

The emissions of an engine will be affected by the ignition timing used, in addition to the air/fuel ratio. Figure 4.18 shows the relationship between the ignition advance and the production of oxides of nitrogen and carbon monoxide. Figure 4.19 shows the specific fuel consumption (ie grams per kW hour) and the emissions of hydrocarbons. All four graphs are for an engine in light-load cruise conditions. Looking first at the emission of oxides of nitrogen, it can be seen that these increase as ignition timing is advanced. Running light-load advances of 40 or more degrees is common, giving good responsiveness off load. However, if emissions standards need to be met, this advance may have to be reduced. On the other hand, the emission of carbon monoxide (CO) is affected very little by ignition timing, being much more influenced by the air/fuel ratio. It can be seen that at stoichiometric and lean air/fuel ratios, increasing the ignition timing can reduce specific fuel consumption substantially. Finally, the emissions of hydrocarbons at stoichiometric and rich air/fuel ratios increase with advanced timing, but timing has little influence at very lean air/fuel ratios such as 19:1.

Ascertaining the best ignition timing by juggling on paper these interrelating factors is nearly impossible. Instead, making real-time changes to the ignition timing while using an exhaust gas analyser and a dynamometer is the only practical way of seeing how the ignition timing being used influences emissions, power and fuel economy.

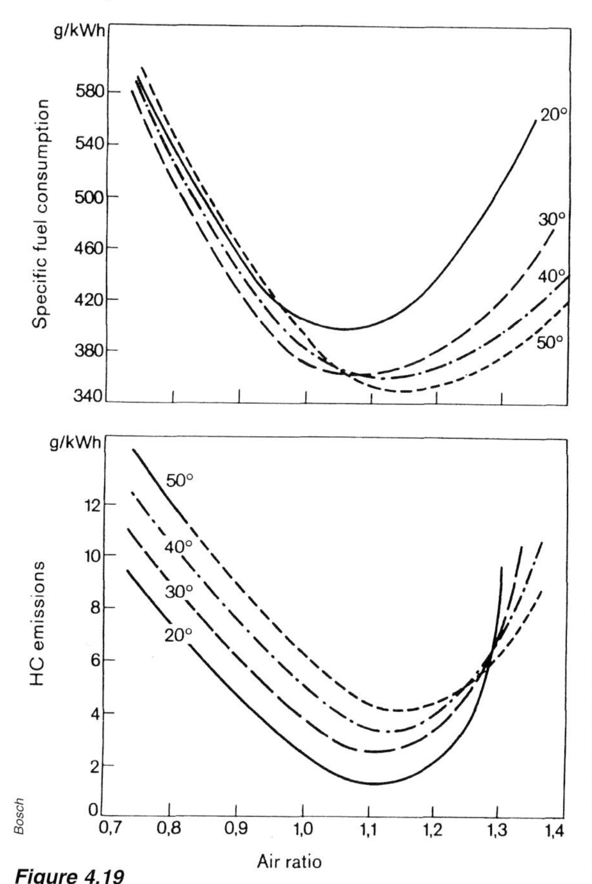

Figure 4.19
The relationship between the ignition advance, specific fuel consumption and the emissions of hydrocarbons. At stoichiometric and lean air/fuel ratios, increasing ignition timing can reduce fuel consumption substantially. The emissions of hydrocarbons at stoichiometric and rich air/fuel ratios increase with advanced timing, but has little influence at very lean air/fuel ratios such as 19:1 (air ratio of 1.3).

Subaru Impreza WRX Management Maps

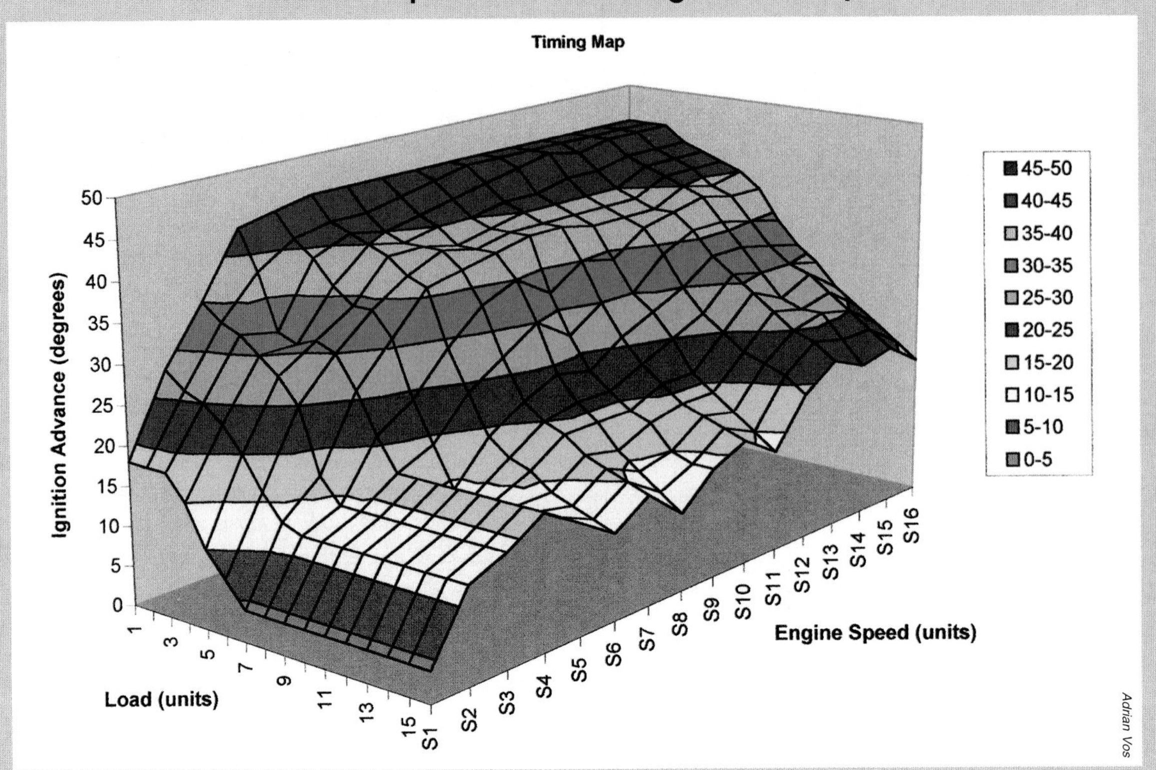

Figure 4.20
The ignition timing map for the standard 1993-96 Subaru Impreza WRX turbo 2-litre. Note that ignition timing at very light loads is very advanced (40-45 degrees) for good economy and off-boost responsiveness (dark blue), while at high loads and high revs the timing is back to 20-25 degrees (browny/purple). At high loads and low engine speeds (lefthand yellow and green) ignition advance is as high as 20 degrees, falling quickly to 10-15 degrees as boost arrives.

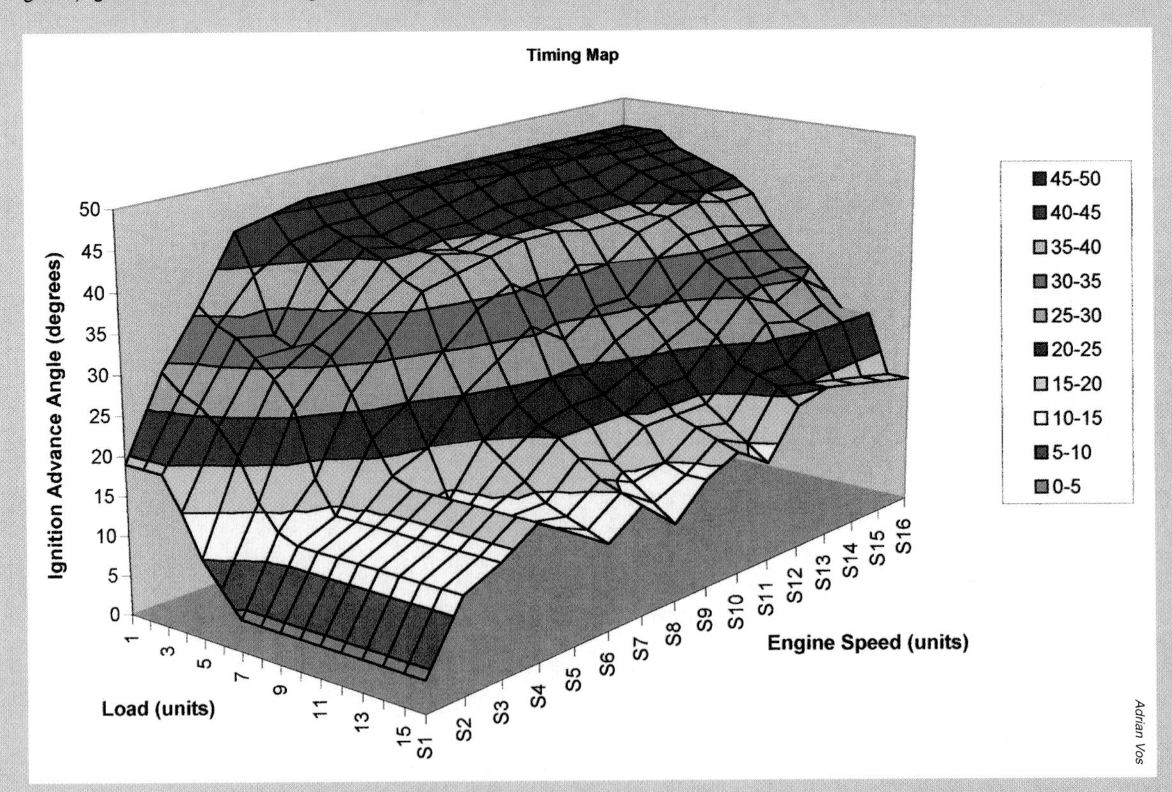

Figure 4.21
A supposedly hot chip version of the standard Impreza WRX timing map shows some light-load timing has been added (pink) and the timing at high loads and rpm has been retarded a little (extended light green). Everywhere else, the timing is near identical with standard.

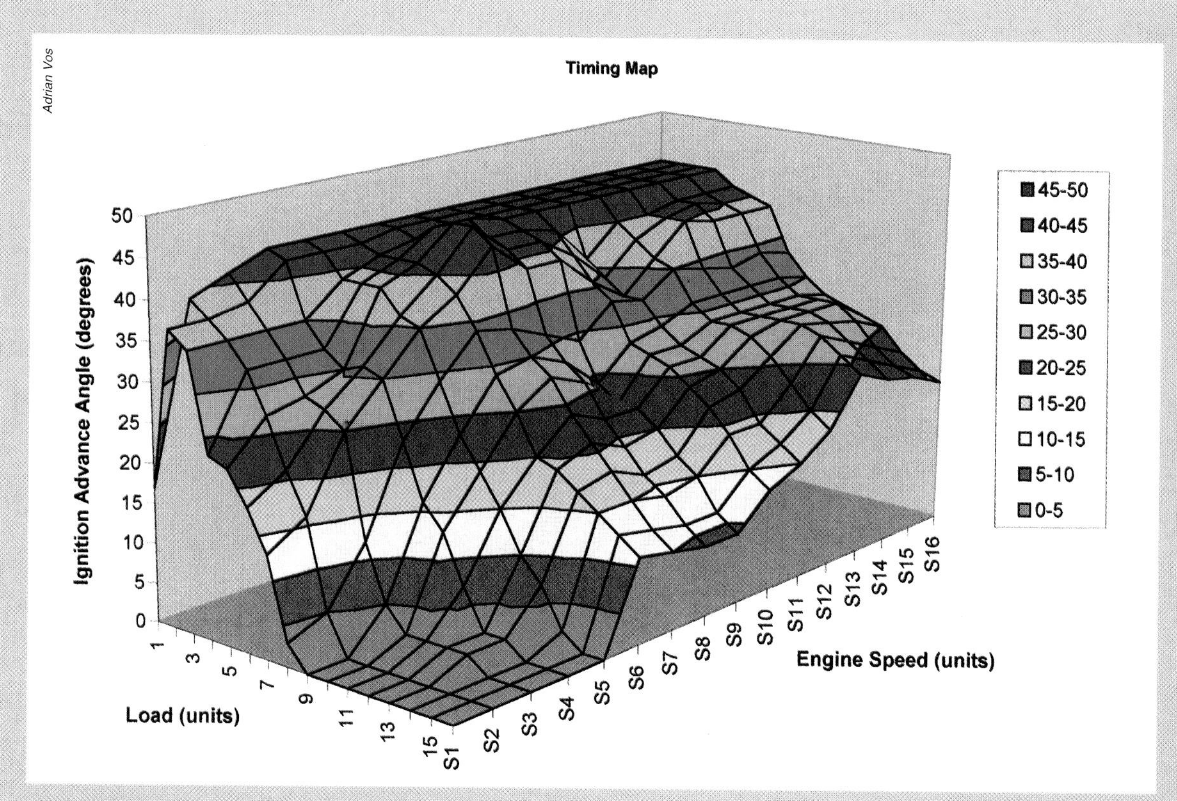

Figure 4.22

A highly modified Impreza WRX timing map, apparently from a Group N factory rally car but quite streetable. This map is radically different from standard! First, at high loads and low engine speeds the timing is massively retarded, probably to allow fuel to pass through the engine without burning (for a turbo anti-lag function). At peak torque, high load, the timing is also more retarded (to prevent detonation with higher boost); however, high rpm/load conditions use the same timing as standard. At low loads and engine speeds the timing is also dramatically retarded, before jumping to more than 35 degrees once some load has been applied.

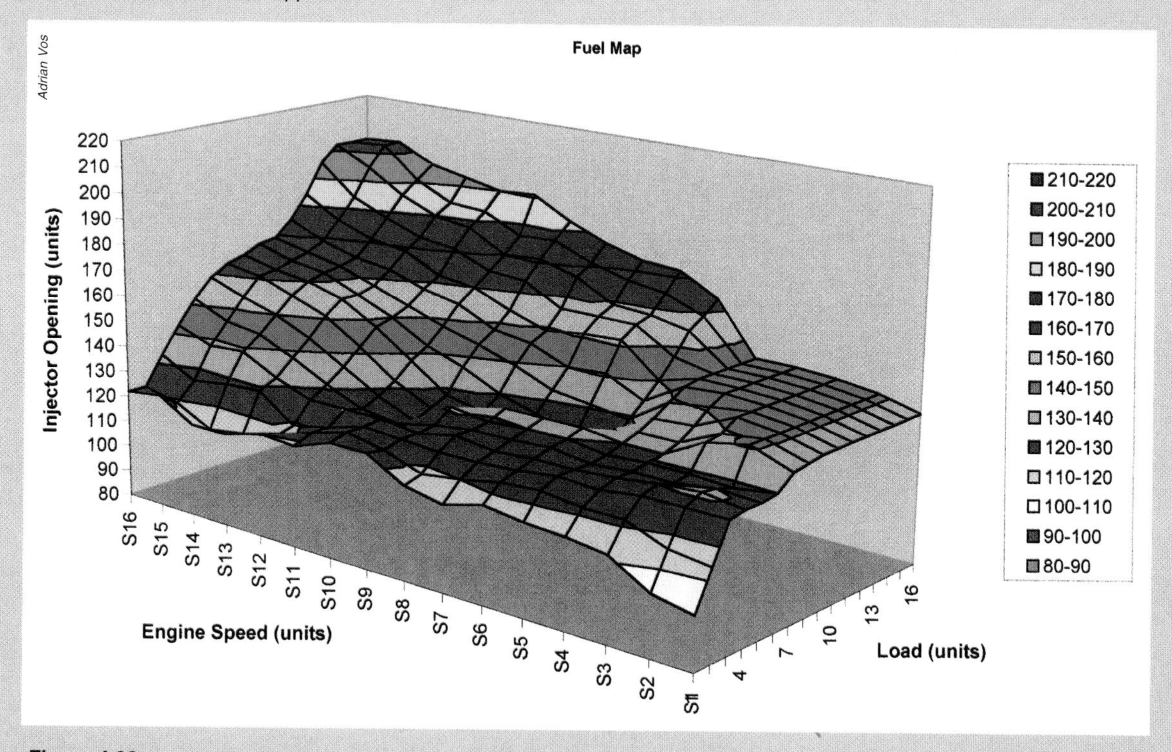

Figure 4.23

The fuel map for the standard 1993-96 Subaru Impreza WRX. (Note the fuel graphs have their axes oriented differently to the ignition maps.) Looking first at the top line (ie high load, from low to high engine speeds), the sudden increase in fuel required as the engine comes on boost can be clearly seen. Also clear are the 'peaks' and 'valleys' required in the fuel delivery to provide optimal mixtures.

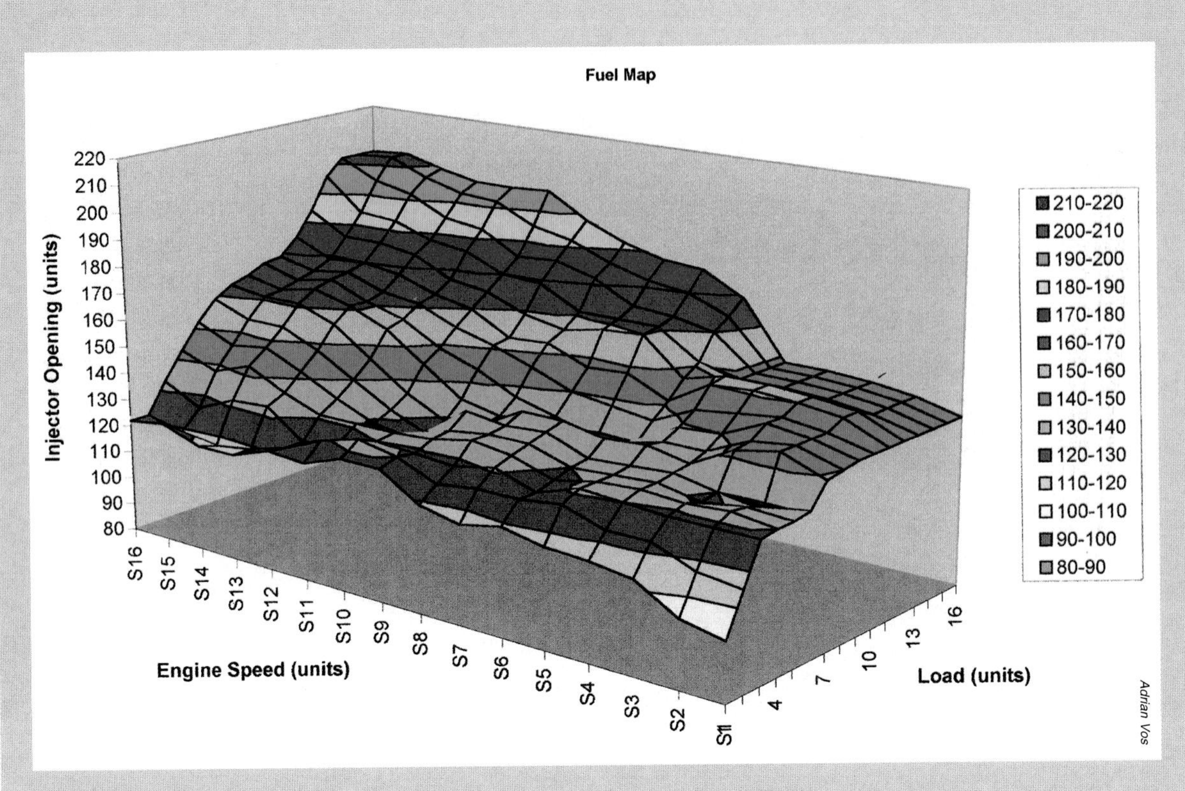

Figure 4.24
The supposedly hot chip version of the standard Impreza WRX fuel map shows the fuel delivery has been enriched, especially in the mid-range. This can be seen by noting the larger areas of blue and orange when compared with the standard map. Top-end fuelling remains identical to standard.

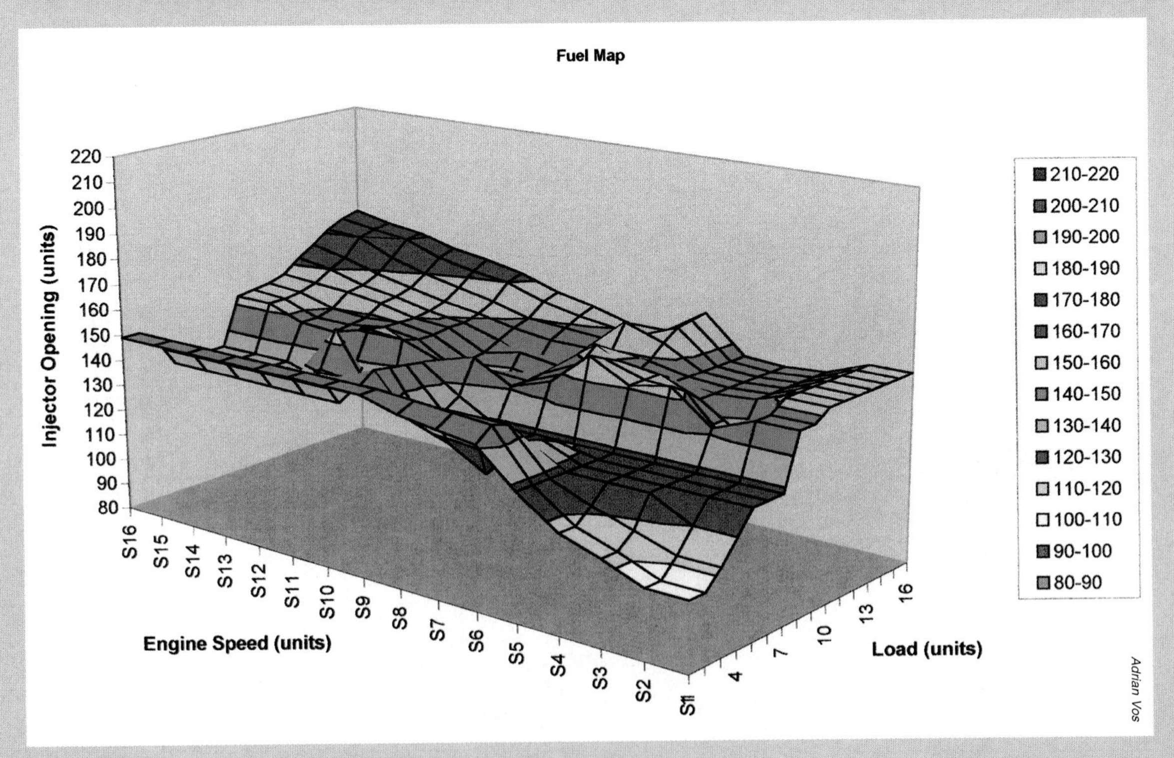

Figure 4.25
Again the factory-developed Group N fuel map shows a radical change over the standard map. The engine has been substantially leaned out at high loads and rpm (note the lack of any injector openings over 180 units). At high engine speeds and low loads, a great deal of fuel is added, which (along with the retarded timing) allows either an active or passive anti-lag function. There is also a peak of fuel (light blue) that occurs at high loads as the engine comes on boost — helping to control detonation.

This BMW utilises MoTeC programmable management to control its modified 3.5 litre engine. While primarily a road car, it acquits itself on the racetrack quite well.

Cranking and Idle

Some programmable engine-management systems have a default cranking advance of 15 degrees, a value about midway through the range of appropriate cranking advances. Smaller engines with faster cranking speeds need a greater ignition advance (up to 20 degrees), while slower cranking speeds of a high-compression engine will require less advance (down to 10 degrees). The compression ratio of the engine will also determine the likelihood of kickback on starting. Engines with a low static compression ratio of 8:1 will accept an ignition advance of anything from 0-20 degrees without kick-back. A 10:1 compression ratio will reduce this to 15 degrees, 11:1 to around 10-12 degrees, while race engines using very high compression ratios of 12-13:1 can sometimes tolerate no cranking ignition advance at all.

Most engines will idle happily with an ignition advance of 15-32 degrees. This is a very wide range — some engines will certainly not be happy at 32 degrees and others won't be at 15 degrees! An overly high amount of ignition advance for a given engine will result in lumpiness at idle, excessive hydrocarbon emissions and sometimes exhaust popping, while too little advance will also cause lumpiness. If the engine runs closed-loop fuel control at idle, too much idle timing advance can disrupt the oxygen sensor reading, causing the self-learning process to overly enrich the idle mixture. Setting the optimal ignition timing can therefore best be done by trial and error variations.

Timing that's more advanced at slightly lower engine speeds than idle is sometimes used to help stabilise idle. This is effective because, when the engine starts to slow down, the greater ignition advance causes the engine to produce more torque, so increasing engine speed. Many factory management systems use ignition timing as a major element in controlling idle smoothness, with an increase or decrease in rpm at idle responded to by a change in timing advance.

Cruise

At light loads, as are used in normal everyday cruise conditions, an ignition advance of 40 degrees or more will improve responsiveness and economy. This advance can be used successfully on many engines — even those with an 11:1 compression ratio, if they're being run on high-octane fuel. One factor limiting the cruise ignition advance that can be used is the maximum ignition timing attack rate provided by the ECU — that is, how fast the timing can change. If very advanced timing is being used with light loads and the attack rate is not high, there may be slight detonation when the engine load suddenly increases.

The hotter the camshaft(s), the less advance that will be able to be used in light load conditions (the limiting factor being driveability rather than detonation in this case), however timing in the range of 35-40 degrees is still usually used. Engines with good combustion chamber design will be able to run up to 45 degrees in these conditions. Fuel economy and engine responsiveness are both very much affected by light-load ignition timing.

High Load

The torque output of a given engine is proportional to average cylinder pressures, so the full throttle ignition timing advance used should relate to the torque curve rather than power curve. The maximum ignition timing that can be used at peak torque is **usually** limited by the occurrence of detonation. A detonation limit is always the case in forced aspirated engines, but not always the case in naturally aspirated engines. As an example of the latter, one Porsche flat six developed best power with a maximum advance of 8 degrees, even though the engine did not detonate at even 27 degrees of advance! A Mercedes V8 engine was able to run 38 degrees at high rpm, peak load without audible detonation. However, best results came from a full load advance of 28 degrees.

On a modified engine having increased compression and hot cams, a peak torque advance of 28-36 degrees can often be used. In a factory forced-induction engine using a little more boost than standard, the peak torque timing will be around 18-22 degrees, while in a naturally aspirated engine converted to forced induction without internal modifications, timing should be well back at about 10 degrees.

Because, as already indicated, most forced-aspiration engines and many naturally aspirated engines develop best performance when the ignition timing is advanced close to the point of detonation, great care should be taken when setting the full-load ignition timing. To assess the maximum ignition advance that can be safely run at a given rpm and load, a dyno is a very useful tool. When the dyno is used in this manner, the engine is held under load at a single rpm and the ignition timing is slowly advanced.

If the rate of power increase tapers off to zero (or in fact power starts to **decrease**), the timing should not be advanced further. If the power development of the engine starts to fluctuate rapidly, the timing advance is excessive. These power fluctuations can be clearly seen when a dyno is used in a steady state, expanded power scale, bar graph mode. Note that the power fluctuations occur well before detonation is audible. The ignition timing should be retarded by 2-4 degrees from the point of power fluctuations. Optimal ignition timing is that which gives a lack of detonation, the lowest exhaust gas temperatures and maximum torque.

From peak torque through to peak power, a modified naturally aspirated engine should increase in ignition advance to 36-40 degrees, a boosted factory turbo car should be running around 25-28 degrees, while an after-market, non-decompressed forced induction engine should be conservatively timed at around 15 degrees.

If the engine uses reliable knock sensing and the ignition timing can be retarded quickly at the onset of detonation (and then re-introduced only slowly), more advanced timing than these figures can be used at high rpm. An intake air temperature correction chart that quickly pulls off timing advance with increased air intake temperatures can also allow the main table's ignition timing to be fairly advanced. For example, with an intake air temp of 120 degrees C, the timing can be retarded by 12-15 degrees, so providing an acceptable level of safety while still allowing good cool weather and short-burst performance. The importance of using a programmable ECU that has tables for the intake air temperature correction of ignition timing can be seen from this example.

Acceleration

The ignition timing used during acceleration transients should have an attack rate that's quick enough to keep up with the timing requirements. This parameter is often specified as the maximum number of degrees per second change that is permitted. One source suggests that an attack rate as high as 650 degrees a second may be needed in some high-performance engines. If the attack rate is not sufficiently high, mid-range detonation can occur in the transition from light-load cruise (perhaps with 45 degrees of advance) to full throttle at peak torque (perhaps requiring only 15 degrees of advance). However, if the attack rate is set too high, slight changes in throttle will cause rapid, undamped jumps in timing, which can cause minor detonation. This is especially the case at low engine revs — at higher rpm, the attack rate can also be higher.

Overrun

On deceleration (with injector cut-off working) most factory cars run retarded timing, such as 10-12 degrees. However, in modified cars this has been found at times to cause an exhaust burble, and if this is unwanted, more advanced timing (20-26 degrees) can be used. The amount of deceleration timing advance used may affect the strength of engine braking available.

TRACTION CONTROL

Traction control is a function available in high-end programmable ECUs. Traction control allows the greatest possible transfer of torque to the road surface, so giving the best possible acceleration. The aim of traction control is to actually allow a small degree of slip between the tyre and the road — on dry road surfaces, best acceleration occurs with slip rates of between 10 and 30 per cent. However, it should be noted that on loose sand and gravel, a slip rate of more than 60 per cent gives best acceleration!

Programmable management ECUs sense wheelspin and reduce engine power by either cutting individual cylinders' fuel or ignition — in some systems the torque-reducing approach that is taken is configurable. Slip can be calibrated as either a percentage or as a speed differential. The speed of the driven wheel is calculated either indirectly from rpm and gear or from the direct input from two or four wheel speed sensors. Filtering of the slip value is user-adjustable, as is the amount of engine torque reduction that occurs when the Traction Control System is operating. This is typically defined as 75 per cent.

Traction Control Systems work within tightly defined parameters. To prevent engine stalling, the rpm below which the Traction Control System will not work is often definable (typically 3000-5000rpm), while the minimum throttle opening at which the Traction Control System will operate is normally set at 10 per cent. The amount of desired slip at various throttle positions is set in some systems by means of a table. Typically a 2km/h slip is used at throttle openings below about 20 per cent, increasing to about 10km/h at full throttle. The minimum speed at which the Traction Control System operates is normally set at 15km/h and the minimum rpm somewhere in the range 2500-4500rpm.

So the driver can override the effect of traction control, a throttle opening above which the Traction Control System reduces its operation can be set. If this figure is set to 100 per cent, the Traction Control System will operate at all throttle openings, while if it's set to a typical value of 78 per cent, at throttle openings higher than 78 per cent the driver can still get some wheelspin.

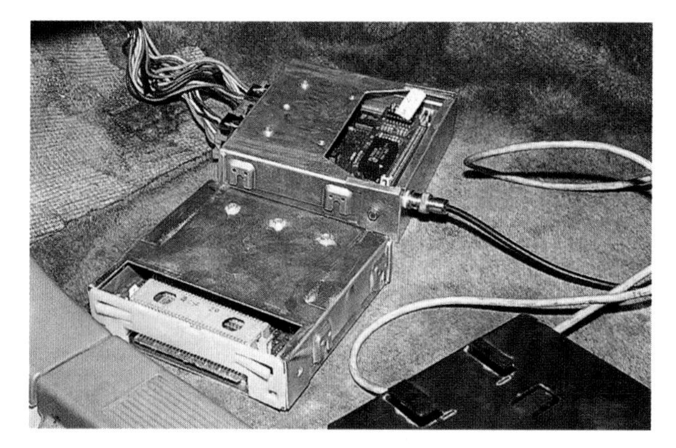

The Kalmaker system uses a modified GM-Delco ECU (rear) during re-calibrating of the factory maps. These can then be transferred to the Mem-Cal in the standard ECU (front) and the standard ECU then plugged back into place.

A form of traction control can also be implemented by decreasing engine power in the lower gears, where wheelspin is much more likely. In turbo cars, the boost can be reduced in first and second gears, or, if the ECU is sufficiently sophisticated, boost can be mapped against road speed or selected gear. In a naturally aspirated car, the ignition timing can be retarded by some ECUs in the lower gears. Note that this form of traction control cannot take into account different road surfaces and different weather conditions.

THE KALMAKER GM-DELCO SYSTEM

The Kalmaker system can be used to re-calibrate standard GM-Delco engine management software, as used in Holden and Chevrolet vehicles. It takes a slightly different approach from that of traditional programmable management in that the factory ECU, sensors, actuators and ECU software maps are all retained intact. This has major advantages in that:

- All engine management system parts are standard and so are available as spare parts from the car's maker.
- Dealership diagnostic machines (and the mechanics operating them!) will function normally.
- Sophisticated functions of the factory ECU software seldom found in aftermarket programmable management systems (eg security, auto trans control, lean cruise) are retained.

Communication with the GM-Delco ECU

The GM-Delco ECU uses a separate plug-in Mem-Cal that contains the EPROM (program and engine data) and resistor packs (calibration of engine fuel back-up values). The rest of the electronics within the ECU includes the injector drivers, counters, timers and so on that are required in any engine management system.

The EPROM (Erasable Programmable Read Only Memory) on the Mem-Cal is not amenable to byte-by-byte re-programming. To allow changes to data to be made on the fly, the Mem-Cal is replaced with a new PC board dubbed a Real Time Board. The RTB contains a static RAM and a series of resistors. The resistors calibrate the engine's fuel back-up values and also tell the ECU how many cylinders the engine has. (Engine cylinder number is specified in both the hardware and software.) The static RAM contains both the program and the engine management maps, with the latter changed during re-programming.

A coaxial cable is used to connect the PC to a new connector located on the modified ECU or, on later vehicles, to the standard diagnostic plug. One conductor is an earth while the other connects to the data port already present in the ECU. One of the aspects of the GM-Delco system that makes it amenable to program manipulation is the data port is already two-directional.

The program addresses (ie which bytes do what) are contained within the individual Kalmaker scripts. The Kalmaker program itself is just a general-purpose editor; the scripts are where all the intelligence is contained. A script has been written for each of the different Holden Mem-Cals (and some Chev) programs, of which there have been many.

As with conventional programmable management systems, use of the two-directional port to write directly to the static RAM gives seamless changes in real time. (This is important as sudden mixture or timing changes can be dangerous if the engine is being run under load on a dyno during the re-programming.) The serial data cable connects the PC parallel port to the ECU via an interface board. The interface is needed because the ECU high-speed serial data link does not conform to RS232 specification and its baud rate is 8192, not the standard 9600. The interface board allows two-directional communications via the PC's parallel port status lines.

During re-calibration, the standard Holden GM-Delco ECU is temporarily removed, being replaced with a development-type ECU. The new ECU is a modified Holden unit, running the Real Time Development Board that allows communication with a laptop PC. Loaded onto the laptop is the Kalmaker software, permitting the readout of all of the fuel, ignition, idle maps (and the hundreds of others!) in the standard Holden engine management software. In addition to being able to read engine operating data, the laptop can be used to re-calibrate any of the engine-management maps in the development ECU.

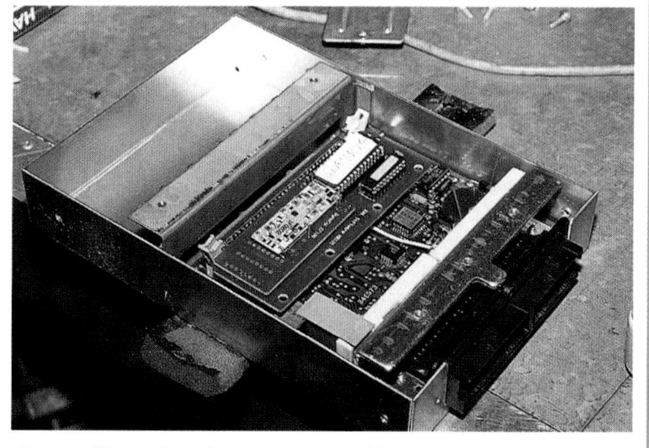

Some Chevrolet Corvettes are able to be re-programmed using Kalmaker. This Corvette ECU has the Real Time Board installed, complete with additional GM-Delco knock-sensing PCB (the small board sitting on top of the RTB).

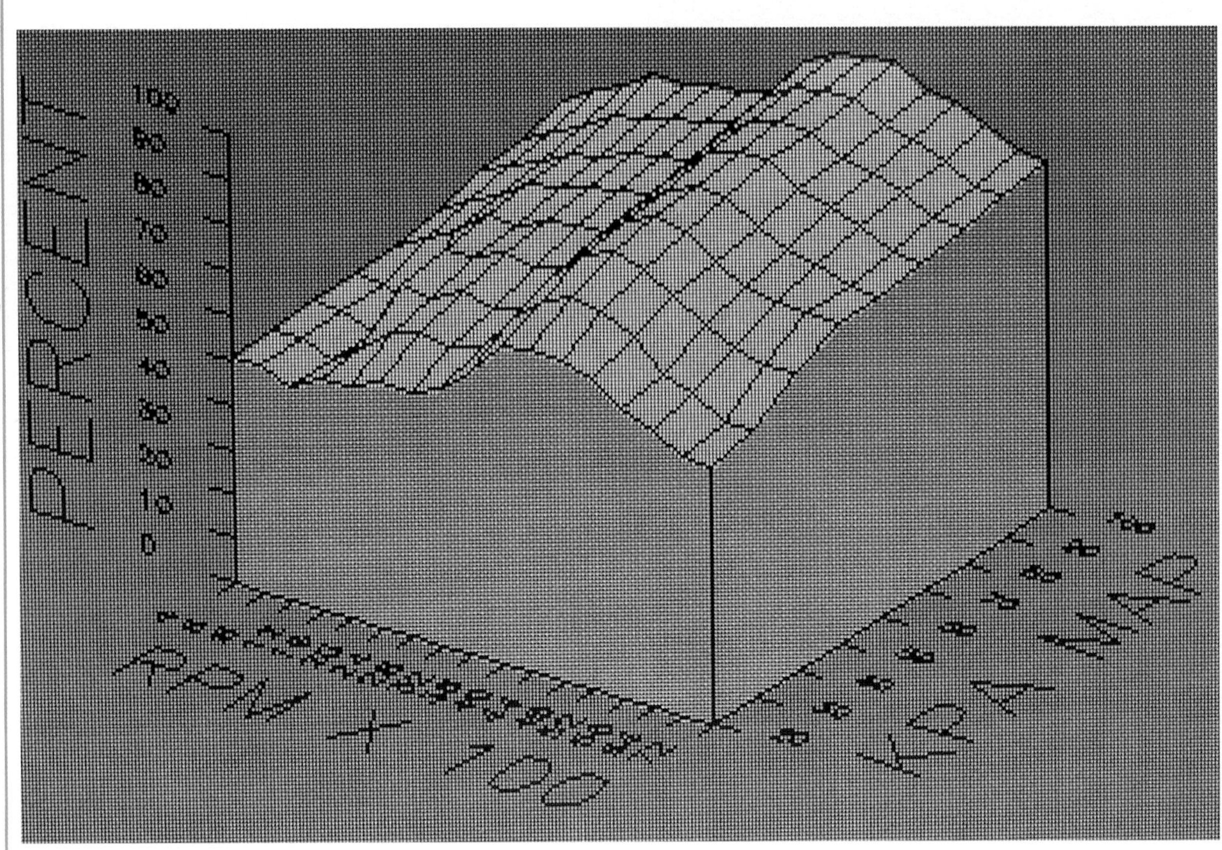

The volumetric efficiency map from a standard GM V8, as displayed with Kalmaker software. The higher the VE, the greater the amount of fuel injected. The strong mid-range torque of this passenger car 5-litre V8 engine can be clearly seen.

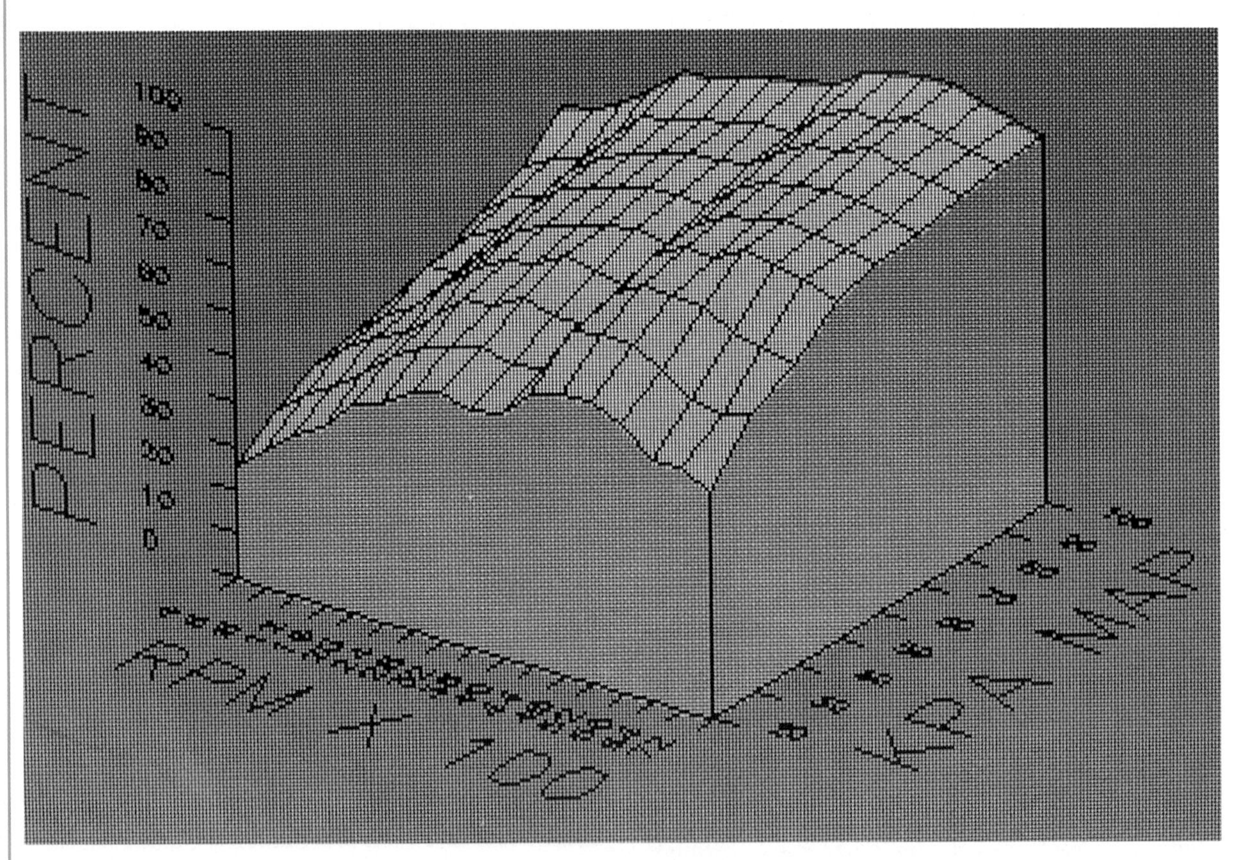

The volumetric efficiency (fuel) map from a GM Group A 5-litre V8. Running a much better flowing intake system and hotter cam than the standard car, the Group A engine has poorer breathing low in the rev range but does not fall off in volumetric efficiency at high rpm.

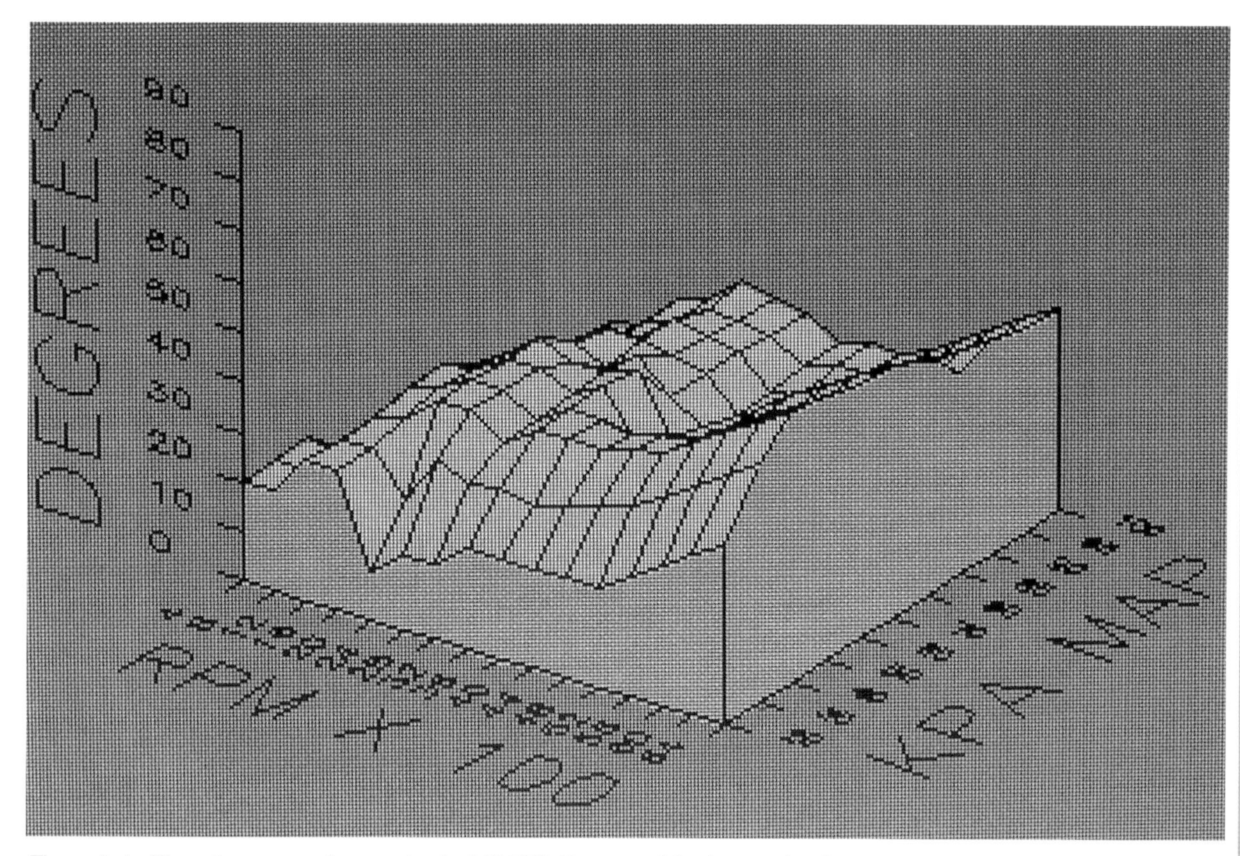

The main ignition advance map from a standard GM V8. Because of the low combustion pressures at high and low rpm, the ignition timing can be relatively advanced (at least when compared with the rest of the map!) without detonation occurring. Here this is especially noticeable at high rpm.

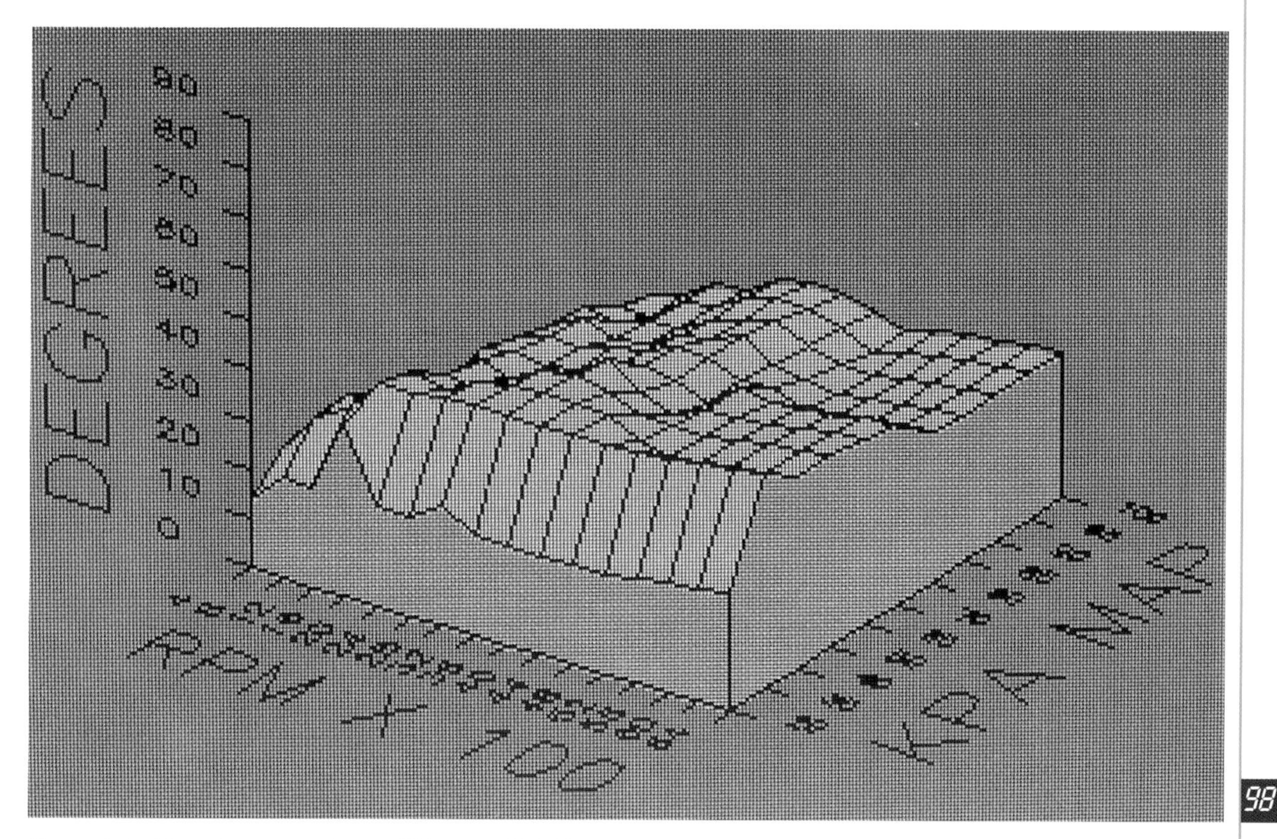

The main ignition advance map from a GM Group A V8. When compared with the standard engine, the sporty version has much more advance at low loads and rpm. However, this tapers off at high loads and revs because of the higher combustion pressures present.

ECU Software

It's worth looking at the factory engine management software maps in some detail, simply because their complexity clearly shows why re-calibrating factory engine management can be so successful. (Of course, before this can be done, the special software must be available!) The examples used here are for a MAP-sensed Holden 3.8-litre V6; however, other GM-Delco systems are very similar.

There are four different types of variables used by the GM-Delco ECU factory software:

1. So-called zero dimension variables that are used to specify the number of engine cylinders, whether there is an auto or manual gearbox installed, etc

2. One-dimensional variables that are used to represent counters, delays, air/fuel ratios, etc

3. Two-dimensional variables that are look-up tables comprising such factors as idle air motor steps versus rpm, air/fuel ratios versus time, etc

4. Three-dimensional variables that are the 3D maps. An example of such a map is air/fuel ratio versus rpm versus manifold pressure

The program logic can be divided into six main areas of operation. These are:

1. Fuel

2. Spark

3. Idle Air Control

4. Diagnostics

5. Output Logic

6. Variables

In addition, the program also determines a number of internal modes, such as whether the engine is cranking or running and whether or not the oxygen sensor is operational. These modes are saved and used in various calculations.

The processes followed in the calculation of the required amount of fuel are:

1. Estimate of the mass of air entering the engine.

2. Look up the desired air/fuel ratio for that engine speed and MAP.

3. Multiply air mass by fuel/air ratio to give fuel mass required.

4. Translate fuel mass to injector pulse width.

The mass of air per cylinder is calculated from manifold pressure, intake air temperature and engine speed. This figure is then multiplied by the volumetric efficiency of the engine. A three-dimensional table is used to specify low rpm volumetric efficiency, which is calculated as a function of engine speed and MAP. Another similar look-up table is used for high rpm volumetric efficiency.

Should coolant temp be below 44 degrees C, air/fuel ratio correction is applied as a function of coolant temp and MAP. Two further single-dimension variables control the decay rate of the enrichment and the minimum to which it can fall. The main air/fuel ratio look-up table uses three dimensions, with air/fuel ratio expressed as a function of MAP and engine speed.

Adding a bias to the injector pulse width compensates for injector opening delays caused by battery voltage variation. For example, at a battery voltage of 11.2 volts, the bias is 1.16 milliseconds. Because of non-linear biases in injector operation at small openings, a two-dimensional table is used to apply further correction. Despite the mechanical fuel pressure regulator maintaining fuel pressure at a fixed headroom over manifold pressure, battery voltage changes apparently cause sufficient variation in fuel pump pressure to require another correction factor. This works as a function of battery voltage. Yet another injector correction is a delay imposed in injector operation. Employed when the ECU is used to control single point injection (so not enabled in the Commodore), it can give better air/fuel mixing.

One of the factors giving cars equipped with this management system such good open-road fuel consumption is the lean cruise mode. Lean cruise is enacted when the coolant is above 80 degrees and the speed higher than 68km/h. After 150 seconds, the air/fuel ratio is increased in 0.1 ratios at 0.2-second intervals. This increase is limited to a value derived from a three-dimensional look-up table specifying the enleanment value as a function of engine speed and MAP.

As you would expect in a factory system, the fuel injector control for engine starting is complex. There are pre-set variables for cranking pulse width and also for the decay rate of this base crank pulse width. The steps by which this pulse width decays are also specified. The 'clear flooding' throttle position is set at 98 per cent opening or more, with the injector pulse width reduced to 7.895 milliseconds during cranking with the throttle in this position.

To give accelerator enrichment/deceleration enleanment, no fewer than 22 different variables are used! These include such an exotic as coolant temperature correction of the deceleration enleanment, which is then decayed by a specified amount over a specified time.

Closed-loop running — where the oxygen sensor controls the mixtures — uses numerous variables. The minimum coolant temperature at which closed-loop running will start is 44 degrees C for idle and 31.25 degrees C for running conditions. The program takes six seconds before it switches from open loop to closed loop after acceleration and it will do this only with a manifold vacuum of more than 5kPa.

The Short Term Fuel Trim is a fast-reacting air/fuel ratio correction system using the output of the oxygen sensor. The oxygen sensor outputs a voltage signal that's categorised either 'rich' or 'lean'. The longer the ECU receives a 'rich' (or 'lean') signal, the greater the correction applied. This results in an air/fuel ratio under closed-loop conditions that oscillates around the stoichiometric point.

The Long Term Fuel Trim uses an array of 24 learning cells. Each cell handles a range of rpm and MAP; the array covering the engine's overall operating conditions. When the engine is operating in closed loop, the fuel term is calculated and then modified by the value of the block learn cell corresponding to the rpm and MAP

While the Kalmaker system for GM vehicles has allowed the functionality of the GM ECU to be revealed, cars such as this Mazda use engine management software which is still much more of a closed book.

conditions present. If the engine has operated with that manifold vacuum and engine speed for a number of seconds, 'learning' occurs. Blocks can affect neighbouring blocks, allowing smooth interpolation. Disconnecting the battery clears these cells — one reason a car may run initially poorly after the battery has been disconnected.

The Long Term Fuel Trim RAM data can be accessed after the event, the ECU having continuous on-board data logging of the air/fuel ratio at 24 different load/rpm sites. For the Long Term Fuel Trim to become active, the oxygen sensor must be working correctly (indicated by its voltage output and frequency of voltage swings) and the engine neither accelerating nor decelerating. During the manufacture of the car, the program is configured to recognise a stoichiometric air/fuel ratio whose value corresponds to the 'switching' voltage of the oxygen sensor being used. An injector constant is also programmed, allowing larger or smaller injectors to be used.

For the calculation of spark timing, the ECU receives from the crankshaft position sensor a reference pulse well before Top Dead Centre (TDC). It then calculates the required spark timing, counts forward and delivers the spark.

If the car is at idle, the main spark is drawn from a two-dimensional map as a function of MAP. If the car is running, the initial spark value is derived from a three-dimensional look-up table that shows spark advance as a function of engine speed and MAP. At low MAP pressures (ie high vacuums), the table has increased resolution. At engine speeds of over 4800rpm, a high-rpm spark figure is added to the main advance rate. The output of the three-dimensional coolant correction chart is then used to modify the timing. This chart shows the correction as a function of coolant temp and MAP. So negative coolant corrections can be made, another variable (coolant offset) is subtracted from the positive value coolant correction chart.

When the exhaust gas recirculation function is enabled (it isn't in the example Commodore), the spark is further advanced as a function of MAP and engine speed. Another correction not currently used (but available) is for barometric pressure variation. This chart has correction data as a function of MAP and barometric pressure.

When shifting the automatic transmission from Park or Neutral into Drive, spark is retarded by 5.98 degrees if the engine speed is greater than 3600rpm. This is to cushion 'shift-shock' — the lurch that occurs when shifting into gear. Spark advance is also increased by an amount proportional to the rate of acceleration. However, the change in spark from one calculation to the next (ie attack rate) is limited by a variable expressed in degrees/millisecond, with a figure of 0.01 degrees/millisecond used in our example.

Depending on coolant temp, the rate of change of throttle position is also used to retard the spark timing. However, the spark retard logic is bypassed if the vehicle speed is less than a pre-set variable, or if the engine speed is higher or lower than other pre-set values. When deceleration fuel cut-off is employed, spark timing is decayed until a minimum value is reached, whereupon the actual fuel cut-off starts.

During starting, spark timing uses an initial spark timing value which is then modified by a two-dimensional chart, being advanced on the basis of cranking speed. However, should cranking speed be below 400rpm, the crankshaft position sensor output can become inaccurate. In this case, spark is generated by the ignition module, using its back-up spark capability. Once rpm exceeds approximately 400, the ECU switches spark timing back to its normal mode. When the engine is idling, a two-dimensional table is used to look up timing as a function of MAP. Interpolation is used if the MAP value falls between two points.

At the end of the spark timing calculations, the spark advance value is checked against a one-dimensional variable limiting the maximum spark advance angle permitted and another limiting the minimum spark timing angle. These are 60.2 degrees and -17.8 degrees respectively. Other spark timing variables set during manufacture include one to establish the engine position at which the distributor reference pulse occurs, and a crankshaft position sensor lag correction factor (a correction for electronic delays in the sensor and pick-up). This is set at 200 microseconds on the V6.

In fact, there are more than **70 variables** (ranging from one-dimensional to three-dimensional) used in the calculation of **just the final spark advance**! As a comparison, some aftermarket programmable engine-management systems have only four or five.

Re-calibration of the ECU Software

With the development ECU fitted to the car, the management maps are tuned in a similar way to other PC-controlled programmable management systems. The Kalmaker calibration data is presented in plain English, being shown as numerical tables or bar graphs. Changes to data can be typed in, or the bars clicked and dragged.

Figure 4.26 shows a screen grab from the Kalmaker program. It is part of the main ignition advance map, with this particular display for 2000rpm. On the vertical axis are degrees of ignition advance and on the horizontal axis is load, shown in kilopascals absolute manifold pressure. (The third dimension of the map comprises the different rpm scales, but this can only be seen on a 3-D graph.) One hundred kilopascals pressure absolute represents full throttle in this naturally aspirated engine. Here it can be seen that the ignition timing at full throttle, 2000rpm, is 14.1 degrees. The system tracks which bar is being accessed by the ECU; it can be seen that the 100kPa bar is dark, indicating that when this screen grab was made, the engine was at full load.

Figure 4.27 (opposite) shows a screen that relates to the knock sensor function. The ignition timing retardation per knock sensor 'count' can be specified on a 2D table. Here it can be seen that at 6400rpm, a retard of 0.032 degrees per count is being used. In an engine running much more advanced timing than standard, this figure would be raised so that if detonation occurred, the timing would be retarded more sharply. This approach can actually be seen in Figure 4.30 (page 103).

One very good feature of Kalmaker is its ability to graphically display all the ECU inputs and outputs. Figure 4.28 (opposite) is a sample display. From top left to right, the display shows:

- Coolant temperature
- Manifold pressure (ie MAP)
- Road speed
- Throttle position opening
- Manifold intake air temperature
- Oxygen sensor voltage
- Air/fuel ratio
- Battery voltage
- Barometric pressure
- Idle air control valve steps
- Ignition retard
- Short Term Fuel Trim
- Fuel injector base pulse width

The blocks across the bottom of the display are functions that can be toggled on/off.

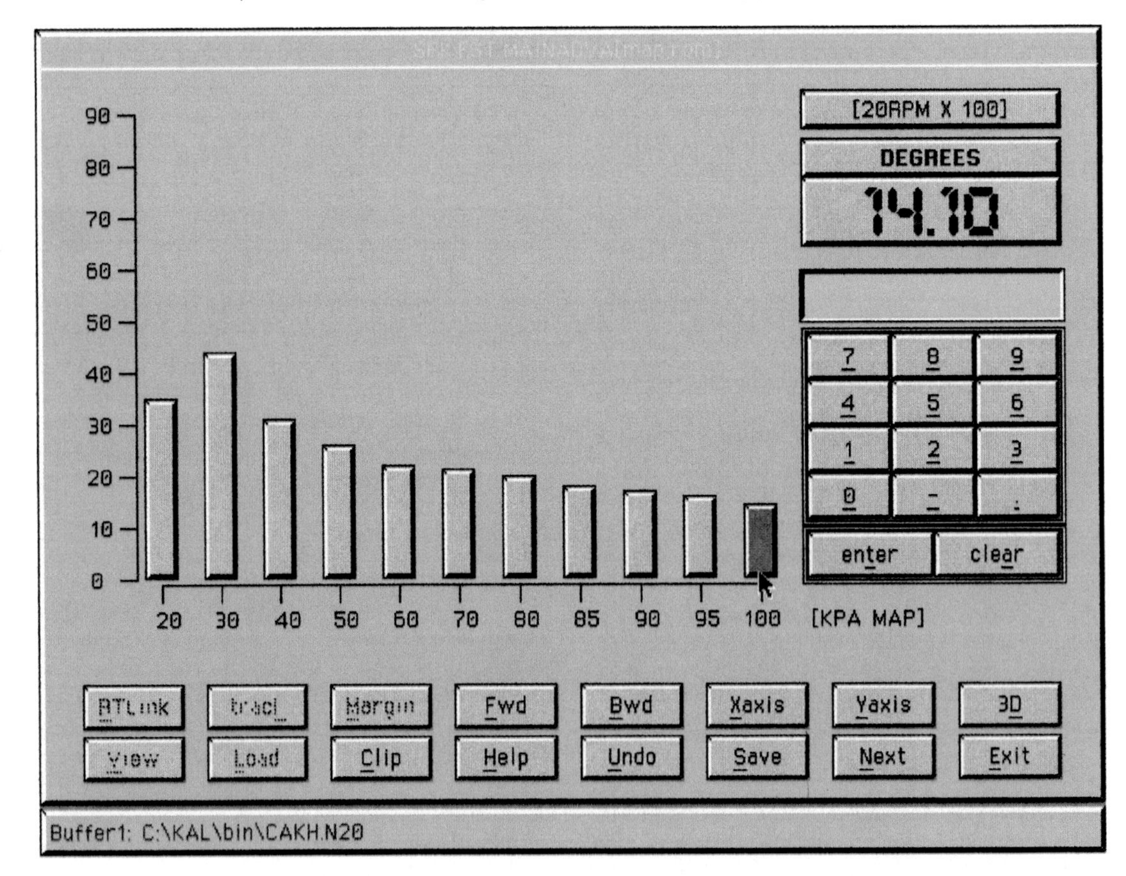

Figure 4.26
A GM-Delco ignition map display, as seen with Kalmaker software. This map is for 2000rpm and shows volumetric efficiency up the vertical axis and manifold pressure on the horizontal axis. The highlighted bar is for full throttle (100kPa absolute) and shows a timing advance of 14.1 degrees.

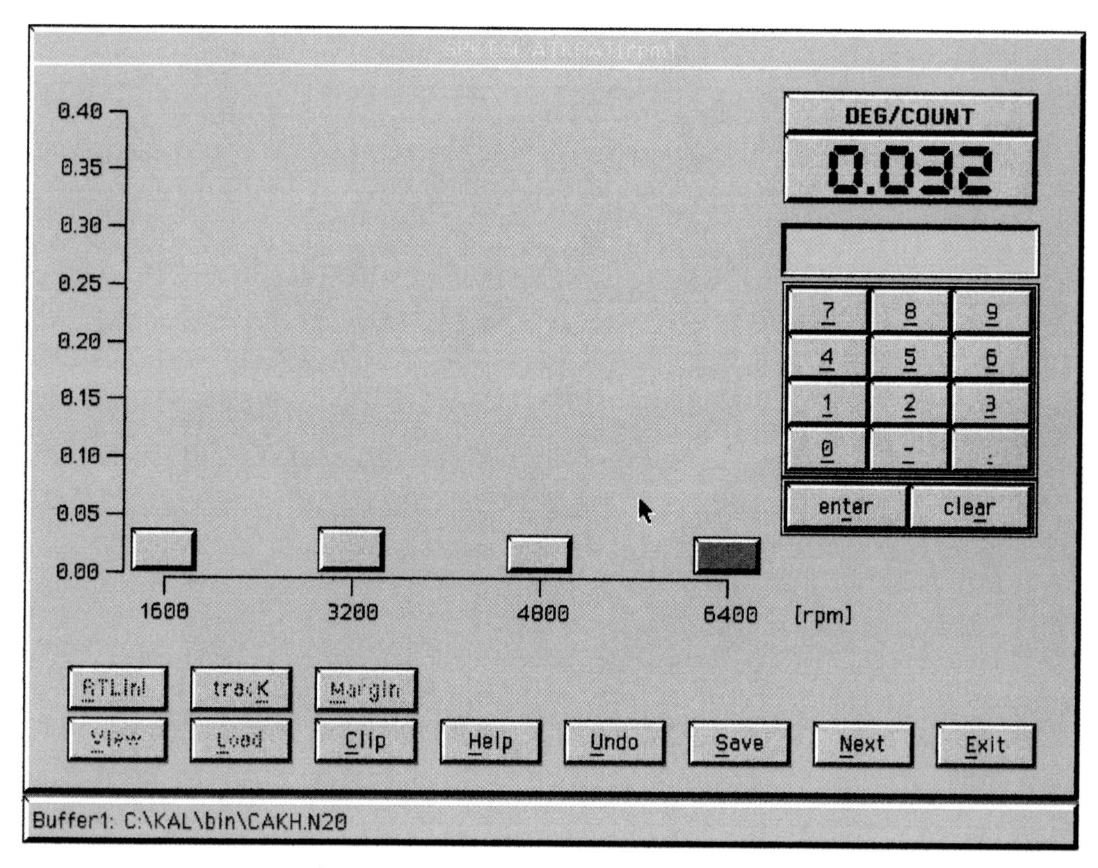

Figure 4.27
The amount of ignition advance pulled off when detonation is detected can be set with this Kalmaker map. The retard is expressed in degrees per knock sensor count and, as can be seen, it can be set for different rpm.

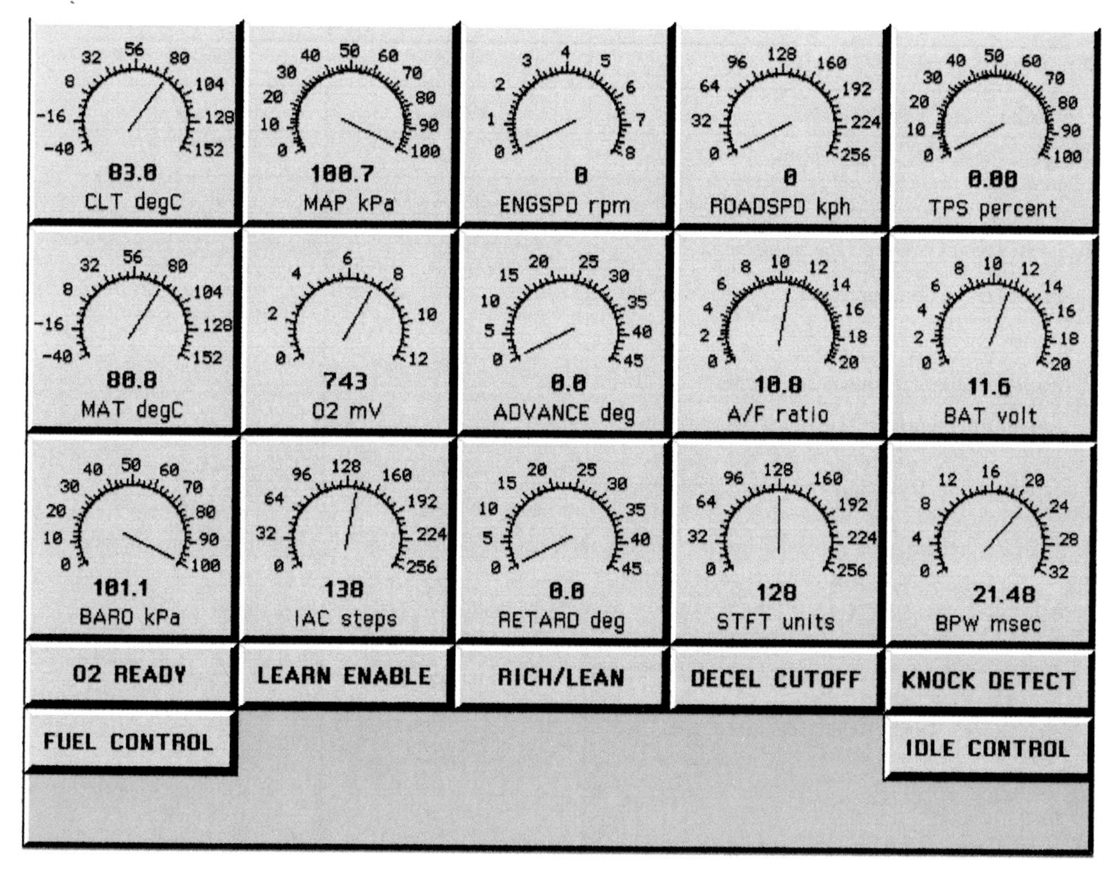

Figure 4.28
The Kalmaker gauge screen allows the viewing of all of the inputs and outputs in real time.

The Kalmaker system for GM logs engine and transmission information in real time by collecting data from standard sensors.

Variable	Modified Holden V6	Standard Holden V6
Spark High Advance Rate (degrees/1000 rpm)	1.98	2.03
Spark Maximum Retard (degrees)	9.84	8.09
Fuel Max Pulse Width (milliseconds)	24.002	10.986
Fuel Cut Low RPM	6300	5715
Fuel Cut High RPM	6400	5817
Fuel Cut Time Delay (seconds)	0.04	0.10
Idle Air Control Max Position (steps)	220	196
Idle Air Control Closed Loop Threshold (kilopascals)	39.86	34.33
Idle Air Control Warm Up Decay (seconds)	10	5
Idle Air Control Deadband (rpm)	18	50
Idle Air Control Sag RPM	206	400
Spark Idle Air Control Advance (degrees)		
(kPa MAP)		
20	20.39	26.02
30	24.26	26.02
40	30.59	26.02
50	31.99	26.02
60	30.94	26.02
70	26.37	26.02
Spark Knock Sensor Retard (degrees/count)		
(rpm)		
1600	0.038	0.030
3200	0.065	0.030
4800	0.069	0.030
6400	0.080	0.030

Fuel Volumetric Efficiency (per cent)		
(kPa at 400 rpm)		
20	13.3	53.5
30	14.8	60.2
40	16.4	62.5
50	17.2	49.6
60	20.3	53.5
70	23.4	55.9
80	26.6	58.2
90	30.1	76.2
100	33.6	71.1
(kPa at 1600 rpm)		
20	31.2	44.9
30	33.6	59.4
40	37.1	67.2
50	38.7	71.9
60	40.2	73.4
70	44.1	77.0
80	50.0	78.9
90	57.8	80.5
100	62.1	81.2
(kPa at 3200 rpm)		
20	44.9	60.5
30	48.4	80.1
40	53.5	82.4
50	59.8	84.4
60	64.5	85.5
70	68.4	86.7
80	70.3	87.1
90	73.8	86.7
100	76.2	84.4
(kPa at 6400 rpm)		
20	66.0	50.8
30	67.2	58.6
40	70.3	62.5
50	73.0	70.3
60	76.2	74.2
70	80.5	76.2
80	89.1	78.1
90	98.8	78.1
100	99.6	78.1

Figure 4.29
*This listing is a **very small** extract comparing the re-calibrated software of a modified 3.8-litre V6 Holden with the standard program. The modified engine featured a new camshaft, higher compression ratio and larger valves. It developed 60 per cent more power than standard.*

The Kalmaker GM Delco System

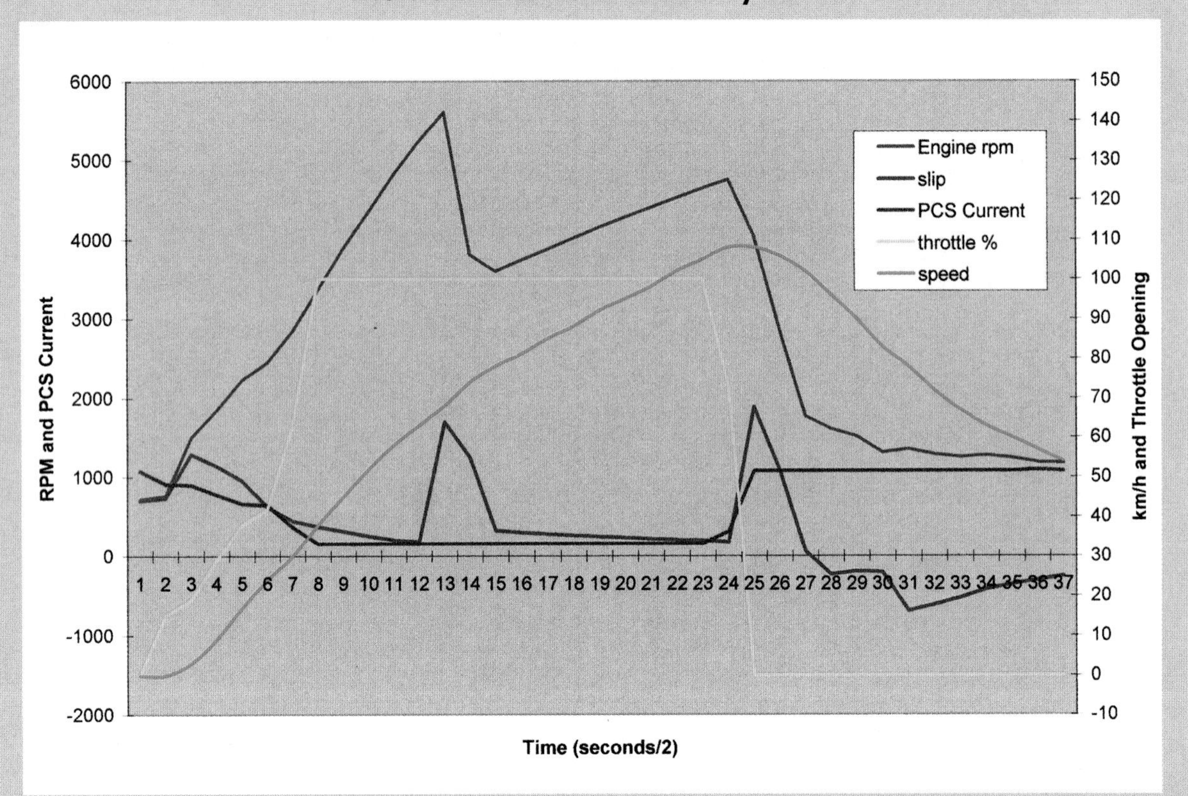

Figure 4.30
The Kalmaker software allows the logging of engine and automatic transmission factors in real time. While the system logs 10 times a second, this graph — showing auto trans behaviour — has been simplified, with data shown at 0.5-second intervals.

Displayed is engine rpm, torque converter slip rpm, throttle opening, vehicle speed and Pressure Control Solenoid current for a 16.5-second period.

During this time, the car was accelerated from a standstill to a speed of 108km/h. The throttle was then closed and the car gradually slowed to a speed of 54km/h. The car was left in 'Drive' during this manoeuvre.

The light blue line shows the car's speed and the yellow line shows throttle position, both being referenced against the righthand axis. Engine revs are shown by the pink line, while torque converter slip is shown by the dark blue line, both referenced against the lefthand axis. It can be seen that when the car is stationary, engine speed and torque converter slip are equal. This is because at an idle rpm of 700rpm, the slip must be 700rpm if the car is not moving! With a throttle opening of 100 per cent, the car accelerates rapidly with the amount of slip decreasing. The first-second gear change occurs at 13 on the horizontal axis and with the newly applied load, the slip within the torque converter rises. It slips by up to 1700rpm before the value drops back to about 300rpm after about 1.5 seconds.

In second gear, the driver keeps his foot flat to the floor until the engine speed reaches 4750rpm, on which he lifts his foot entirely (23 on the horizontal axis). The transmission immediately changes from second to third to fourth. Again, there is a major increase in slip through the torque converter, which then goes into negative numbers as the engine brakes the car.

During these procedures, the hydraulic fluid Pressure Control Solenoid is varied in its duty cycle, controlled by the current flow through it (shown by the brown line). A high current flow results in a low fluid pressure, while a low current flow increases fluid pressure and thus the clamping forces. With the application of full throttle the current rapidly reduces, staying at 156mA for the 1-2 full throttle gear change. It rises to 1074mA as the throttle is lifted, responding to the reduced torque load on the transmission. Even when engine braking, the PCS keeps pressures low.

The Kalmaker system can collect this type of data from the standard sensors, not only for the automatic transmission but also for the engine.

When the tuning process is complete, the development ECU is removed from the car, its job is finished. The new calibration data contained in the laptop PC's memory is then transferred to the original ECU's Mem-Cal using a normal EPROM burner. Following that, the standard ECU is replaced in the car.

Figure 4.29 (Pages 103-104) is a **very small** extract from the comparison of the re-calibrated software of a modified 3.8-litre V6 with the standard program. The modified engine featured a new camshaft, higher compression ratio and larger valves. It developed 60 per cent more power than standard.

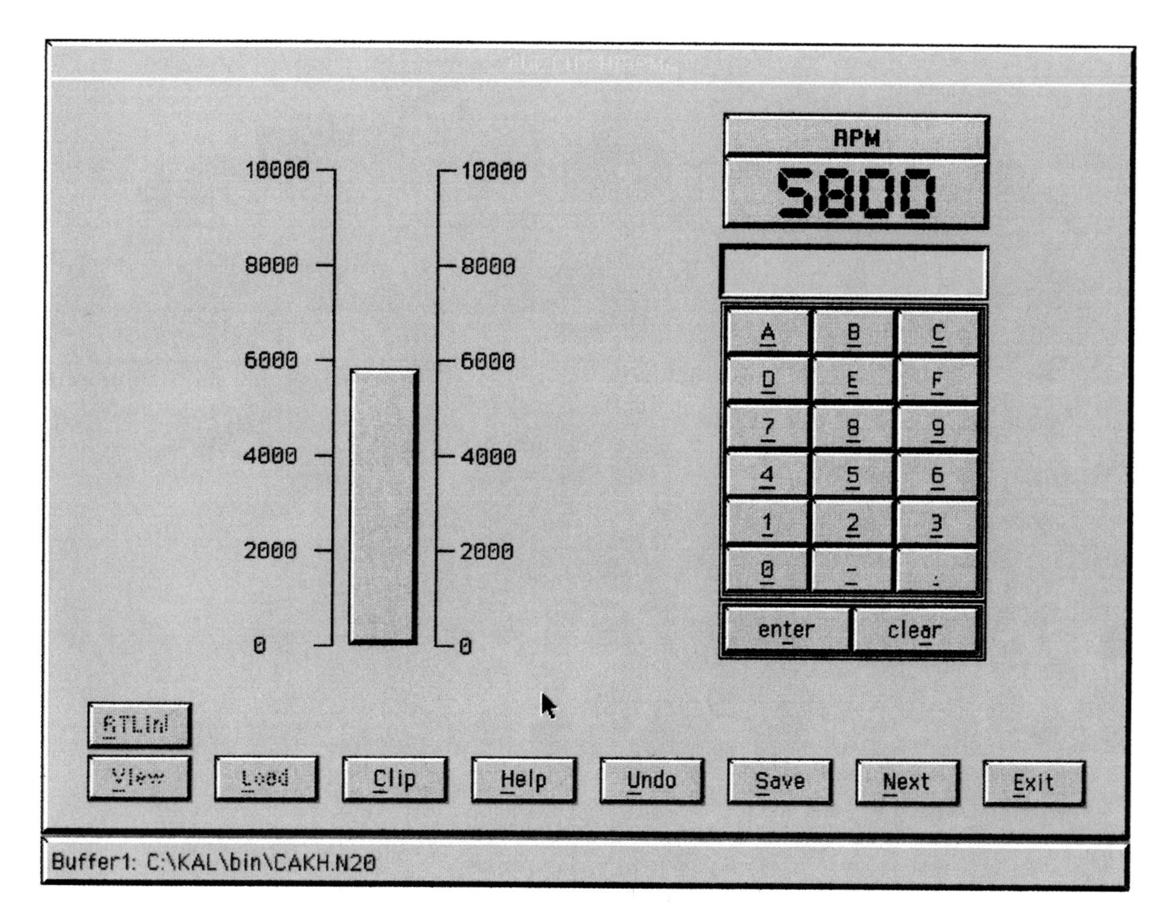

Figure 4.31
Using Kalmaker, the rev limit of the engine can be very easily altered.

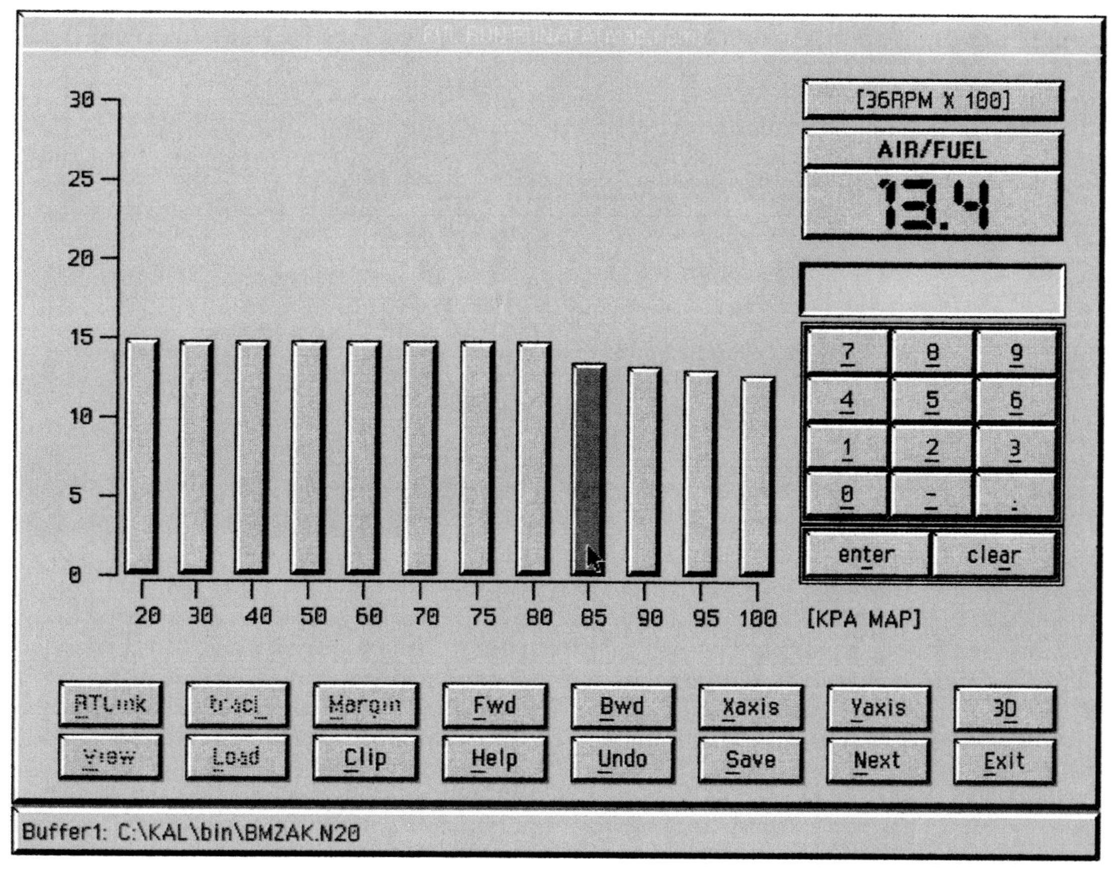

Figure 4.32
In addition to the volumetric efficiency (fuel) charts, Kalmaker allows the selection of target air/fuel ratios.

Air Intake Systems

Holden

Most EFI cars use a boxed air filter to prevent dust and other impurities from being drawn into the engine. Air enters the box through a pick-up snorkel and is filtered through a pleated paper filter before passing to the throttle body. There's often an airflow meter positioned between the airbox and the throttle body, and on turbo and supercharged cars, the inlet air is usually directed to the compressor rather than to the throttle. Following the throttle blade is a complex intake manifold using a plenum chamber and tuned-length runners for each cylinder.

The design of this system means most EFI engines breathe through just the one inlet. The air is pushed into the engine intake by atmospheric pressure, with the flow caused by the partial vacuum that's created as either the pistons descend or the turbo or supercharger draws air. The engine can breathe only as much as can pass into this duct, so it's vital that it flows very well. If the intake is restrictive, power, throttle response and economy will all suffer. After the throttle, the resonant tuning of the intake system will help determine how much air is breathed at each engine speed. While seldom utilised in modified EFI engines, changing the tuning of the intake plenum and runners can make a huge difference to the shape of the power curve.

The intake system can be divided into two sections: before the throttle body and after the throttle body. To make the nomenclature easy, I'll call the system in front of the throttle the 'intake system' and the throttle and the parts that follow it the 'intake manifold'. First, let's look at the system before the throttle body.

MEASURING INTAKE SYSTEM PERFORMANCE

Many people look at the standard intake system only as long as it takes for them to work out how to unbolt it! They then fit a bare cone filter in place of the standard box, slam the bonnet shut and hope they have improved performance. Sometimes they have, but a significant number of times they have actually created more problems than have been solved. The first step in any logical development of the intake system is to measure the flow and temperature performance of the standard system. If it doesn't **need** improvement, why spend money trying to do so?

It's quite easy to measure the flow restrictions of the standard system and doing this will take only a few hours and cost less than the price of a pizza! Once you realise that any flow restrictions will make themselves known as

The restriction of any of the parts of the intake system can be easily measured on the road using a fluid manometer. An off-the-shelf item like this can be bought or a temporary manometer can be made for nearly nothing.

This intake system consists of a snorkel, airbox, internal filter and an airflow meter. Note the small standard snorkel used here.

pressure drops it becomes straightforward. Measuring how much total pressure drop is occurring will tell you how well the system as a whole flows; measuring where the major drops are occurring means you can pinpoint the actual restrictions.

Pressure Drops

But what's all this about pressure drops, anyway? If you have a vacuum/boost gauge fitted to your car, you will already be familiar with the idea. When the engine is idling with the throttle nearly closed, there's a strong vacuum present in the plenum chamber. It occurs because the engine is trying to draw in air, but that air isn't being made available to it. In this case, the closed throttle blade is the restriction. In the same way, any other restrictions in the intake system will cause a pressure drop, though not nearly as big as the one shown on the engine vacuum gauge. In fact, the pressure drops in the intake system are so small they need to be measured in inches (or centimetres) of water.

Before you can measure pressure drops, you will need to make a manometer, an instrument very sensitive to small pressure variations. The DIY instructional comic (page108) included here shows you how to make and use the manometer.

The use of a water manometer means pressure drops as small as 1cm of water (0.014psi) can be easily measured. This makes the instrument very, very sensitive to any flow restrictions on the intake system. While extremely simple to make, no other cheap off-the-shelf instrument can

Some cars have a blocked airfilter indicator, such as this one from a VW Transporter. In the same way as this indicator detects a pressure drop, a manometer will show which parts of the intake flow well and which parts pose restrictions.

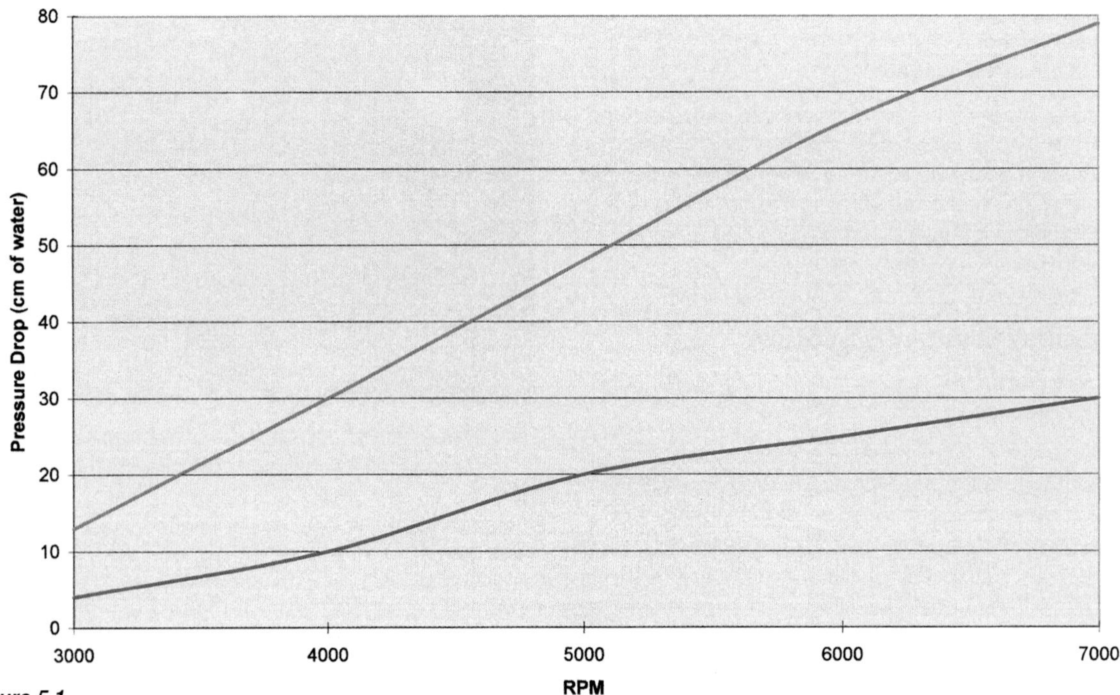

Figure 5.1
The pressure drop of the airbox intake of a Subaru Liberty (Legacy) RS, before and after the fitting of a new snorkel. The pressure drop was reduced by up to 70 per cent and the 0-100km/h time dropped from 6.5 to 6.1 seconds!

measure such small pressure drops with such a high degree of accuracy.

Testing that was carried out on a Subaru Liberty (Legacy) RS showed there was **more than** 80cm of water (1.1psi) pressure drop before the throttle butterfly at full power. The actual magnitude of the pressure drop could not be recorded because an 80cm manometer was the tallest that could be fitted in the car's cabin! Even before the airfilter, the pressure drop was still 78.7cm of water. The cause of this major intake restriction was a resonating box normally hidden inside the guard. Unlike some boxes that are tee'd into the intake, in this design all the induction air passed through the box. The inlet to the resonating box was a small sharp-edged bellmouth, positioned quite close to the metalwork of the inner guard. The resonating box was removed and a new intake to the airbox made. This substantially improved flow with Figure 5.1 showing the 'before' and 'after' pressure drop measurements. With some other minor intake mods, the 0-100km/h time dropped from 6.5 to 6.1 seconds!

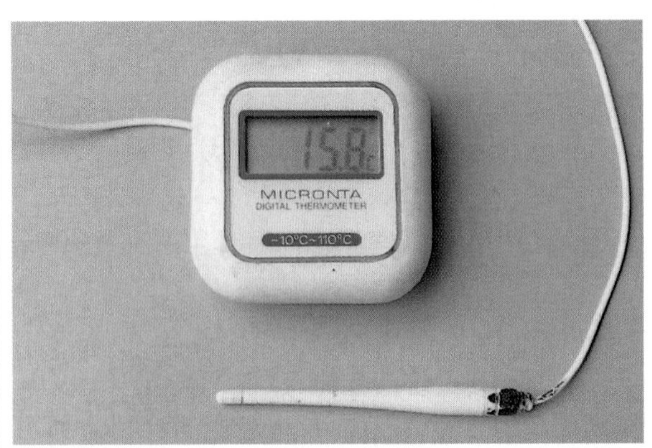

In addition to having good flow, the intake system must also breathe air that is as cool as possible. A cheap LCD thermometer with a remote probe can be used to measure the actual intake air temps.

Temperature Measurement

Not only does the intake system need to flow with the least restriction possible, it also needs to pick up the coolest air. The term 'cold air induction' is often misused — the air being breathed can never be 'cold' unless the day is cold as well! However, the air under the bonnet is very frequently extremely hot. This is the case for a couple of reasons. First, most of the air under the bonnet has passed through the engine cooling radiator and air-conditioning condenser. The air may also have been preheated by an intercooler, engine oil cooler and transmission cooler. And then when the air **does** reach the underbonnet area, the hot exhaust manifold and the engine itself will further warm it. This means the coolest air for induction will **not** be found inside the engine bay!

The temperature of the induction air can be measured with a remote-probe digital thermometer. These can be found cheaply at discount stores and supermarkets. Try to get one that measures up to 70 degrees C — a maximum measuring temp of 50 degrees C is also common. Placing the probe temporarily inside the airbox is a simple way of measuring the temperature of the intake air. The lower the temperature of this air, the better. This is because cooler air is denser, so the same volume of cooler air contains more oxygen. When added to fuel, that extra oxygen can be turned into more power. As a rule of thumb, expect the peak power to improve by 1 per cent for every 4 degrees C temperature decrease of the intake air temperature. Throttle response also benefits from lower intake air temps, and especially in forced induction cars, the lower the inlet air temp, the less likely the engine is to detonate. It's possible to have inlet air temperatures only 1 or 2 degrees C above ambient, but anything less than 5 degrees C higher than ambient is very good.

The intake air temperature is likely to increase with the car stationary but idling, as occurs at traffic lights. If the

Measuring Intake Flow Restrictions

BEFORE YOU MAKE MAJOR INDUCTION CHANGES (LIKE REPLACING A FILTER OR AIRFLOW METER) IT MAKES SENSE TO FIND WHERE THE INTAKE FLOW RESTRICTIONS ACTUALLY ARE. HERE'S HOW...

ANY FLOW RESTRICTIONS IN THE INTAKE SYSTEM SHOW UP AS DROPS IN AIR PRESSURE TO BELOW ATMOSPHERIC. TO MEASURE THESE YOU FIRST NEED TO MAKE A MANOMETER.

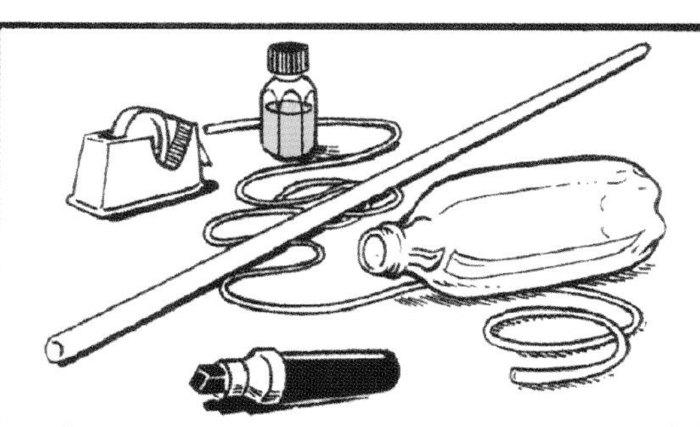

YOU'LL NEED 3 METRES OF CLEAR PLASTIC TUBE, A PLASTIC SOFT DRINK BOTTLE, A METRE LONG BROOM HANDLE, A MARKING TEXTA, SOME FOOD COLOURING AND SOME CLEAR STICKY TAPE.

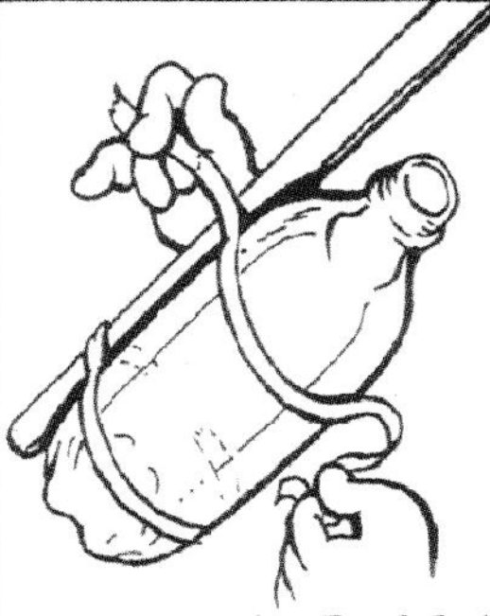

TAPE THE BROOM HANDLE TO THE SIDE OF THE PLASTIC DRINK CONTAINER.

PUSH THE OPEN END OF THE PLASTIC TUBE TO THE BOTTOM OF THE DRINK CONTAINER, AND THEN RUN IT UP THE BROOM HANDLE, TAPING IT INTO PLACE.

DRAW A LINE THREE QUARTERS OF THE WAY UP THE SIDE OF THE BOTTLE. MEASURING FROM THIS LINE, MARK OFF 1 CM INCREMENTS UP THE BROOM HANDLE.

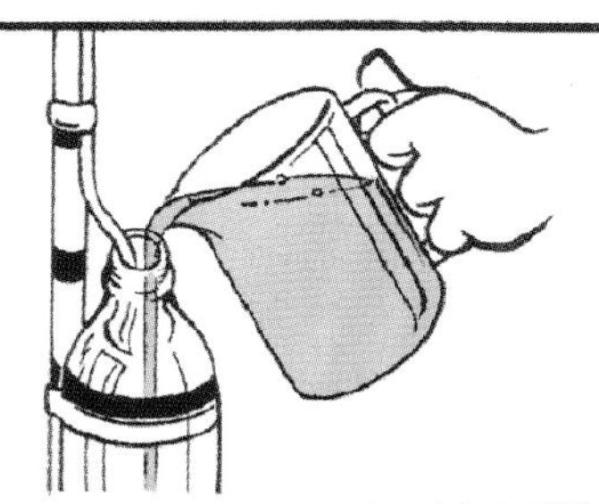

FILL THE DRINK CONTAINER WITH COLOURED WATER TO THE MARKED LINE AND MAKE SURE THAT THE TOP OF THE BOTTLE ISN'T SEALED SHUT.

..YOU'VE JUST FINISHED MAKING A ***MANOMETER.***

NOW THAT YOU'VE MADE A MANOMETER, HOW DO YOU USE IT?

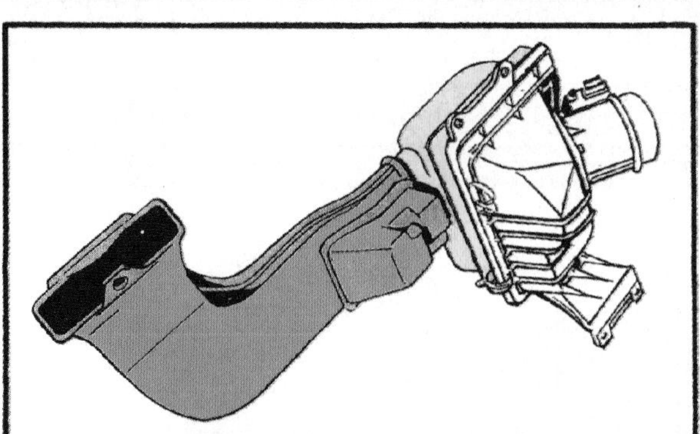
THE FIRST POSSIBLE FLOW RESTRICTION IS THE INTAKE TO THE AIRFILTER BOX.

DRILL A SMALL HOLE INTO THE AIRFILTER BOX ON THE ATMOSPHERE SIDE OF THE FILTER AND THEN SCREW A TINY PLASTIC IRRIGATION FITTING INTO THE HOLE.

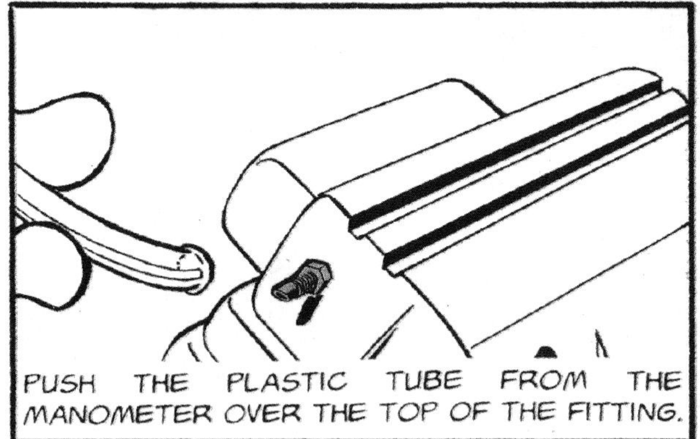
PUSH THE PLASTIC TUBE FROM THE MANOMETER OVER THE TOP OF THE FITTING.

2·5cm
VRROOOM!
REV THE CAR IN NEUTRAL AND YOU SHOULD SEE THE WATER MOVE UP THE MANOMETER TUBE SLIGHTLY.

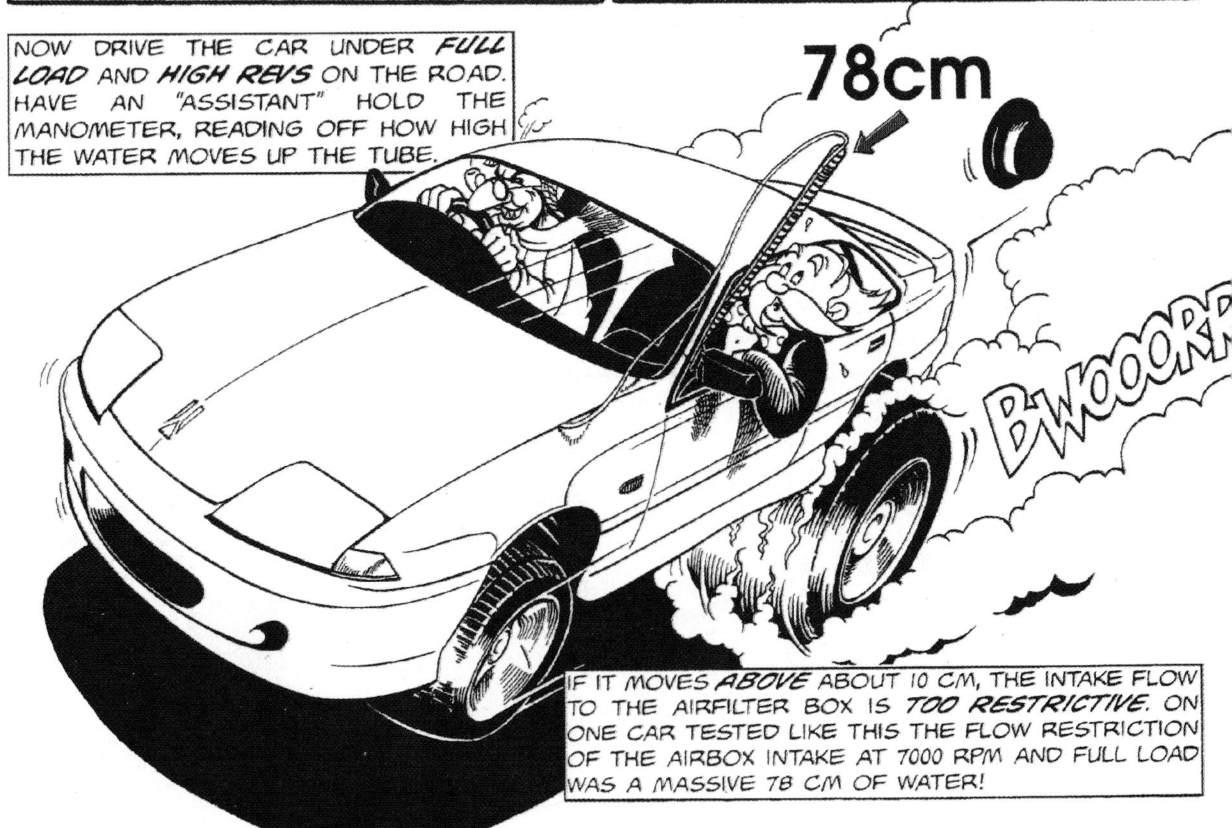
NOW DRIVE THE CAR UNDER FULL LOAD AND HIGH REVS ON THE ROAD. HAVE AN "ASSISTANT" HOLD THE MANOMETER, READING OFF HOW HIGH THE WATER MOVES UP THE TUBE.
78cm
BWOOOR
IF IT MOVES ABOVE ABOUT 10 CM, THE INTAKE FLOW TO THE AIRFILTER BOX IS TOO RESTRICTIVE. ON ONE CAR TESTED LIKE THIS THE FLOW RESTRICTION OF THE AIRBOX INTAKE AT 7000 RPM AND FULL LOAD WAS A MASSIVE 78 CM OF WATER!

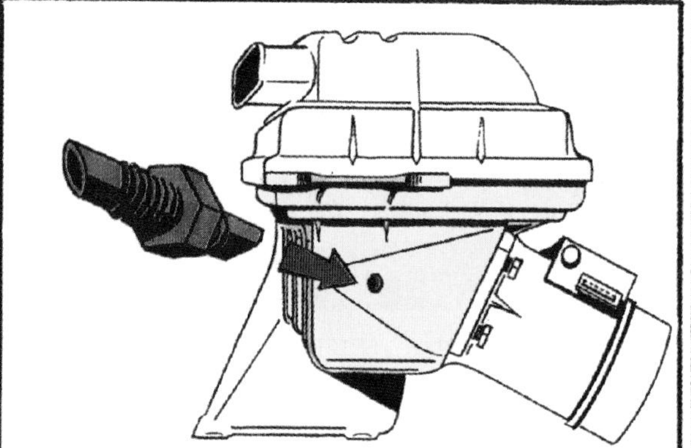

NEXT ATTACH THE PROBE TO THE ***ENGINE*** SIDE OF THE AIRFILTER BOX. YOU'LL NOW BE MEASURING THE FLOW RESTRICTION CAUSED BY THE INTAKE TO THE BOX, THE BOX ITSELF, AND THE AIRFILTER.

TO WORK OUT HOW MUCH PRESSURE DROP IS BEING CAUSED BY ***JUST*** THE FILTER AND BOX, SUBTRACT THE FIRST FIGURE FROM YOUR SECOND MEASUREMENT.

IF THERE'S MORE THAN ABOUT 25 CM OF PRESSURE DROP ACROSS THE FILTER, CONSIDER GETTING A BETTER FILTER!

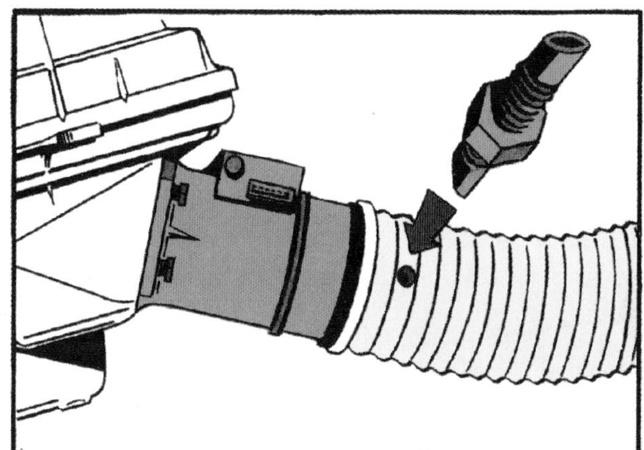

NEXT IN LINE IS THE ***AIRFLOW METER***. CONNECT THE PROBE TO THE ***ENGINE*** SIDE OF THE METER AND SEE WHAT PRESSURE LOSS IS OCCURRING IN THE SYSTEM UP UNTIL HERE.

IT'S THEN EASY TO WORK OUT HOW MUCH OF THE PRESSURE DROP IS BEING CAUSED BY ***JUST*** THE AIRFLOW METER.

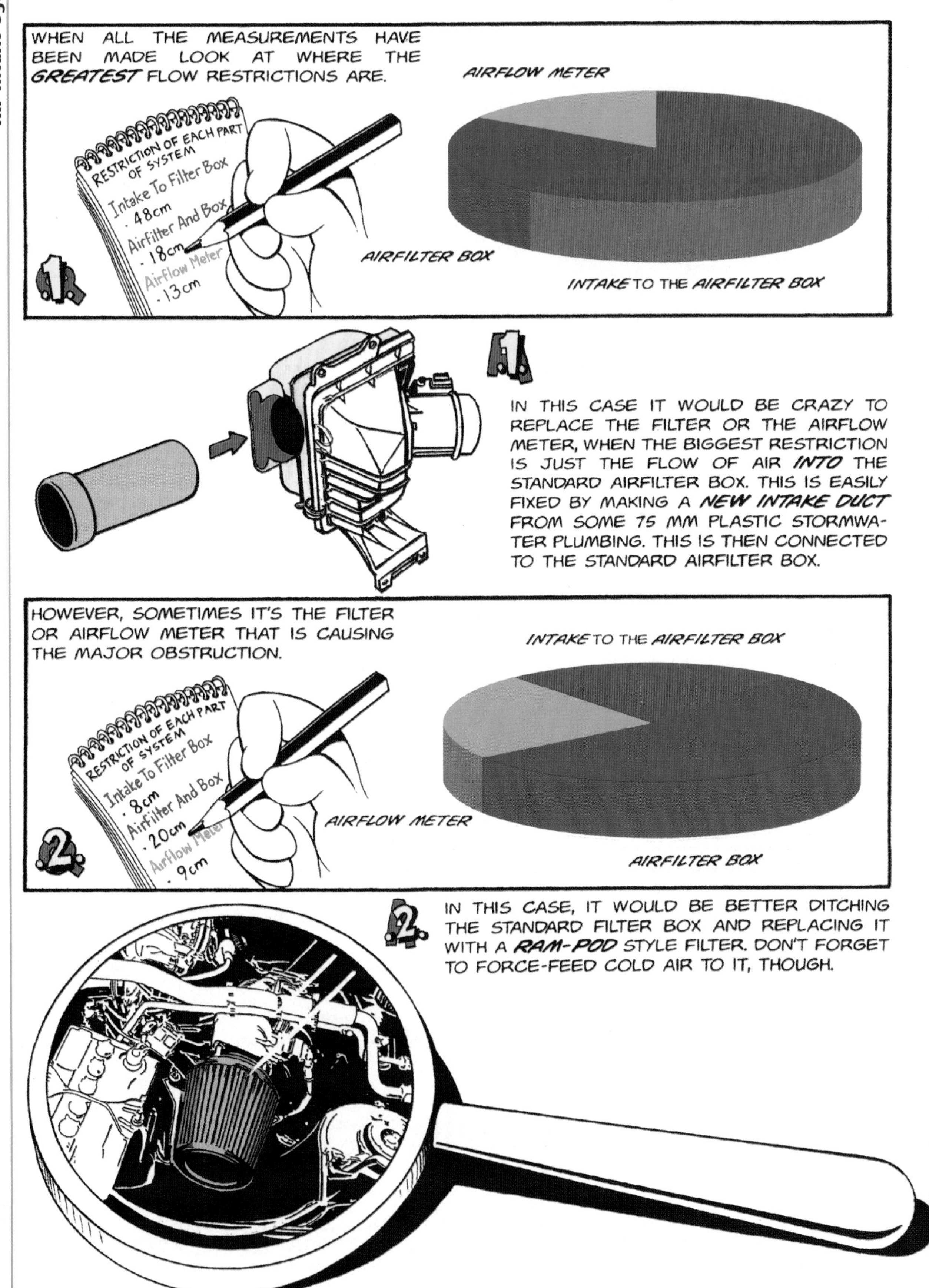
WHEN ALL THE MEASUREMENTS HAVE BEEN MADE LOOK AT WHERE THE GREATEST FLOW RESTRICTIONS ARE.
RESTRICTION OF EACH PART OF SYSTEM
Intake To Filter Box
- 48cm
Airfilter And Box
- 18cm
Airflow Meter
- 13cm
1.
AIRFLOW METER
AIRFILTER BOX
INTAKE TO THE AIRFILTER BOX
1.
IN THIS CASE IT WOULD BE CRAZY TO REPLACE THE FILTER OR THE AIRFLOW METER, WHEN THE BIGGEST RESTRICTION IS JUST THE FLOW OF AIR INTO THE STANDARD AIRFILTER BOX. THIS IS EASILY FIXED BY MAKING A NEW INTAKE DUCT FROM SOME 75 MM PLASTIC STORMWATER PLUMBING. THIS IS THEN CONNECTED TO THE STANDARD AIRFILTER BOX.
HOWEVER, SOMETIMES IT'S THE FILTER OR AIRFLOW METER THAT IS CAUSING THE MAJOR OBSTRUCTION.
RESTRICTION OF EACH PART OF SYSTEM
Intake To Filter Box
- 8cm
Airfilter And Box
- 20cm
Airflow Meter
- 9cm
2.
INTAKE TO THE AIRFILTER BOX
AIRFLOW METER
AIRFILTER BOX
2.
IN THIS CASE, IT WOULD BE BETTER DITCHING THE STANDARD FILTER BOX AND REPLACING IT WITH A RAM-POD STYLE FILTER. DON'T FORGET TO FORCE-FEED COLD AIR TO IT, THOUGH.

intake air temperature rises rapidly in these conditions, the response off the line will be reduced and the car may initially detonate. As an example, the intake air temperature was measured at the throttle body of a standard six-cylinder car. With the engine up to operating temperature and the car moving, the air temperature at the throttle body was 31 degrees C on a 20 degrees C day. After three minutes of stationary idling this had risen to 38 degrees, after five minutes 40 degrees C and after seven minutes 45 degrees C. Figure 5.2 shows these results.

On this car, the standard airbox snorkel breathes very hot air that has passed through the radiator. The hot air reduces power and dulls throttle response.

MODIFYING THE INTAKE SYSTEM

Armed with the knowledge gathered from your pressure and temperature testing, you can decide which parts of the system need changing.

Replacing the Airbox

First, if the measured pressure drop of the standard airbox and filter (not including the snorkel) is the highest of any part of the system, it makes sense to discard the complete airbox. One approach is to replace it with the airbox from another car.

I had 12 standard airboxes flow tested. Four of these boxes used the same Holden/Nissan/Subaru filter element while the others used a variety of filters. Tested at one inch of water on the flowbench, the boxes using the same Holden/Nissan/Subaru filter element had flows that varied from 78cfm to 129cfm. In other words, the box design (not the filter) caused a flow variation of more than 65 per cent! This means it's much more important to select an airbox that flows well rather than to buy a fancy filter that fits into the standard box. The flow of the other design boxes (all tested with a new filter element) varied from 52cfm to 129cfm.

When assessing alternative airboxes, consider a number of factors, starting with the cross-sectional area of the filter (bigger is better) and how the inlet and outlet ducts have been integrated into the design of the box. By the latter, I mean that a box that has right-angled inlet and outlet ducts butt-jointed against the box walls is likely to have poorer flow than one where the walls of the box gently flow into the ducting. Both the inlet and outlet ducts should have as large a cross-sectional area as possible. Finally, keep in mind you will need to buy replacement filter elements — so don't pick an airbox from an ultra-rare car!

Replacement of the standard airbox can also be made with a cone-type aftermarket filter or ram pod that has been suitably adapted to the airflow meter or inlet duct.

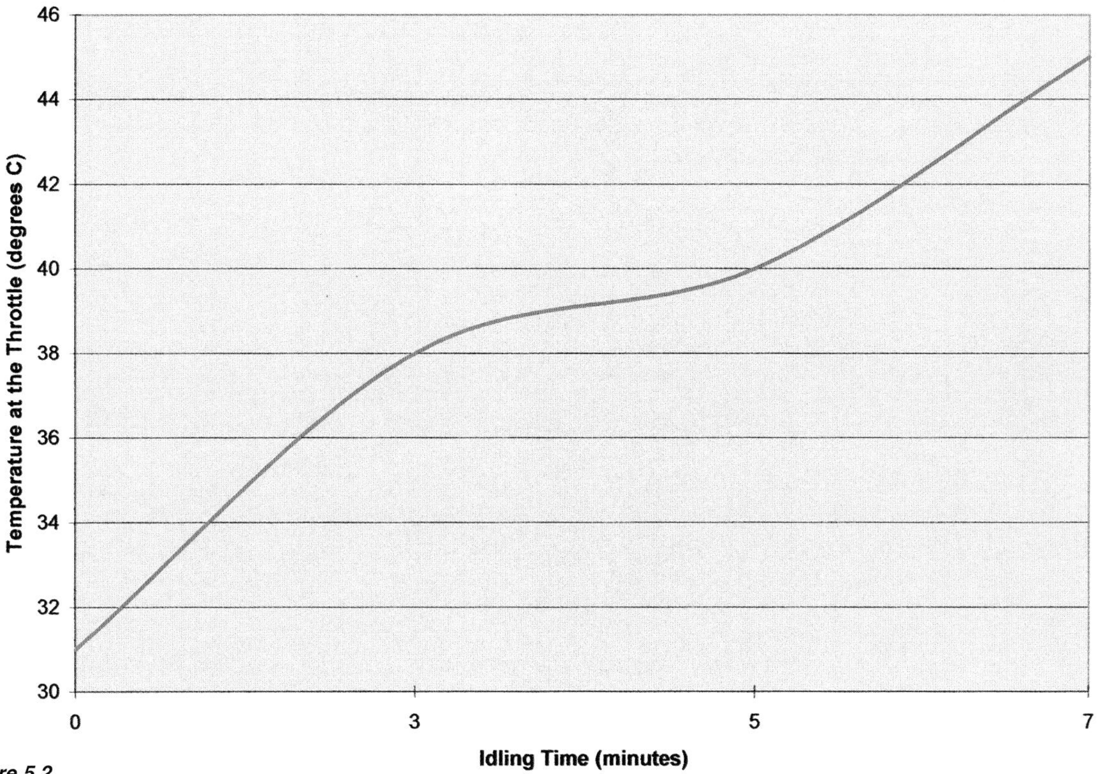

Figure 5.2
The intake air temperature of most cars increases rapidly when the car is idling. This was the rise in intake air temperature of a standard six-cylinder car, measured on a 20-degree Celsius day.

This will improve intake flow over even a good airbox. At the same time as the airbox testing mentioned above was taking place, a three-inch K&N barrel-shaped filter was flow-tested. It flowed 146cfm at one inch of water — better by 13 per cent than even the best box. However, to be effective in a car, the cone filter **must not** be exposed to hot air — simply placing it under the bonnet where the standard airbox previously sat is a very poor idea. This is because any factory cool air duct leading to this area will **not work** unless the new filter is enclosed and the new box sealed to the original duct. With the filter simply plonked into position, air will be drawn from the general engine bay, not specifically from the original supply duct.

This means aftermarket filters need extensive heat shielding (normally sealing against the closed bonnet) and careful ducting of air from outside the engine bay. Be wary of dyno runs that show major power gains from newly installed exposed filters — the runs are often made with the bonnet up and the dyno cooling fan blowing fresh air across the engine bay...

Figure 5.3 shows a 'before' and 'after' dyno run made on a car where an exposed underbonnet filter had been installed. The runs were made with the bonnet closed. You can see that at all areas of the power curve but the extreme top end, power actually decreased. Note also the sudden drop in power low in the rev band — almost certainly the result of the knock sensor retarding timing when detonation occurred as a result of the engine breathing hot air.

Another point to keep in mind when considering a replacement is that cars using hot-wire airflow metering can have their air/fuel mixtures upset if the airflow meter is fed turbulent air. This has been known to occur on some cars where the airbox has been replaced with an aftermarket filter attached directly to the airflow meter. Induction noise can also become intrusive on quiet cars equipped with an exposed filter. I'm not talking about the mild noise increase you get with a set of extractors and a big exhaust, but a booming, hollow-sounding 'baaaarrrff' every time you change gears or sharply back off! This is especially the case in turbo cars.

Finally, the length of pipe that connects the factory airbox with the throttle body has normally been carefully sized in diameter and length to provide effective intake resonant tuning. If this pipe is removed so a replacement filter can be mounted directly on the throttle body, power is quite likely to fall. Dyno testing carried out on a Daihatsu Charade showed that power variations of more than seven per cent were possible across the mid-range when intake pipe lengths that ranged from 30cm to 100cm were fitted.

Replacing the Factory Filter

Detailed testing of air filters that takes into account dust filtration, airflow and how quickly performance declines when in service, is rare. However, UK magazine *Fast Car* carried out a well-researched test in their June 1996 issue. In that test, the filters' pressure drops were measured at an airflow of 250cfm. A standard fine dust was then

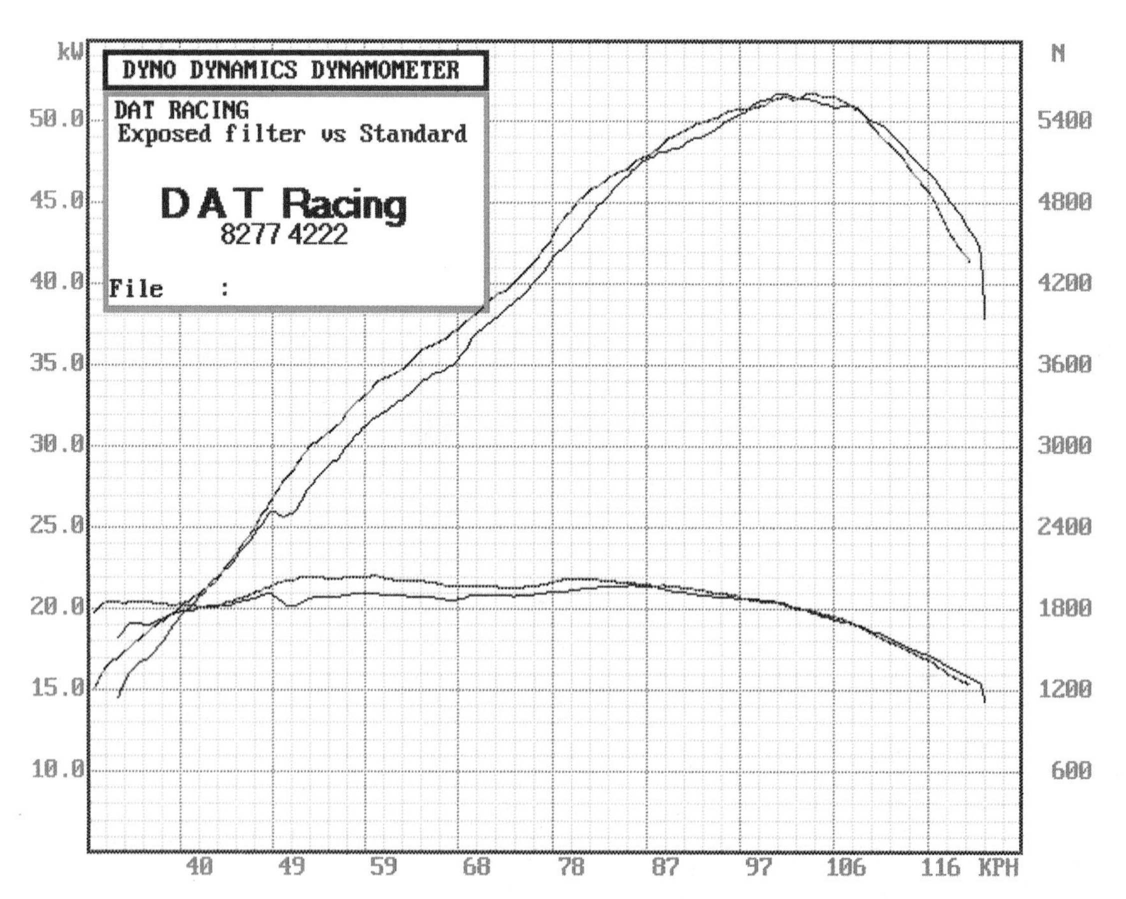

Figure 5.3
An exposed underbonnet filter does not always work well! The red line indicates the power with the new filter fitted to a Hyundai . It can be seen that through most of the rev range power actually decreased over standard.

Filter	Construction	Filter Efficiency (%)	Dust Capacity (grams)	Pressure Drop at 250cfm (cm of water)
Peugeot	Single layer foam on wire mesh	43	100+	16
Foam #1	Three layer foam on perforated metal	85	259+	14
Foam #2	Three layer foam on perforated metal	68	188+	16
Foam #3	Three layer foam on perforated metal	73	186+	16
Foam #4	Three layer foam on perforated metal	87	207+	17
Cotton gauze #1	Pleated cotton gauze not supported	94	62	15
Cotton gauze #2	Pleated cotton gauze not supported	96	52	14
Cotton gauze #3	Pleated cotton gauze not supported	96	115	13

introduced and the test continued until the pressure drop exceeded 60cm of water (0.85psi) or the test went on for more than an hour. The proportion of dust trapped by the filter was measured and expressed as a percentage. The mass of the dust adhering to the filter when either the pressure drop exceeded 60cm of water or an hour had elapsed was also measured. If it was the time interval that limited the test, the dust capacity is indicated with a (+) sign after the figure. The tests were carried out on the standard and aftermarket filters suitable for a Peugeot 205 GTI, but it's the performance of the different filter types that's of most interest.

As can be seen above, those expecting a major performance gain from changing the filter will be disappointed — the pressure drops of the different filters varied by just 4cm of water (0.06psi)! However, the filtering ability of the filters and how much dust they could hold before the flow was restricted varied dramatically. It can be seen that the pleated cotton filters work very efficiently but become blocked quickly, while the foam filters do not filter as well but can hold a lot more dust before becoming blocked. Most unusually, in this test, the standard manufacturer's filter performed poorly. These results show that a pleated cotton filter that's cleaned frequently provides the best combination of flow and filtering.

If you still believe a significant power increase can come from replacing the drop-in factory filter element, here's something else to ponder. A standard car featuring a 3.8-litre V6 was dyno tested while equipped with a dirty standard paper filter. The measured power peaked at 99kW. Next, a run was made with a slightly cleaner (but still fairly dirty) factory filter. The power curve peaked at just the same number — 99kW! With the baseline firmly established, a brand-new aftermarket oiled cotton gauze element was tried. This time there was a peak reading of 98kW, with a second run showing 100kW. These peak figures and also the complete dyno power curves showed there was nothing in it — and, remember, that's comparing a **brand-new** aftermarket filter with two **dirty** factory elements!

Building a Blocked Filter Alarm

To maintain the best possible power and fuel consumption, the air filter should be changed when it becomes dirty enough to pose a major airflow restriction. Knowing when this is the case is easy if you install a pressure switch that sounds an in-cabin buzzer. A suitable switch is available from RS Components (stores in most states and countries) with the catalog number 317-443. The switch

If an exposed airfilter is used, it must be sealed off from the hot underbonnet air. In this Subaru Impreza WRX rally car the shield uses a rubber strip that seals against the inside of the bonnet.

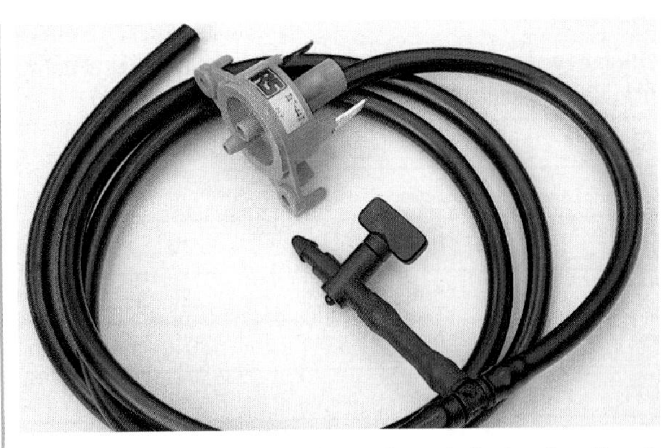

This switch can be fitted to the intake to indicate when the air filter has become restrictive. It triggers at only two inches of water pressure differential!

closes when subjected to a pressure differential of just two inches of water. It can be installed so the vacuum port of the switch is connected to the engine side of the filter and the other port left open. With the switch plumbed like this, it will detect the total pressure drop of the filter, the filter box and the inlet duct to the box. However, if the alarm is over-sensitive, plumb the switch so its second port is connected to the other side of the filter. That way, just the pressure drop of the filter element is being monitored. If it's still over-sensitive, tee an adjustable bleed into one of the hoses leading to the switch.

The exit duct from the airbox should use a bellmouth. Subaru, Nissan and Holden use bolt-on bellmouths like this.

Making a New Snorkel

If your measurements reveal that the factory airbox and filter are fine, but the intake snorkel to the airbox seems configured to breathe hot, restricted air, good improvements can generally be made. Enlarging this duct and leading it outside the engine bay dramatically improves flow and reduces inlet air temperature, the only downside being the need to change the filter element more frequently.

An exposed aftermarket filter like this will breathe very hot air, especially when the car is stationary. This can cause detonation and power loss problems when the car is accelerated hard off the line.

The easiest way to make a new inlet is to use plastic three-inch (75mm) PVC stormwater fittings and pipe, available from your local hardware store. If you get the right mix of 15-degree, 45-degree and 90-degree bends, you'll find you can soon put together a duct that works well. You will probably need to cut a hole in the inner guard to provide access for the new duct, but this hole can sometimes be made directly under the airbox, so remaining invisible to a casual observer. Note that the in-guard resonant boxes used before the airbox in some cars can be discarded, with a barely noticeable increase in induction noise usually the result.

At the airbox you can either cut a new hole in the right portion of the 'box to take the new duct or adapt the duct to meet with the factory airbox opening. This can be easily done using 'gutter adaptors' (which turn the round pipe into rectangular plumbing) or one of the other plastic plumbing fittings available. If the adaptor doesn't quite match the box opening, file the airbox hole until

A simple bellmouth can be easily formed in a piece of PVC pipe if a few centimetres of one end of the pipe is softened with a heat gun and then the pipe pushed down over a funnel.

Here's a new airbox outlet that has been flared using the heat gun and funnel technique. One edge needed to be ground off so it would slide into the exit duct of the airbox.

everything is a neat fit. Airboxes are made from strong, hard plastic, so this can be done fairly easily. Note the new duct must be sealed to the box — if you simply direct the pipe in the vague direction of the box intake, there's no guarantee the intake air will actually be drawn from this pipe!

Forming the whole new inlet from one piece of plastic PVC pipe can make an even smoother, better flowing duct. Use either three-inch or four-inch diameter thick-walled pipe (both available from your local hardware store) and a heat gun to soften the pipe until you can shape and bend it. Do one section at a time using small tools (like a spark plug socket) and gloved fingers to form the curves and bends. You need to be really careful that the pipe doesn't close up on any bends and that the inside surface stays smooth, but with care, a superb airbox intake duct can be formed.

The new flared outlet duct in place. After it was fitted, the measured pressure drop through the airbox was reduced.

The shape of the duct where it's open to the atmosphere is a very important factor in determining its flow. To help find which type of intake pick-up flows the best, I had three different designs flowbench-tested.

First was a straight piece of three-inch tube, cut off square at the end. The actual ID of the tube was 73mm, giving a cross-sectional area of 41.8 cm^2. At a pressure difference of three inches of water, the 215mm long tube flowed 255cfm. Next was a three-inch pipe running a straight flare. Again, the tube was 215mm long, with the length of the 11-degree flare being 75mm. The mouth opening was 115 mm in diameter, giving an intake area of about 104cm^2. The airflow with this design was noticeably improved, with a measured flow of 318cfm recorded. This represents a 25 per cent increase over the straight pipe.

Finally, a custom-made bellmouth was trialled. Again 215mm long, the bellmouthed intake used the same flare angle as the previous pipe. This time, though, a 22mm

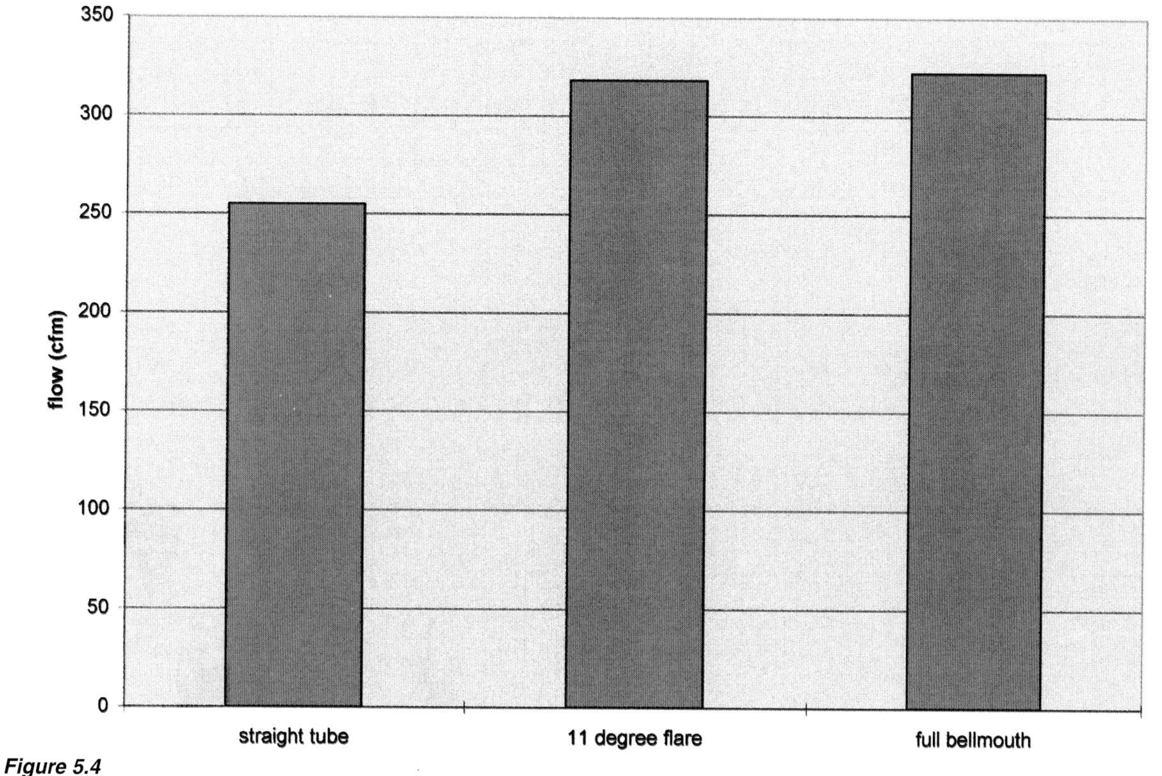

Figure 5.4
The results of flowbench tests (carried out at a test pressure of three inches of water) show that a simple flare on a pipe can improve the flow of air into it quite substantially. The extra gain with a full rolled-edge bellmouth is then quite small.

tube was rolled around the top, being brazed to the edge of the flare. Plastic body filler was used to fill any imperfections, giving a very smooth radius entry into the tube. Incidentally, the shape of this design was based on published tests that indicated this would flow very well indeed. However, the gain of this sophisticated inlet over the straight flare was only tiny. The measured result was 322cfm, up over the straight flare by only 1.2 per cent! Figure 5.4 (previous page) shows the results of this testing.

This testing indicates it's very worthwhile placing a straight flare on the end of the intake duct, but a more sophisticated bellmouth probably isn't worth the effort. A straight flare is easily made if you're using plastic pipe to form the duct. Heat the last few centimetres of the plastic pipe with a heat gun until it becomes pliable. While it's still hot, slip it over the outside of a plastic funnel and push down firmly. The pipe will be formed into a gentle flare. When it has cooled, use fine sandpaper to smooth the lip until there are no sharp edges to create air turbulence. When you have finished making the new intake duct, paint it black with a can of spraypaint.

The front-end styling of the car more than anything else determines the location from which the new duct can pick up air. As indicated earlier, a hole cut in the inner guard will allow a duct to inhale air from outside the engine bay, though note that the wheel arch area in many cars is a low-pressure area so is not ideal. If a front spoiler is fitted, there may be openings in it that can be connected to an intake duct. However, the duct should not be susceptible to picking up water, and any vehicle used in very wet climates or in conditions where flowing waterways need to be forded should avoid low-mounted intake ducts. In fact, rural cars should always use a wire grille across the mouth of the intake duct to prevent flying insects filling the airbox!

Facing the intake duct forward so it's pressurised by the car's forward movement works quite effectively. This is a little odd because the physics indicate that it's only at very high speeds (above 200km/h) that this **should** become effective. However, many sedan-based racing cars take elaborate steps to use forward-facing intake ducts of really massive proportions and it has also proved to be effective on road cars operating at quite modest speeds. So where this can be achieved, a forward-facing duct should be used.

How well, then, can a new inlet duct into the standard airbox actually work? A car featuring a 5-litre V8 was dyno tested. The engine was breathing through just the standard over-radiator intake and in this form power peaked at 125kW (168hp) at the wheels. Next, this intake was blocked off and a new three-inch PVC plastic stormwater pipe passing through the inner guard was used to feed air to the box. Power immediately jumped to a peak of 128kW (171hp) — a gain of two per cent. Next, both the new intake **and** the standard intake were opened up together. Power rose to 131kW (176hp) — a five per cent increase. This is a very good result for such a cheap and simple modification.

The standard intake to this airbox is very small. Aim to make the cross-section of the snorkel duct at least 45cm² — about the equivalent of a three-inch (76mm) pipe.

Measurements were also carried out on a 2-litre four while the car was being road tested. At peak power, the standard car had a total intake pressure drop of 25cm of water. Further, on a 15-degree C day, the inlet air temperature was as high as 47 degrees C. A new three-inch plastic pipe inlet was made for the airbox, picking up air inside the guard from behind the slotted bumper. A small flared bellmouth was also made from plastic pipe (using the heat gun technique described above) and this was placed in the exit tube of the standard airbox. These measures decreased the pressure drop by 40 per cent to 15cm of water and dropped the intake air temperature by as much as 29 degrees C! Throttle response improved noticeably as a result of these changes.

If you decide to replace the airbox with a completely new design, look for a box like this. The inlet and outlet ducts are beautifully contoured and the box flows very well.

The intake air system cannot be over-sized. This Audi 2-litre Touring Car uses an enormous forward-facing duct with a huge duct connecting it straight to the airbox.

After the Filter

An important point to keep in mind is that flow losses are cumulative in nature — they add up one after the other. This means **any** inlet restriction is bad. Some cars follow the airfilter box with a resonant chamber and, while it generally causes little pressure drop, if you're after every single improvement possible, you can replace it with a length of pipe. Note you should not use PVC pipe within hot areas of the engine bay — the pipe will soften and sag at temperatures over about 80 degrees C.

The duct that connects the airflow meter or filter box with the throttle butterfly in naturally aspirated cars is usually quite long. It may also have a tight bend or two along the way, so can cause an airflow restriction when the engine is developing high power. If manometer testing **does** prove it restrictive, replace it with a larger-diameter tube using more gentle bends. Exhaust tubing and mandrel bends can be used for this job.

An example of this sort of modification was carried out on a six-cylinder Ford, standard but for extractors and a hi-flow exhaust. The car used a dual-path plastic duct to connect the airbox outlet with the throttle body. Manometer testing indicated that this duct caused about half of the total pressure drop prior to the throttle body. The duct was replaced with a three-inch mandrel-bent steel tube. The measured peak power immediately increased by four per cent. It's interesting to note that the next model of this car replaced the dual-tube intake with a single large-diameter duct! However, as indicated earlier in this chapter, changing the design of the duct between the airbox and the throttle body may have unforseen consequences if this duct is part of the tuned-length intake system.

If you fit a large intake duct that picks up air from the front of the car, the filter will become dirty faster than with the standard intake. This is the result of driving through a swarm of locusts with a four-inch free-flowing intake fitted to the airbox!

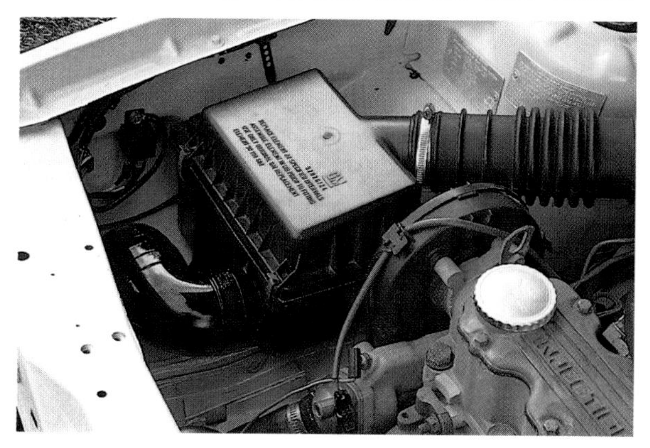

Here a new intake duct has been fitted to an airbox. The new intake draws air from within the guard. It was made from three-inch (76mm) PVC bends and pipe, and both improved the intake flow and dramatically dropped intake air temps.

Beware of normally hidden restrictions such as this resonating chamber (nearest to camera) within the guard of a Subaru Liberty (Legacy) RS. In this car, all air is breathed through the very restrictive box.

Finally, the airflow meter may prove to be one of the most restrictive parts of the entire induction system — especially if it's of the vane type. The best approach in this situation is to replace it with one that's larger or of a different, less restrictive design. This topic is covered in Chapter 3.

If it can be integrated into the car's styling, a bonnet scoop makes a good source of cool intake air. A NACA scoop like this works well when it's positioned in an area of attached airflow.

THE INTAKE MANIFOLD

The system that directs air to each cylinder and regulates how much air is breathed by the engine can have huge implications for performance. While modification in this area is rarely done, with good research and development, excellent gains can be made.

Throttle Body

The throttle body controls the amount of air that can enter the engine. Most cars use a single throttle placed at the beginning of the plenum chamber but some use dual or even triple throttles. In the latter case, the individual throttle blades are often contained within the one casting. In multiple-throttle engines the blades are usually opened in sequence, with the second (and/or third) starting to open after a pre-determined opening angle of the first blade. The exceptions to this are those engines that use one throttle blade per cylinder. These are normally located close to the cylinder head and open simultaneously.

If the engine power has been lifted considerably, the throttle may be posing an undue restriction even when it's fully open. The easiest way to measure whether this is the case is to check the manifold pressure at full power. In a turbo car, measuring the boost pressure either side of the throttle will show if there is a restriction — the two figures should be nearly identical. In a naturally aspirated car equipped with a Manifold Absolute Pressure (MAP) sensor and programmable engine management, the manifold pressure will be able to be read off on the laptop PC screen. It should be very close to atmospheric pressure — a maximum of 3kPa lower than atmospheric pressure. In a car running standard factory engine management, direct measurement will need to be made. This can be done using a manometer as described above, but with one important difference. The manometer sensing tube should pass through an in-cabin tap so the manometer can be isolated from the plenum chamber except when full throttle is used. Otherwise, manifold vacuum will suck all the water out of the instrument!

Intake Manifold Design

An engine constantly starts and stops its intake airflow as the intake valves open and shut. When the piston descends, it creates a low pressure in the combustion chamber that causes a negative wave to race along the intake port and manifold runner. In a naturally aspirated

The connecting duct from the airbox to the throttle body can sometimes be restrictive. This Ford used a dual tube intake that was manometer measured as having a significant pressure drop...

...so it was replaced with this three-inch (76mm) mandrel-bent steel pipe. (Paint came later!) Peak power improved by 4 per cent.

Interestingly, the next model of the Ford had as standard the single large-diameter duct you can see here! Each of these cars has the superb dual-length, changeover intake manifold shown on page130.

This Honda Civic race car runs exposed intakes with cool air being supplied by the convoluted tube ducts. Taking this approach loses the chance of making use of the Helmholtz resonant intake tuning that's possible with an enclosed plenum.

This Ford Cortina drag car uses six throttles and an enclosed plenum chamber fed by the large hose that goes to a removed headlight. Closed, it looks like this...

...while with the plenum cover removed, it looks like this! Note how the length of the intake runners has been extended as far as possible and the superb bellmouths used on each intake.

car, the wave is below atmospheric pressure while in a forced aspirated car the wave is at a pressure of less than manifold pressure. When the wave reaches the plenum chamber, it's reflected back towards the engine. This returning wave has the potential to help ram more air into the combustion chamber, but only if it reaches the valve when it's again open! (As you'd expect, the motion of the pressure waves within the intake is considerably more complex than this, with this description more a general model than the reality.) The trick is to design the system so the reflections arrive back at the intake valves at the right time — helping to push air into the engine.

The gains available from a well-tuned intake system should not be underestimated. Pioneering work done by Jaguar on their mechanically injected racing engines showed it was possible to gain **more than** 100 per cent volumetric efficiencies (VE) by using very long intake runners. This means the cylinders actually breathed in more than their swept volume! Many manufacturers have since developed efficient tuned intake systems that give very high volumetric efficiencies. As an example, Ford in Australia developed a superb dual-length intake manifold for their straight six, a system that achieves a VE of 100 per cent at the peak torque of 3000rpm. Power and torque gains of up to 30 per cent have been seen in some engines — and that's as much as is achieved by low-boost turbocharging!

The three variables that can be changed in designing an intake system are runner length, runner diameter and plenum chamber volume. Jaguar found with their experiments that the longer the intake runners, the better were the peak torque outputs, and other manufacturers have seen similar results. However, very long runners cause an overly great pressure drop and are not properly tuned in length for peak power, so explaining the changeover characteristics used in some sophisticated intake systems.

A starting point for working out the length and diameter of intake runners can be gained from the following equations. In a Helmholtz Resonance system (one with runners connected to a common plenum), US-based engineering guru David Vizard suggests that a runner length of seven inches at 10,000rpm makes a good starting point.

This Ford Pinto engine had radically different power curves, depending on the length of intake trumpet fitted. Lower rpm power variations of 30 per cent were experienced! Fig's 5.5, 5.6 and 5.7 overleaf show some of these variations.

(Note the runner length is the distance from the plenum chamber right to the intake valve seat.) Add to this another 1.7 inches for each 1000rpm **less** than that which the system is being tuned for. Tuning for peak torque (not peak power) is the norm, so if the engine were being tuned for 4000rpm, a runner length of 17.1 inches (43.6cm) would be required.

The very good book *Desktop Dynos* suggests the intake system can be tuned for the second, third, fourth or fifth reflected pulse. So this is done, the following intake runner lengths should be adopted:

Second pulse length (inches) = 108,000 divided by rpm

Third pulse length (inches) = 97,000 divided by rpm

Fourth pulse length (inches) = 74,000 divided by rpm

Fifth pulse length (inches) = 54,000 divided by rpm

You will soon see that for an average-sized engine bay, often the longer the runner that can be fitted in, the better!

One equation for runner diameter is to multiply the engine volume in litres by the engine's volumetric efficiency, then by the tuned rpm, then divide this sum by 3330. The final figure is square rooted, giving the runner diameter in inches. As an example, a 5-litre engine with an 80 per cent efficiency (expressed as 0.8) and tuned for 3000rpm, will have a runner diameter of 1.9 inches (48mm). The volume of the plenum should be around 80 per cent of the volume of the cylinders to which it's connected. Most intake systems use runners that slightly taper down in size as they approach the intake port.

There are also several computer programs around that can help you design a new intake system. These include *Engine Analyzer*, which can calculate runner length and diameters. Note, though, that a lot of information about every detail of the engine needs to be input before this program can start to work. *Controlled Induction* and *Controlled Induction Junior* can also be used in this area, while probably the best software for designing intakes is *Dynomation*.

However, as with extractor design, nothing beats dyno experimentation with different intake designs. The complexities of an internal combustion engine are very difficult to model, and with factors such as engine bay room, pressure drop if intake runners are overly long or small in diameter and individual engine-to-engine variations, the equations are only a rough guide. If making prototype intakes, use mild steel tube and sections of rubber hose. The lengths of runners can then be readily changed and the results tested on the dyno or against the stopwatch on the road. When it comes time to build a 'proper' prototype, use exhaust tube and MIG welding — so much the better for mocking up a trial intake than TIG'd alloy!

Extensive testing was carried out on the intake system of a Ford 2-litre SOHC Pinto four-cylinder engine. The very modified engine (cam, compression, head work) used quad Weber-style throttle bodies and Haltech engine management. For the first series of intake tests, the engine was fitted with 4-into-2-into-1 extractors (headers) and an open exhaust. In all tests, the engine was revved from 3000rpm to 8200rpm in third gear.

Figure 5.5 shows the dyno results of tests undertaken with different-length intake trumpets. The red line indicates the power output with a length of 12 inches (30.5cm) from the intake valves to the end of the Weber-style trumpets, which were open to the atmosphere. In this form, power peaked at 98kW (131hp) at the wheels, with a slight dip in the mid-high rpm power curve. Next, trumpets 14.5 inches (36.8cm) long were fitted. Power dropped substantially, peaking at 88kW (118hp), as shown by the blue line. Finally, a trumpet length of 13.5 inches (34.3cm) was tried. This gave the best results (green line), with power peaking at 100kW (134hp) and with no dips in the power curve. This test indicated a change in peak power of 14 per cent with a trumpet length variation of just 2.5 inches (6.3cm)!

The car was then fitted with 4-into-1 extractors (headers) and further intake system testing was carried out. Figure 5.6 shows the results (with the power curves at the top, tractive effort curves at the bottom). The red line indicates the power achieved with trumpets 12.5 inches (31.8cm) long. Power was improved over the previous test achieved with (approximately) this length of trumpet probably because the 4-into-1 extractors were working better than the 4-into-2-into-1 design. A trumpet length of 20.6 inches (52.3cm) was then tried. This required the use of mandrel-bent 90-degree bends so the runners could be fitted in the engine bay. In this form, the engine developed the power curve shown by the brown line in Figure 5.6 (opposite).

It's worth looking at the power curves of Figure 5.6 closely. The long trumpets gave a staggering increase in bottom-end/mid-range power, lifting it by as much as 32 per cent! However, at high revs, power was down by as much as 8 per cent over the short trumpets. The lines crossed at about 5500rpm, meaning that for a road (or rally) car, the long trumpets would have been ideal. For a drag car (the purpose of the buildup) top-end power is required, meaning the short trumpets were preferred. Obviously a dual-length intake manifold that changed from long to short trumpets at 5500rpm would have given the best of both worlds.

Bare intake trumpets are not very practical on a road car (or a road car/drag car!) so consideration was given to the filtering system to be used. Previous testing had shown that slip-on oiled foam filters can disrupt the trumpets' intake flow significantly, so the use of a remotely filtered airbox placed over the trumpets was a likely scenario. But how would this affect the intake tuning? With the long 20.6 inches (52.3cm) runners still in place, a cardboard airbox with a 5.6-litre volume was fabricated over the intake Weber bellmouths. This was initially left open at one end.

Figure 5.7 (opposite) shows the results. The green line is the power curve before the airbox was placed over the intakes. The blue line shows what happened with the airbox in place — there was a power increase at about 5500rpm (110km/h in third gear) but also a major dip in the power curve at about 6500rpm. The presence of the

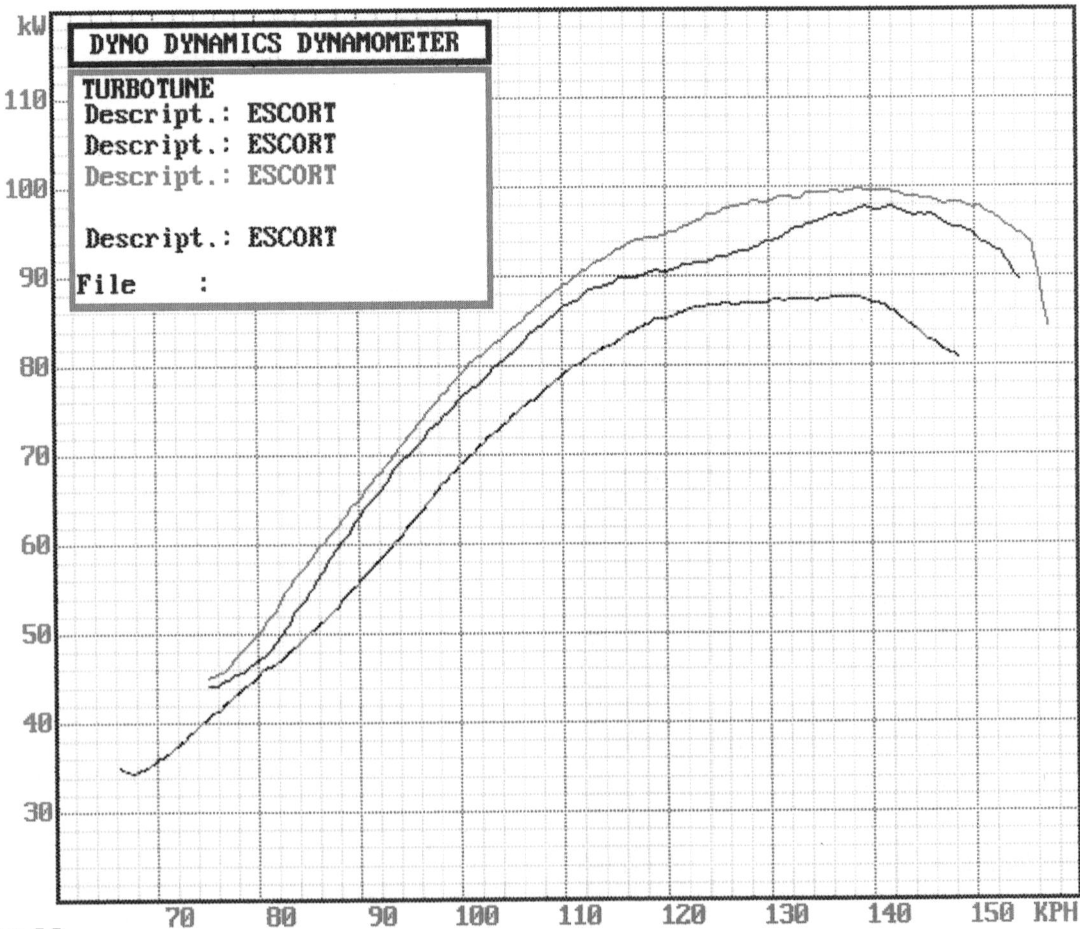

Figure 5.5
A power variation of 14 per cent was experienced when intake trumpets 12, 13.5 and 14.5 inches long were fitted to a multiple throttle body 2-litre Ford Pinto engine. In this case, a trumpet length of 13.5 inches (34.3cm) gave the best results, with power peaking at 100kW (134hp) and with no dips in the power curve.

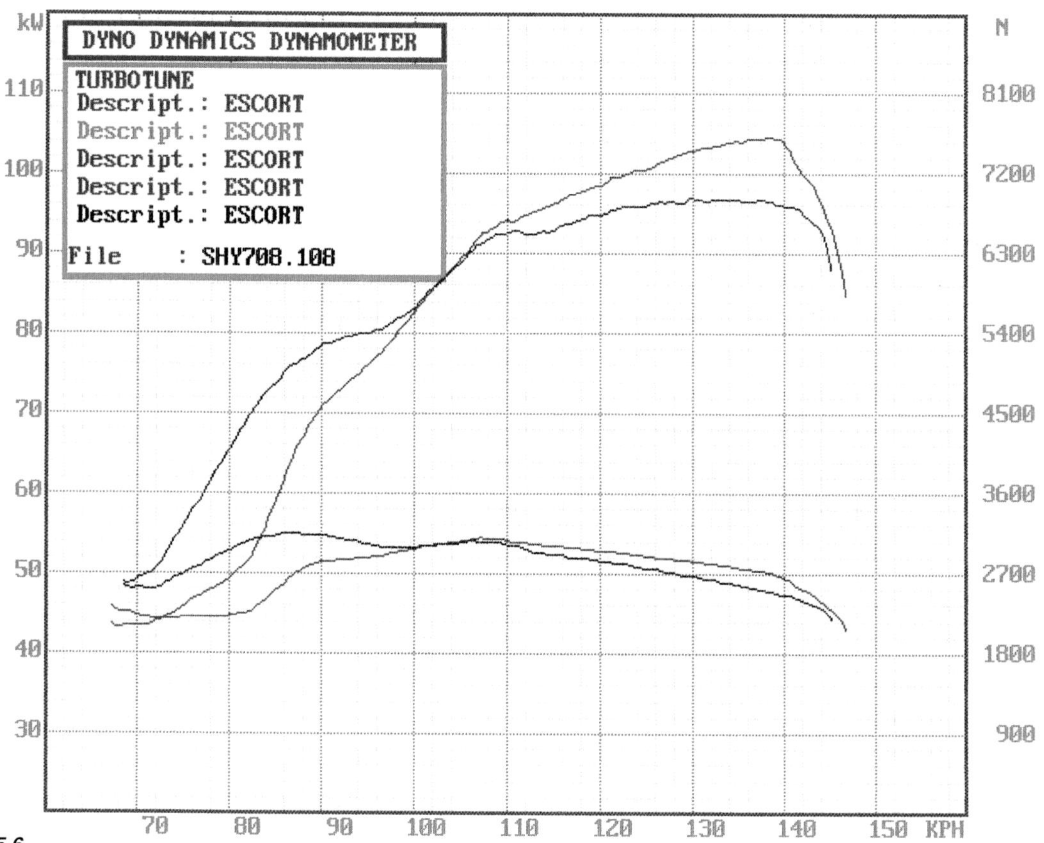

Figure 5.6

Fitting long trumpets to the 2-litre Ford gave a staggering increase in bottom-end/mid-range power, lifting it by as much as 32 per cent! However, at high revs, power was down by up to 8 per cent over the short trumpets. The lines crossed at about 5500rpm, meaning that for a road (or rally) car, the long trumpets would have been ideal.

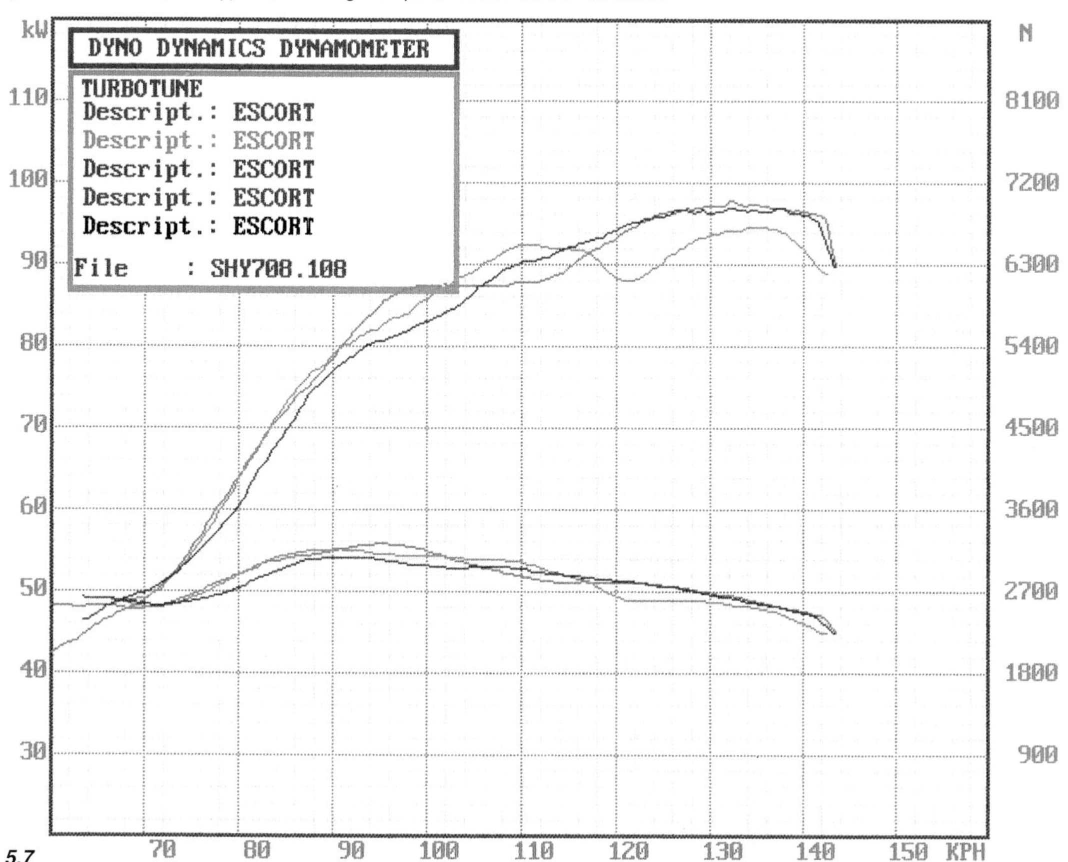

Figure 5.7

These dyno power curves (the top three lines) show the result of placing an airbox over intake bellmouths. The green line is the power curve before the airbox was placed over the intakes, while the blue line shows the result with the airbox in place. There was a power increase at about 5500rpm (110km/h in third gear) but also a major dip in the power curve at about 6500rpm. The end of the airbox was then closed off, with a 10.6-inch (27cm) three-inch diameter tube used to feed air to the airbox. The red line shows the power curve that resulted.

airbox definitely changed the tuning of the system. The end of the airbox was then closed off, with a 10.6-inch (27cm) long, three-inch diameter tube used to feed air to the airbox. A bellmouth was placed on the exposed end of the feed tube. The red line in Figure 5.7 (previous page) shows that installing the feed tube smoothed the peaks and dips out of the power curve. Using the long runners, enclosed airbox and 10.6-inch intake duct gave a smooth power curve with — for this type of engine — good mid-range torque.

Importantly, each of the intake tuning changes had a significant effect on the power curve.

With the Lotus bodywork removed, the huge flexible intake air duct can be seen running to the large plenum.

On the Lotus, the carbon-fibre plenum works with a quad throttle intake system. For an engine that develops only 142kW (190hp), all these intake air system components would seem to be oversized, but this isn't the case.

Variable Intake Manifolds

It can be seen from the above example that variable intake systems are likely to give major gains throughout the rev range. Variable intake systems change runner length or plenum volume, depending on engine rpm or load. This allows the intake to have more than one tuned rpm — giving better cylinder filling at both peak torque and peak power, for example.

The system can be variably tuned in a number of ways, including (especially on six-cylinder engines) connecting

The Lotus Elise uses a mid-mounted engine that's fed combustion air from the side duct positioned just behind the door. Its air intake system is shown in the pictures above.

Saab

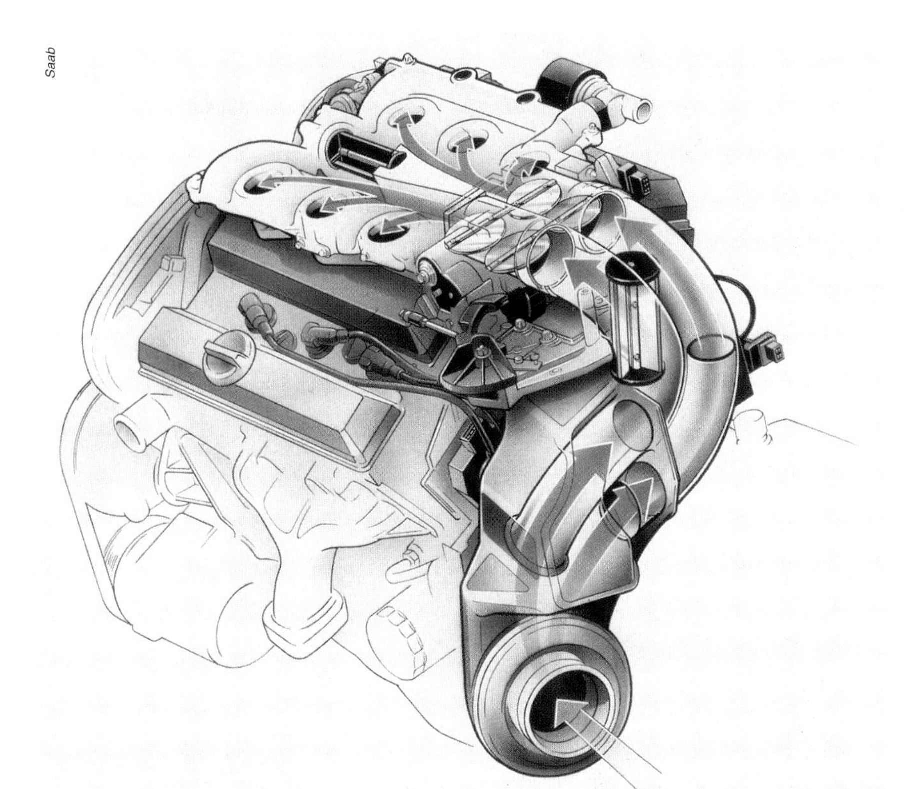

Figure 5.8
The variable intake system used by Saab on the 3-litre V6 9000. Two butterfly valves are positioned in the intake system, effectively changing the length of two main intake passages.

twin plenums at high rpm but having them remain separate smaller tuned volumes at lower revs. The introduction of a second plenum into the system at a particular rpm is another approach taken. However, the most common method is to have the induction air pass through long runners at low revs and then swap to short runners at high rpm. Note this doesn't mean the long runners need to be positively closed — opening parallel short runners is often sufficient to change the effective tuned length of the intake system.

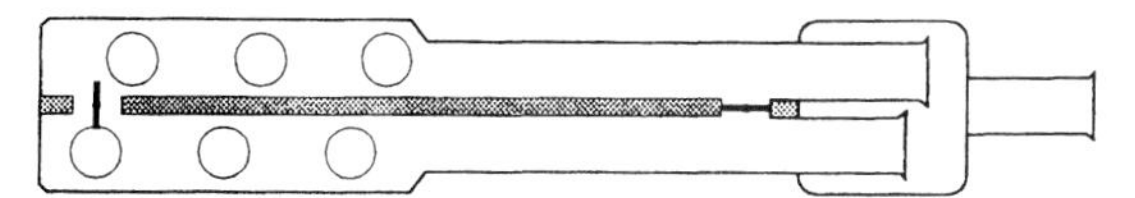

At low rpms: #1 valve open for idle air equalisation, #2 valve closed.

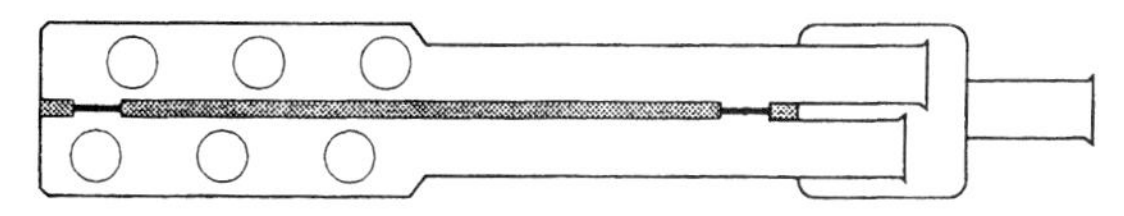

Up to 3200rpm: Both #1 & #2 valves closed. Longest induction length.

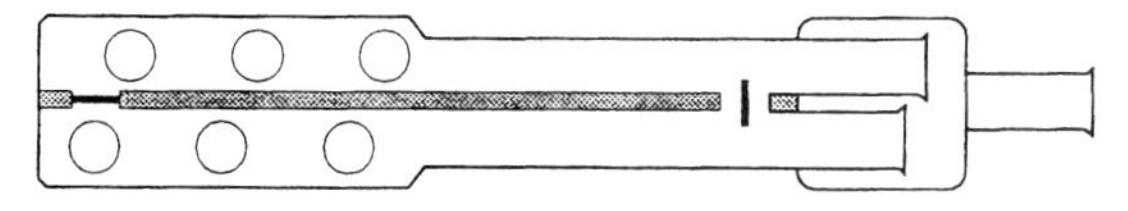

From 3200 to 4100rpm: #1 closed & #2 open. Shorter induction length.

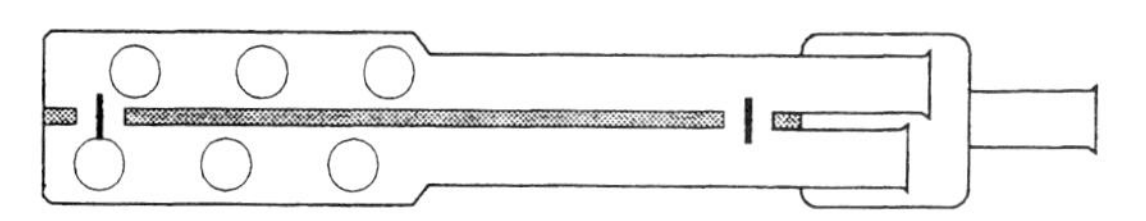

Above 4100rpm: Both #1 & #2 valves open. Shortest induction length.

Saab

Figure 5.9
Valve #1 in the Saab system joins the two intake passages near to the cylinder runners, while Valve #2 effectively shortens them.

One example of a variable intake system is that used by the 3-litre V6 Saab 9000. The system feeds each bank of cylinders with a separate plenum, with each plenum in turn fed by a long intake. The plenums can be acoustically joined by the opening of a valve, with a second valve effectively shortening the dual runners. Figure 5.8 shows a cutaway view of the engine, while Figure 5.9 shows the layout of the system. The valves are operated in the following manner:

- Idle to 2000rpm: Valve #1 is open (to allow equalisation of the air from the idle air control valve) and Valve #2 is closed.
- 2000-3000rpm: both Valves #1 and #2 are closed, giving the longest induction length.
- 3000-4100 rpm: Valve #1 is closed and Valve #2 is open, giving a shorter induction length.
- Above 4100rpm: Both Valves #1 and #2 are open, giving the shortest induction length.

Figure 5.10 (overleaf) shows another view of the system and the relationship between the intake system valve positions and the engine's torque curve.

Intake Manifold Layout

Given the intake runners need to be long and the plenum chamber generally large, some thought needs to be given to how everything will be fitted in under the bonnet. Different engine configurations need different approaches, with V engines the hardest. Because of potential underbonnet clearance problems and the

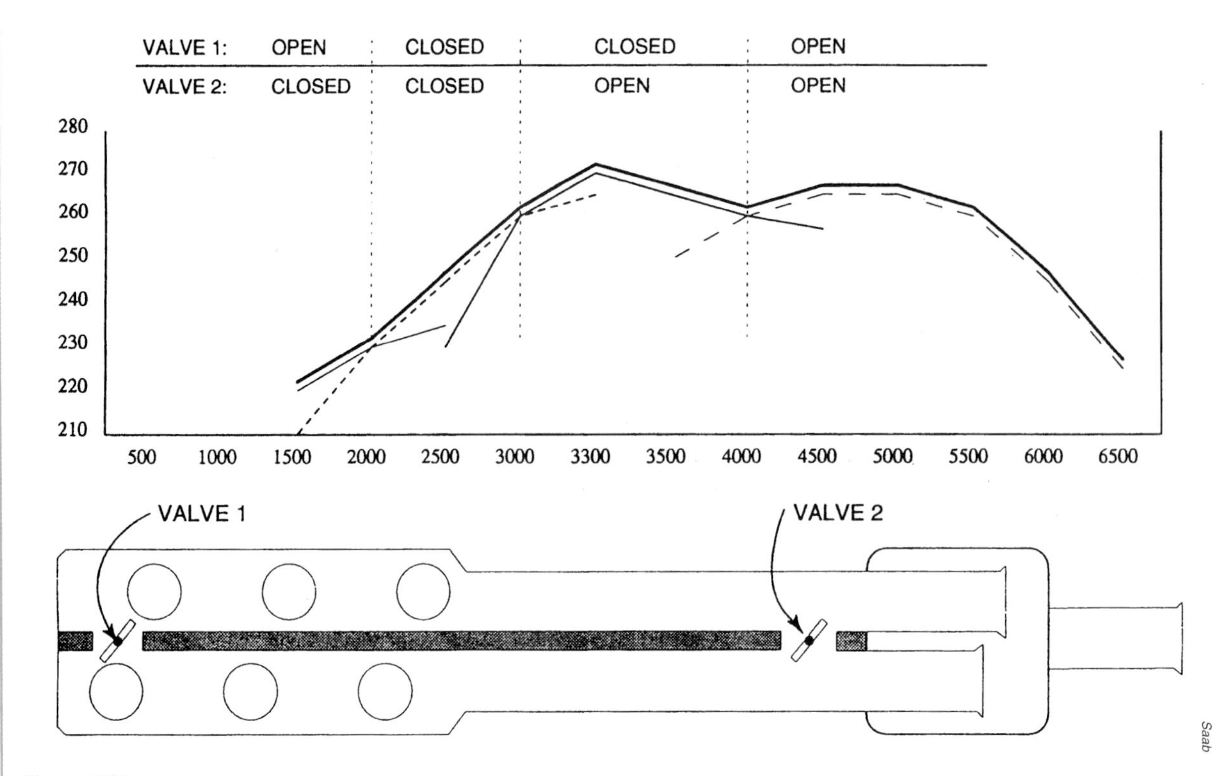

Figure 5.10
The torque curve shows clearly how the opening and closing of the valves have been matched to rpm so the maximum torque possible is realised at all engine speeds. Variable intakes like this can be controlled by programmable management systems, with dyno testing used to show the best valve opening and closing revs.

large number of intake runners required, the runners for a V8 are very limited in length if they're to remain straight. As a result of this, many manufacturers bend their V8 intake runners through 180 degrees, placing the plenum at the base of the runners in the valley of the V. For example, both Holden and Mercedes do this. However, while this is good for mass-market passenger cars that require a lot of bottom-end torque, peak power suffers because of the pressure drop caused by the long, relatively small runner diameters.

'Straight' engines can use curved runners that pass over the rocker cover to give a very long intake design. This approach was first used by BMW on their injected sixes in the early seventies. However with crossflow heads, this also means the plenum ends up on the hot (exhaust) side of the engine so heat shielding is needed. Using runners that dip well downwards before turning around to come up to the plenum can lengthen the effective runner length while keeping the plenum chamber on the cool side of the engine. You need only look at the tricks manufacturers go to in the pursuit of long intake runners to realise their importance!

The throttle body in most cars consists of a single throttle blade mounted at the entrance to the plenum chamber. This is the throttle body on a Nissan RB20DET six-cylinder engine.

If measurement indicates that the throttle body is restrictive, it can be replaced by a larger unit. Here an XE Ford Falcon unit has been fitted to a Nissan RB30ET 3-litre turbocharged six. The engine developed over 335kW (450hp).

Where the runners enter the plenum, bellmouths should be used to smooth the flow of air into the tube. A metal spinner can spin custom-fabricated bellmouths out of aluminium, with these then being inserted into the ends of the runners.

Many people suggest that in a turbo car the entrance to the plenum should not be directly opposite any of the runners; it should face a blank wall so the airflow collides with it, spreading evenly within the plenum. If the intake pipe is aimed straight down one of the runners, that cylinder may receive too much air, giving it a leaner mixture. Most manufacturers use shorter runners on the turbo versions of their cars, often with a larger plenum chamber.

The air intake system on this Hyundai race car would be perfect on a road car — large-radius bends, big-diameter ducts and a well-designed airbox.

Intake Manifold Construction

Both the runners and plenum chamber should be constructed of heavy-duty materials. This is because the very powerful pressure wave activity occurring within the system can cause fractures and cracks in lightweight designs. The runners can be made from mild steel exhaust tubing, with the plenum folded up from sheet steel and the head flange made up like an exhaust flange. A more expensive — but better looking — approach is to fabricate the system from aluminium sheet and tube. The finished item can be polished, powder-coated or painted, depending on the materials used and the budget available.

Standard road cars always have the injectors placed close to the intake valves, but many race cars move the injectors much further back along the intake runners. It's suggested that the greater evaporation of the fuel that occurs when the spray is positioned well back from the head aids intake air cooling, so increasing its density. In fact, I have seen a race engine on an engine dyno that used this approach. After running hard, its carbon-fibre intake runners were so cold that condensation was streaming down them! However, in a road car this distant injector location is likely to lead to poor driveability at low loads and the greater possibility of an engine fire if a backfire occurs.

Ford

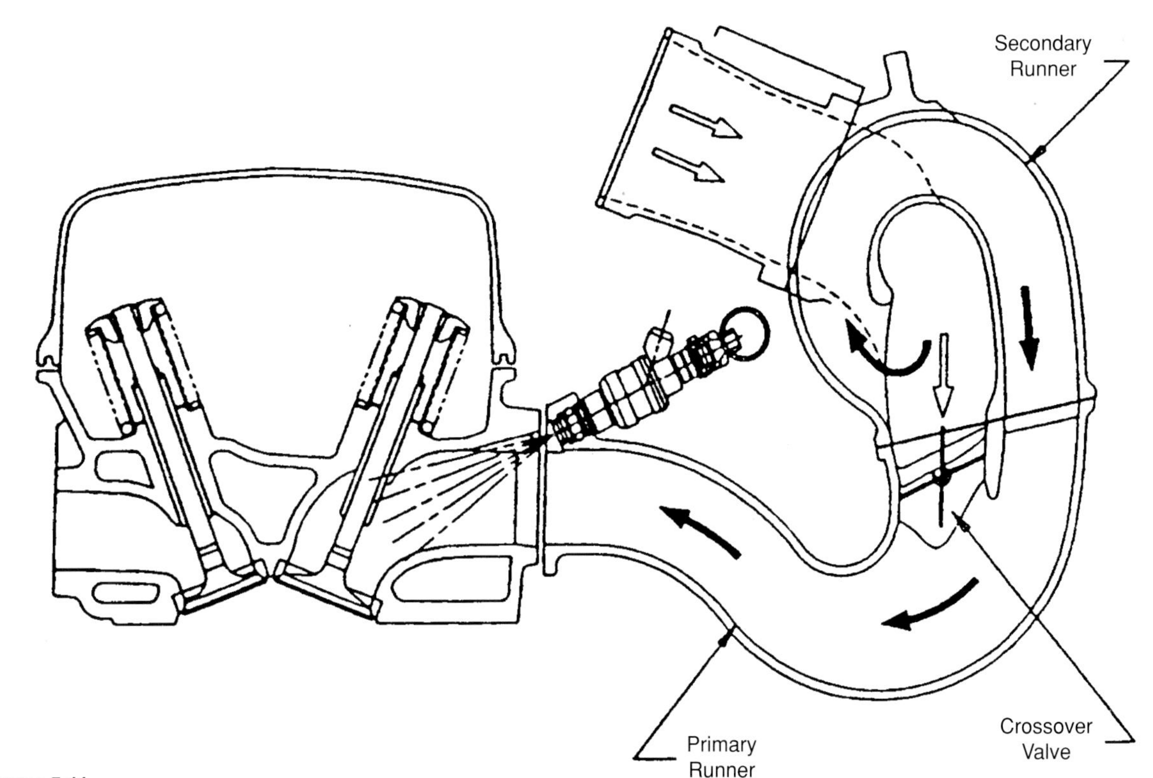

Figure 5.11
The variable-length intake manifold fitted to the Ford Falcon six-cylinder engine is a superbly compact design. The secondary (long-length) runners actually wrap around the plenum chamber.

The BMW M3 uses six throttles and short, large-diameter runners. This inlet system is designed very much for high power outputs rather than an emphasis on mid-range torque.

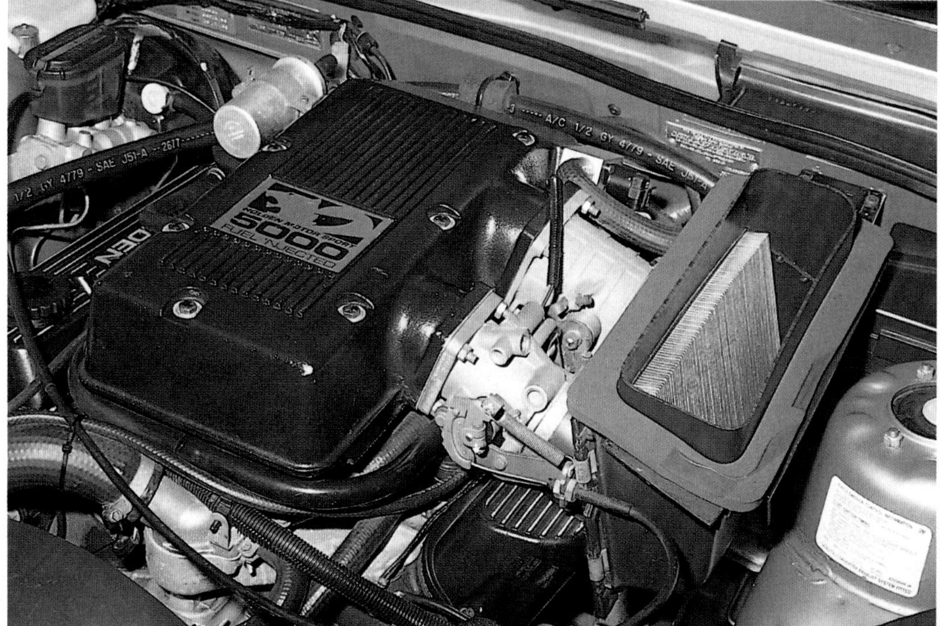

The Walkinshaw Group A VL Holden Commodore uses a very large plenum chamber, straight intake runners with bellmouthed trumpets, and dual sequential throttle bodies. For sheer power, the intake system is superb, but engines equipped with it have less bottom-end torque than those using the standard long-runner intake.

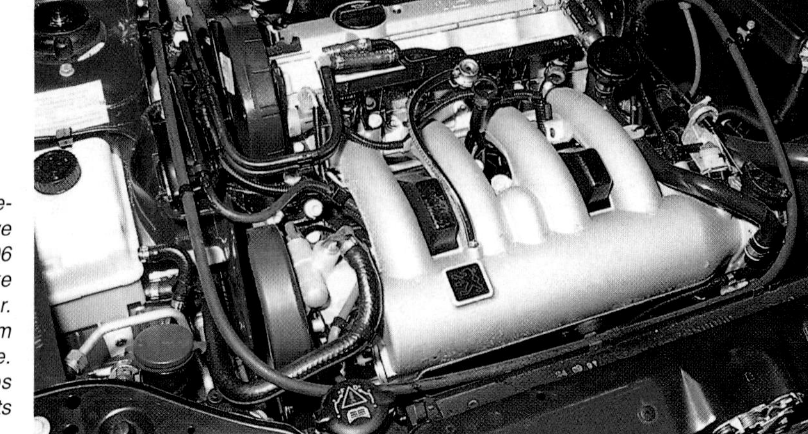

Unlike many transverse-engine front-wheel-drive cars, this Peugeot 306 GTI6 places the intake at the front of the car. The very large plenum chamber is noticeable. The engine develops 124kW (166hp) from its two litres.

For maximum power, the slide throttles used on this 2-litre Audi Touring Car are unparalleled. This is because when they are fully open, nothing remains to obstruct the inlet air path.

This 2-litre turbo Nissan V6 uses twin throttles, each of which feeds a separate small plenum chamber. In some engine operating conditions, a third butterfly connects the two plenums. The complex design of this type of tuned intake can really only be proved through experimentation.

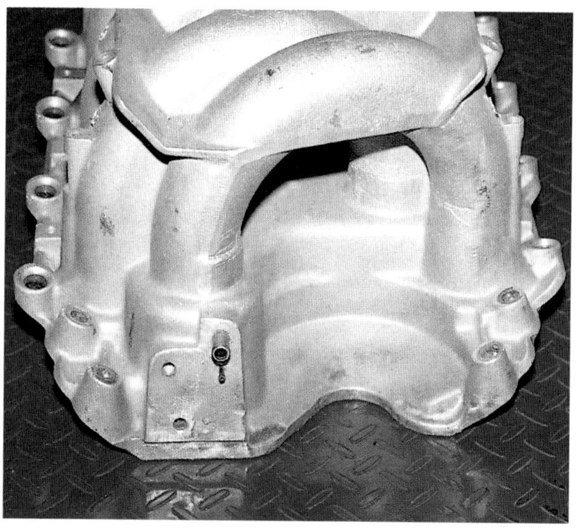

The intake runners on this V8 are about 20 inches (51cm) long. They help give the engine excellent low-down torque but also limit top-end power when modifications are made.

The Porsche flat six uses dual plenums and very large-diameter, beautifully curved runners. One of the two plenums is shown here.

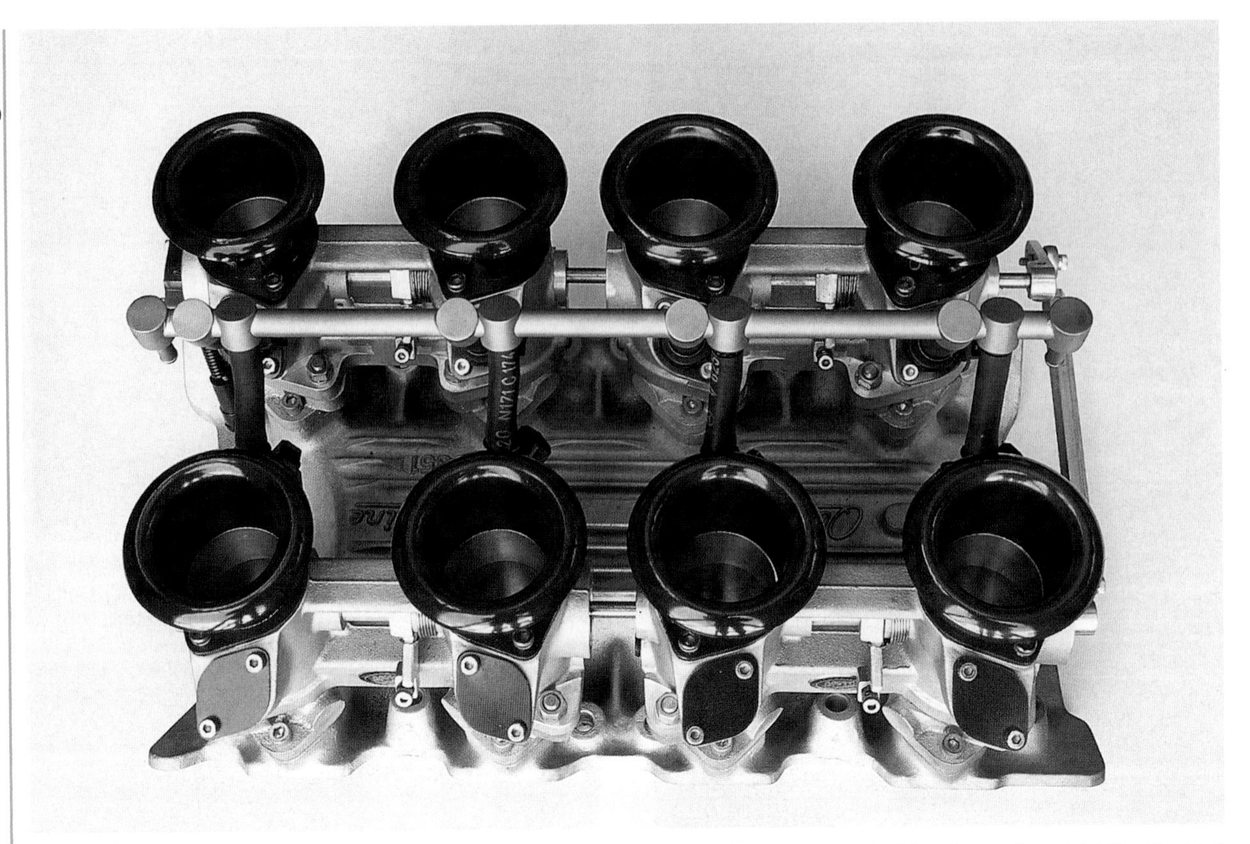

Multiple throttles allow the greatest amount of airflow while still retaining fine throttle control. This aftermarket eight-throttle inlet manifold is for a V8. Note how the manifold has provision for two injectors per cylinder.

Some factory manifolds use bolt-on injector mounts, making it easy to adapt these to custom intake manifolds. A turned-up fitting welded to the runner can also be used, but taking this custom-everything approach can be very time consuming. Holding the injectors in place with the fuel rail is a neat and easy way of avoiding the need for multiple tapped fittings on each runner.

MODIFYING THE STANDARD INTAKE MANIFOLD

The standard intake manifold and plenum can be modified. Extrude Honing, a special process where an abrasive solution is forced through the intake runners, can be used to smooth and enlarge the intake system. Another approach is to cut open the plenum chamber and port the runners with a die grinder, before welding the plenum

Car manufacturers sometimes have different intake systems available for the one engine. This Alfa Romeo flat four intake uses just a single throttle body...

...while this version uses a larger plenum chamber and quad throttles. Selecting the best intake system made by the manufacturer gives you a head start.

back up again. However, I have seen some expensive failures occur when the intake system has been ported by hand in this way. Gains of only a few per cent in power mean the time could have been better spent making a completely new intake system.

One of the most time-consuming tasks in making a new intake is the marking and cutting out of the new manifold plate, and then the welding of port-matched intake runners to it. This step can be avoided if the engine uses a two-piece intake manifold, where the plenum chamber and upper portion of the runners separate. In this case, a new plenum chamber (perhaps with revised throttle bodies) can be easily made and attached to the shortened runners. Airflow into the runners can be encouraged by the use of push-in spun alloy bellmouths. This type of modified standard intake will usually achieve better top-end power than the standard system.

Not all OE intake manifolds use cast alloy. This Toyota engine uses a tiny plenum chamber and tubular steel intake runners.

This 5-litre V8 racing engine uses a crossover design with almost straight inlet runners and eight throttles. Note the spun alloy bell-mouths and the foam rubber that seals to a large airbox.

An underbonnet view unlike most! The owner of this turbocharged Mini Moke has made his own inlet manifold using a simple small tubular plenum and short runners. The use of mild steel makes for an easy fabrication.

In order that intake air velocity be high at low loads (so increasing cylinder filling and therefore torque), this Nissan RB20DET six uses butterflies to control airflow through one of the two intake ports used for each cylinder. At low loads, the butterflies are shut, while at high loads both intake ports are open.

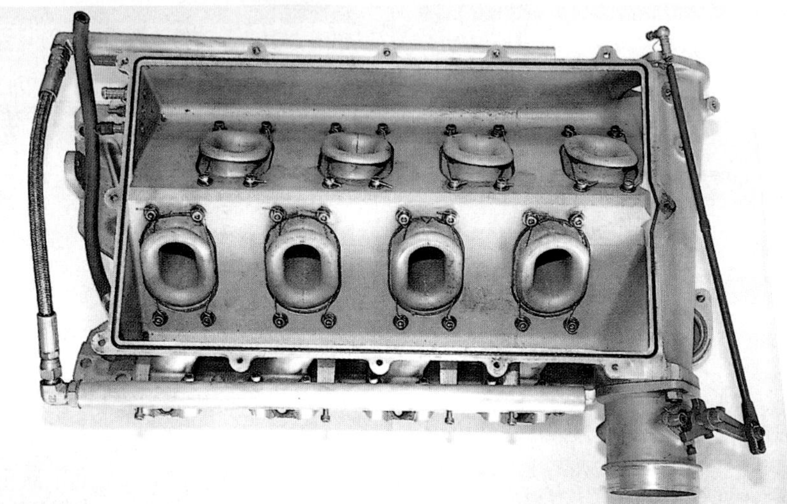

The MoTeC intake manifold for the Holden V8 uses short straight runners and a single plenum largely divided by a central ridge. Twin throttles (one is missing in this view) are opened sequentially.

The multiple throttle bodies from a 20-valve Toyota 4A-E 1.6-litre engine have been adapted to a Nissan A14 engine that uses mechanical injection.

The fitting of a new plenum to the original intake runners has been done here. Taking this approach makes the production of a new intake system straightforward.

Another approach is to adapt a carby manifold to EFI use. This will result in an intake system that uses large-diameter, relatively short runners. A multiple throttle assembly can be fitted where the carburettor normally sits. The injectors can be placed either in this throttle assembly, or the manifold can be drilled and tapped to accept injectors on each intake runner.

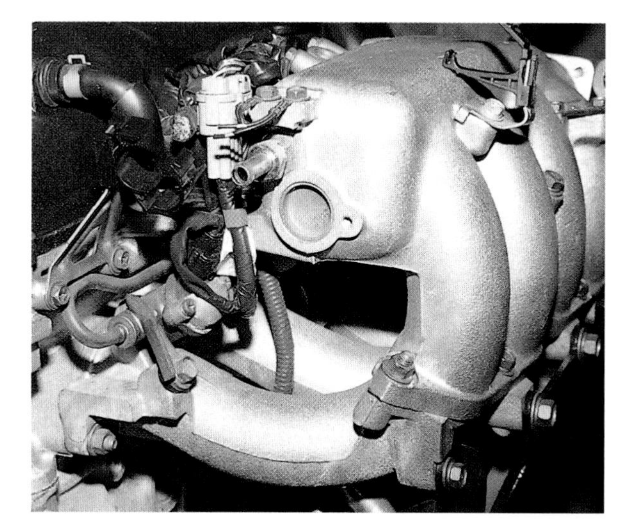

The long runners used in a typical engine can be clearly seen here. This is a Nissan SR20DET turbo 2-litre four developing 147kW (197hp) in standard form.

If a higher torque design is needed than that provided by the new car manufacturer, you're probably not in luck! This is because nearly all road car engines use intake systems designed to bolster as much as possible bottom-end and mid-range torque. Making improvements in this area over the standard manifold will be extremely difficult.

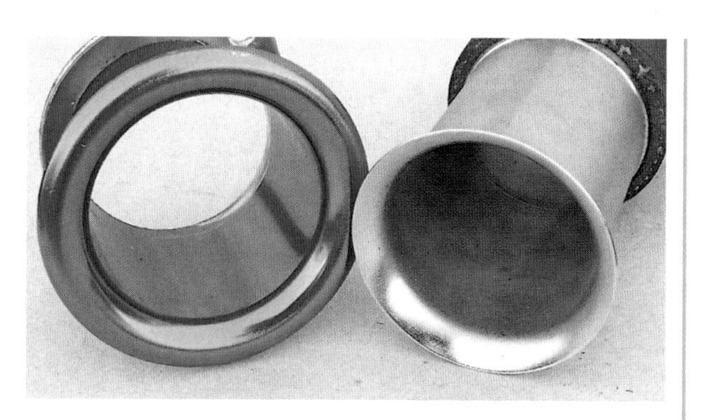

The shape of the intake bellmouth can affect the power developed by the engine. Testing carried out on a Ford Pinto engine showed the Weber-style trumpets on the right to be more effective than the bellmouths with the full rolled edge shown on the left.

This is a V8 carby intake manifold adapted to EFI using a new throttle body. Note how the runners have been modified to accept injectors, making it a multipoint manifold.

If the standard plenum can be unbolted from the intake runners, a new plenum and throttle system can be fabricated much more easily. This engine has a plenum that easily separates from the runners.

Exhaust Systems

The fitting of a new exhaust is one of the most common power upgrades made to cars. It is also the most effective, with good results coming from the fitting of a free-flow exhaust on almost every car. So why do the car manufacturers make such bad exhausts in the first place? It's because they need to satisfy the conflicting criteria of low noise, light weight, low emissions and low cost. The first means that original equipment exhausts are very quiet, but as a result, often restrictive to flow. Good emissions performance requires the use of relatively small (and so restrictive) catalytic converters placed near the engine; and low cost means press-bent pipes are used and other corners are cut.

EXHAUST FUNCTION

An exhaust system has three main functions:

- To transfer the noxious gases resulting from combustion to the rear of the car, so preventing the poisoning of the car's occupants
- To quieten the exhaust note
- To aid wherever possible the development of engine power

The quietening of the exhaust is required because the pulsating gas flow would otherwise expand rapidly when it left the tailpipe, creating noise. The energy remaining in the exhaust gas after it has left the engine or turbo is in the form of heat and velocity. As each exhaust valve opens, there is a sudden rush of gas into the pipe. A six-cylinder engine rotating at 3000rpm has 150 pulses per second passing into the exhaust pipe. These pulses travel down the exhaust at the speed of sound and reflect from any pipe junctions or sudden changes in pipe diameter.

If the natural frequency of any length of the exhaust system is excited by these pulses, the whole pipe will ring — or resonate. Bad resonances create vibration and noise inside the car at just a few combinations of engine speed and load. In other words, a resonance is the sort of thing that drives you mad when you're on a long journey and the exhaust is booming away at exactly 100km/h in fifth gear! As the name suggests, a resonator is a muffler designed to reduce resonances. But the exact difference between mufflers and resonators has become blurred — some manufacturers use the same box construction for both tasks.

Many standard exhausts are made quite poorly — this pipe has been flattened by the new car manufacturer to clear a suspension component. In this case it's easy to improve the flow with an aftermarket exhaust!

The noise reduction requirements of a muffler vary from engine to engine but the maximum silencing effect is normally required over the 50-500Hz frequency range. The noise reduction possible from a good muffler working in this range may reach 30dB — a 1000-fold decrease in sound power.

Mufflers can be divided into two main types. Reactive mufflers work by reflecting the pressure waves within interconnected chambers until they cancel themselves out. Dissipative (or absorption) mufflers operate by allowing the gas to expand through perforated tubes into a silencing material such as fibreglass mat or stainless steel wool. Many mufflers use both techniques within the one design, with absorption mufflers most effective at high frequencies. The louvres or holes used within a muffler help determine the silencing characteristics, the area of perforations determining the frequency of maximum noise reduction, and the type of perforations determining the magnitude of the loss. The proportion of open to closed areas of the perforated pipe is one characteristic that can be changed so different noise reductions occur.

Supercharged drag cars use short exhaust stubs from each cylinder. The pipes are not connected together and so flow from one cylinder does not help scavenge another.

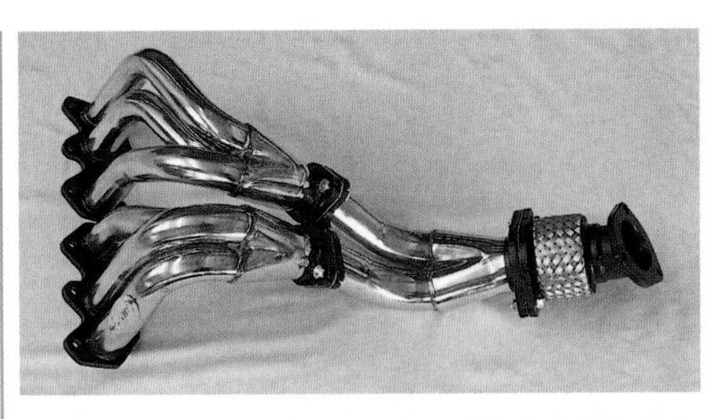

A tuned-length extractor system for a VW Golf VR6. Quite short primary and secondary pipes have been used in a 6-2-1 configuration. The diameter of the pipes is a trade-off between making the greatest use of exhaust pulsing and keeping back pressure as low as possible.

The function of the exhaust in aiding the development of power by the engine is a very important part of exhaust design. A naturally aspirated engine benefits from an exhaust system tuned to make effective use of exhaust pulsing. It does this by using individual pipes for each cylinder, with the length and diameter of each pipe carefully designed. The intention is that a low-pressure wave is reflected back to the exhaust valve at the exact time it's opening, creating a momentary lower pressure in the primary pipe and so aiding exhaust flow out of the combustion chamber. In addition, if the individual pipes are joined together through gently converging collectors, the flow of gas from one cylinder will help draw gas from the other cylinders. This scavenging effect will be improved if the speed of exhaust gas flow is high.

Contrary to popular opinion, any exhaust back-pressure harms performance. There is no properly tuned engine where increasing exhaust back-pressure causes an improvement in power, torque or fuel economy. However, the maximum effect of exhaust pulsing and scavenging may come from an exhaust system small enough for some exhaust back-pressure to be developed! This means that, for naturally aspirated cars, as big an exhaust as possible should be used **once the tuned-length part of the exhaust is passed**. Turbocharged engines simply require the biggest exhaust that can be fitted after the turbo!

This butterfly is used within an exhaust to vary back-pressure. Normally, the butterfly is automatically controlled to maintain only a small amount of back pressure, keeping the exhaust quiet.

Few tests have been done that clearly show the effect of changing back-pressure. Most muffler and exhaust comparison tests change more than one parameter simultaneously, making the identification of exhaust back-pressure as a culprit difficult. However, Wollongong (Australia) mechanic Kevin Davis has done extensive testing of varying back-pressure on a number of performance engines. These range from turbocharged Subaru Liberty (Legacy) RS flat fours to full-house traditional pushrod V8s. In **not one** case has he found any improvement in any engine performance parameter with increased exhaust back-pressure.

Reducing back-pressure to a minimum while still keeping the exhaust pipe dimensions practically small is most easily done using dual pipes. This factory system uses dual cat converters (pictured) in addition to dual mufflers further towards the rear of the car.

The tests came about because Kevin has developed a patented variable-flow exhaust that uses a butterfly within the exhaust pipe. He initially expected to use the system to cause some back-pressure at low loads "to help torque". However, he soon changed his mind when any increase in back-pressure proved to decrease torque on a properly tuned engine. What increasing the back-pressure **does** do is dramatically quieten the exhaust.

One of the engine dyno tests carried out by Kevin was on modified 351 4V Cleveland V8. Following the extractors, he fitted a huge exhaust that gave a measured zero back-pressure. Torque peaked at 573Nm (423ft-lb) at 4700rpm, with power a rousing 329kW (441hp) at 6300rpm. He then dialled-in 1.5psi (10.4kPa) back-pressure. As you'll see later, very few exhausts are capable of delivering such a low back-pressure on a road car. Even with this small amount of back-pressure, peak torque dropped by 4 per cent and peak power by 5 per cent. He then changed the exhaust to give 2.5psi (17.3kPa) back-pressure. Torque and power decreased again, both dropping by 7 per cent over having zero back-pressure!

Figure 6.1 shows the power curves gained in the tests. These results were achieved on a large engine with a large overlap cam — one of the type some people suggest is 'supposed' to like back-pressure.

If, in fact, power **does** increase with increased exhaust back-pressure, it is most likely the air/fuel ratio and/or ignition timing are no longer optimal for the altered state of engine tune.

MUFFLER TYPES

Once, a typical performance muffler used multiple baffles inside it. The exhaust gas was forced to meet blank walls inside the muffler, making its way out through holes punched in the tube. Often it would then have to squeeze through even more holes before it could continue on its way. Each section of the muffler allowed expansion and pulse reflection, decreasing noise. These mufflers were reasonably quiet but very restrictive to flow. This type of baffled muffler is still currently fitted to some new cars.

Next on the scene was the Turbo or reverse-flow muffler. Even this muffler dates back to the 1960s, where it was first used on the turbocharged Corvair Spyder in the US. The Turbo muffler doesn't use baffles to block off flow, so it has less flow restriction. Instead, the Turbo takes the

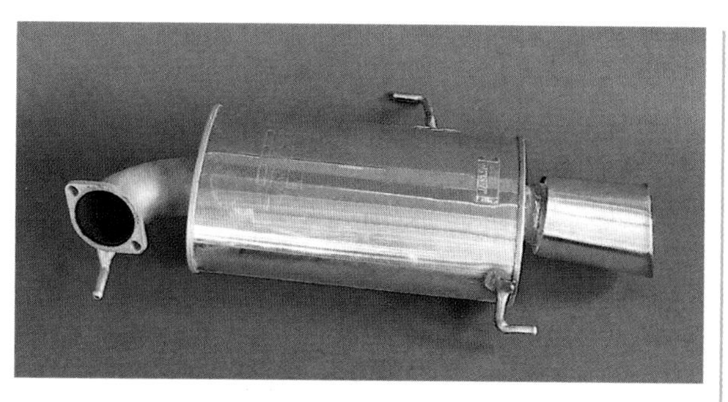

The Japanese aftermarket mufflers are among the best in the world with their combination of good flow and noise reduction. They use straight-through designs, sometimes with multiple chambers within the main body of the muffler.

exhaust gas on an S-shaped path through the muffler The gas enters the muffler, travelling straight down to the other end of the muffler, where it is forced to turn through 90 degrees. It then heads back in a different tube the way it has come, before it's forced to turn around again. Finally, it flows out of the muffler.

The benefit of the Turbo muffler is that it's effectively three times longer inside than outside! The disadvantage is that each of those 90-degree turns causes a flow restriction. Some of these mufflers use little curved internal guides that are designed to help the gas make these U-turns, but restriction is invariably caused. The Turbo muffler works by allowing the pressure waves to reflect and dissipate in the chambers at each end of the muffler and to pass through perforations in the tubes that carry the gas. The best Turbo mufflers use bends external to the muffler body to turn the gas through 90 degrees.

Finally, the most modern type of muffler is the straight-through design. The straight-through muffler uses a single perforated tube that takes the exhaust gases directly

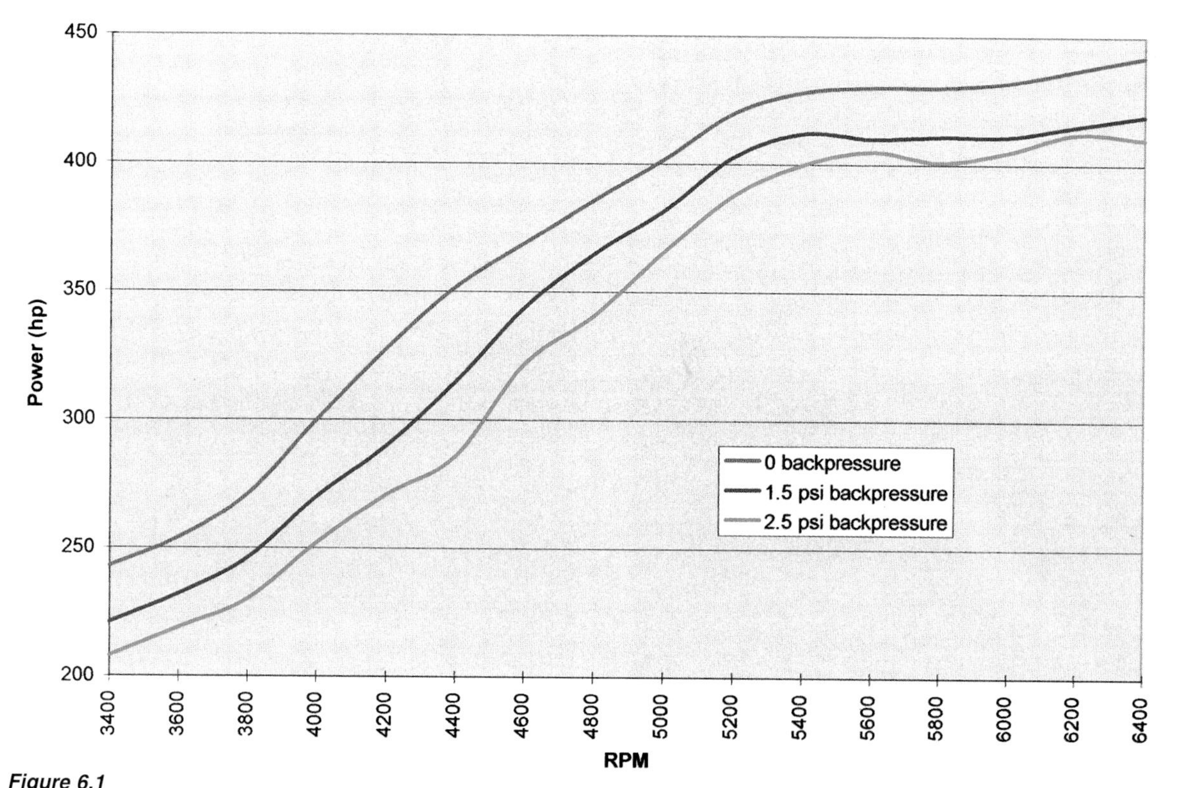

Figure 6.1
On a well-tuned engine, increasing back-pressure reduces power. This graph shows the reduction in measured power on a modified 351 Ford V8. Even 1.5psi (10.4kPa) back-pressure reduced power!

Flowbench testing has shown that mufflers vary enormously in flow. And, no, the most restrictive ones aren't always the quietest!

from the inlet to the outlet. The exhaust gas can travel through the muffler with almost no restriction at all. The sound waves expand through the holes in the pipe and are absorbed in the muffler packing. Most straight-through mufflers use holes that have been punched cleanly, but poor designs have a 'louvre' type where there are projections of lots of little bits of metal into the gas flow.

These are the three basic types of mufflers but there are also variations on the designs. Some sophisticated straight-through mufflers use two chambers, with the exhaust gas expanding into the second chamber after it

Variable flow mufflers are becoming more common as standard equipment. They allow good flow and quiet performance.

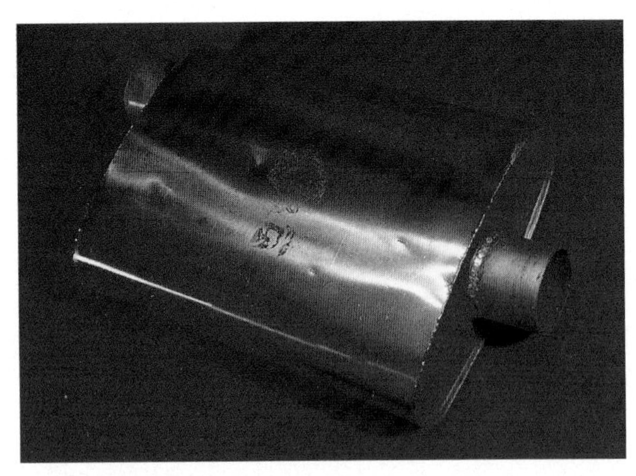

A muffler that's poorly made often performs badly as well! The rippled body of this stainless steel muffler shows the manufacturer doesn't have good standards of quality control and workmanship.

has squeezed through the perforations in the main tube. Others use a dog-leg design, with a central open chamber and offset inlet and outlet pipes that pass through their own respective chambers.

VARIABLE-FLOW DESIGNS

Some car manufacturers are now fitting variable-flow mufflers and variable-flow exhausts. Such active devices have huge advantages in exhaust design in that the flow capability, silencing ability and effective pipe lengths can all be altered to give the best outcome at different engine loads and vehicle operating conditions. Variable-flow mufflers alter the exhaust gas path through the muffler. This type of muffler has the ability to provide excellent silencing in cruise and light-load conditions while at the same time allowing (noisier!) low-restriction flow at high loads. An external solenoid or vacuum actuator can be used to control a movable flap within the muffler, or the flap can be sprung to move when exhaust back-pressure reaches a pre-determined level.

Variable-flow exhausts most frequently use an adjustable butterfly within the exhaust. The butterfly can be located after the mufflers and varied in opening so a constant (but relatively low) exhaust back-pressure is maintained irrespective of the flow. This allows the mufflers to silence

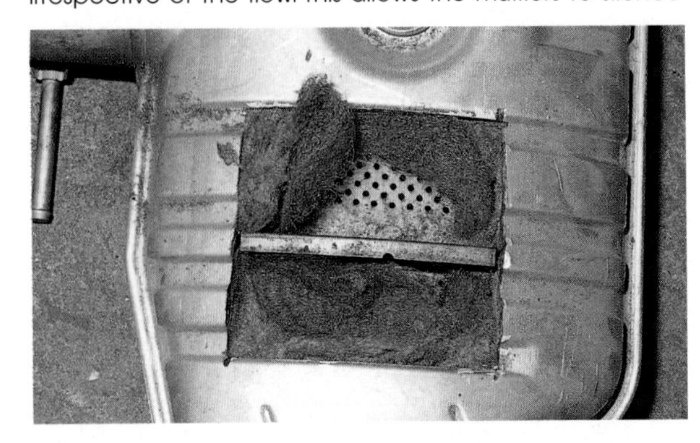

Most mufflers use stainless steel wool and/or fibreglass packing as the silencing material. Note how this Original Equipment muffler uses a large central chamber to allow expansion of the exhaust gases. Many muffler shops are happy to let you take away discarded standard mufflers to cut open. This way, a lot can be learned!

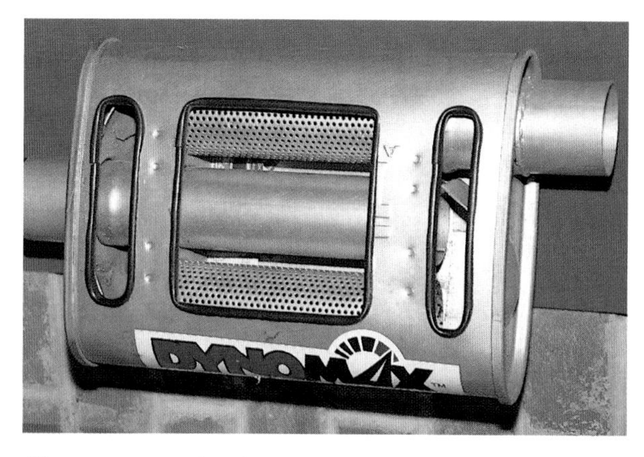

The tortuous path taken by the exhaust gases through a reverse flow ('turbo') muffler can be seen in this cutaway view. As a result, reverse-flow mufflers flow quite poorly when compared with straight-through designs.

very effectively, even when a very large diameter exhaust pipe is used. Alternatively, the butterfly can be placed closer to the engine and used to alter the effective tuned length of the exhaust pipe. When partially closed, the butterfly will cause wave reflection to occur from that point, while when open, the pressure waves will reflect from a point further along the exhaust (from the first muffler, for example). Variable exhaust actuators are most likely to be ECU-controlled.

Both variable-flow mufflers and variable-flow exhausts will become much more common because they allow a far better compromise of flow and silencing than fixed exhaust design. In modified road cars, such an approach is likely to yield very good results, although much experimentation (especially when altering wave reflection points) will need to be undertaken to gain best results.

COMPARING MUFFLERS

A genuine straight-through muffler outflows any other type. When compared with the same length of empty pipe, a good straight-through design flows 92 or 93 per cent of the maximum possible. That is exceptionally good, and can be compared with the poor flow of atraditional Turbo reverse-flow design that is typically down to

Aluminised mild steel mufflers have much shorter lives than stainless-steel mufflers. However, if you want cheap performance, a mild steel muffler is quite acceptable. Mild steel mufflers are available in the same range of internal designs as stainless mufflers.

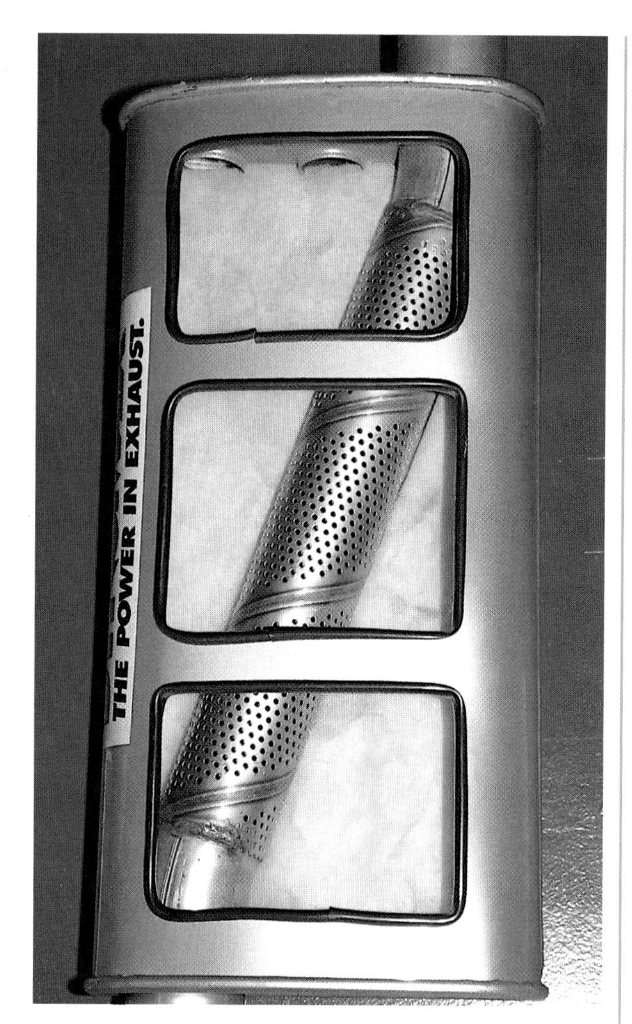

A straight-through muffler is literally just that! When you look into a straight-through design, there should be no protrusions into the gas path to impede flow. Some mufflers have welding dags or even punched louvres that project into the perforated tube.

59 per cent. Those mufflers using a dog-leg internal design with offset chambers have a flow of about 65 per cent, while traditional baffled mufflers can be as low in flow as 38 per cent. These figures are the result of extensive muffler testing carried out on a flowbench.

It's important to realise that these muffler flow figures are not just theory — the choice of muffler **will** affect the power being developed by your car's engine. An example can be seen in a strong naturally aspirated V8 being dyno tested. The owner had fitted a twin 2.5-inch (64mm) exhaust system with two well-known reverse-flow mufflers. From the mufflers to the back of the car, smaller two-inch (51mm) tailpipes had been fitted. This exhaust did not use catalytic converters.

Tested in this form, the power peaked at 213kW (285hp) at the wheels. The two-inch tailpipes were then dropped off, with power immediately responding with an increase to 218kW (292hp). The restrictive reverse-flow mufflers were then replaced with a single, large straight-through design. With the required changes to maintain the air/fuel ratio as it was in the previous tests, the power rose to 239kW (320hp)! This means the reverse-flow mufflers were dropping 26kW (35hp), or 11 per cent from the engine's output.

More and more standard cars are using active exhausts, where the exhaust path is changed depending on load. The flow control vacuum canister can be seen in this Ferrari.

The amount of noise developed with a particular design of muffler is very dependent on the car to which it is fitted. When used with a resonator or on a turbo car, straight-through designs can be very quiet. In fact, in a muffler comparison test carried out on a 3.8-litre V6, the two straight-through designs were the quietest of all the mufflers. The restrictive reverse-flow designs were among the noisiest — so much for the theory that you cannot have a quiet **and** free-flowing muffler! In a second test carried out on a turbo four, the same applied — the quietest and best-flowing mufflers were the straight-through designs.

However, no fewer than four of the small straight-through mufflers measured as being very quiet on the small (but powerful) turbo engine were later fitted to a 5-litre V8. The power gain over the standard system was excellent, but the noise was unbearable. I suggest the only **certain** way of ascertaining the noise level is to actually fit the muffler to the vehicle. Obviously, it helps if you can find someone else with your type of car who has already done that!

There is, however, a quick and easy way of comparing the noise output of mufflers before they are fitted to the car. It both looks and sounds odd, but it is effective. If you hold a muffler up to your mouth and yell or scream through it while someone listens at the other end, the sound attenuation (silencing effect) of the muffler will be quite apparent. A 'yell test' like this was carried out using a sound level meter to record the SPL at a distance of one metre from the muffler outlet. Comparison was then made with the noise output of the mufflers when they had been tested on a car. The order and relative magnitude of the yell test figures closely matched the results achieved during full-load dyno testing.

SELECTING MUFFLERS

Inspect any muffler you are thinking of buying. Look at the construction of the muffler — well-made mufflers are often quieter because the company has done a good job in all aspects of the design and construction. Using a torch or the sunlight, carefully examine what you can see of the inner workings. If it's a straight-through design, there should be no welding dags or other bits of metal protruding into the path of the exhaust gas. Even a small lump of weld can make a big difference to flow. If possible, look through the perforations in the tube. If you can see white fibreglass packing, you should be concerned. Fibreglass is an excellent silencing material, but the exhaust gas flow can easily suck the fibres through the perforations. To prevent this, there should be a layer of stainless steel wool up against the holes to protect the fibreglass. When you look through the muffler, make sure the tube size doesn't decrease within the muffler. At one stage, a very well-known 2.5-inch (64mm) muffler actually had internals that were closer to two-inch (51mm), perhaps explaining its poor flowbench results!

The larger the muffler canister, the more likely it is to be effective at silencing. The Japanese aftermarket exhaust manufacturers are probably the best in the world at producing very quiet exhausts with excellent flow, and one common denominator of their systems is the use of large mufflers. You want the body of the muffler to be as large as can be fitted under the car. However, the muffler shouldn't have big flat panels, because these can vibrate and create resonances. Instead, they should be stiffened by being curved or even by having the manufacturer's name deeply stamped on them.

Most mufflers are made from either aluminised mild steel or stainless steel. Some one-off or low-budget designs use only painted mild steel and should be avoided. Mufflers are prone to corrosion because the engine produces acids that cause the pipe to corrode from the inside out. This is especially the case if lots of short trips are undertaken. Stainless-steel mufflers use one or a combination of two types of material: 304 or 409 grades. Of the two, 304 is the more expensive and is a similar grade of material to that used in kitchen sinks. The lower-grade 409 is easily picked because it attracts a magnet. Both types of stainless steel can be polished to a high lustre, but 304 is more durable, with mufflers made from this material having warranties of up to 10 years. In comparison, aluminised mild steel mufflers last four or so years, depending on the type of trips undertaken.

An exhaust is much more likely to be quiet if it uses three major components: a catalytic converter, resonator and muffler. Each of these performs a silencing function, so the noise level will increase if any is left out. Note that even if you wish to reduce restriction to a minimum, a straight-through, well-designed resonator will cause almost no impediment to flow while at the same time substantially dropping noise levels.

EXTRACTORS

Extractors — called headers in some markets — use individual pipes for the exhaust ports of each cylinder. Extractors are available in two different designs: 'interference', where the pipes are of unequal lengths, and 'tuned-length', where the (usually longer) pipes are of a similar length. The interference design provides better results than a typical cast manifold because the less tightly bent pipes flow better and there's also less pressurising of the exhaust ports of adjacent cylinders. In addition to these factors, the tuned-length designs take advantage of the reflected negative pressure pulses to aid flow through the exhaust valve. However, the pipes can only be tuned for one engine speed (just one rpm), so will be 'out of tune' for other revs. It's normal to tune the pipes for the revs at which peak torque is developed. Second, there often isn't enough room for four, six or eight very long primaries in most engine bays! All this means the benefits of tuned-length extractors aren't always realised in road cars.

These extractors use a 4-into-1 approach with a 'twisted' collector. Which extractor design works best varies from car to car and depends on the amount of room available as well as the characteristics of the engine.

While many standard cast-iron manifolds are not very efficient, it should be noted that some **are** very good indeed. On a four-cylinder car, a well-designed 4-into-2 cast manifold that's followed by dual, long exhaust pipe secondaries can work very nearly as well as extractors. Other cars use standard 4-into-1 extractors that are also very effective. This is especially the case in front-wheel-drive cars with the engine mounted transversely. In these cars, there's normally plenty of room for a substantial manifold using long, equal-length primaries and gentle bends. The gains made by replacing this type of design are therefore going to be only small — don't assume all cars automatically benefit from the fitting of extractors!

Extractors vary substantially in their quality of manufacture. This will affect the power gained and also the ease of fitting. The latter is very important because good extractors are usually a tight squeeze within the engine bay. If a hammer is needed to squash a tube to gain clearance, some of the potential power benefits will be lost! In good-quality designs, the header flange should be at least 8-10mm thick to prevent heat distortion. The pipes should fit **within** the holes cut in the header plate (not be butt-jointed to the outside of the flange), so the flange holes should have been cut larger than the port size to accommodate the wall thickness of the tube. The welding should be neat (neat welds are usually good welds) and the bead should have good penetration into both the flange and the material of the pipes. There shouldn't be any welding dags on the inside of the tubes, and the mouths of the tubes should be beaten out to match the port shape. Mandrel bending rather than press bending should be used.

Extractor design is complex. If all details of the engine are known (including cam characteristics, port size and so on) a software package such as *Engine Analyser* can be used to develop some preliminary extractor designs. However, even with this data available, the extractors still need to be made and tested for their effectiveness before you can be confident that they will provide the desired results. A Graham Bell in his book *Performance Tuning in Theory and Practice* suggests the following equation can be used to determine the best primary pipe length, and others have agreed with him.

$$\text{Pipe Length} = \frac{850 \times ED}{RPM} - 3$$

Where rpm = engine speed for which the extractors are being tuned, ED = 180 plus the number of degrees the exhaust valve opens before BDC, and with the pipe length expressed in inches. This equation applies to both 4-into-1 and 4-into-2-into-1 designs, but with the latter it's the total length of the primary and secondary pipes that's being determined.

Companies developing aftermarket extractors build and dyno-test at least four or five sets of extractors before

A lot of room is left for extractors when the cast-iron manifold is removed from this FWD Celica. With the extractors and a two-inch (51mm) cat converter fitted, the car picked up 11 per cent in peak power.

they're confident they have produced a good design. In other words, even those who have built perhaps hundreds of different types of extractors over the years still cannot be sure of the end result when they start work on a new design. When selecting extractors, you should ask to see the dyno test results of each of the designs you're considering buying.

Manufacturers place the exhaust oxygen sensor close to the engine, normally in the exhaust manifold. This means when extractors are fitted, the oxygen sensor is moved. Two conflicting requirements need to be met by the new location — the oxygen sensor should be placed close enough to the engine that it quickly attains operating temperature, but it should also be located where it can monitor the exhaust output of all of the cylinders. Unheated sensors are often mounted on a single primary pipe close to the engine, while the heated types are usually placed after the collector. It's not difficult to replace an unheated oxy sensor with one that uses a heater, so if a digital multimeter shows the output of the oxygen sensor is poor (in cruise, the output voltage should oscillate between 0-1 volt in most designs), a heated type should be fitted.

EXHAUST COATINGS

Ceramic-based coatings can be applied to extractors, exhaust pipes and turbo exhaust housings. The coatings are durable and, unlike chrome plating, will not 'blue'. A number of colours are available and the coatings are claimed to be stable to 690 degrees C. Note that exhaust gas temps in many engines can exceed this figure at high loads, though! The coating slows the radiation of heat, both keeping the exhaust gases hotter and the rest of the engine bay cooler.

The coatings can be applied to the pipes both externally and internally, with the latter probably helping to smooth the inner surface and so improve flow. The manufacturers of the coatings claim the hotter exhaust gases flow better, causing their velocity to remain higher and so promoting better cylinder scavenging. But when they're examined closely, the claims are only for very small power increases, of the order of 1 per cent. However, the reduction in underbonnet temperatures can improve power by allowing the inhalation of colder, denser air. One manufacturer's test saw a turbo car decrease in

These superb 2-litre open-wheeler race car extractors show how it's done! Note how the pipes at the exhaust ports change from rectangular to round in cross section, the pairing of cylinders 1 and 4 and the pairing of 2 and 3, and the very gentle bends used. The oxygen sensor is placed after the last join so it can monitor the output of all cylinders.

Extractors are used on some factory high-performance cars. This Lexus GS300 uses large-diameter, beautifully bent primary pipes. The 166kW (222hp) naturally aspirated six is only three litres in capacity.

underbonnet temperature from 58 degrees C to 40 degrees C, while the car experienced an 8 per cent power increase. In this case, the coating was applied to the factory cast-iron exhaust manifold.

Exhaust coatings of this type do not have the problems associated with thermal wrapping tapes, which can cause rapid corrosion of the exhaust pipe. However, if it's desired that a thermal wrap be used, boiler tape is a cheaper alternative to the tape sold specifically for wrapping automotive exhausts.

PIPE DIAMETERS

One of the few sources of independent information on appropriate exhaust pipe diameters can be found in the literature of those companies that manufacture truck mufflers. For example, Donaldson produces a chart that matches recommended pipe diameters with engine power. At a 2psi (13.8kPa) maximum back-pressure, Donaldson recommends in the table overleaf the following for four-stroke petrol engines:

A high temperature coating like that used on these extractors provides some minor power gains but the real advantage is in reducing the temperature of other underbonnet components. Note the very long primaries being used on this 450kW (600hp) Chev V8.

Recommended Maximum Power (kW, petrol engine with 2 psi back-pressure)	Pipe Diameter (inches/mm)	
75	2	51
120	2.5	64
165	3	76
230	3.5	89
375	4	102
550	5	127
635	6	152

Note that kilowatts can be converted to horsepower by multiplying kW x 1.34. From the table it can be seen that the recommended pipe sizes are much larger than often considered necessary on a car. However, back-pressure readings support this need for large-diameter pipes. Many high-performance cars have exhausts that are too small for their peak power exhaust gas flows.

When fitting extractors or manifolds, oxygen sensor-safe sealant should be used if damage to the oxygen sensor is to be avoided.

BEND FLOWS

Exhaust pipes can be bent using two different techniques — mandrel and press bending. Press bending machines are commonly found in exhaust workshops, while mandrel bending machines are much rarer. Press bends are made with tooling that remains external to the tube, causing some flattening of the tube as it's bent. Mandrel benders use a mandrel of similar diameter to the inner diameter of the tube. The mandrel is pulled through the tube as it is bent, forcing the tube to keep very nearly the same inner diameter as a straight piece of the same tube.

Mandrel bends are generally regarded as being much superior to press bends but, depending on the tightness of the bend, the flow difference can actually be quite small. Flowbench testing was carried out on a selection of 2.5-inch (64mm) diameter bends, with the bends tested being:

- 45 degrees press and mandrel
- 90 degrees press and mandrel
- 120 degrees press and mandrel
- 180 degrees mandrel

A press-type pipe bender is used in most exhaust shops. The resulting bends do not retain their full internal cross-section and so press bends flow less than that achieved by mandrel bends. However, on gentle bends the difference is not great.

This Audi two-litre Touring Car uses very large-diameter primary runners, especially considering that with 'only' about 225kW (300hp) available, they are not extremely powerful racing cars.

Where an exhaust is made from welding together short mandrel bends, the system may in fact flow more poorly than if a continuous length of press bends had been used. This is because there will almost always be steps and penetrated welding beads at the joins.

Note that both the press and mandrel bends were of very good quality, something often hard to achieve with a press bender! The flowbench testing was carried out using three inches of water pressure differential. The mandrel 45-degree bend decreased flow by just over 1 per cent when compared with the same length of straight tube, while the press bend dropped it by 2 per cent. The tighter 90-degree mandrel bend was down over a straight length of tube by 0.6 per cent, with the press-bent tube 3 per cent worse in flow than the straight tube. The mandrel-bent 120-degree bend dropped flow by 2 per cent, while the press-bent 120-degree bend dropped flow by a significant 5.6 per cent. The press bender could not produce a 180-degree bend, while the 180-degree mandrel bend was down in flow by 1.9 per cent over the same length of straight tube. Figure 6.2 shows these results.

This shows that the flow of both the mandrel- and press-bent tubes is generally very good, but the mandrel bend typically has less than half the flow restriction of a press bend. However, there is a very important point to consider if specifying mandrel bends. Because the vast majority of exhaust shops do not have a mandrel bender, their 'mandrel-bent' exhausts consist of many pre-formed mandrel bends cut and welded together to form the exhaust. Unless the welding is done with great care so that a bead does not penetrate the pipe and there's no offset at the joins, the flow may well be worse than that which would have been achieved with a continuous length of press bends! Note also that many exhaust shops that make an exhaust in this way grind back the welds and then paint the pipe so the method of exhaust construction may not be at all obvious.

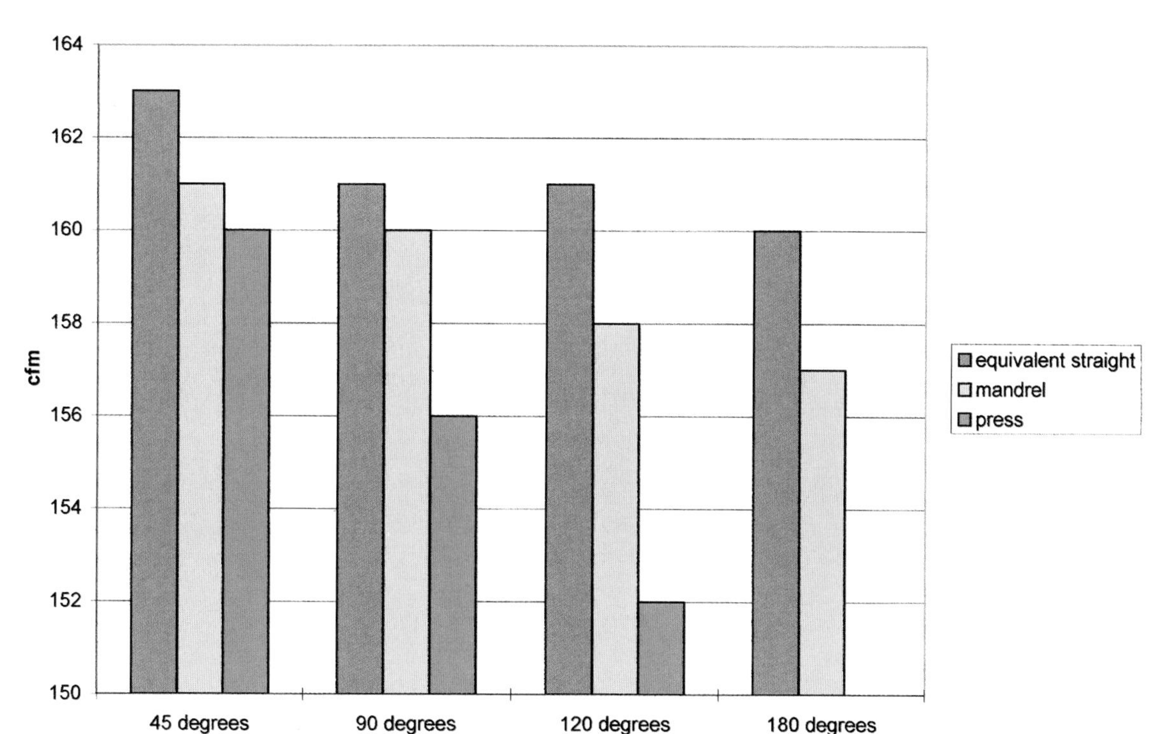

Figure 6.2
The flow of both mandrel- and press-bent tubes is generally very good, but mandrel bends have typically less than half the flow restriction of press bends.

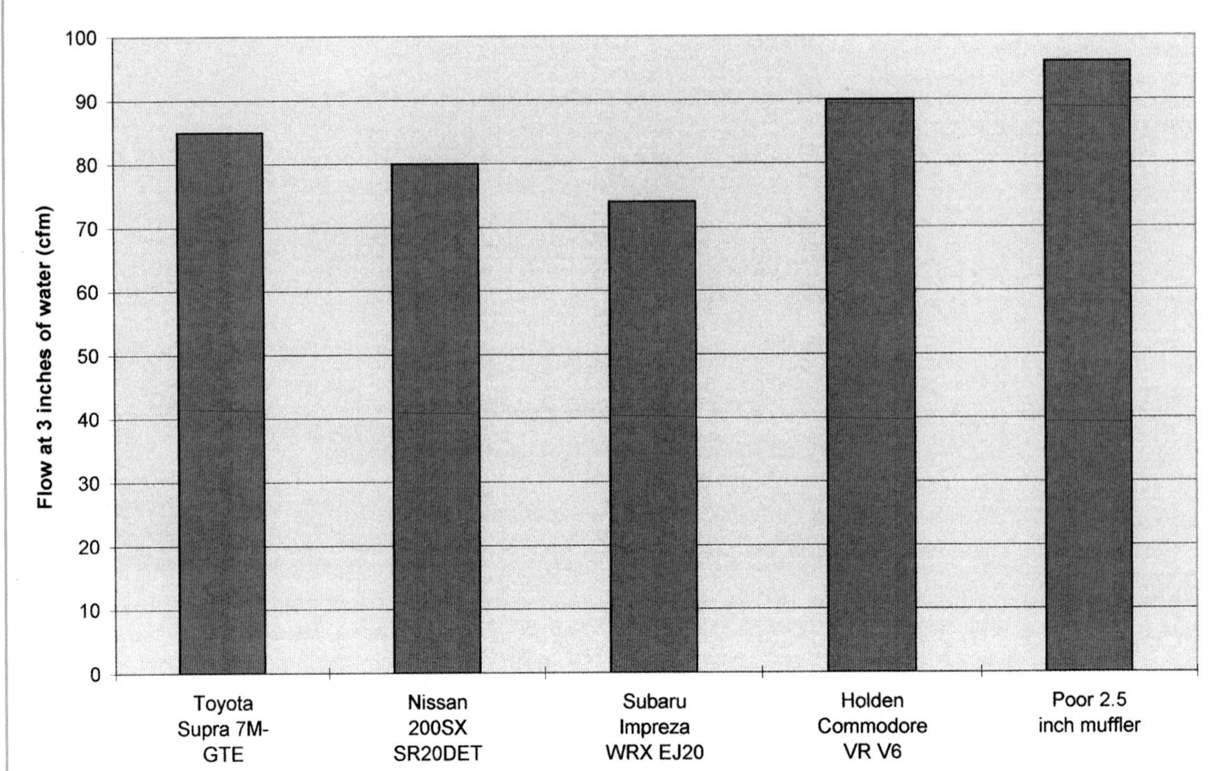

Figure 6.3
Catalytic converters are generally very restrictive. Here are the flows of a variety of original equipment cat converters, compared with a ***very poorly flowing*** *2.5-inch (64mm) muffler. In most exhaust systems, the cat converter poses the greatest restriction to flow of any single part of the exhaust.*

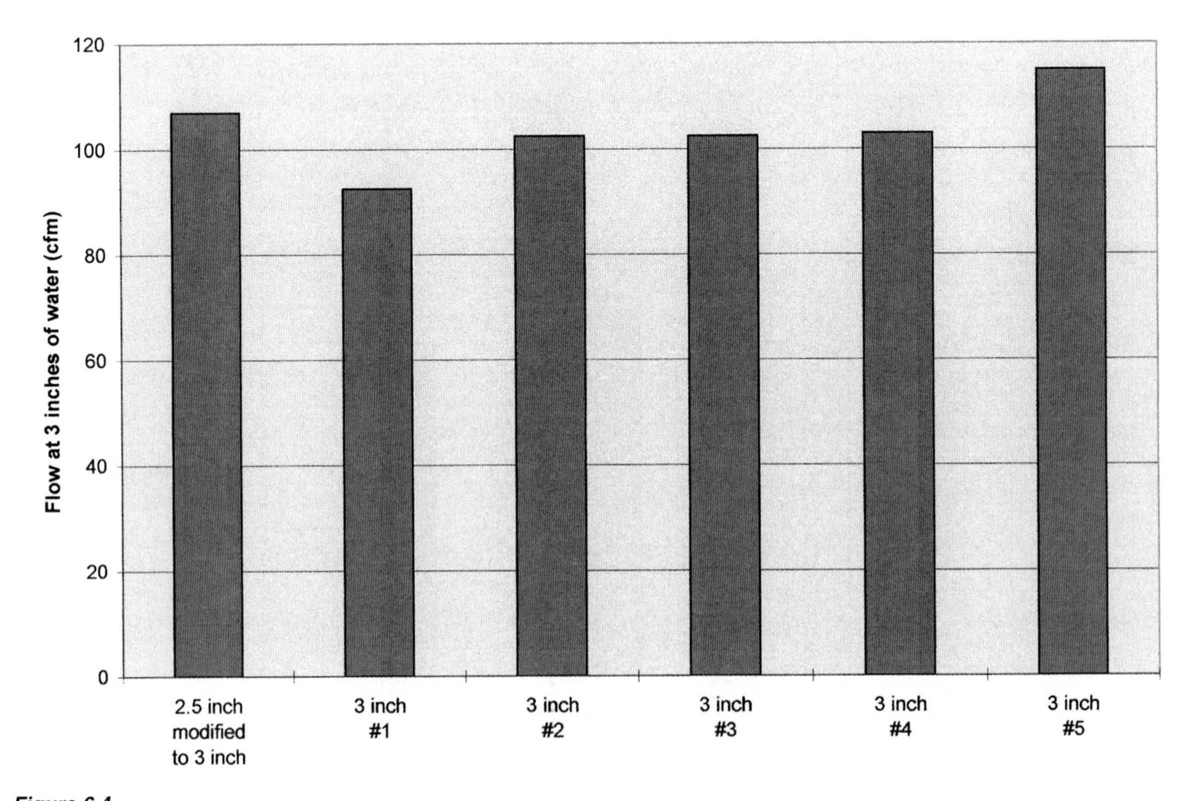

Figure 6.4
Aftermarket cat converters vary substantially in flow. This flowbench testing showed that a 2.5-inch (64mm) cat modified to take three-inch (76mm) nipples outflowed many genuine three-inch (76mm) designs.

CATALYTIC CONVERTERS

Catalytic converters are devices that reduce toxic emissions. They do this by passing the exhaust gas through a ceramic honeycomb that has been very thinly coated with precious metals such as platinum and rhodium. These metals accelerate the degradation of the toxic substances, promoting the transformation of carbon monoxide and hydrocarbons to carbon dioxide and water. The oxides of nitrogen are also reduced to neutral nitrogen.

Of the restriction to flow posed by the complete standard exhaust system, the cat converter is almost certain to be the most restrictive element. Flow testing carried out on four standard catalytic converters clearly shows this. Three of the cats were in used but good condition. Standard catalytic converters from a 7MGTE Toyota Supra Turbo, SR20DET Nissan 200SX Turbo and EJ20 Subaru Impreza WRX Turbo were selected. On the flowbench, the best of the three was the 62mm Supra Turbo cat with a flow of 85cfm at three inches of water pressure differential. The Nissan 200SX 65mm cat was next best at 80cfm, while the 55mm WRX cat trailed the field at 74cfm. A brand new 64mm cat converter for a Holden Commodore VR V6 was then tested at 90cfm. In comparison, a **very poor** 2.5-inch (64mm) muffler flowed about 96cfm at the same pressure differential! Figure 6.3 (opposite) shows these results.

Aftermarket big-bore catalytic converters generally flow much better than these factory designs. Flow testing carried out on five three-inch (76mm) aftermarket cats showed the worst flowed 93cfm (at three inches of water) and the best, 115cfm. Interestingly, a small-bodied 2.5-inch (64mm) cat converter was adapted to take three-inch nipples (76mm) and then flowed a good 107cfm, showing that the core cross-sectional area isn't the primary determinant of flow. Figure 6.4 (opposite) shows these results.

In terms of emissions, cats vary significantly in performance — enough that an emissions test could be failed with a poor-quality (but brand-new!) cat converter fitted. Six aftermarket cats, one new factory cat and one 'empty' cat were fitted to a car equipped with a 3.8-litre V6. The emissions were measured on a chassis dyno with the car in Drive, 'travelling' at 60km/h, with a manifold vacuum of 17 inches of mercury (-58kPa) and with a power of 6.5hp (4.9kW) being measured at the wheels. This is typical of light cruise conditions. Before testing, the cats were brought up to full operating temperature.

The testing indicated that the brand-new factory replacement cat and one of the aftermarket cats failed to work, except in reducing the output of oxides of nitrogen! The aftermarket cat had an inlet air pipe and it eventuated that for the cat to work effectively, it was a requirement that air be pumped in through this pipe. However, this inlet pipe was capped when the manufacturer sold the cat converter, and there was nothing with the cat that suggested the performance would be adversely affected if it were fitted to a vehicle not equipped with a suitable air pump! I can only conclude that the brand-new factory cat converter was defective.

A large exhaust, especially on a turbo car, will give excellent performance gains.

The importance of cat converters in actually reducing emissions is likely to increase dramatically in the years to come. This means care needs to be taken in the placing of the cat converter within the exhaust system if the cat converter is to still work effectively. The cat needs to be located so it heats quickly following engine start-up. Manufacturers achieve this by placing the cat relatively

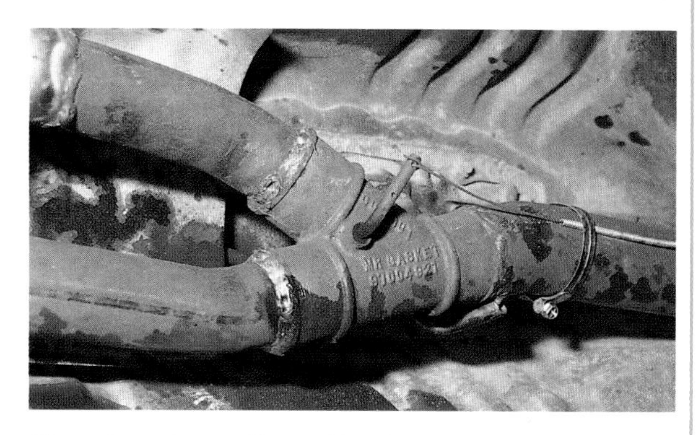
The exhaust 'cut-out' is as old as the performance car itself. It allows you to select between two different pipes — one with many mufflers and the other with perhaps none! Note, though, that a cast-iron cut-out like this leaks a little, so the unmuffled pipe may have some flow even when it is not selected.

close to the engine (actually attached to the end of the exhaust manifold in some cases) and by using pipework in front of the cat converter that has a low thermal mass.

One way the latter is done is to use a double-skin pipe between the exhaust manifold and the cat. The inner skin has little thermal mass, so most of the heat is carried

A catalytic converter uses an internal ceramic matrix coated with precious metals. If spread out, the area of the matrix exposed to the exhaust gas would be similar in size to a football field.

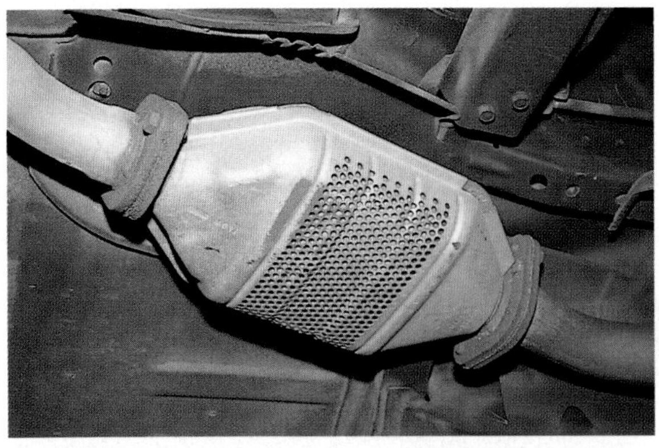

The factory-fitted cat converter is a major source of restriction on most cars. In any performance exhaust, dual standard cats should be fitted or the single cat upgraded in size.

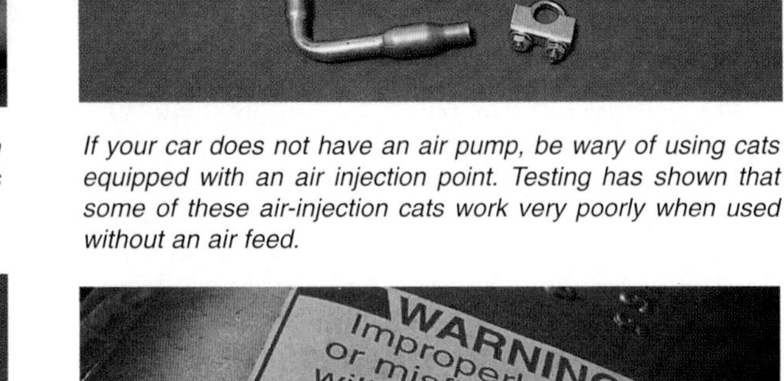

If your car does not have an air pump, be wary of using cats equipped with an air injection point. Testing has shown that some of these air-injection cats work very poorly when used without an air feed.

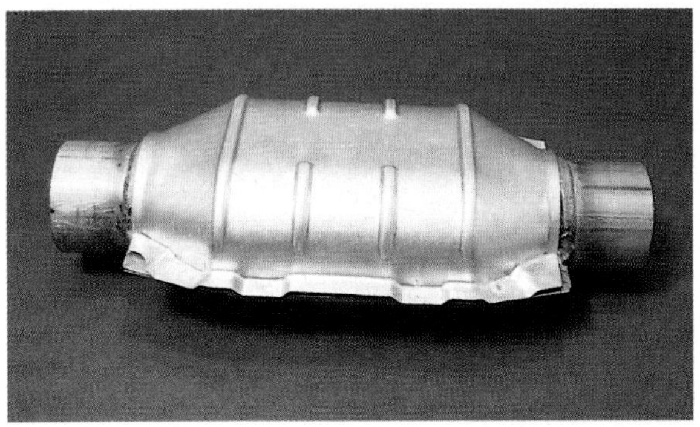

A three-inch (76mm) cat converter like this flows far better than most standard cats. Note that even if the exhaust pipe size is only 2.5-inch (64mm), using a large cat still makes good flow sense.

While cats don't start working properly until they reach a temperature of about 400 degrees C, a cat converter can also be overheated. Overly high temperatures (over 800 degrees C) are especially likely to occur if a misfire causes raw fuel to pass into the cat, where it then burns.

through to the cat converter, quickly bringing it up to operating temperature. When the manifold and piping in front of the cat are replaced, the characteristics of this thermal transfer may be changed. However, moving the cat much closer to the engine than standard may cause overheating of the cat converter when maximum power is being developed. Predicting the emissions performance of the car with a changed exhaust is very difficult, and where legal emissions requirements must be met, 'before' and 'after' emissions testing should be carried out.

The silencing effect of cat converters varies substantially. As part of the above cat converter test, noise was measured with a sound pressure level meter held at a 45-degree angle, 1.2 metres from the exhaust pipe outlet. The car was positioned outdoors and the engine was held at 3000rpm in neutral while the noise measurements were being made. The dB(A) weighting curve was used. With an empty cat in place, the measured noise level was 84dB(A), the same figure produced by the best flowing aftermarket cat. However, typically, the noise level was dropped to 80dB(A), with the standard cat converter also giving this noise output figure.

Muffler	Noise at 3000rpm, no load [dB(A)]	Peak noise during dyno power run [dB(A)]
No muffler	90	122
Best straight-through muffler	81	115
Best turbo muffler	88	116
Best dog-leg muffler	86	116
Best baffled muffler	90	120

Figure 6.5
The noise readings obtained when a number of three-inch (76mm) mufflers were tested on a Nissan 200SX turbo shows that full power noise level testing is much more demanding than the measurements made at 3000rpm with no load!

TESTING EXHAUSTS

Exhaust systems can be tested in three ways — noise output, back-pressure and the engine power developed.

Noise Testing

Exhaust noise is typically tested at just a single engine rpm with the engine under no load. In practice this means the car is in neutral and engine revs are held at 3000rpm (or a similar engine speed). The dB(A) noise meter weighting curve is selected and a reading made with a noise meter positioned behind and slightly to one side of the tailpipe. Other tests include drive-by simulations at certain speeds and throttle settings. However, there's a number of problems with taking these approaches.

First, the dB(A) weighting curve that's used significantly underrates the importance of the low-frequency noise in the measuring process. This is a problem because it's the deep notes that are usually more objectionable, whether you are in or out of the vehicle. Second, if you hold a noise meter behind a stationary vehicle and watch the meter's display, you will see the sound pressure level rises and falls as the driver changes engine rpm. And higher revs don't always mean a louder noise! A test being conducted at 3000rpm may give an artificially low figure — the exhaust may be louder at both 2800rpm and 3200rpm. This means single-rpm low-load noise readings can be treated with some disrespect.

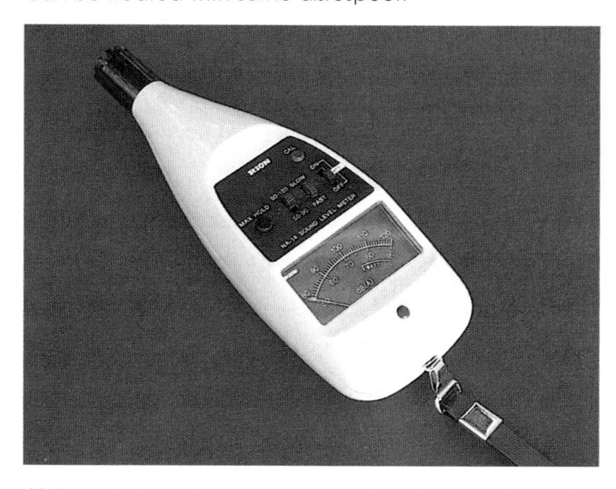

Noise meters can be used to assess the silencing effect of the exhaust system. However, note that it's easy to get a reading which is much quieter (or noisier) than the system sounds to your ears. This is because a noise meter measures only a small part of what we call noise.

Peak noise readings made either inside the car or within a dyno room during a full power run are far more meaningful — though muffler manufacturers don't like to use these numbers because they always seem so high! Figure 6.5 (opposite) shows the noise readings obtained when a number of three-inch (76mm) mufflers were tested on an SR20DET Nissan 200SX turbo. In addition to each of the test mufflers, the car's exhaust system used an empty cat and a small resonator. The piping of the system comprised three-inch (76mm) tube from the turbo back. As you can see from the table, full power noise level testing is much more demanding than the measurements made at 3000rpm with no load!

Before installing a cat, check it carefully to see if there's a marking showing the required direction of flow.

There's no simple answer to the problem of easily and cheaply measuring exhaust noise. At the minimum, to do the job properly a chassis dyno needs to be employed, along with a spectrum analysing noise meter — and the only people likely to have that equipment are the manufacturers of new cars! In the final analysis, if the noise level is below the legal requirements (normally established with just a simple test like those described above) it's your ears that need to be the final judge. Note that an exhaust that flows very well does **not** have to be noisy if the right mix of components is used within it.

Back Pressure Testing

Back-pressure is easily tested. In fact, some people (like me!) have a gauge measuring exhaust back-pressure permanently mounted on the dashboard. But first, what exactly is the back-pressure that has been spoken about in so much detail? When the engine is running, the rising pistons force the end result of combustion out through the open exhaust valves. If the exhaust gas cannot escape readily down the exhaust pipe, a pressure will build up. This is exhaust back-pressure. The higher it is at the head of the exhaust, the more restrictive to flow is the exhaust system.

To measure exhaust back-pressure, all you need to do is drill a small hole into the exhaust pipe and attach a pressure measuring pipe. In cars with oxygen sensors, the

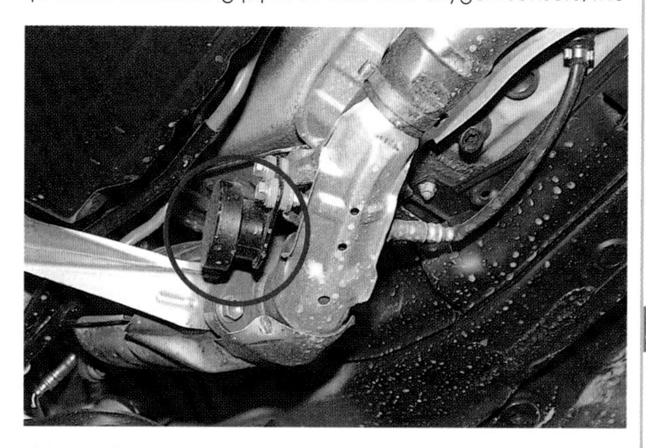

Aftermarket exhausts of whatever cost and construction are rarely as quiet and smooth as manufacturers' OE systems. One of the reasons is manufacturers do a huge amount of testing, resulting in the fitting of vibration dampers such as this one on a Lexus exhaust.

A plug-in silencing module can be used to quieten a loud exhaust when necessary. This one is an A'PEXi design.

sensor can be unscrewed, with a pressure measuring tap replacing it. For a permanently mounted sensor, or where you wish to measure the back-pressure at different points along the exhaust, a small steel tube can be inserted through a drilled hole and then brazed to the exhaust pipe. Another technique is to drill the pressure-tapping holes and then use a special tool to insert a flared nut. A steel tube that has been brazed to a threaded fitting can then be screwed into each test point, and when removed, the hole can be blanked off with a short bolt. A small-bore rubber hose is pushed over the end of the steel tube and its other end connected to a normal pressure gauge — a boost gauge is fine.

On turbocharged cars, the sensing pipe is normally attached straight after the turbo, while on a naturally aspirated car you can connect the pipe either at the end of the tuned-length system (at the outlet of the extractors, for example) or near the cylinder head. On a naturally aspirated engine still equipped with a cast-iron manifold, the pipe is most easily attached at the beginning of the light-gauge steel exhaust tube.

One example of back-pressure testing was that carried out on a turbocharged three-litre six-cylinder. In standard form this engine developed 150kW (~200hp). With the 2.5 inch (64mm) exhaust near to standard in configuration, the peak measured back-pressure was a very high 7psi (48kPa)! The effect of this back-pressure was to slow turbo spool-up time, cause more combustion chamber contamination and create greater pumping losses. After measuring the back-pressure of the complete exhaust, the probe was then tapped in at various points along the exhaust to ascertain which components were causing the greatest restriction.

This test showed that the first short section of standard wrinkled pipe connecting the turbo to the catalytic converter was causing 26 per cent of the total back-pressure! This illustrates the importance of using large dump pipes off the turbo, where the gases are still hot and expanded. Across the standard catalytic converter 19 per cent of the restriction was occurring, while the length of 2.5-inch (64mm) pipe between the cat and the rear muffler was causing a hefty 26 per cent of the total restriction. The tailpipe and the aftermarket reverse-flow muffler made up only 30 per cent of the overall restriction — 70 per cent of the flow restriction was occurring before the single muffler. This helps to explain why many factory exhaust systems drop in diameter towards the rear of the car — the extra size isn't required and some money can be saved without penalty. Figure 6.6 shows the pattern of back-pressure.

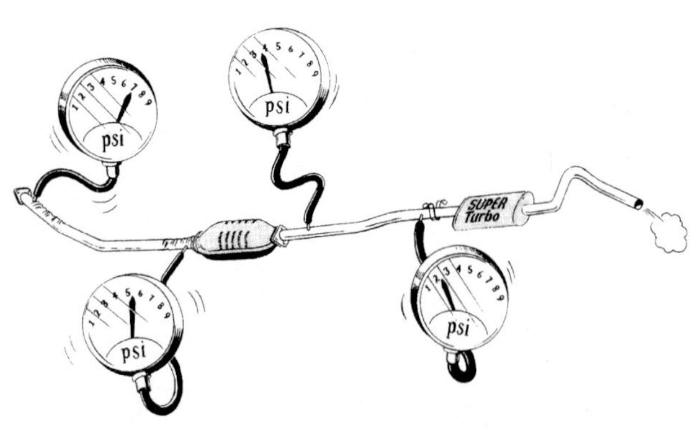

Figure 6.6
The measured maximum back-pressure at various points in the exhaust system of a turbocharged 3-litre, six-cylinder car. Measuring the back-pressure in this way lets you isolate the sections of the exhaust system causing the greatest restriction to flow.

Using a dyno during exhaust system design allows quite strange things to be discovered. This peripheral port rotary race engine exhaust system gained a small but distinct power improvement with an oversize tip fitted!

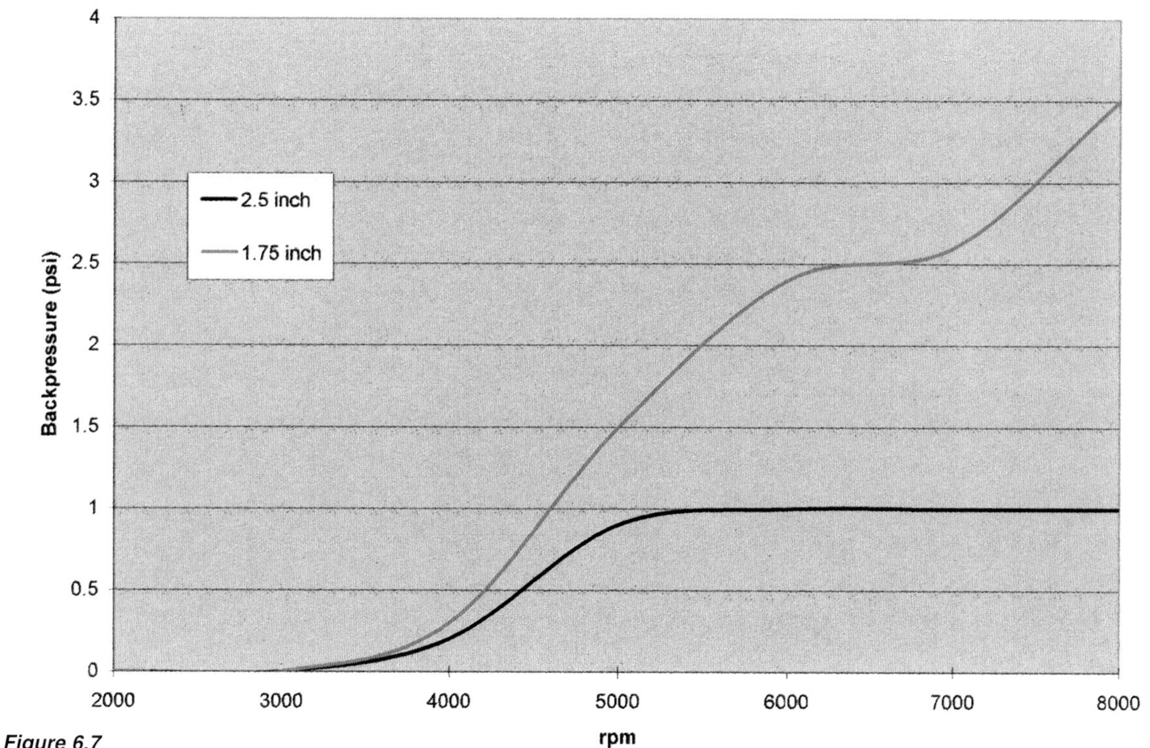

Figure 6.7
The measured back-pressure of a 1.5-inch (44mm) exhaust versus a 2.5-inch (64mm) exhaust on a turbo 660cc three-cylinder car. Keeping back-pressure as low as possible will benefit both power and economy.

In this case, a cost-effective outcome was required, so the pipe from the turbo to the cat converter was replaced with three-inch tube, the catalytic converter was upgraded to a three-inch (76mm) design, and a three-inch (76mm) pipe was used between the cat and the retained reverse flow muffler. In this revised guise, the peak back-pressure fell to 5psi (35kPa), down from the previous system's 7psi (48kPa). The drop in back-pressure of 29 per cent resulted in a maximum improvement in power of 9 per cent, interestingly, recorded at a low 3000rpm.

It's important to realise that back-pressure and exhaust sizing are related to engine power and so to exhaust flow, not engine size. A tiny 660cc Daihatsu Handi three-cylinder car equipped with an IHI RHB5 turbo and EFI provides an example. In one configuration the engine developed 87kW (116hp) at the wheels — around 200hp per litre at the flywheel. For the initial back-pressure test, the car used a three-inch dump pipe out of the turbo, decreasing in diameter through a custom-made cone down to an exhaust made from 1.5-inch (44mm) press-bent tube. A straight-through central resonator and a small muffler at the rear of the car were fitted.

With a turbo boost pressure of 15psi (104kPa or about 1 Bar), the exhaust back-pressure gauge indicated that above 5000rpm the exhaust was flowing very poorly. In fourth gear the pressure was zero below 3000rpm, 1.5psi (10.4kPa) at 6000rpm and 3.5psi (24.2kPa) at 8000rpm. In third gear at 5000rpm, back-pressure was 1.5psi (10.4kPa) at 6000rpm 2.6psi (17.9kPa), with a peak of 3.6psi (24.8kPa) at the 8000rpm red-line. In first and second gears a peak of 5psi (34.5kPa) and 4.3psi (29.7kPa) was recorded respectively — both at 8000rpm. The higher peak back-pressure recorded in first and second gears is interesting, probably indicating the effect of a 'traffic jam' within the first section of pipe when there's much less time for the exhaust gases to escape.

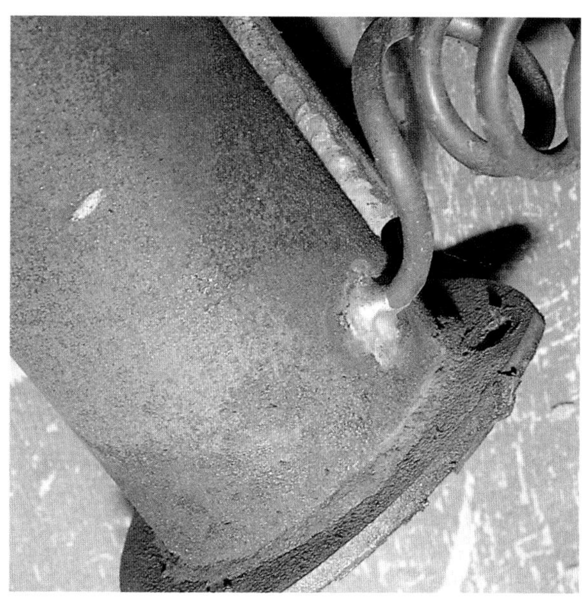

A tapping into the exhaust pipe can be used to measure exhaust back-pressure. This copper tube has been brazed into the dump pipe off a turbo. It's wound into a spiral to shed heat, with the free end attached to a rubber hose run to a pressure gauge.

A new exhaust was made for the car using mandrel-bent 2.5-inch (64mm) tube, a straight-through muffler and a surplus Subaru Liberty (Legacy) RS front resonator. After this exhaust was fitted, the car felt like it went as hard on 15psi (104kPa) boost as it had previously on 22psi (152kPa) while the measured back-pressure was reduced to a peak of just 1.5psi (10.4kPa). Figure 6.7 shows these results. The minimum rpm at which boost was seen also decreased, showing that quicker turbo spool-up was occurring.

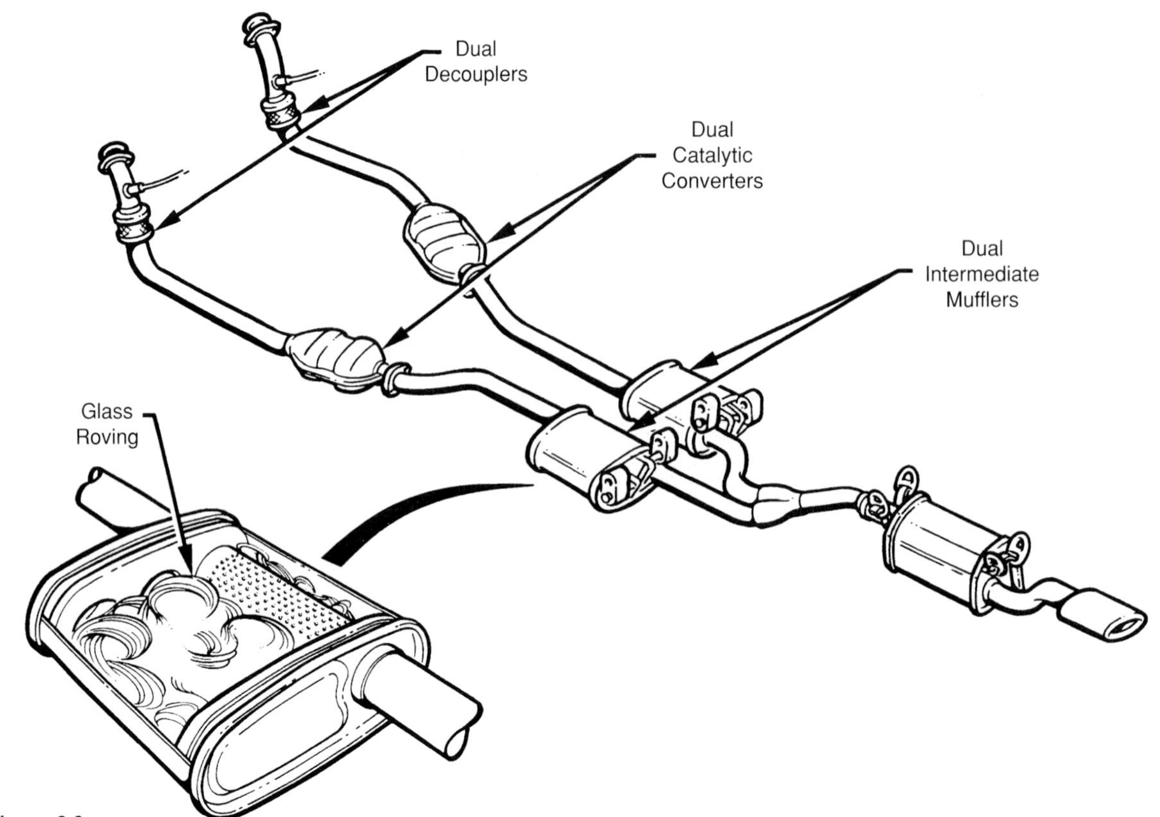

Figure 6.8
Using dual pipes, dual cat converters and dual mufflers gives good flow without requiring a large-diameter pipe that can be difficult to fit under the car. Dual systems like this lend themselves well to V-engine configurations.

Dyno Testing

Rather than measuring exhaust back-pressure, the following two cars were compared for power output on a chassis dyno before and after the exhausts were fitted.

The first car was a 2.0-litre SOHC four-cylinder. As standard, the car used a cast-iron exhaust manifold with relatively long 21cm (8.3-inch) runners formed into a 4-into-2 design. The standard system used twin 41mm pipes from the manifold, joined in a single 41mm tube under the engine. Following this, the pipe increased in size to 43mm for the metre or so before it reached the catalytic converter. After the cat converter, the pipe increased in diameter to 50mm, passing through a resonator and then to the rear muffler where it finally shrank to 45mm. This inspection indicated there were two apparent major restrictions to flow — the 41mm bottleneck at the end of the primaries and the 45mm restriction at the tailpipe.

A new exhaust was fitted. It retained the standard cast-iron manifold, but with the primaries increased in diameter to 44mm. This 3mm increase in the diameter of the secondaries was an educated guess — no calculations

A normal boost gauge can be used to measure exhaust back-pressure if it's plumbed to the exhaust pipe.

This 2-litre four-cylinder system retained the standard cast-iron manifold but used increased-diameter exhaust pipe secondaries. A 2.5-inch (64mm) pipe was used for the rest of the system. Along with some minor intake mods, peak power improved by 14 per cent with no torque loss anywhere in the rev range.

or modelling of the exhaust system were made. The rest of the system was then formed from 2.5-inch (64mm) tubing, using mostly press-bends except over the tight confines of the rear axle, where two mandrel bends were used. A 2.5-inch (64mm) cat converter, 2.5-inch (64mm) straight-through mild steel muffler and a small straight-through resonator were also fitted. A short length of spiral-wound flex pipe was used at the end of the primaries to prevent exhaust breakage when the transverse engine rotated within the engine bay on the application of power. Note this truck-style flex pipe is much, much cheaper than automotive braided exhaust flex, but the exhaust system must be very well supported if it is used.

With the new exhaust fitted, the engine revved far more freely and was noticeably stronger above 5000rpm. On the dyno, the new exhaust and some simple airbox mods lifted the peak wheel power from 58kW to 66kW (78hp to 88hp), a 14 per cent gain. Despite what many believed to be 'too large' an exhaust, there was also absolutely no loss in power at the bottom end of the rev range.

The second example was a 2.2-litre DOHC four. In very near to standard form, the car had a peak power at the wheels of 71kW (95hp). A set of 4-2-1 extractors was then fitted. These used 37cm long 1 5/8-inch (41mm) primaries meeting with 38cm long, two-inch (51mm) secondaries. A two-inch (51mm) cat converter replaced the 1.5-inch (44mm) standard cat, while the rest of the system used the stock 1 7/8-inch (48mm) tube and Toyota resonator. Dyno testing of the car in this configuration showed a 4kW (5hp or 6 per cent) gain in peak power. In many (but not all!) cases, extractors on a late-model car give about a 5 per cent improvement in peak power.

A new exhaust which uses a section of spiral-wound flexible tube. This is usually used on trucks and is much cheaper than braided flex couplings sometimes used in exhausts.

Next, the 1 7/8 inch (48mm) pipe was replaced with two-inch (51mm) tube and a single two-inch (51mm) reverse flow muffler was fitted. This resulted in a further gain of 4kW (5hp), so fitting the two-inch (51mm) system and the extractors had therefore lifted peak power by 8kW (~11hp), or just over 11 per cent. The pipe diameter following the extractors was then increased in size again, with 2.5-inch (64mm) tube, a 2.5-inch cat converter and a 2.5-inch (64mm) reverse-flow muffler fitted. This resulted in a decrease in power at the bottom end of the rev range of 3kW (4hp), a similar performance to the two-inch (51mm) system in the mid-range, and then a 1kW (1.3hp) further gain at the top end. Peak power was therefore lifted by 13 per cent with a 6 per cent decrease at very low revs. I would suggest this low rpm power could be regained through changing the low rpm ignition timing map. However, with the standard engine management, the two-inch (51mm) system was best for this car.

Peugeot

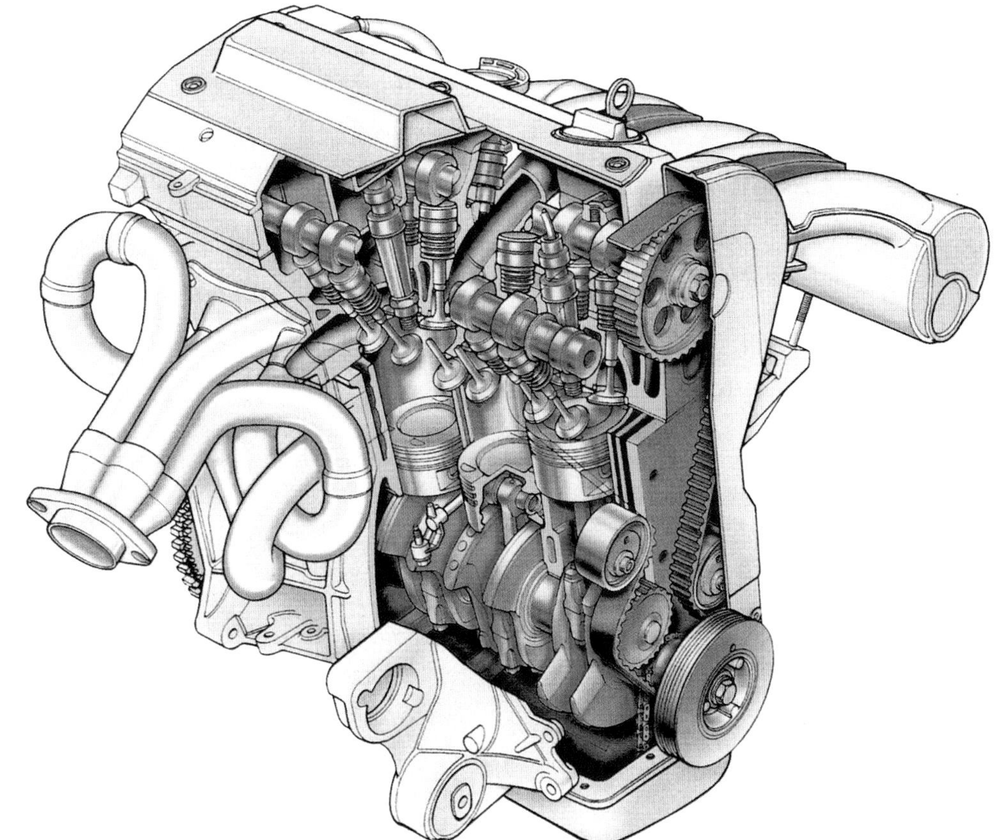

Figure 6.9
The long, large-diameter primary pipes are standard on this Peugeot 306 GTI6 engine. The 2-litre four-cylinder engine develops 124kW (166hp) at 6500rpm and 196Nm at 5500rpm.

Turbocharging & Supercharging

When a naturally aspirated engine is modified for greater performance, great pains are taken to smooth the path of the intake air. Exhaust and inlet valves are often changed; the valve seats and the ports are modified; a high-lift, long-duration camshaft is used; and special attention is paid to the design of the filter and the air pickup. And all this is done so the atmosphere can push as much air as possible all the way into the cylinders.

Now look at the alternative approach — to simply increase the pressure pushing air into the engine. With no internal changes, the same-capacity engine can easily double in peak power! (Of course, you also need to make sure the engine is strong enough to reliably develop all this power.) But compared with the difficulties in developing more power from a conventional engine, it's almost ridiculously easy to get grunt from an engine with forced aspiration.

An engine is force-fed by either a turbocharger or supercharger. A turbocharger uses the flow of the exhaust gases to spin a turbine. The turbine is mounted on the same shaft as a compressor, which in turn forces air into the engine. The more exhaust gas the engine produces, the faster the turbo spins, forcing even more air into the engine. On the other hand, a supercharger is belt-driven directly from the crankshaft; it is an engine-driven air pump or fan that forces air into the engine's intake.

Both approaches to forced induction attempt to push more air into the engine than the engine is capable of breathing. This results in 'boost' — the development of a higher pressure in the intake manifold than exists in the atmosphere. Boost pressures are expressed in pounds per square inch (psi), Bar, or kilopascals (kPa). Sometimes, the intake pressure on forced-aspiration engines is expressed in absolute pressure, ie the boost pressure plus atmospheric pressure (which is about 14.5psi, 1 Bar or 100kPa). An engine with a boost pressure of 7psi will have a gauge pressure (ie reading on a boost gauge) of 7psi and an absolute intake pressure of (14.5 + 7) = 21.5psi. Another approach is to express boost as a 'pressure ratio'. The pressure ratio is found by adding the gauge boost pressure to atmospheric pressure and then dividing this sum by atmospheric pressure. For example, a boost of 7psi is a pressure ratio of about 1.5. (7 + 14.5 divided by 14.5 = 1.48.) In this book, unless otherwise mentioned, all turbo and supercharger pressures are expressed as gauge pressures.

The advantages and disadvantages of turbochargers and superchargers are summarised in Figure 7.1 (opposite). However, the actual approach to forced induction taken often depends more on other factors than the advantages and disadvantages listed. For example, if a factory-produced turbo engine is available, it's not worthwhile buying a naturally aspirated version and then supercharging it. (However, there can be arguments for removing the standard turbo and fitting a supercharger in its place!) If you have an engine that's not available in a factory-forced induction version — but a well-engineered supercharger kit is available — then taking this supercharged approach is likely to be much cheaper and easier than developing your own turbo version of the engine.

Two of the most important differences between the various types of supercharger and the turbocharger are efficiency and the shape of the boost curve. Efficiency can be measured in terms of adiabatic and volumetric parameters, but, basically, the lower the efficiency, the higher the outlet air temperature and the more power required to drive the compressor. As can be seen from

	Turbocharger	Roots-type Supercharger	Centrifugal-type Supercharger	Screw-type Supercharger
Efficiency	High	Medium	High	High
Drive power loss	Low - via exhaust gas back-pressure pumping losses	High – via crankshaft drive	Medium – via crankshaft drive	Medium – via crankshaft drive
Engine torque curve	Low engine torque at low rpm. Depending on turbo matching, may also fall off at high rpm.	Strong at low and medium rpm. Fall off can occur at high rpm due to high power consumption and high outlet air temperature.	Greatest gains at high rpm unless matched to produce maximum airflow at lower rpm – then pressure relief valve needs to be used which reduces efficiency.	Higher efficiency at higher speeds extends speed range over Roots-type. Lower leakage at low speeds improves torque over Roots-type.
Installation	Requires major changes to exhaust and intake systems	Requires installation of drive system and changed intake	Requires installation of drive system and changed intake	Requires installation of drive system and changed intake
Lubrication	Requires high quality engine oil feed	Self-contained lubrication	Self-contained lubrication or engine oil feed	Self-contained lubrication or engine oil feed
Engine matching	Very difficult – both turbine and compressor must be matched.	Easy to match and easy to fine-tune output.	More difficult to match but fine-tuning of output easy.	Easy to match and easy to fine-tune output
Driveability	Turbo lag	No lag, on-tap torque	No lag but maximum torque increase is only high in rev range	No lag, on-tap torque
Fuel consumption	Better than a naturally aspirated engine producing similar power	Better than a naturally aspirated engine producing similar power	Better than a naturally aspirated engine producing similar power	Better than a naturally aspirated engine producing similar power
Emissions	Presence of thermal mass of the exhaust housing increases cat converter light-off time	No effect	No effect	No effect

Figure 7.1
The advantages and disadvantages of different forms of forced induction.

This turbo Centre Housing Rotating Assembly (CHRA) shows the compressor wheel (right), which is held on the turbine/centre shaft assembly by a nut. One of the two coolant connections can be seen and the oil feed and drain holes are at the top and bottom in this view. The compressor and turbine housings attach directly to this assembly.

the table, the Roots-type supercharger has the lowest efficiency of the common forced-induction compressors — but this hasn't stopped this type of supercharger being widely used in current factory-engineered blown vehicles.

The other important characteristic is the shape of the boost curve developed by the compressor. A turbo requires a good flow of exhaust gas before it spins quickly enough to develop boost. This means that at low loads and/or low rpm, boost will not be instantly available. Instead, after quickly opening the throttle, it will take a second or two (or much longer if the turbo is poorly matched) before boost builds and the torque output of the engine increases. This delay is known as turbo lag and

On the left is the turbine wheel and shaft from a HT18 Hitachi turbo, normally fitted to a 13B Mazda rotary engine. On the right is a T34 60-series Sierra turbine wheel and shaft. The HT18 turbine wheel has blades that are much more curved than those of the T34 turbine, giving good bottom-end response. The T34 has less blade curvature and the tips of the blades are short, giving less exhaust back-pressure but also being slower to spool-up.

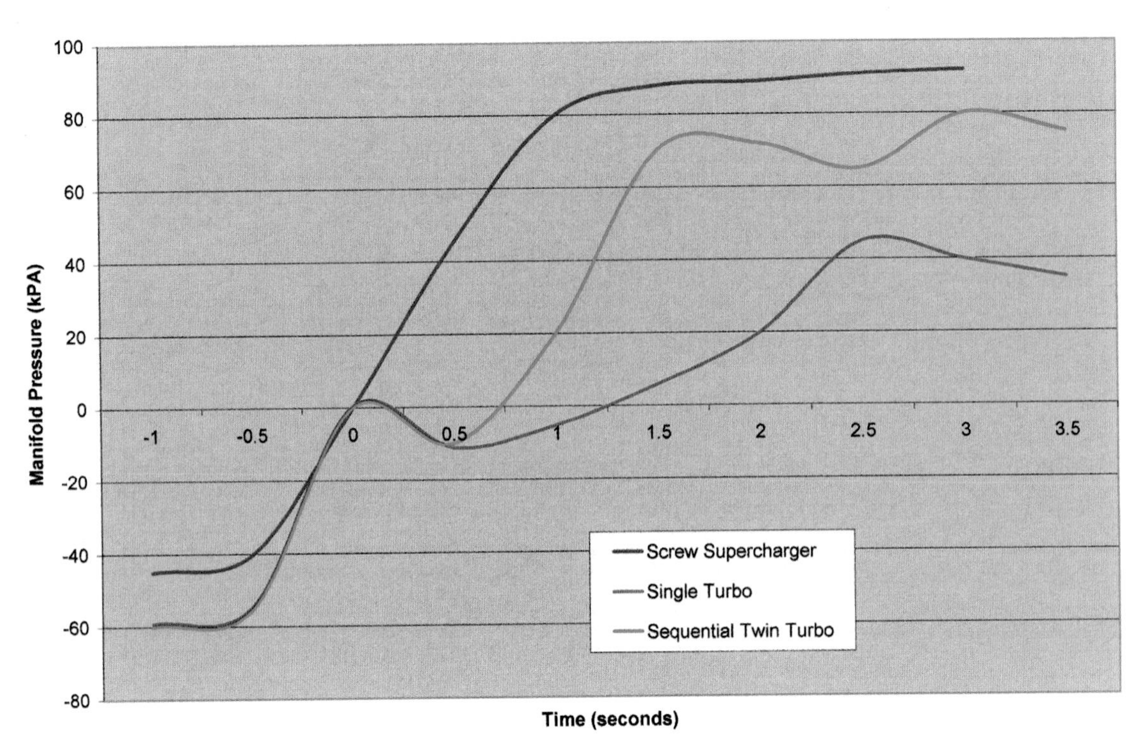

Figure 7.2
The boost response of an engine when it was equipped with a screw-type blower, single turbo and twin turbos. When the throttle was fully opened at -0.5 seconds, the positive displacement supercharger boosted far more quickly than either turbo configuration.

is present to a greater or lesser degree in all turbocharged cars. However, when a turbo car with a well-matched turbo is driven hard, the faster it goes, the faster it feels like it wants to go!

A positive displacement supercharger (Roots or screw types) can maintain a constant boost at all engine rpm, so there is effectively no lag at all. Figure 7.2 shows the results of testing that Mazda carried out when developing their Miller Cycle screw-type supercharged engine. (Note the screw-type supercharger is also known as the Lysholm design.) The graph shows the rate of boost increase in an engine equipped with a screw supercharger, single turbo, or twin turbos. As can be seen, a single turbo takes seconds longer to reach maximum boost, and even a twin turbo configuration is much slower to peak in boost level than the positive displacement supercharger. When being driven, a car equipped with a positive displacement supercharger feels very much like a car with a larger, naturally aspirated engine — there is no rush of boost, just strong and instantly available torque.

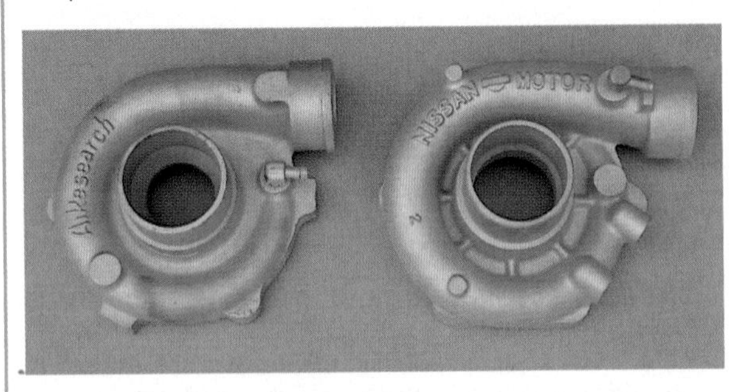

These compressor housings are both from the T3 family of turbos. However, the Saab unit on the left has an A/R ratio of approximately 0.5, while the Nissan compressor housing on the right has an A/R of 0.6.

A Garret T4 V-trim compressor wheel and its matching turbine. Most turbos used in road applications have a similar-sized compressor and turbine.

An A/R ratio of 0.63 is typical of those exhaust housings used on 2- to 3-litre engines, with smaller motors having a lower A/R, such as 0.36 or 0.48. Housings are available with different A/R ratios for the same turbine, allowing the fine-tuning of the turbo response. This housing has an internal wastegate, as is common on factory turbo cars.

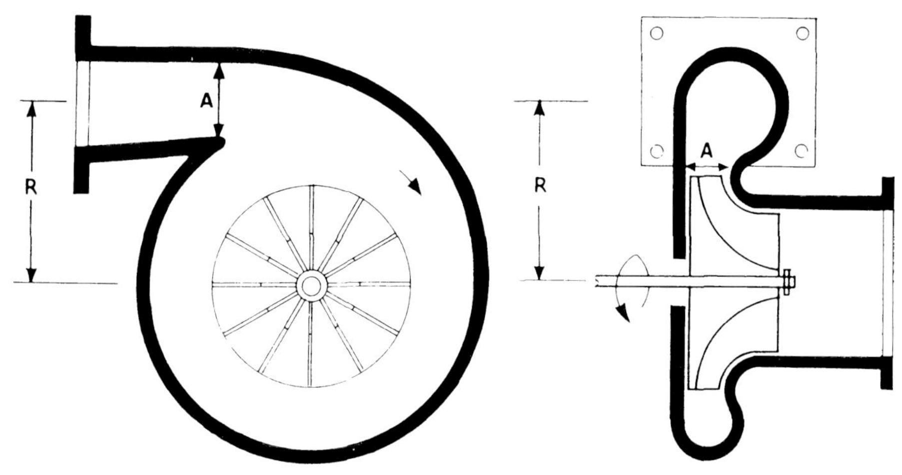

Figure 7.3
The A/R ratio of a turbo housing will determine how quickly the turbo needs to spin to pass a given volume of air. Changing the turbine housing A/R is an effective way of fine-tuning the turbo to its application.

A centrifugal supercharger develops peak boost at maximum engine rpm (or if not, vents excess manifold boost that would otherwise be developed). This usually means that at lower than maximum rpm, boost is also less than maximum. This makes a centrifugal supercharger better suited to larger-engined cars that have a surplus of bottom-end torque but breathing problems at high rpm. A centrifugal supercharger in this sort of application feels strong through the whole rev range, with no lag.

TURBO SPECIFICATIONS

There are four basic variables in the specifications of turbos. These are:

- The flow of the compressor wheel
- The A/R ratio of the compressor housing
- The flow of the turbine wheel
- The A/R ratio of the turbine housing

The main shaft and the turbine wheel comprise a single unit. This unit is used with a dedicated centre (which contains the bearing assembly), to which a variety of compressor wheels, compressor housings and turbine housings can be attached.

The exhaust housing on the left is an IHI VJ20, and the one on the right is a Hitachi HT12. Unlike this pair, the turbine housings from the same turbo family generally have the same stud pattern, making swaps within the family much easier to achieve.

The way the A/R ratio is defined is shown in Figure 7.3. The larger the A/R, the slower the turbo needs to spin to pass a given volume of air. Changing the A/R ratio therefore changes the flow characteristics of the turbo, even without a change of the compressor or turbine sizes. But these wheels can also be changed — their flow being mainly determined by their diameter and the shape of their blades. The A/R ratio of the exhaust housing is very

The range of compressor housings available for the IHI A'PEXi RX6B roller bearing turbos. From left to right — TCW77L, TCW15 and TCW10.

Family	Model	Compressor				Turbine		
		Trim	Wheel Diameter (Inducer, Exducer in inches)	A/R ratio	Flow (lbs/min)	Trim	Wheel Diameter (Inducer, Exducer in inches)	A/R ratio
Small Frame	GT15	56	1.29/1.73	0.33	18	58	1.26/1.62	0.34
		56	1.29/1.73	0.33	18	58	1.26/1.62	0.35
		60	1.49/1.93	0.48	22	72	1.38/1.62	0.35
	GT17	55	1.43/1.92	0.53	21	72	1.48/1.75	0.56
Standard Frame	GT25	48	2.09/3.00	0.70	44	76	1.85/2.12	0.64
		48	2.09/3.00	0.70	44	76	1.85/2.12	0.86
	GT30	56	2.25/3.00	0.70	55	84	2.16/2.36	0.63
		56	2.25/3.00	0.70	55	84	2.16/2.36	0.82
		56	2.25/3.00	0.70	55	84	2.16/2.36	1.06

Garret

Figure 7.4

Some of the specs for the Garret GT Series ball bearing turbos. These turbos are available with a variety of compressors, turbines and turbine A/R ratios. Note that Garret uses the rule of thumb that each pound/minute of air is enough to develop nine horsepower.

important in determining the boost curve, so changing the turbine housing for another with a different A/R is often done to fine-tune a turbo to its application.

Reducing the mass of the turbo's rotating assembly will quicken transient response. The replacement with ceramic of the austenitic nickel steel alloy usually used to form the turbine is sometimes done to reduce the mass of the spinning assembly. Testing by Nissan of ceramic turbos has indicated up to a 21 per cent decrease over a conventional turbo in the time required to achieve 5.8psi boost. However, the ceramic blades have also been known to shatter when the engine is working hard. If this occurs, very hard particles enter the engine through the exhaust ports and cause major internal damage.

Ball (rather than plain) bearings are also used to improve the transient response of some turbos, and composite material compressor wheels are also being used in some applications. Another approach to quickening response is to use a variable nozzle turbo that has multiple vanes arranged around the periphery of the turbine housing. The angle of the vanes can be changed, altering inlet area and exhaust flow angle onto the turbine wheel. Variable nozzle turbos are currently being used extensively on OE diesel cars.

Figure 7.4 shows the available specs for some of the Garret GT Series ball bearing turbos. As can be seen, the complete turbos are available with different compressors, turbines and turbine A/R ratios. Note that Garret uses the rule of thumb that each pound/minute of air is enough to develop nine horsepower.

The exhaust housings available for the A'PEXi RX6B. clockwise from upper left — P20, P22, P25, P27, P18, P16 and P13. These numbers refer to the housings' A/R ratios.

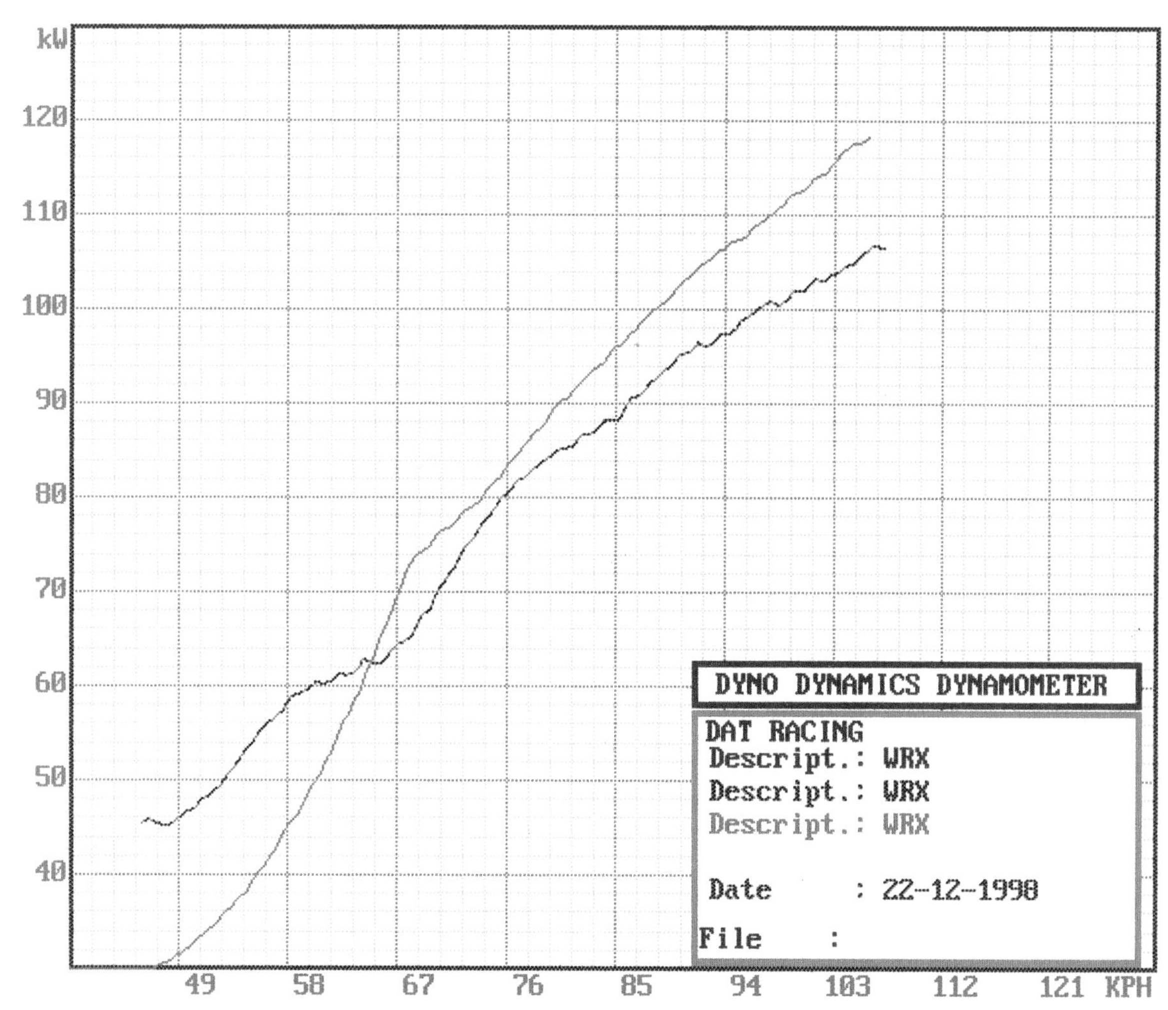

Figure 7.5
The power curves of a Subaru Impreza WRX equipped with a TD04 turbo (blue line) and a TD05 turbo (green line). Note the car was tested on a two-wheel-drive dyno, limiting measurable power to 120kW. At different stages in the model's life, the WRX was available standard with each of these turbos. The smaller turbo provides better bottom-end response at the cost of higher rpm power.

MATCHING TURBOS

The matching of a turbocharger to a particular engine and application is complex, time-consuming and very easy to get wrong! The latter is partially the case because there aren't any fixed criteria against which to measure success. Manufacturers that sell turbo road cars spend thousands of person-hours using sophisticated software, building and testing engines, and evaluating different turbos. And the end result of that process might be described by one person as "hopeless" and by another as "perfect"!

In fact, some manufacturers change the turbo that's fitted to a car during its model life. Figure 7.5 shows the power output of a MY99 Subaru WRX Impreza. The constant four-wheel-drive turbo car was tested in rear-wheel drive on a two-wheel-drive dyno, and in this dyno configuration was limited to a maximum power of 120kW at the wheels. The MY99 used a TD04 Mitsubishi that features both a smaller compressor and a smaller turbine than the TD05 used in the older MY98 model. The blue line on the graph shows the power output with the standard TD04, while the green line is the same car fitted with the previous model's TD05. As can be seen, the later-model turbo provides better bottom-end response at the cost of higher rpm power. In this case, the owner of the car was very happy swapping from the smaller to the larger turbo, much preferring the extra top-end power of the TD05. So which is the better-matched turbo? Subaru obviously decided the smaller turbo was better, but most enthusiasts disagree...

The matching of a turbo to its application will be affected by:

- The peak power and torque developed by the engine
- The shape of engine's torque curve
- The efficiency of the turbo
- The turbo's airflow characteristics
- The turbo's dynamic characteristics
- The car's gearing, mass and aerodynamics
- The required driving characteristics of the car

To say this is extremely complicated is understating the case, especially when it's realised that many of these factors interrelate. In effect, it's simply impossible for anyone working outside of a major manufacturer to take into account all these factors.

The turbo which best matches an engine and application is always open to discussion! These different-sized turbos were both fitted by Subaru to the WRX. On the left is the TD04 Mitsubishi from the MY99, while on the right is the larger TD05 used in the older MY98 model. The partial power curves of the engine equipped with each turbo are shown in Figure 7.5 (previous page).

Further, those matching a turbo to an engine by the use of extensive dynamometer testing face another hurdle. It is the bottom-end response that's the hardest factor to optimise when matching a turbo to its application. And on a dyno, quite an incorrect impression can be gained in this very area. On either a chassis or engine dyno it's very easy to hold high loads at low engine speeds — something that almost **never happens** on the road! For example, on a 200kW (~270hp) 3-litre turbo six-cylinder engine, full load on the dyno at 1500rpm is easily held for three or four seconds. However, this does not accurately reflect road use — the engine would only be in this load/rpm situation if towing a trailer up a steep hill in first gear! This load doesn't occur at full throttle when accelerating through the gears from a standstill, because the car is at 1500rpm only once (in first gear) and then only for a fraction of a second. This is an important point, because artificially holding a high load creates lots of exhaust gas flow, which in turn quickly accelerates the turbo.

In other words, the type of dyno testing that holds the engine at high loads at low rpm creates an overly favourable boost curve. For example, on the dyno, a particular engine/turbo combination might be able to develop 3psi boost at 2000rpm, but it won't on the road when the car is actually being driven! To test the boost curve accurately on a dyno, the engine should be ramped through its rev range using the loads and ramp speeds characteristic of the engine when it is powering the car through each of the gears. This is seldom, if ever, done.

The best theoretical matching of a turbo to an engine starts with the use of compressor maps. These show the volume or mass flow at the different pressure ratios (ie boost levels) the compressor is capable of flowing. However, the mathematical process of working out from these maps the most appropriate turbo configuration is generally so involved and complex that, in the real world, almost no DIY mechanic takes this approach. Another reason for the scarcity of this approach is that many turbo manufacturers will not even release this type of information — unless it's to major car manufacturers! And even if such technical information **is** available, factors such as the mass airflow of the engine also need to be known. It's extremely difficult to **accurately** calculate the:

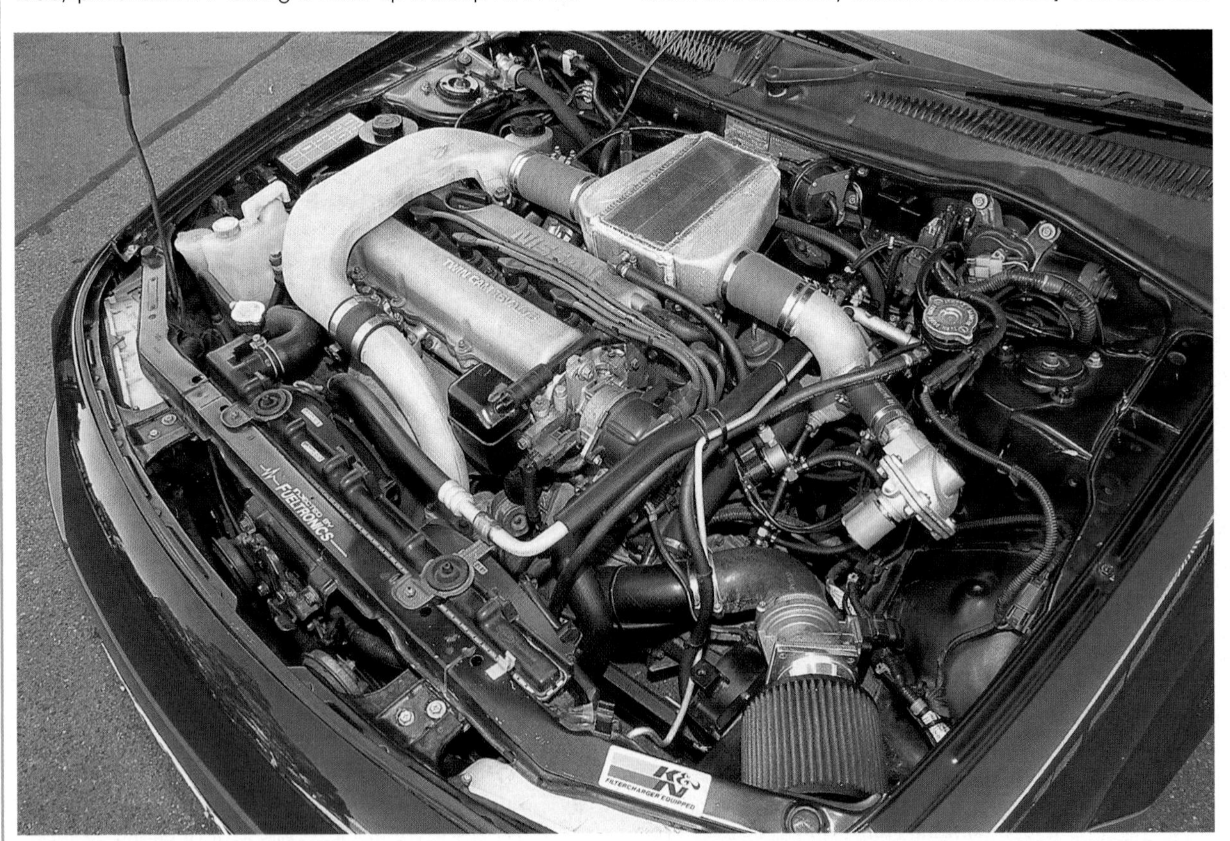

If the compression ratio is kept high, the off-boost performance will be stronger, in turn giving more exhaust flow and allowing the turbo to spool up more quickly. This Nissan SR20 2-litre four retains the standard naturally aspirated compression ratio and uses a relatively large T3 usually fitted as standard to a 3-litre, 150kW engine.

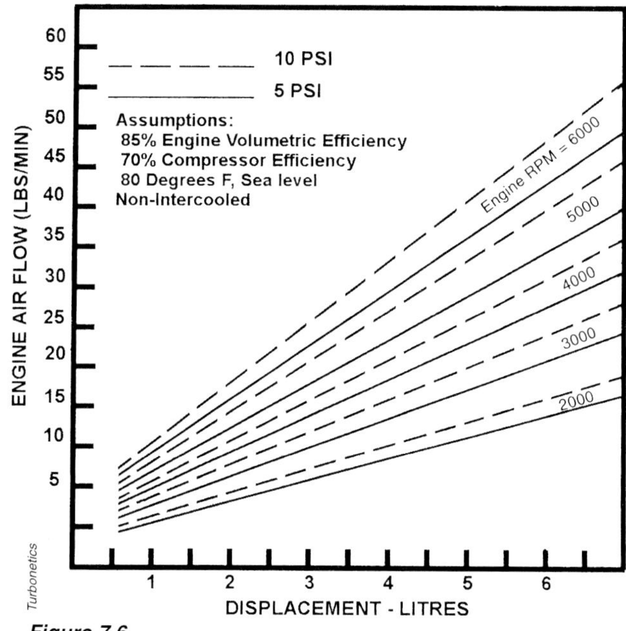

Figure 7.6

The mass airflow for a turbocharged, unmodified, four-valves-per-cylinder engine or a heavily ported two-valve engine, estimated to have a volumetric efficiency of 85 per cent, a compressor efficiency of 70 per cent, no intercooler, at sea level with an ambient temp of 27 degrees C.

- Engine volumetric efficiency at all rpm
- Turbocharger efficiency at various airflows
- Intercooler efficiency in various airflow and weather conditions

without access to a great deal of engineering data!

When you look at the lack of detailed turbocharger specifications, the difficulty of applying that information, and the expense of buying on spec the brand-new turbo you roughly calculate you require, it's understandable

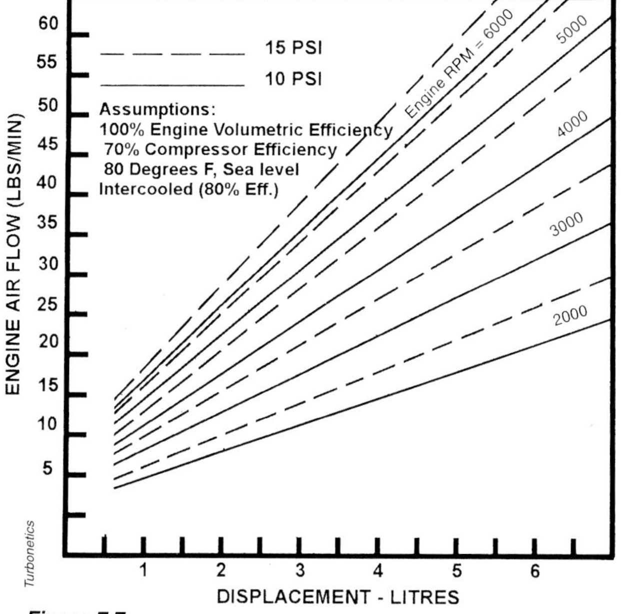

Figure 7.7

The mass airflow for a turbocharged, unmodified, four-valves-per-cylinder engine or a heavily ported two-valve engine, estimated to have a volumetric efficiency of 100 per cent, a compressor efficiency of 70 per cent, an intercooler, at sea level with an ambient temp of 27 degrees C.

that this is a path taken by very few people indeed. However, that doesn't stop all the textbooks pretending this is the approach taken by everyone! But with enough information available, and by making some educated guesses, a simplified version of this approach can still be viable. Turbonetics is one company that releases compressor maps and also information on turbo matching. The approach taken by Turbonetics is shown as follows.

The first step is to estimate the maximum mass flow of the engine. Figure 7.6 shows the mass airflow for a turbocharged unmodified four-valves-per-cylinder engine or a heavily ported two-valve engine, estimated to have a volumetric efficiency of 85 per cent, a compressor efficiency of 70 per cent, no intercooler, at sea level with an ambient temp of 27 degrees C. Figure 7.7 is for the same type of engine, but this time with 100 per cent volumetric efficiency and intercooled. Figure 7.6 shows the mass airflow requirement for 5psi and 10psi boost levels, while Figure 7.7 shows the requirements for 10psi and 15psi boost. These boost pressures correspond to pressure ratios of 1.3, 1.7 and 2.0 respectively.

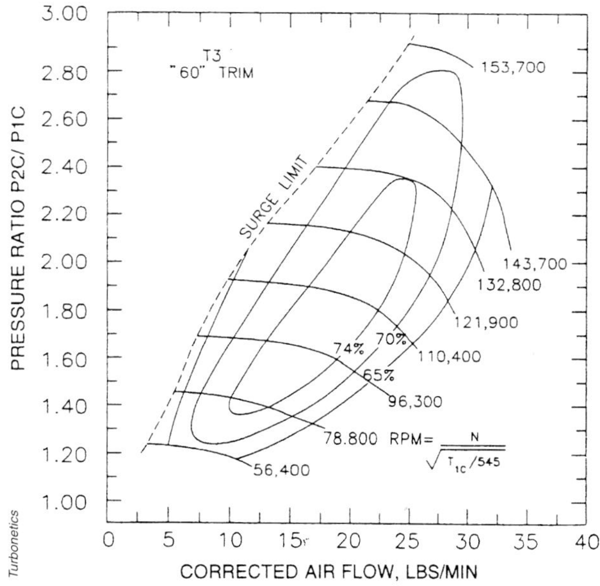

Figure 7.8

The map for the Turbonetics T3 '60' trim compressor. If the mass airflow of the engine can be calculated, a compressor map can be used to ascertain the suitability of a compressor for the engine application.

For example, a 3-litre engine with 85 per cent VE, a 6000rpm redline, no intercooler and running 10psi boost (a pressure ratio of 1.7) has an estimated mass airflow of about 26 pounds/minute. At peak torque (estimated to be 4000rpm) the airflow will be about 16 pounds/minute. The compressor should be sized so the airflow required **at peak torque** is near to the best efficiency 'island' on the compressor map, while the airflow at maximum output falls into an efficiency area of 60 per cent or higher. Figure 7.8 shows the compressor map for the Turbonetics T3 '60' trim. At a pressure ratio of 1.7 (boost of 10psi) the peak torque airflow of 16 pounds/minute falls into the island of highest efficiency, while the peak rpm airflow occurs at better than 65 per cent efficiency.

Note the 'surge limit' marked on Figure 7.8. Compressor surge occurs when the compressor wheel is attempting

to flow pressurised air that cannot be accepted by the engine. In other words, if the compressor is too large for the application, surge can occur. Surge is characterised by sudden reversals of airflow, causing a loud resonating sound to be emitted by the turbo and the boost pressure to fluctuate rapidly. The noise emitted is far louder than, for example, the sound of an engine detonating, so is able to be detected by ear alone! Compressor surge can quickly destroy the turbo. Sometimes, a large compressor will be on the verge of surge at low throttle angles or when dyno testing causes unrealistic boost conditions. However, if surge does not actually occur on the road, this is not a concern.

Another way of estimating the turbo compressor size needed is to work backwards from the naturally aspirated and desired turbo engine power figures. For example, a 2-litre engine develops 100kW (134hp) and it's desired that 180kW (~240hp) be developed. Boost pressure is expected to be in the range of 10-15psi. Using Garret's rule of thumb that nine horsepower can be developed for each pound/minute of air, at 240hp a mass flow of about 27 pounds per minute at an estimated pressure ratio of 1.7-2.0 will be needed. Consulting Turbonetics compressor maps shows that a high-flowing T3 compressor would therefore be suitable.

Once the compressor has been sized, the turbine and its housing need to be chosen. Manufacturers' charts list appropriate turbine sizes for each compressor wheel, and this guide is generally fine for road cars. However, there's usually a variety of different A/R ratio turbine housings available for each wheel — and the chosen turbine housing A/R will make a large difference to the way in which the turbo boosts. The larger the A/R ratio, the slower the turbine will spin for the same passage of exhaust gas and the lower will be the exhaust back pressure. Overly large A/R ratio turbine housings will therefore cause the turbo to boost too late; overly small A/R ratios will cause the turbo to potentially over-speed or to exert excessive exhaust back-pressure on the engine. OE manufacturers usually specify a relatively small turbine wheel and a low A/R so the turbo boosts early in the engine rev range and has quick response.

The table below shows the range of turbine housings available from Turbonetics for the Garret T3 and T4 turbo families. Note that Garret and IHI use a different numbering system to show the A/R ratio, but for each, the higher the number, the larger the A/R. A split pulse exhaust housing uses two parallel intakes, fed by two branches of the exhaust manifold.

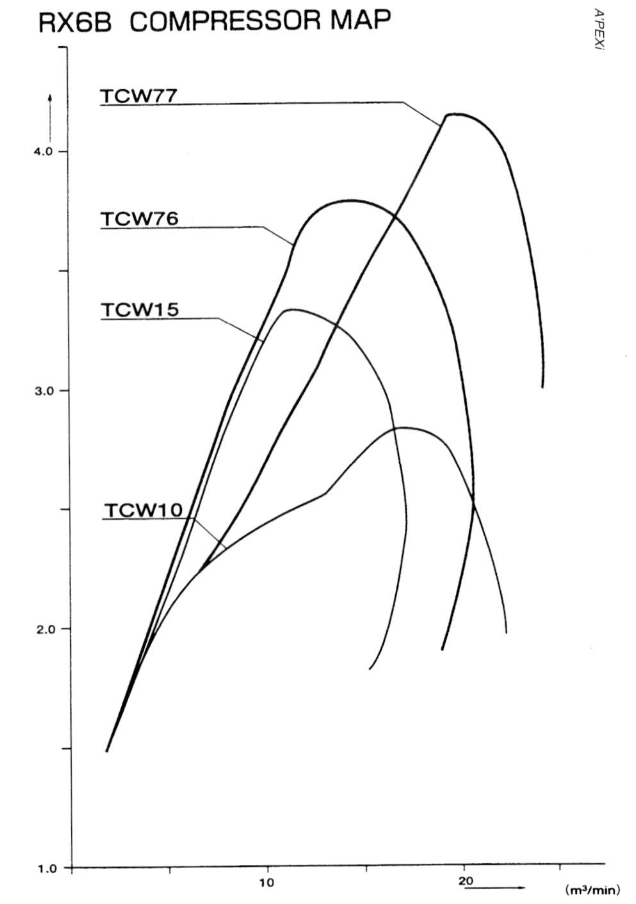

Figure 7.9
The partial compressor maps for the A'PEXi RXB6 compressors. Note airflow is expressed in cubic metres per minute on this map.

A'PEXi recommended turbo configurations for the Nissan RB26DETT, all the way up to 670kW!

However, if having read all of the above, you desire an easier way of matching a turbo to an engine, simply find a very similar OE turbocharged engine to the one you wish to turbocharge and look at the turbo used by the manufacturer. For example, if you wish to turbocharge a naturally aspirated 2-litre 16-valve four-cylinder engine that develops about 100kW (134hp) in standard form, the turbo used by Nissan on their 2-litre 16-valve SR20DET engine to develop 147kW (197hp) is likely to be a very good choice. (The naturally aspirated SR20 Nissan develops 105kW.) Manufacturers strive to achieve good low-rpm response and are scrupulous in not exceeding turbocharger specifications that could lead to over-speeding or compressor surge. Given this turbo on modified SR20DET engines is good for just over 200kW (~270hp), there would also be room for some power increases through the use of higher boost pressures and breathing improvements.

Turbo Family	Turbine Housing A/R ratios
T3	0.36, 0.48, 0.63, 0.82
T4	
(standard)	0.58, 0.69, 0.81, 0.96, 1.3
(split pulse)	0.58, 0.70, 0.84, 1.00, 1.15, 1.32, 1.52

The IHI A'PEXi RX6B roller bearing turbo is available with a range of compressor and turbine housings. Three compressor housings are available (TCW77L, TCW15 and TCW10) and seven exhaust housings (P13, P16, P18, P20, P22, P25 and P27). A'PEXi has recommendations for various Japanese engines; for a 375kW (~500hp) Nissan RB26DETT engine the use of a single TCW76LSP25 is suggested. The maps for the RX6B compressors are shown in Figure 7.9, while Figure 7.10 (opposite) shows the other

The same approach can be used when upgrading the size of a turbo on an engine already turbocharged. For example, let's say the standard turbo engine develops 80kW

Configuration	Turbo		Appropriate Maximum Power (kW)
	Compressor	Exhaust Housing	
Twin	Standard	Standard	270
Twin	TCW15	P16/P18/P20	410 – 630
Single	TCW10	P18/P20	450 – 670
Twin	TCW10	P20	450 – 670
Single	TCW76LS	P25/P27	340 – 460
Twin	TCW76L	P13/P16/P18	460 – 670
Single	TCW77LS	P25/P27	390 – 500
Twin	TCW77L	P13/P16/P18/P20	500 – 670

Figure 7.10
The A'PEXi-recommended turbo configurations for the Nissan RB26DETT engine, all the way up to 670kW (890hp)! Note the range of power outputs over which each turbo works is quite large; however, the smallest combination that yields the required maximum power will give best driveability.

(107hp). The first step in hunting for more power is to increase boost pressure (along with upgraded intercooling, improved engine management and so on). This will typically result in a peak power of about 105kW (141hp), but a further power upgrade will need a larger turbo. So do you now go for the turbo from an engine that in standard trim develops 105kW (141hp), or from an engine developing what you want the **new** peak power figure to be — perhaps 120kW (161hp)? Many people jump too great a distance, resulting in a turbo that takes too long to come on boost and is dull and unresponsive unless the engine is working very hard. Therefore, in this case, the turbo from a turbocharged engine developing in standard form perhaps 110kW (147hp) would be best.

While this approach looks quite simplistic, it takes into account the relative exhaust and intake flows, turbo speed and turbo boost pressure and approximates the engine volumetric efficiencies. If you find the new turbo isn't as suitable as you had hoped, you also have a saleable item in that the turbo is standard. Turbos that have been 'high flowed' to suit a particular oddball application are much harder to shift on the second-hand market...

However, when you start talking about much higher power applications, there are many fewer single-turbo role models available for you! But don't OE manufacturers produce 200kW and 300kW turbo engines? They certainly do, but generally they have decided the only way sufficient bottom-end response can be gained is to use twin turbos. In itself that's a **very** significant point — especially in the light of claims by some that there will be plenty of bottom-end torque (and no problems with 300kW) from their custom-built, one-off, very special turbo that they're prepared to assemble just for you!

Sometimes it's possible to get away with the use of a single large turbo if the engine has a large swept capacity and/or uses an auto transmission with a high-stall torque converter. The large engine gives good low rpm power even before the turbo starts to boost, while the high-stall torque converter lets the engine rev a little harder than it otherwise would before the car launches, helping to spool the turbo up. However, in most high-power road applications, twin turbos (run in either sequential or parallel forms) are required. Even in drag racing applications, twin turbos are often employed to maintain a flat torque curve across as wide a range of engine speeds as possible. If you choose to use a single very large turbo, be prepared to trade off a lot of bottom-end response.

Twin turbos give better response and quicker boosting than using a single, large turbo.

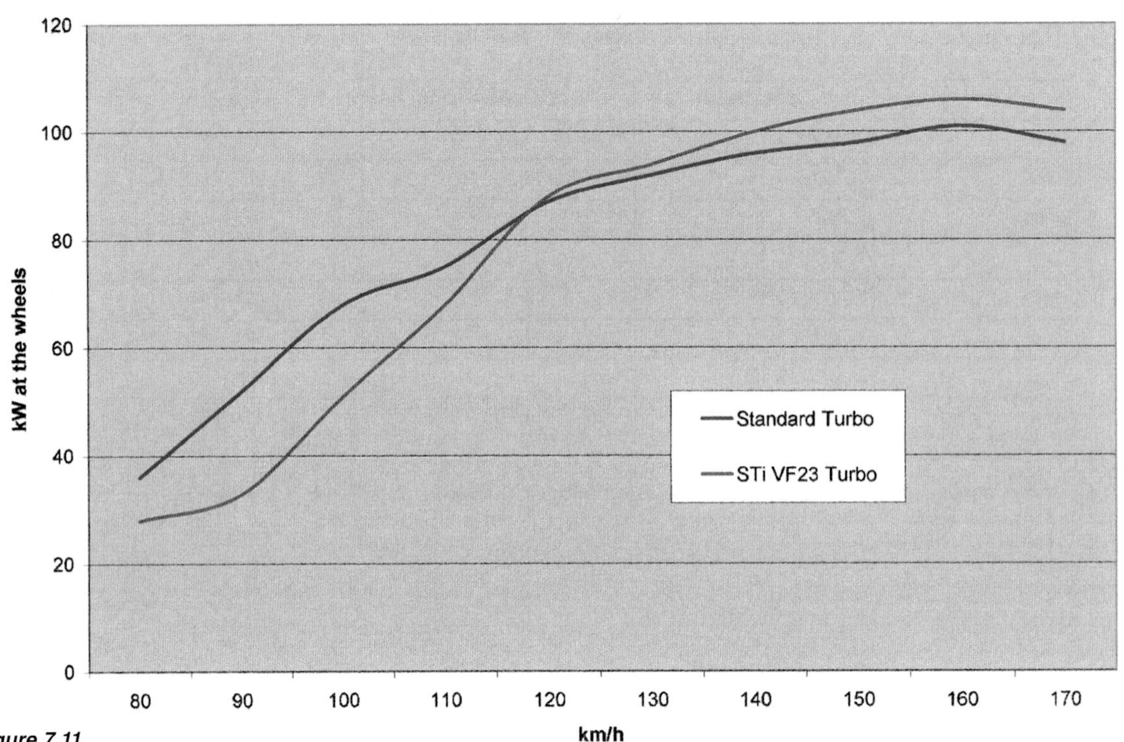

Figure 7.11
On this MY99 Subaru Impreza WRX, swapping the standard turbo for a larger STi VF23 turbo gave a peak power gain of only 5 per cent. However, in the lower portion of the rev range, the decrease in power was substantial.

MATCHING TURBO SWAPS

Many of the preceding comments apply to turbo swaps. Increasing the size of the turbo will lead to a decrease in power at lower rpm with a corresponding increase in the potential power available at higher rpm. Many people will suggest that a larger turbo can be fitted with no trade-off in low rpm power or response, but this is **very** rarely the case.

Figure 7.11 shows a dyno comparison performed on two Subaru WRX Imprezas. The constant four-wheel-drive cars were tested on a four-wheel-drive dyno in third gear. With the standard Mitsubishi TD04 turbo, the MY99 Impreza (fitted with a free-flow rear muffler) had a peak power of 101kW at the four wheels. The car with the larger STi VF23 turbo developed 106kW at the wheels — a gain of 5 per cent. However, the shape of the two power curves is very different. In the lower portion of the rev range, the car with the standard turbo had up to 58 per cent more

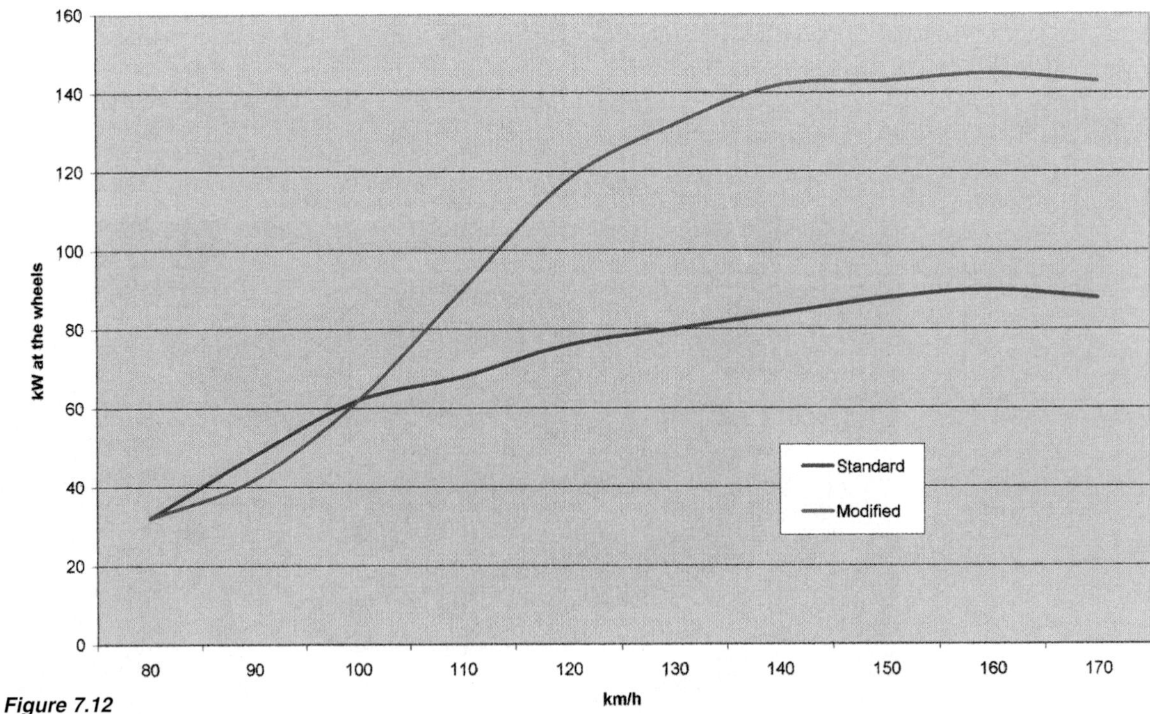

Figure 7.12
The dyno results of a modified Subaru Impreza WRX compared with a standard car. Both cars were MY98 models, but the modified car had been fitted with an A'PEXi exhaust, electronic boost controller, front-mounted large intercooler, airfilter, programmable management system and a new turbo. Peak power is up by just over 60 per cent with almost no loss at low revs.

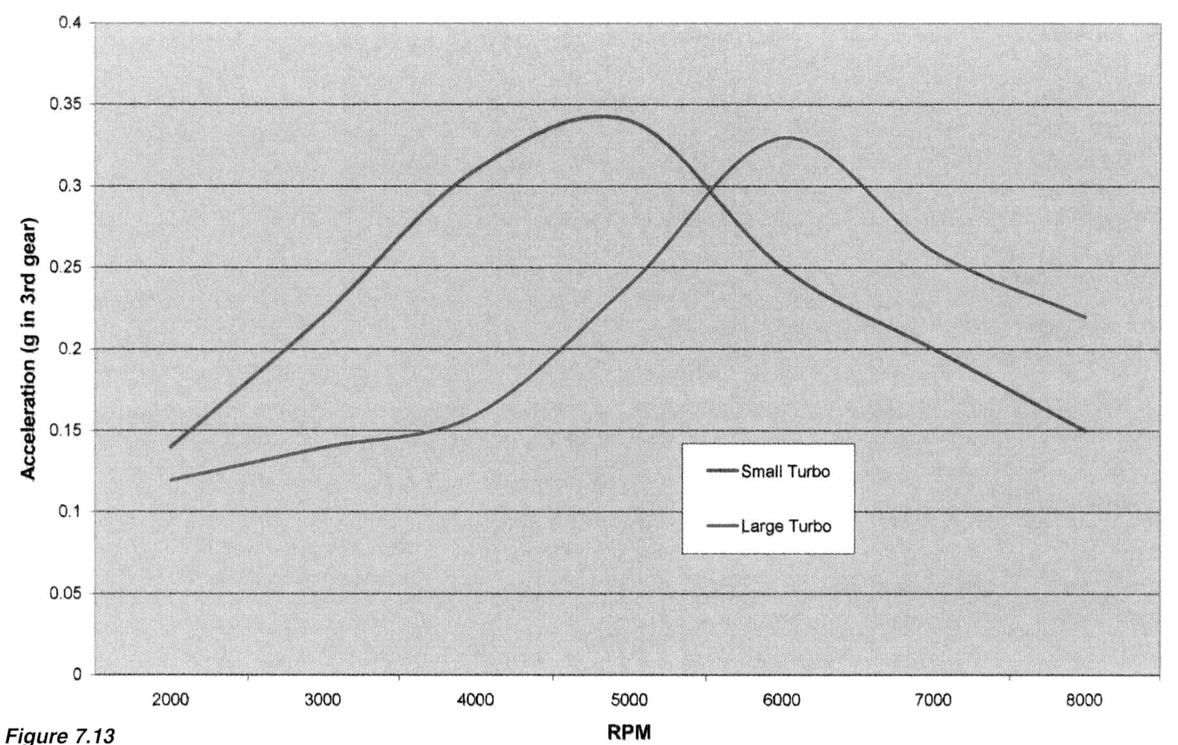

Figure 7.13
A comparison of the on-road acceleration of a Daihatsu Mira Turbo, when equipped with the standard RHB31 turbo and with a larger RHB5 turbo. There was a very substantial reduction in on-road acceleration at engine revs of less than 5500, with acceleration down by 48 per cent at 4000rpm! However, other modifications did improve this situation.

power! In this particular case, it would appear that the larger turbo isn't doing much for performance at all — although with more boost, it would probably lift peak power considerably.

However, swapping in a large turbo doesn't **have to** result in a major compromise between lower and upper rpm power. Figure 7.12 (opposite) shows the dyno results of yet another modified Impreza WRX compared with standard. Both cars were MY98 models, but the modified car had been fitted with a long list of top-quality A'PEXi parts — exhaust, electronic boost controller, front-mounted large intercooler, air filter, programmable management system and a new turbo. As can be seen, peak power is up by just over 60 per cent with almost no loss at all at low revs. A good car!

Figure 7.13 is another turbo swap comparison, this time carried out by measuring the acceleration achieved on the road. The car was a Daihatsu Mira Turbo with a

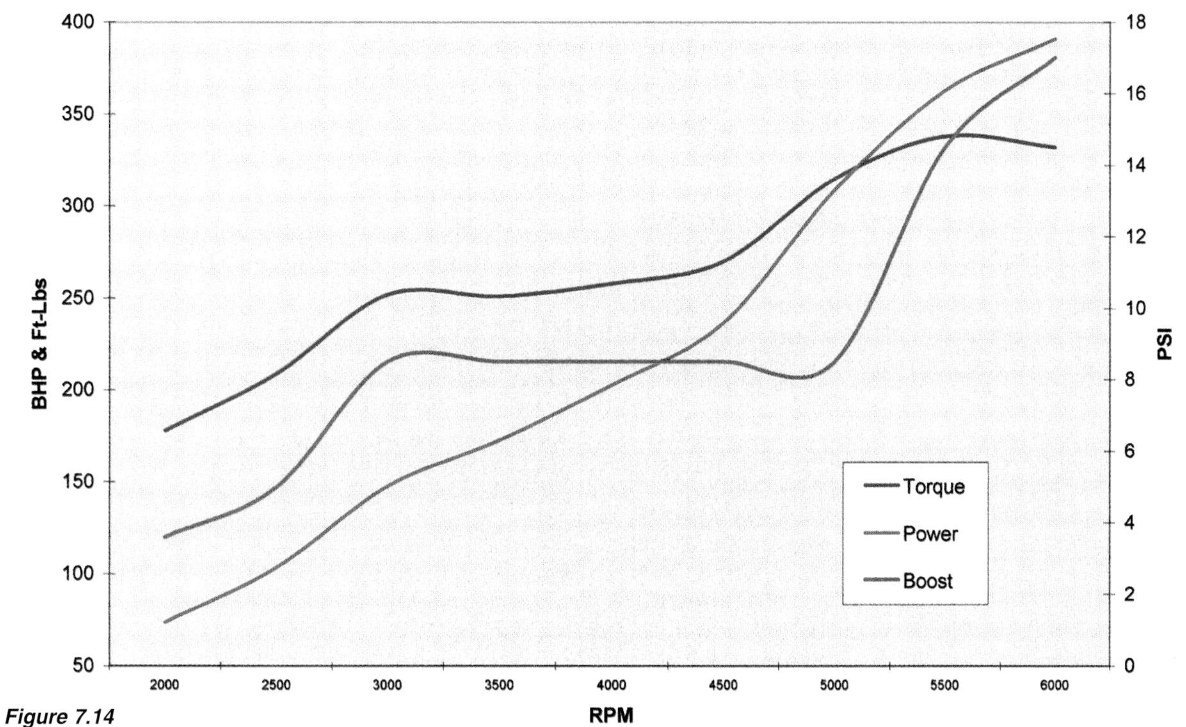

Figure 7.14
The engine dyno data for a Nissan RB25DET 2.5-litre DOHC turbo six, fitted with an internal wastegate hybrid T3/T4 turbo. The turbo used a Sierra T3 turbine wheel in an RB30ET exhaust housing, and a T4 compressor wheel. The internal wastegate was able to control boost only between 3000rpm and 5000rpm, with boost then rising as high as 17psi.

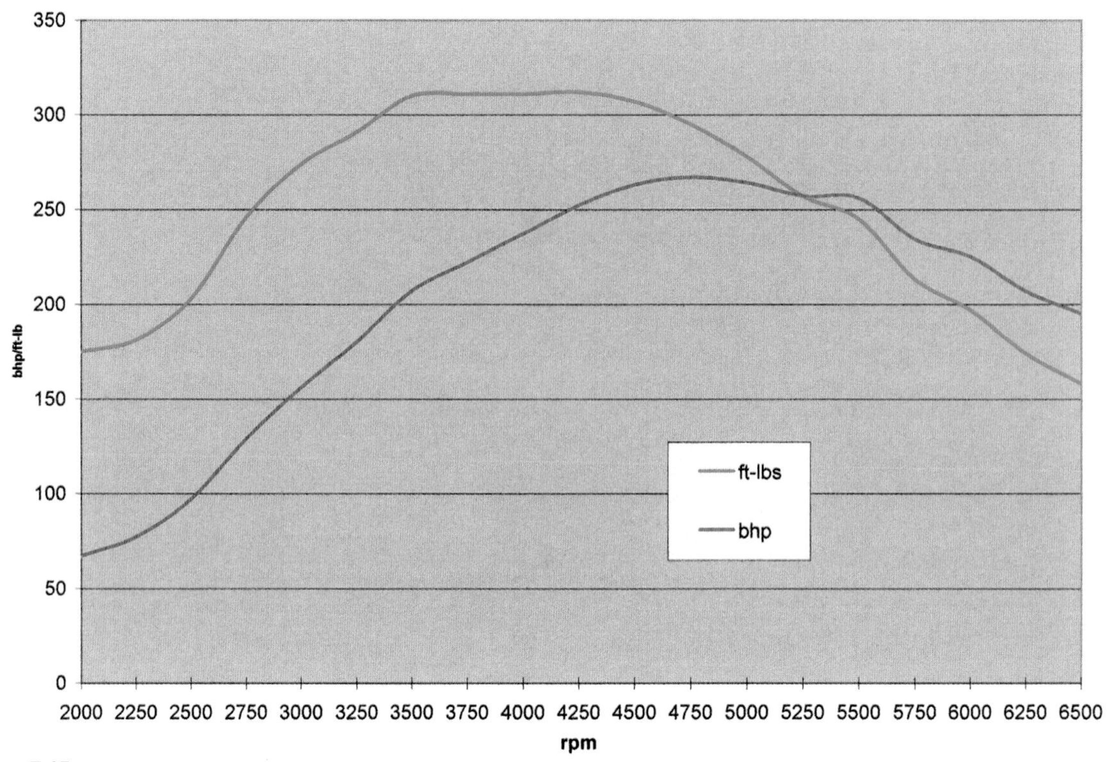

Figure 7.15
The power and torque curves of a 2.5-litre Honda V6 equipped with the turbo and manifolds from a 2-litre Honda V6. While power is not high, the engine develops more than 200ft-lb (270Nm) of torque from 2500rpm to 6000rpm, giving a very wide power band. A higher peak power could almost certainly have been realised with a larger single turbo (or a turbo on each cylinder bank) but in this form, the driveability is excellent.

three-cylinder, 660cc engine. In standard form, this engine develops 49kW (66hp). The turbo used was from a Daihatsu Charade GTti, which normally develops 75kW (101hp) from its 1-litre three-cylinder engine. The standard turbo was a RHB31 and the Charade turbo a RHB5. At the time of these acceleration runs, no changes other than the turbo swap had been made. The acceleration was measured in third gear using an accelerometer (see Chapter 11), a technique that's more valid than using a dyno because all the real-life variables are present. As can be seen, there was a very substantial reduction in on-road acceleration at engine revs of less than 5500. In fact, acceleration was down by 48 per cent at 4000rpm!

Interestingly, in this case, there was a substantial pressure drop across the standard (very small) intercooler. When a larger intercooler was fitted, peak power improved considerably and the turbo also spooled up more quickly. While the new turbo was still less responsive down low, the situation depicted in Figure 7.13 (previous page) improved substantially!

Figure 7.14 (previous page) shows the engine dyno data for a Nissan RB25DET 2.5-litre DOHC turbo six. This was fitted with an internal wastegate hybrid T3/T4 turbo, using a Sierra T3 turbine wheel in an RB30ET exhaust housing, and a T4 compressor wheel on the induction side. As can been seen, the internal wastegate was able to control boost only between 3000rpm and 5000rpm, with boost then rising as high as 17psi. On the road, the car required use of the gearbox to keep the turbo on the boil — fifth gear at lower rpm gave no boost at all. In this case, the on-road performance was significantly worse than you would imagine when looking at the engine dyno curves.

Figure 7.15 shows a quite different approach. If anything, in this case, the turbo was a little small for the application. The Honda 2.5-litre C-25A naturally aspirated V6 was decompressed to 7.3:1 and turbocharged using the turbo and manifolds from a 2-litre Honda V6. On 14.5psi boost the engine developed 267hp (199kW) at only 4750rpm. However while the power was not high, the engine developed more than 200 ft-lb (270Nm) of torque from 2500 all the way to 6000rpm, giving a very wide power band. A higher peak power could almost certainly have been realised with a larger single turbo (or a turbo on each cylinder bank), but in the above form the driveability was excellent.

This type of approach can also be seen in Figure 7.16 (opposite), which shows the manufacturer's power and torque curves for the Calibra 4WD turbo. The torque curve very clearly shows the quick boosting of a small turbo — maximum torque is achieved at just over 2000rpm — and how this boost is held right through to 5000rpm. However, once past 5000rpm, the torque starts to drop quickly. To gain a great deal more power from this engine, a larger turbo would need to be fitted. But as a very driveable turbocharged car, the Calibra in standard form has excellent engine characteristics!

Figure 7.17 (opposite) shows the chassis dyno power curves of a Nissan Gazelle owned by mechanic Anthony Wilson and powered by a CA18DET 1.8-litre turbo engine. The standard turbo is a T25 with a 0.49 A/R exhaust housing and a medium/small compressor spinning in a 0.48 housing. The red line on the graph shows that in standard form, power at the wheels peaked at just over 100hp (75kW). Increasing boost to 15psi and temporarily removing the aircleaner increased this to 125hp (93kW)

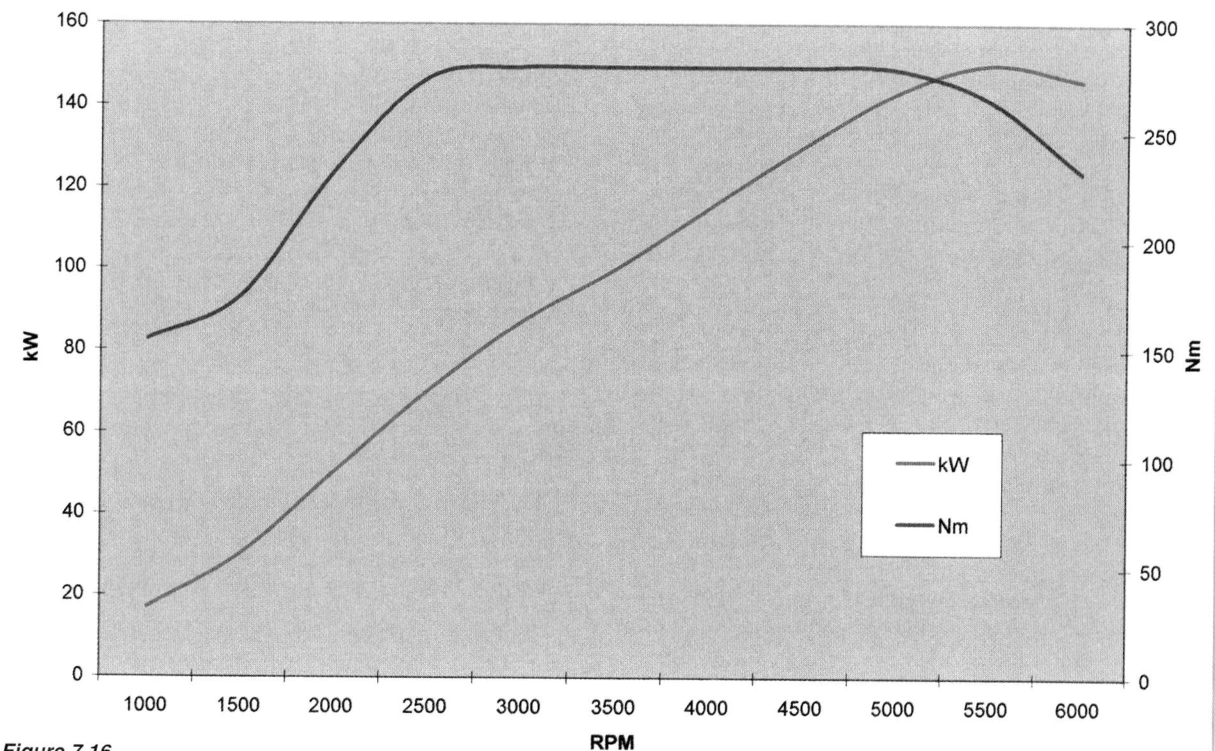

Figure 7.16
The manufacturer's power and torque curves for the Calibra 4WD turbo show the result of using a small turbo. Maximum torque is achieved at just over 2000rpm and is then held to 5000rpm, where the torque then starts to drop quickly. This turbo sizing gives excellent driveability.

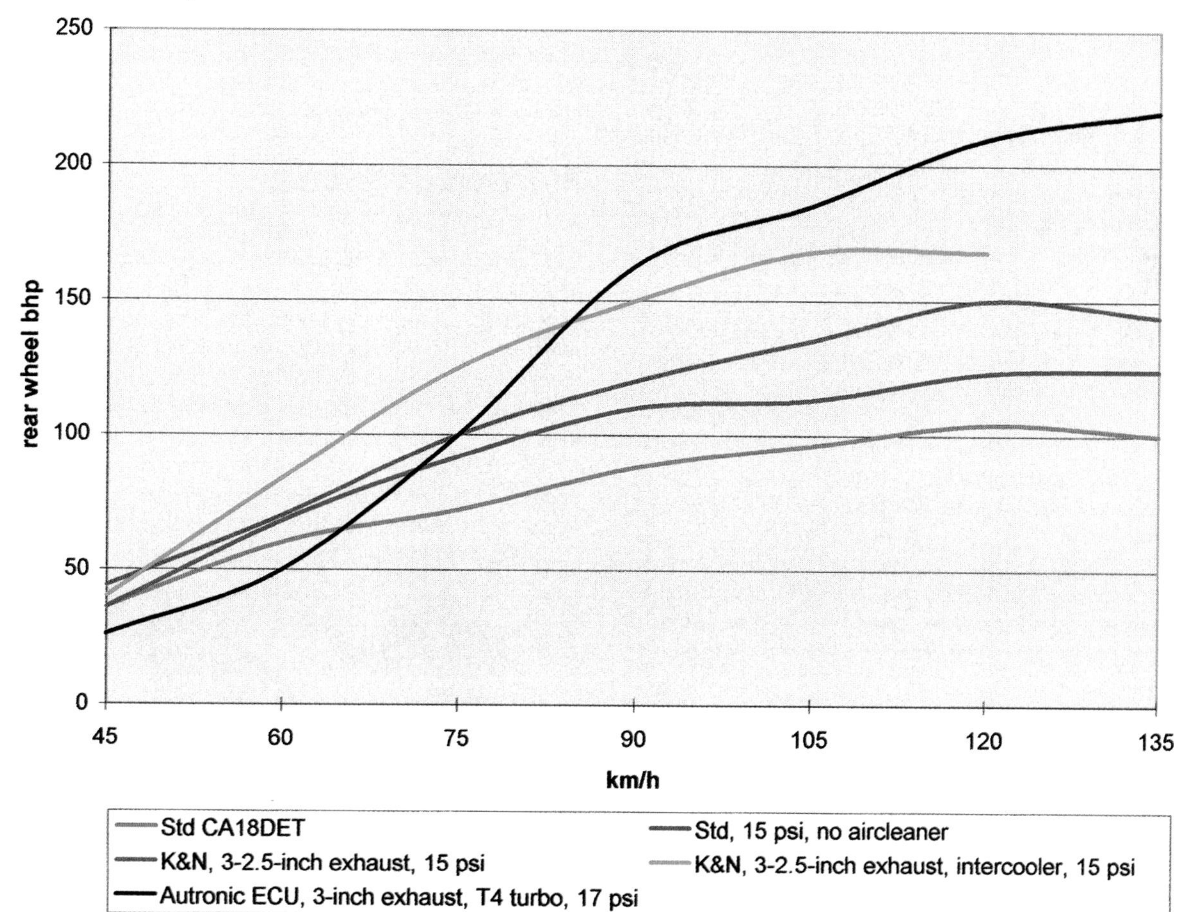

Figure 7.17
The chassis dyno power curves of a Nissan Gazelle, fitted with a CA18DET 1.8-litre turbo engine. The red line shows the rear wheel power in standard form. The blue line shows the result of increasing the boost to 15psi and temporarily removing the aircleaner. The result of adding a new intake filter and 3-inch-into-2-inch exhaust is shown by the purple line, while the green line shows the results of a new filter and a large intercooler. Programmable management and a large T4 turbo were then fitted, with this power curve shown by the black line.

— a very cheap 25 per cent lift in power! Adding a new intake filter and 3-inch-into-2-inch exhaust improved power to a peak of 150hp (112kW) — so far, a 50 per cent power gain! Next, a new filter and a large intercooler were fitted, giving a peak of about 160hp (119kW). Note the standard turbo and engine management were both still being used at this stage. As you will subsequently read, Anthony then tried another five(!) different turbos, but for **all-round** good performance, he rates the standard turbo as the best.

Next, Anthony fitted programmable Autronic management and a much larger T4 turbo featuring a 0.84 split-pulse exhaust housing and a P-trim turbine. The compressor used was a very old design, apparently from an International tractor. Using a V-band housing, it had no marked A/R but had a six-blade compressor with a 50mm 'eye' (the eye is the diameter of the opening in the compressor cover). He also fitted a full three-inch exhaust and lifted boost to 17psi. This step is shown by the black line on the graph. You can clearly see how the large turbo caused a major change in the shape of the power curve, dropping low rpm power very considerably. Power now peaked at about 220hp (164kW) at the wheels. But was the change in turbo size worth it? Anthony was adamant that it was a step forward, but as a passenger in the car, I thought the result was appalling. The transition from no boost to a great deal of boost happened across only 500rpm and power below 4000rpm was almost non-existent.

However, not one to accept anything less than optimal, Anthony kept on trying different turbos. The next turbo was also a T4, featuring the same exhaust end but using a V-trim compressor that had eight blades and a larger 53mm eye. It was used with a 0.6 A/R compressor cover. This combination gave improved bottom-end performance, came on boost less suddenly and had a better top-end than the previous T04.

Following this, a Nissan turbo from a VG30 DOHC-per-bank 3-litre V6 was fitted. This used a T3 60-trim compressor in a 0.6 A/R housing and a T3 ceramic turbine working in a 0.82 housing. However, this turbo proved to have violent surge when fitted to the 1.8-litre engine. Initial acceleration would provide adequate performance, but then power would stop increasing. The engine and car would violently shake, a high-pitched resonating noise coming from the turbo. The boost gauge needle would also flick rapidly (moving too fast to see at times!) from 0-15psi. Even when past the surge zone, the turbo would not develop more than 10psi boost and gave poor top-end performance. Anthony believes the flow of the T3 compressor was simply inappropriate for the application, but makes the point that on the basis of flow maps, this compressor should have suited the application very well...

This turbo was obviously not the best, so it was removed and replaced with a ball-bearing T3, again from a VG30 Nissan 3-litre V6. This turbo is identical in compressor and turbine specs to the one discussed above, except it uses a compressor 3mm larger in its base circle (but with the same-sized eye). But this turbo also had bad surge problems, although it could hold a higher boost (15psi) at the top end, so developing more power.

At the time of writing, the final turbo Anthony has fitted to his car is the unit from a Nissan RB25DET. This uses a very large compressor wheel (similar to a T4 M-trim) and a turbine wheel the same as the VG30 design but spinning in a housing with the smaller A/R of approximately 0.4-0.6. This turbo gives very good top-end power (although still not as high as the T4) but with excellent boosting and no sign of compressor surge. In fact, 17psi boost is available from just 3000rpm. However, Anthony has inveigled a friend (also with a CA18DET Gazelle) to become involved in the experiment. The turbo fitted to that engine is a T4 using a very small S-trim compressor in a 0.6 A/R compressor cover, with a 0.58 A/R N-trim turbine. This turbo gives a similar rate of boost increase to the turbo on Anthony's car, but has a stronger top end — 360hp (269kW) on the engine dyno.

Each of the turbos fitted by Anthony Wilson appears on paper to be a good choice — perhaps the first T4 excepted. However, the results gained have varied enormously in boost delivery, power and driveability.

As can been seen from all these examples, the manner in which boost builds, the amount of turbo lag and the general driveability will be dependent on the selected turbo. Getting a great deal of power by increasing turbo size almost always results in a major decrease in lower rpm torque. For these reasons, by far the best way of selecting a turbo upgrade appropriate for you and your car is to drive another car of the same sort already equipped with the new turbo. That way, you will be able to assess whether the results match your expectations. That's not a glib answer, nor one that is necessarily immediately helpful. But it does reflect the reality of the situation...

HI-FLOWING TURBOS

The term 'hi-flowing' is used to describe modifications made to the turbo itself. This can involve changing the compressor cover and wheel but leaving the rest of the turbo intact (a 'hybrid' turbo in some vocabularies), or the machining of housings, or the back-cutting of turbine wheels. Other modifications often made to turbos while they are in pieces include the fitting of a 360-degree thrust bearing to replace the more common 180-degree unit, and the use of a reverse-threaded nut to hold the compressor wheel in place. The latter both add durability rather than outright performance.

If a larger-diameter compressor is being fitted in the standard housing, the housing will need to be machined. This machining is carried out on a lathe.

The contour of the inner part of the housing needs to be matched to the shape of the compressor. This can be done in a NC lathe or — at a pinch — by hand with a scraper, as is being done here.

The hi-flowing of a turbo can result in a turbocharger totally unsuited to its application — or one that works superbly. In other words, if someone says they have a turbocharger that's hi-flowed, it tells you nothing other than that the turbocharger itself has been modified. It's a little like describing an engine as 'hot'.

The final clearance between the compressor and its housing should be about 0.030-0.040 inches. This is being established here with abrasive paper.

In my experience, a hi-flowed turbo is even less likely to provide improvement for a road car in all performance aspects than a straight turbo swap. A hi-flow turbo often has mis-matched compressor and turbine ends, resulting in poor bottom-end torque characteristics. However,

The wheel/housing clearance being checked using plasticine.

once again, until you have driven a car with the new turbo in place, you will not be able to say with certainty whether or not you are happy with the end result.

In all cases, if you are dealing with a factory turbo car, boost and breathing improvements (ie exhaust and intake mods) **should precede the changing of the turbo**. Only when excessive inlet air temperatures or exhaust back-pressures occur should a turbo change be considered.

A heavy-duty thrust bearing (left) and the standard thrust bearing. Upgrading the thrust bearing aids turbo durability in high-boost applications.

CHANGING TURBOS

Changing the turbo fitted to a car may be as simple a procedure as unbolting the old and bolting in the new! However, if the turbo is from a different manufacturer or is a different 'family' upgrade (from T3 to T4, for example), it's likely the turbine housing will not be a direct fit to the exhaust manifold. If this is the case, there are three ways of overcoming this problem.

If a turbo swap with another family is being made, the exhaust manifold may need to be modified to suit the new turbine housing. This Mazda 323 unit's outlet passage has been considerably enlarged over standard.

First, the exhaust manifold may be able to be modified. Enlarging the outlet orifice to match the new exhaust housing's inlet and/or changing the location of the mounting holes can be done. Alternatively, an adaptor plate can be made. This is a thick steel plate that's bolted to the original exhaust manifold, with new threaded holes (or studs) to take the new exhaust housing. Finally, a new exhaust manifold can be fabricated. If the new turbo needs to be moved in its mounting position (because it fouls bodywork, for example), making a completely new exhaust manifold has significant advantages.

This turbo has been fitted in a non-standard application. Note the modification to the wastegate rod to clear one of the fittings.

On the compressor side, it's likely that changes in the diameter or location of the hoses will need to be made. However, as this side of the turbo is relatively cool, simple rubber hose adaptors and tubes can be used. Finally, the oil pressure feed, oil drain and water fittings will probably need to be altered. Changing these fittings can be a fiddly operation, and, if the turbo swap needs to take place with the car off the road for as short a time as possible, these hose and fittings should be made up before the actual swap occurs.

This adaptor plate has been used to allow the fitting of a non-standard exhaust housing to the exhaust manifold. Note the use of thread inserts in the plate to strengthen the attachment to the manifold.

TURBO EXHAUST MANIFOLDS

Fabricated exhaust manifolds can be used in three different situations:

- Where an engine not normally available with a turbo is having one fitted
- Where the standard turbo is being changed and the new turbo does not match the standard exhaust manifold
- Where gains in performance are required over that available from the standard turbo exhaust manifold

Two different approaches can be taken to the fabrication of an exhaust manifold. The most common

Most factory turbo exhaust manifolds are adequate (given the limitations of space, cost and gaining quick cat converter warm-up) for all but the most extreme power-ups. Note how this engine uses a turbo with a split-pulse exhaust housing.

approach is to make the manifold from heavy-gauge, mild-steel, steam pipe. This is available in pre-formed bends with a wall thickness of 4mm or so; it can be bought in a variety of internal diameters and bend angles. To form the manifold, the bends are MIG or arc welded together. The other approach is to use much thinner gauge stainless-steel pre-formed bends, usually welded together with TIG gear.

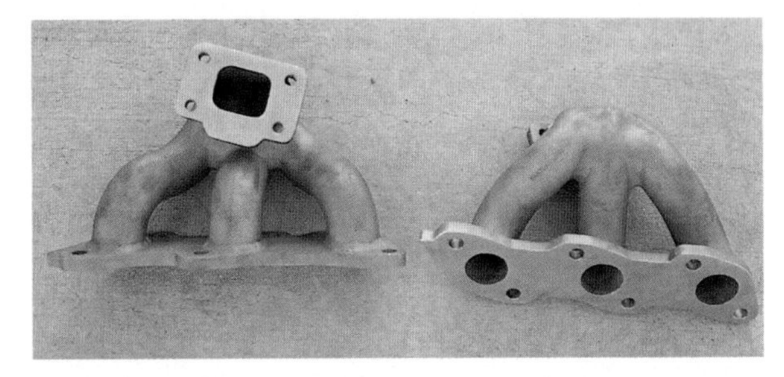

A pair of superbly fabricated exhaust manifolds for a twin-turbo straight six. Thick-walled steam pipe has been used in the construction of these manifolds.

In either case, the exhausts of the cylinders should be grouped in a similar way to that done with extractor exhausts, and attention should be paid to smoothing internal contours and using gentle bends. Relatively small-diameter, short runners will help give quick turbo response. However, despite appearances, factory-style cast-iron exhaust manifolds work quite well, even with power increases of 100 per cent. Fabricating a manifold is normally an expensive proposition, and a fabricated manifold is also much more likely to crack than a cast-iron manifold.

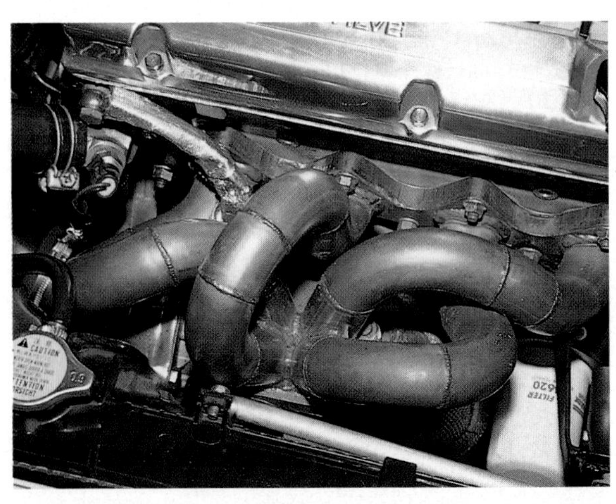

Stainless-steel, thin-gauge bends have been used to construct this manifold. Note the excellent TIG welding and the equal-length, gently curved runners that have been used.

Simple and effective, this manifold has been constructed from steam pipe and matches the very tight confines of the engine bay into which this turbo engine was fitted. Note the cylinder pairing and split-pulse design.

When it's being decided whether or not to replace a factory cast exhaust manifold with a fabricated unit, it's a good idea to have the standard manifold and head flow-tested together on a flowbench. Each exhaust manifold runner should have flow similar to the others, and the presence of the exhaust manifold should not substantially decrease the head's exhaust port flow.

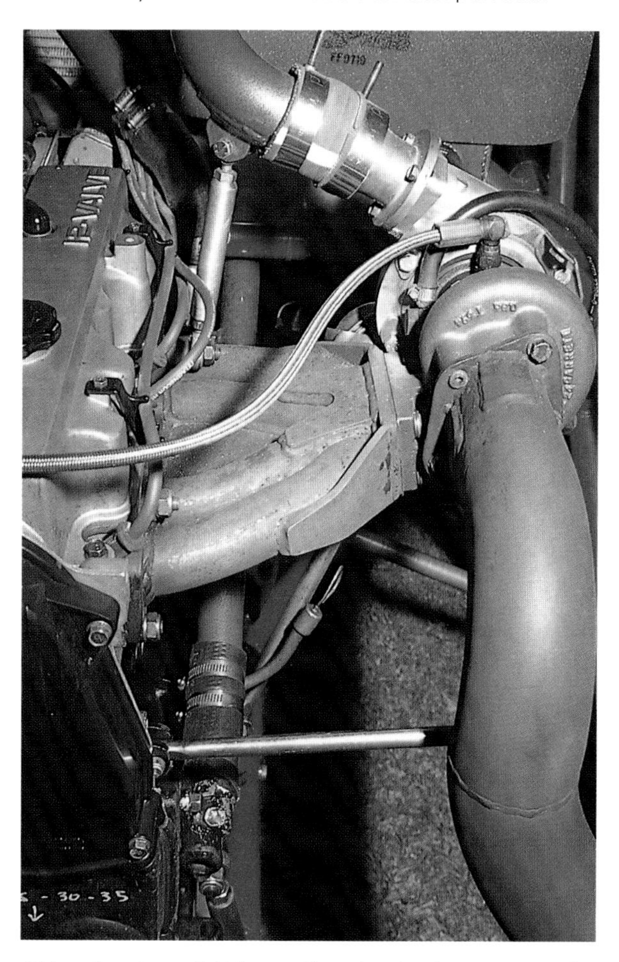

This exhaust manifold in an off-road racing buggy uses relatively straight, long runners and extra strengthening to withstand the high off-road loads. Remember that in most installations, the turbo is supported by only the exhaust manifold.

TURBO WASTEGATES

By far the majority of automotive turbos use internal wastegates. These take the form of a swing valve mounted within the exhaust housing. When it's open, exhaust gases can bypass the turbine, so being 'wasted'. The

A large external wastegate can bypass a substantial amount of exhaust gas, giving good boost control on high-power engines.

swing valve is operated by a rod connected to a wastegate actuator, a canister containing a diaphragm that has a spring mounted on one side and intake manifold pressure pushing on the other side. Boost pressure deflects the diaphragm against the spring, opening the valve and allowing the bypassing of some exhaust gases around the turbo. The magnitude of the boost developed is dependent on the spring pressure and the pressure acting on the diaphragm.

External wastegates often have the facility to apply boost or vacuum to both sides of the diaphragm. However, here only a boost pressure feed is used, deflecting the diaphragm against the internal spring.

The alternative to an internal wastegate is an external wastegate. An external wastegate is physically much larger than an internal design, being an assembly entirely separate from the turbo. As with an internal wastegate, an external wastegate is operated by a diaphragm and a spring. However, in this case, vacuum or boost pressures can be applied to both sides of the diaphragm. Within the wastegate is a large poppet valve. An external wastegate is plumbed into place prior to the turbo. For this reason, the exhaust manifold may need to be modified to accept the wastegate plumbing, or an adaptor block for the wastegate inserted before the turbo. With an external wastegate fitted and appropriately operated, very accurate control of boost can be carried out.

An internal wastegate is fine for the vast majority of turbo road cars. It also has several major benefits over fitting an external wastegate: it's already present so costs nothing,

A Turbonetics Racegate that uses a 42mm stainless-steel valve and is appropriate for use on engines developing up to 670kW (900hp). No internal wastegate can pass the amount of exhaust gas that a large external wastegate like this is capable of flowing.

it's compact and is easily controlled. However, where the volume of exhaust gases has been increased substantially through the development of a lot more power, the standard internal wastegate may not be able to flow sufficient gas. If this is the case, excess gas will be channelled through the turbine, causing boost pressure to rise uncontrollably. Furthermore, exhaust gas back-pressure on the engine will be higher than that which would have been present if a large external wastegate had been fitted. Also, some larger turbos are not available with internal wastegates.

An internal wastegate can be modified for more flow. This involves the boring-out of the wastegate hole and the replacement of the swing valve with one of a larger diameter. The path through the wastegate hole can also be smoothed, and the use of a more efficient exhaust following the internal wastegate will aid the flow passing through it.

TURBO BOOST CONTROLS

The airflow output of a turbo compressor rises as the square of its rotational speed. This means that doubling the turbo's shaft speed increases by a factor of four the air output. This characteristic is quite different for an engine, where a doubling of engine speed will (more or less) double the engine's appetite for air. A turbo that can develop 5psi boost at 3000rpm engine speed may

Before changes are made to turbo boost, a boost gauge should be fitted. Making boost changes without clearly seeing the results is very dangerous indeed!

develop 20psi boost at 6000rpm! Therefore, some form of boost control is needed when using a turbo. An optimal boost control system allows the turbo boost pressure to rise to the designated maximum level as fast as possible and then stay steadily at that boost level even as exhaust gas flow increases.

Boost control can take various forms:

- A restriction in the intake system, sized so not enough air can be inhaled to allow the turbo to exceed the designated maximum boost pressure
- An exhaust restriction, sized so not enough exhaust can pass through the turbine to cause it to exceed the designated maximum boost pressure
- The use of a small turbo that cannot flow sufficient air to exceed the designated maximum boost pressure
- The venting of air from the intake manifold above a designated maximum boost pressure
- The use of a wastegate that bypasses exhaust gas around the turbine when boost pressure reaches the designated level

The latter form of boost control is most commonly used; all factory turbo cars have a wastegate.

The wastegate actuator, wastegate pressure hose and wastegate actuator rod can all be seen here. All variable boost systems bleed off pressure from the hose.

Factory Turbo Boost Control

The control of the wastegate is normally carried out in a simple manner. A hose senses boost pressure from a connection close to the turbo compressor's outlet. Boost pressure travels down the short connecting hose to the wastegate actuator, deflecting the actuator's diaphragm against the spring. If the factory wastegate actuator is set for 7psi boost, the diaphragm will be deflected so the wastegate valve will bypass enough exhaust gas to hold boost at close to 7psi. Figure 7.18 (opposite) shows this type of factory wastegate control. Where boost control is carried out by the ECU, the approach is very similar except that a pulsed solenoid is used to bleed from the wastegate actuator hose.

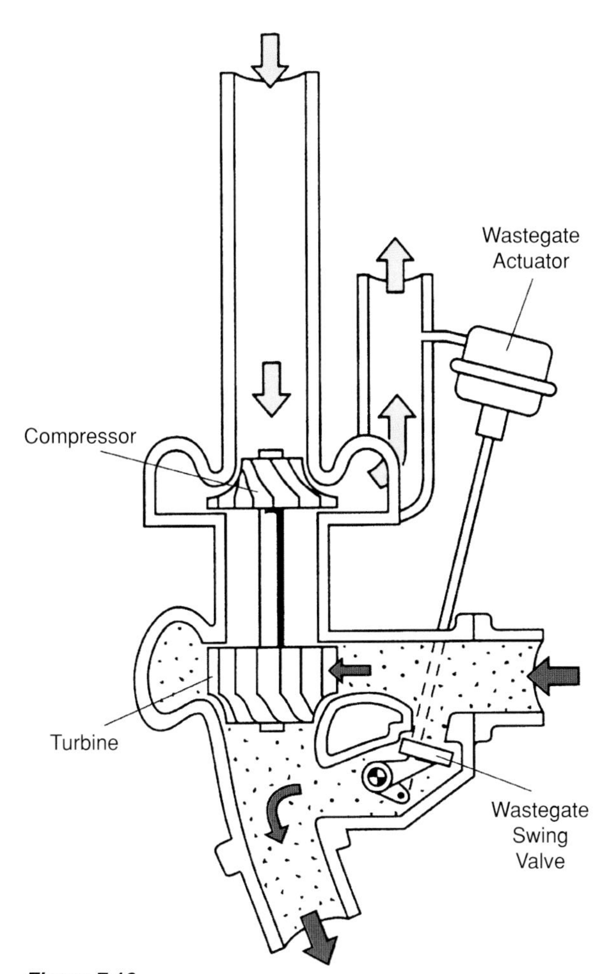

Figure 7.18
The vast majority of factory turbo cars use an internal wastegate to control boost. Boost pressure from the compressor outlet is fed to a wastegate actuator which consists of a diaphragm backed by a spring. The diaphragm is deflected by the boost, so opening a swing valve and bypassing exhaust gases around the turbine.

Factory boost control systems are designed to bring on boost relatively gently. They do this by allowing 'boost creep', where the wastegate actuator starts to open well before maximum manifold boost pressure is reached. This approach is used to allow good throttle control through the development of a flatter torque curve. Wastegate creep reduces straight-line acceleration but makes the car far more user-friendly, especially when cornering hard.

Aftermarket Turbo Boost Controls

A great deal of misinformation is spread in the area of aftermarket turbo boost controllers. There are many who believe that to make boost adjustable, and to get fast turbo response, there's no alternative but to buy an electronic aftermarket boost controller. This is simply not true. In fact, electronic aftermarket boost controllers are extremely expensive when you consider what you get for your money. At the time of writing, a stand-alone electronic boost control cost about half that of a full engine-management system. The engine-management system gives programmable control over ignition, fuel, idle speed **and** turbo boost, among other factors! Clearly, it makes sense to save the extra money to buy a complete engine-management system.

Furthermore, all electronic boost controllers are electronic only in their control of the pneumatically operated wastegate. In other words, an electronic boost control should really be termed an electronic/pneumatic boost controller. It sounds a minor difference, but it's not when it's realised that pneumatic systems can be easily controlled by pneumatic means — not just electronic/pneumatic! I have no doubt that if you pay absolutely top dollars for an electronic boost controller with all the bells and whistles, you may be able to see a tiny improvement in on-road performance. But it will be only a very minor gain that involves the spending of major amounts of money!

Quickly-rising boost is billed as one of the major advantages of electronic boost control. However, **no** boost control can allow the turbo to develop boost more quickly (or to greater levels) than will be achieved if the wastegate is completely shut. To put this another way, if you make sure the wastegate stays shut until the required boost is reached, you will be bringing up boost as quickly as **any** boost control system is able to (anti-lag systems excepted!).

So why is standard electronic boost control used in many factory turbo cars? The major advantage this approach gives the manufacturer (who has integrated boost control into the engine management system) is that boost can be increased or reduced when certain engine conditions are met. For example, during the occurrence of detonation, boost can be reduced, as it can when the engine is not yet up to operating temperature. In some cars, boost can be increased for short periods while the intake air temperature is below a certain threshold. An aftermarket electronic boost control with these features is certainly worthwhile, but at the moment this type of control complexity is found only in full programmable engine-management systems. If you want electronic control of boost, I suggest you buy a full programmable management system that incorporates this feature.

Pneumatic Boost Control Systems

The easiest way of getting maximum boost is to simply pull the hose off the wastegate actuator! That way, **all** the exhaust gas from the engine will pass through the

An electronic turbo boost controller and four pneumatic pressure regulators. Despite common opinion, an aftermarket electronic boost controller is not needed for good boost control.

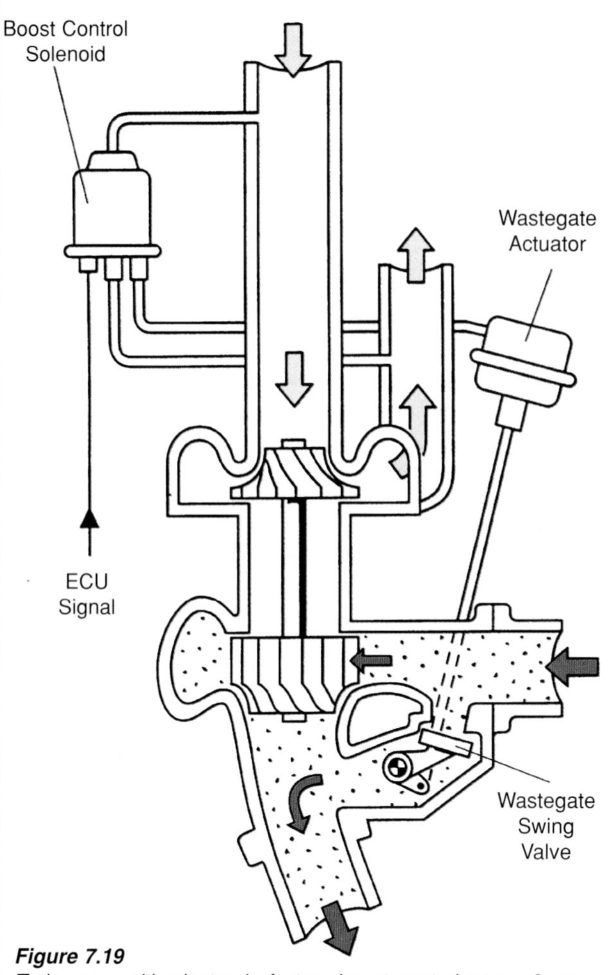

Figure 7.19
Turbo cars with electronic factory boost control use a 2- or 3-port solenoid valve that is pulsed by the ECU, allowing air to be bled out of the wastegate control line and so reducing wastegate opening.

turbine — none will be wasted. You can't get any more boost out of the turbo than that — but the peak boost pressure will almost certainly be too high. Too much boost results in detonation and broken pistons ... **not** what's wanted. However, if the turbo is relatively small for the application and the correct air/fuel ratios and intake air temperatures are maintained, some cars can run without the wastegate connected.

But it's usually much better to build a boost control system, so boost can be varied to suit the occasion. In very high-powered two-wheel-drive cars, the boost may need to be turned down in wet weather, and it is always useful to be able to drop boost pressure (even if you need to lift the bonnet to do so) when only lower-octane fuel is available.

It's quite simple and cheap to build your own adjustable boost control system. Such a system can be configured in a number of ways:

- To give a factory-style gentle increase to boost, which gives excellent throttle control
- To bring on boost very hard, which reduces fine throttle control but gives maximum acceleration
- To give short-term higher-than-normal boost levels when the throttle is floored, which gives excellent short-duration acceleration

Each of these systems is pneumatic in operation. It's very important to be aware that not all pneumatic boost control systems are alike. If someone says, "Oh yeah, I tried a bleed and it didn't work", that doesn't tell you a great deal unless you know the **exact** layout of their system, the car on which it was used and how the results were disappointing. For this reason it's unwise to tar all pneumatic systems with the same brush!

Boost Control #1

If you own a car that has ECU-controlled boost as standard and you wish to retain excellent throttle control and a factory feel to the boost increase, the following system works very well.

Turbo cars with electronic factory boost control use a two- or three-port solenoid valve that is pulsed by the ECU. This allows air to be bled out of the wastegate control line, decreasing the amount of boost seen by the diaphragm. If the ECU directs an increase in boost, it pulses the valve so it's open longer than it is shut, giving a great bleed. If the ECU wants to lower boost, it lets less airflow through the bleed valve. Figure 7.19 shows a schematic of this sort of system.

The boost pressure to move the diaphragm of the wastegate actuator comes via a hose that connects the turbo outlet (or plenum chamber) to the factory boost control valve. If a new adjustable flow control valve is

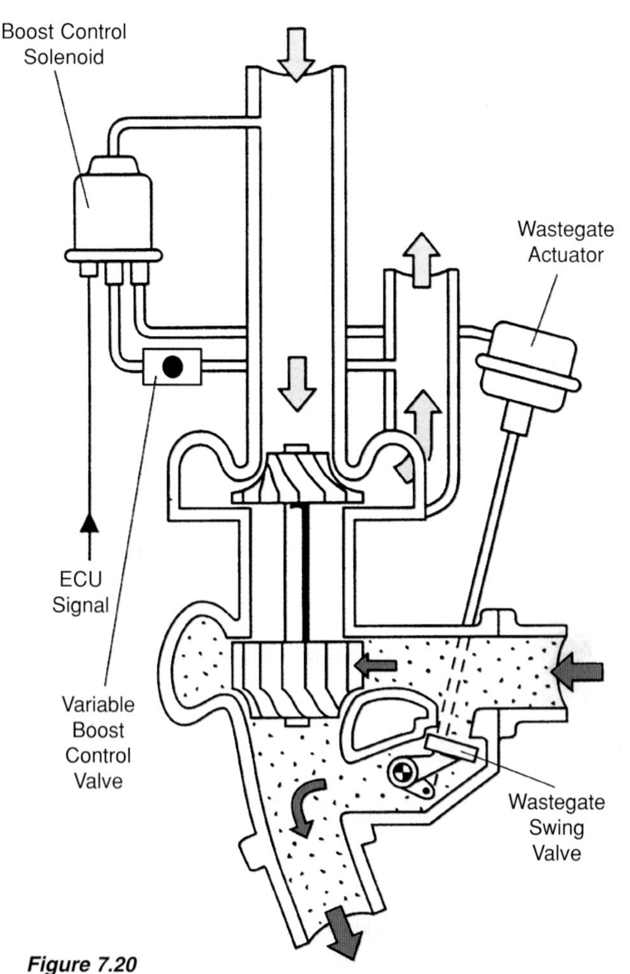

Figure 7.20
Boost Controller #1 uses a new flow control valve inserted in the boost pressure feed to the solenoid. This allows adjustment of the amount of pressurised air reaching the factory boost control solenoid, so altering maximum boost.

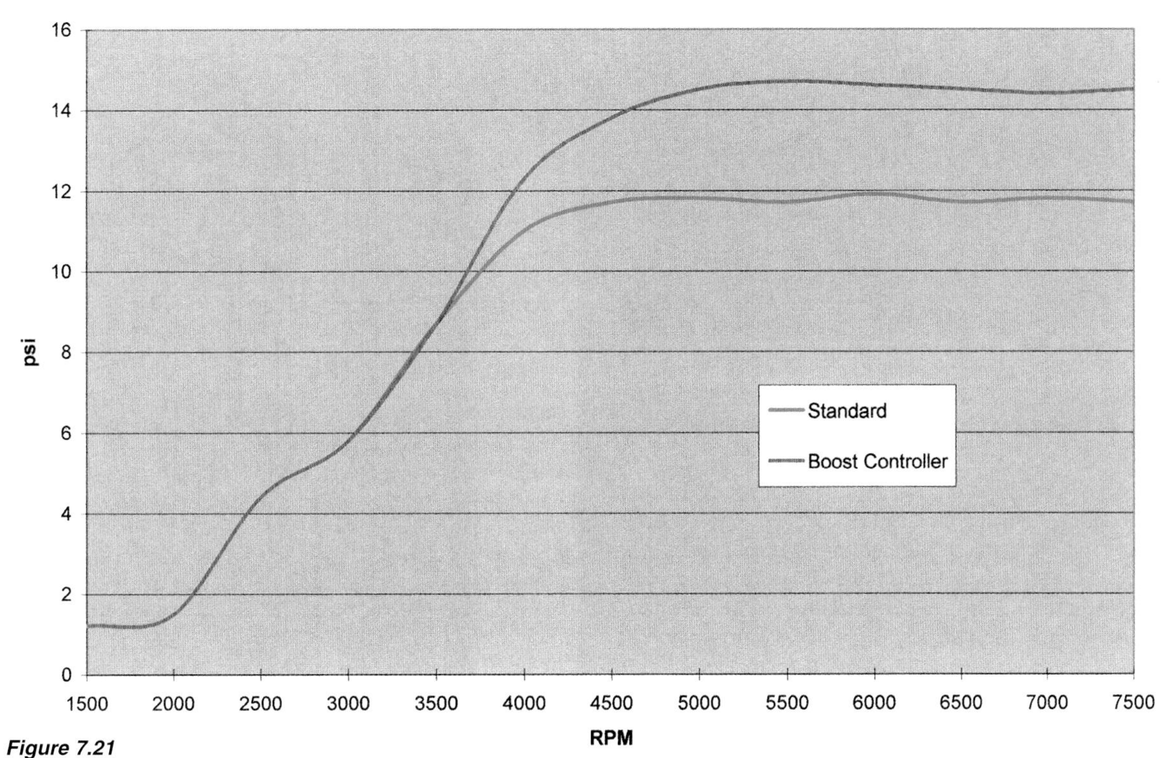

Figure 7.21
The standard and new boost curves achieved with the Boost Control #1 fitted to a Nissan R32 Skyline GT-R. The boost measurements were taken on the road at full throttle in second gear. The boost delivery up to about 3600rpm is the same with or without the control fitted, but from there, boost rises smoothly to the new 14.5psi maximum setting.

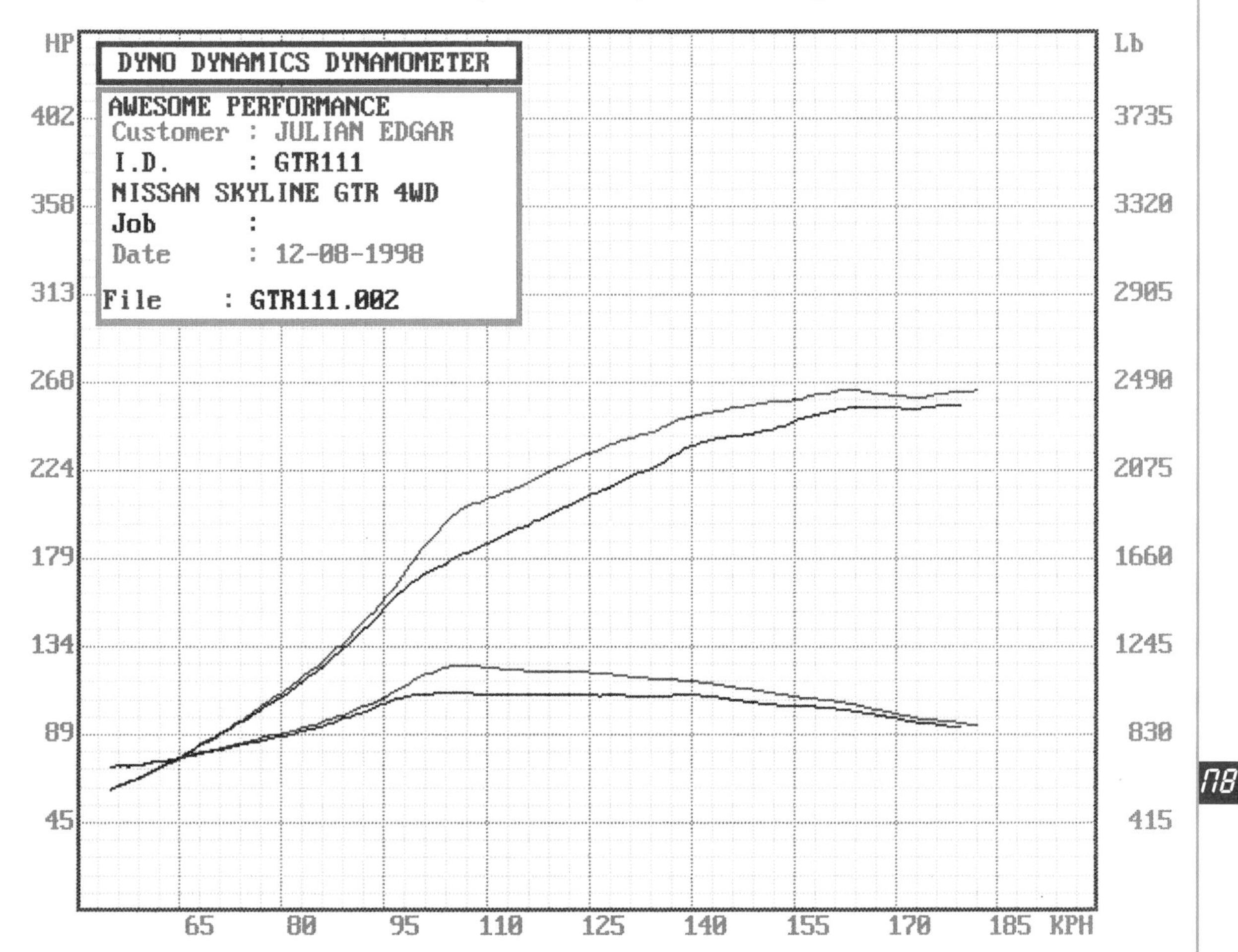

Figure 7.22
The dyno results of increasing the boost on a Skyline GT-R by 2.6psi, using Boost Control #1. The amount of extra power developed was limited by the very high (6.5psi!) back-pressure the standard exhaust was causing.

Suitable for use on cars that have factory ECU-controlled boost, this control valve allows the retention of the ECU control but increases the maximum boost level. It's cheap and very effective.

placed in this hose, the amount of air that reaches the factory boost control solenoid can be regulated. Adjusting this valve then changes the maximum level of boost. Figure 7.20 (page 177) shows where the valve fits into the system. With this approach, you cannot decrease boost from the factory setting, but you can increase it. Increasing boost is as easy as reducing the flow of air that can pass along this hose. The ECU will still pulse the valve, giving much the same shaped boost curve as standard, but boost will reach a higher level before holding at the new value.

Most turbo cars that use ECU-controlled boost do so in an open-loop manner. That is, even though there is an input

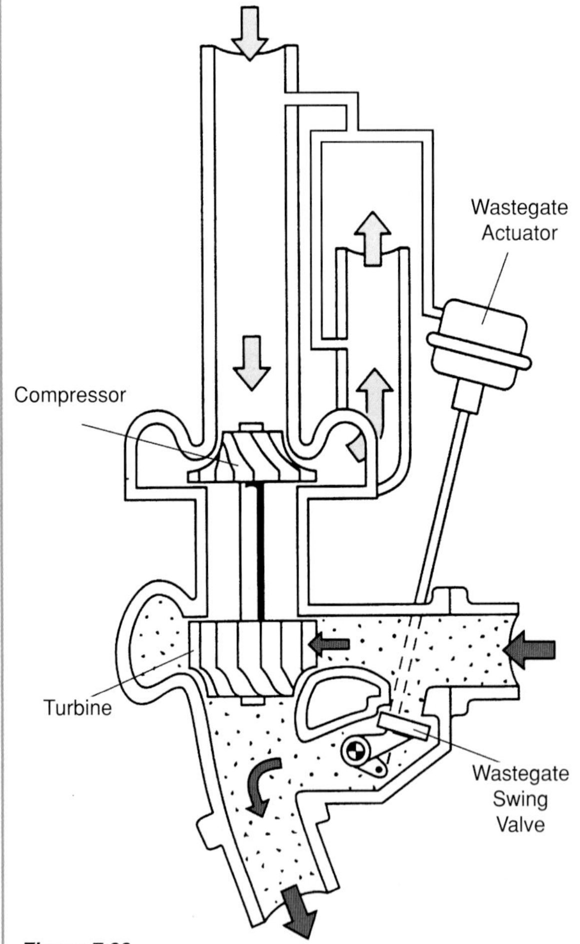

Figure 7.23
A typical wastegate bleed uses a T-piece inserted in the wastegate hose. Air is then bled out of the wastegate line. Taking this approach often gives unstable boost levels.

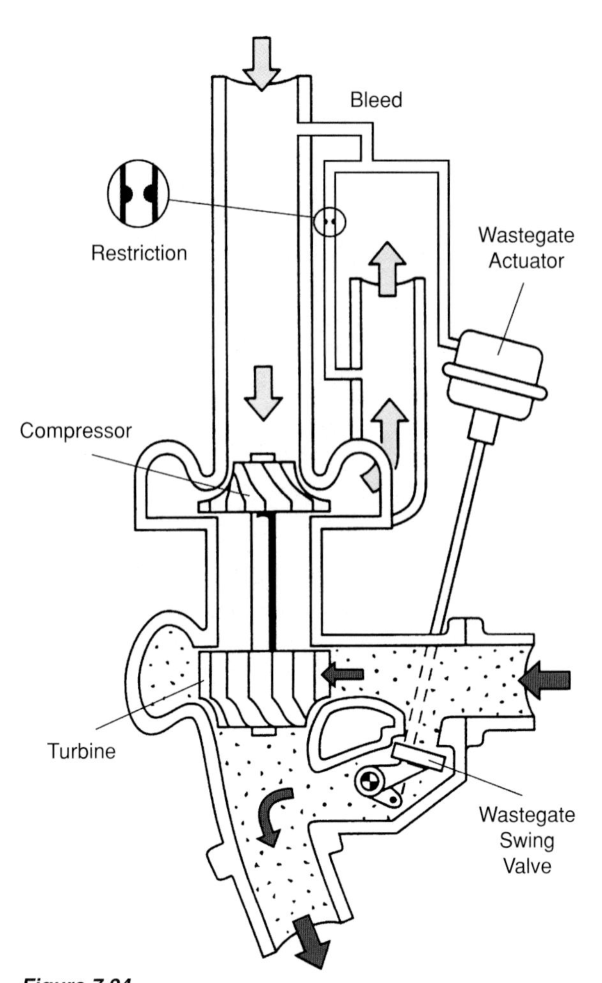

Figure 7.24
By inserting a restriction on the compressor side of the T-piece, the bleed needs to flow only a small amount of air to make a large difference to the pressure seen by the wastegate diaphragm.

sensor that measures intake manifold boost pressure, the ECU does not respond to this input signal in the setting of boost. Accordingly, when the flow of air to the boost control solenoid is decreased (causing an increase in boost) the ECU doesn't respond to this changed situation by increasing the pulse width of the bleed solenoid. However, if the system in your car **does** use closed-loop boost control, the type of boost control system shown here may not work. (Instead, it may be worthwhile bleeding off some boost pressure to the pressure sensor.)

Figure 7.21 (page 178) shows the standard and new boost curves achieved with the control fitted to a Nissan

The boost control valve fitted to a Nissan Skyline GT-R. Closer to the camera is the standard boost control solenoid; the new valve restricts the airflow to this solenoid.

R32 Skyline GT-R, with the boost measurements taken at full throttle in second gear. This engine used twin parallel turbos and the car was standard but for a cool-air duct to the airbox. As you can see, the boost delivery up to about 3600rpm was the same with or without the control fitted. But from there, boost rose smoothly to the new maximum setting, which was set to 14.5psi (1 Bar). Note these are actual measured values — not estimates! On the dyno, the Skyline (tested in rear-wheel-drive mode) showed how the power output and tractive effort responded to the 2.6psi boost increase. Figure 7.22 (page 178) shows these dyno results. However, in this case, the amount of power being developed was being limited by the very high (6.5psi!) back-pressure that the standard exhaust was causing.

This boost control system will give the same peak boost in different gears (varying boost levels are a common criticism of pneumatic-based systems) and if the valve is mounted under the bonnet with short hose runs, good throttle control is retained.

Probably any variable orifice valve could be used as the boost control valve, but I use a Festo Pneumatics valve that has two advantages. First, it features a lock-nut that prevents accidental boost changes. Second, the valve allows adjustable flow in one direction but free flow in the other direction. This characteristic can have advantages in some cars. The actual valve used is the Festo 6509 GRA -1/4- B flow control valve. You will also need quarter-inch barbed hose fittings and quarter-inch ID hose, hose clamps and cable ties. Festo valves are available from Festo and other industrial pneumatics suppliers in each state and in many countries.

The first step in fitting this system is to identify the hose that supplies boost pressure to the factory boost control valve. It will come from either the turbo outlet or plenum chamber in most cars. Insert the new valve in this hose, making sure the valve is installed with the arrow on the valve body pointing in the direction of the normal flow to the factory boost control solenoid. Try to keep the hose runs short, otherwise you'll build some wastegate delay into the system. This will give quicker boosting but reduced fine throttle control.

After that, follow these steps to set the boost level:

1. Set the valve to the fully open position (wound anti clockwise). Drive the car, and check the boost. It should be dead-standard or at the most up by just a very small amount.

2. Undo the lock-nut and close the valve a little. Drive the car again and check the boost. It should be up a little in peak boost but the car should otherwise behave normally.

3. Close the valve a little more until you get the boost level you want. Tighten the lock-nut.

Boost Control #2

If you would like boost to come on hard and fast, or your car does not have factory ECU-controlled boost, here's another system that's almost as cheap to put together. I have used this system on many different turbo cars.

This system bleeds air from the wastegate line so the pressure seen by the wastegate actuator is reduced.

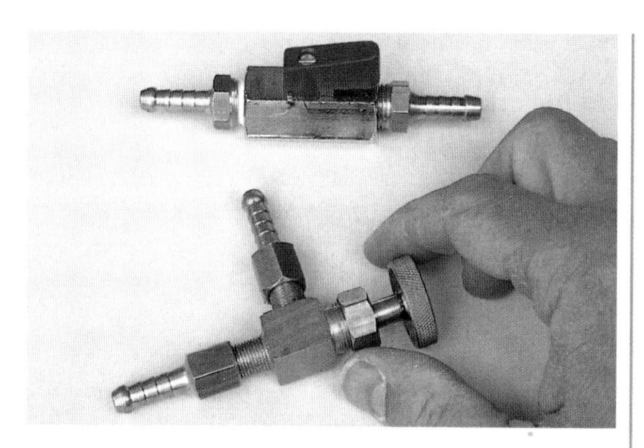

In cars without ECU controlled boost (or where you want boost to rise very quickly), a simple but extremely effective adjustable boost control system can be made using these valves. At the top is a ball valve and beneath, a needle valve.

However, as stated earlier, all pneumatic bleed systems are **not** the same! Figure 7.23 (opposite) shows a typical wastegate bleed. A T-piece is inserted in the wastegate hose and air is bled out through a variable restriction. One of the problems with taking this approach is that often a **lot** of air needs to be bled from the line before the wastegate sees a lower pressure. That's because there's a great deal of air available from the turbo, so bleeding off just a bit often makes little difference to the pressure seen by the wastegate actuator.

The way to overcome this problem is to insert a restriction on the compressor side of the bleed's T-piece, as shown in Figure 7.24 (opposite). Adding the restriction means the bleed needs to flow only a small amount of air to make a large difference to the pressure seen by the wastegate diaphragm. Introducing the restriction is fundamental to the way this system works.

To build the system, you will need two valves, one or two T-pieces, some hose and some hose clamps. The main in-cabin (or underbonnet) control valve is a quarter-inch needle valve. A needle valve is a multi-turn (often five or eight turns) brass valve that gives fine control over flow — it's like a well-made tap. For the restriction, it's best to use a quarter-inch ball valve. A ball valve also allows adjustment of the flow through it, but it's a valve that with just a 90-degree turn either shuts off the flow or fully opens it. Both of those valves are available from companies that specialise in pneumatics and hydraulics — check your phonebook. While you're at the valve supplier, buy two quarter-inch T-pieces and some hose clamps. The only

An over-boost reservoir used as part of the boost control system can eliminate wastegate creep, allowing a rate of boost increase as fast as the turbo can develop it.

other part needed is a couple of metres of quarter-inch fuel hose — you can easily work out exactly how much you need before you go to buy it.

The system is organised as shown in Figure 7.25. The location of the ball valve restriction, T-piece and needle valve are important — don't get their positions within the system confused.

To set the system up:

- With the system connected, fully open the ball valve and fully shut the in-cabin needle valve. Drive the car and check the peak boost. It should be standard.
- Next, fully open the needle valve. Again, drive the car — in most cars boost should be standard or up only a little. However, in some cars boost will now be raised, so be careful!
- With the needle valve fully open, close down the ball valve by a **very small amount**. Boost should now be lifted a little. Keep closing it down and test-driving until boost reaches your new maximum.
- With the needle valve fully open, the boost level should reach the new maximum. With it closed, boost should fall to standard. An in-between setting of the needle valve should give an in-between boost level.
- Put a daub of epoxy glue on the ball valve handle so it can't be inadvertently knocked open or closed.

This boost control system was fitted to a Nissan Pulsar EXA 1.5 and the new and old boost curves are shown in Figure 7.26 (opposite). Note the new boost curve falls in level as revs climb. This characteristic of the boost control system causes concern to some, but it has a major benefit. The control system has a built-in 'scramble' or 'over-boost' capability that allows the engine to develop a higher peak boost pressure for a short time. **This occurs because the volume of air contained within the hoses needs to fill before the wastegate actuator will start to open.** In other words, the hysteresis of the system means there is a wastegate anti-creep mechanism built in.

Since most cars can tolerate a peak boost level higher than can be sustained, this works to the advantage of

Using a bleed near the turbo (here switched on or off by a solenoid) is not nearly as effective as a properly set-up pneumatic boost control. Not all pneumatic boost controls are the same!

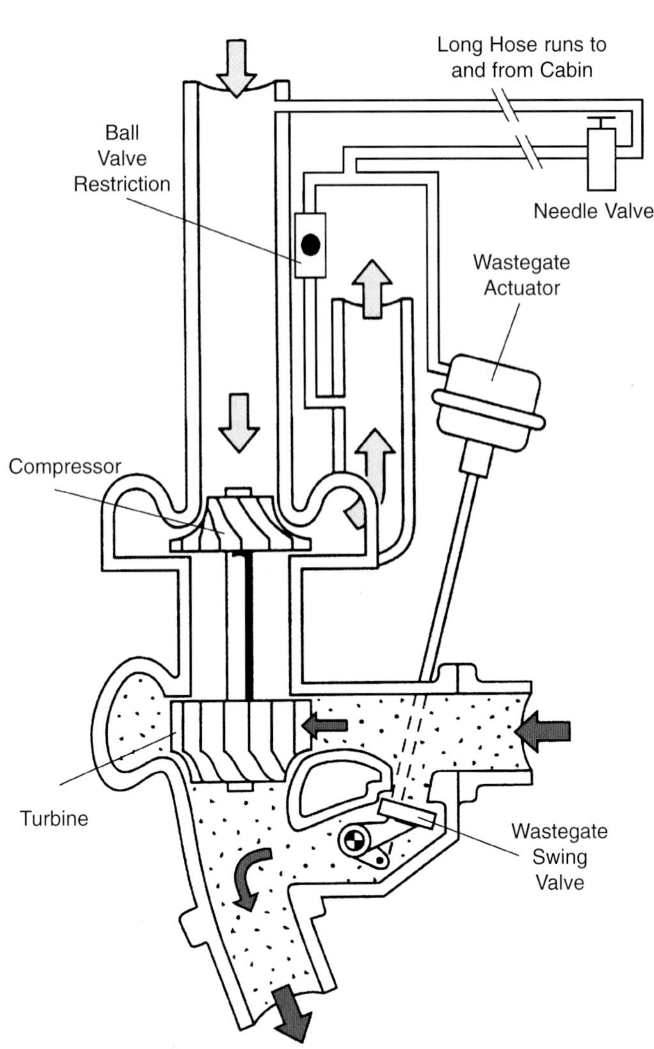

Figure 7.25
Boost Control #2 can be used to give adjustable, in-cabin control of boost. The T-piece and ball valve are located close to the turbo, with the needle valve located in the cabin. The volume of air contained within the hoses running into the cabin provides some anti-creep wastegate delay and is an important ingredient in getting quick turbo boost.

the vehicle's performance. One reason for the ability to cope with short-term 'over-boosting' in the mid-range is that injector duty cycle is still low, so there will be less likelihood of running out of injector capability at this combination of boost and revs. Another reason is that the heat-sink effect of the intercooler (see Chapter 8) will give initially cooler intake air temps. However, all things in moderation! If you let the boost spike too high, you may run into detonation problems, especially around peak torque.

If you would like the over-boost effect to be increased, T into the hose leading to the wastegate a small reservoir such as a small (empty!) aerosol can or a small stainless steel drinking flask. This will allow boost to peak at a higher level whenever you quickly put your foot down, with the boost then dropping as the reservoir fills with pressurised air. If you put your foot down slowly, the boost will rise to its normal value only. As an example, one small FWD turbo car would break into wheelspin on the 1-2 gear change with just this modification performed to it! The amount and duration of this over-boost can be varied by changing the volume of the over-boost can and the diameter of the hose leading to it.

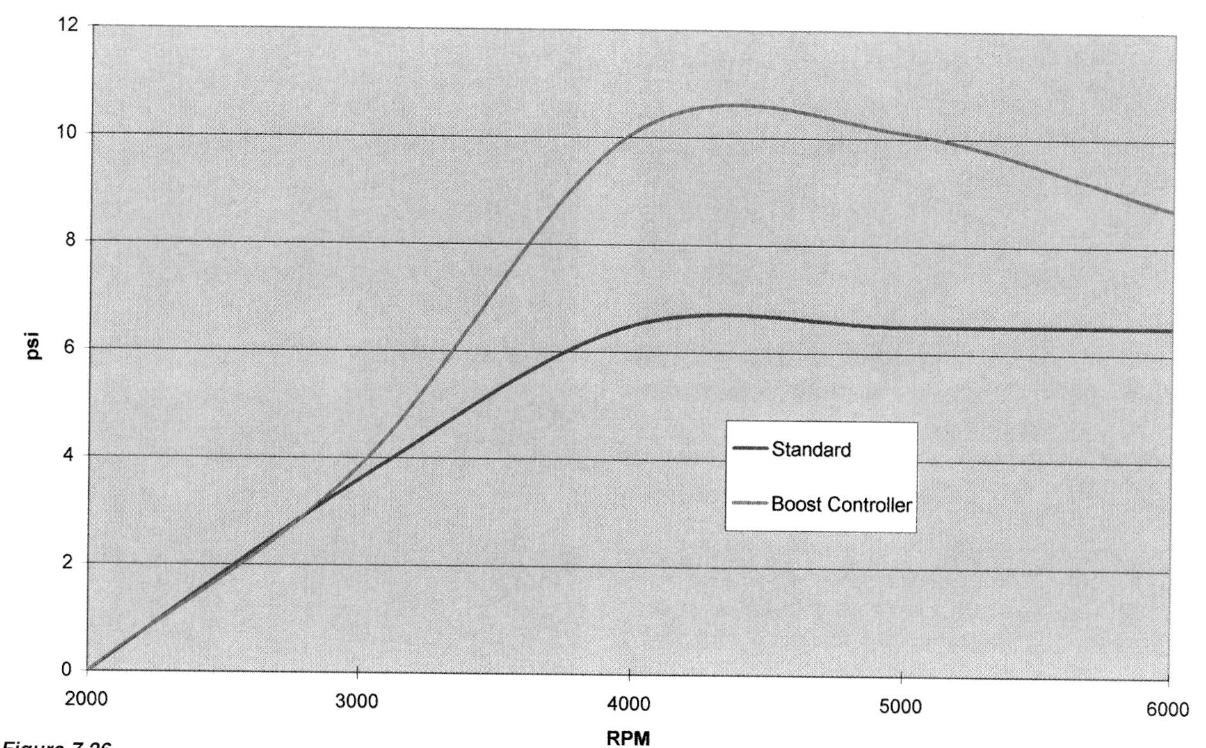

Figure 7.26
The standard and new boost curves achieved with the Boost Control #2 fitted to a Nissan Pulsar EXA. Note that a characteristic of this control system is that boost on some cars falls a little in level as revs climb. This over-boost capability allows the turbo to develop a higher peak boost pressure for a short time.

So how do these boost control systems compare on the road against stand-alone electronic boost controllers? During the development of the second boost control system, I experimented with an electronic system I had developed. That system prevented **any** boost pressure at all from getting to the wastegate until the desired boost level had been reached. This was done using a solenoid that shut off wastegate boost pressure entirely when the car was first accelerated, starting to pulse open only when the selected boost pressure was achieved. However, there was no discernible performance advantage in on-road performance to the well set-up pneumatic system.

Another important advantage the pneumatic systems have over pulsed electronic is they are essentially analog in operation. In other words, the air is bled from the wastegate pressure line in a constant flow. In most electronic systems, a valve pulses to allow the escape of air in short discrete bursts. A sensor measuring boost pressure indirectly controls the pulsing of the valve. It's easy in electronic boost control systems to get into a situation where the system is inadequately damped, leading to a surging of boost pressure. This occurs in the same way that the oxygen sensor oscillates the air/fuel ratio — but unlike the cycling air/fuel ratio, you can certainly feel it when boost is fluctuating by a few psi!

You should note that in pneumatic systems, quite small changes to the layout of the system can dramatically change the way the system works. For example, pneumatic boost control systems are sensitive to the orientation with which T-pieces are placed, the volume of air contained in hoses and other seemingly minor aspects. So if the system doesn't give the driving characteristics you desire, it's wise to experiment further rather than immediately discard the system.

TWIN TURBOS

As indicated previously, using a single large turbo normally results in poor response and poor low rpm boost. The lack of response to changes in throttle position occurs because the larger turbo's rotating assembly has more inertia so is slower to speed up when exhaust gas flows change. A large single turbo also requires substantial exhaust gas flow before it will rotate fast enough to develop boost, so that boost will occur later in the power band. Using two smaller turbos overcomes these problems. In fact, nearly all the world's high-power automotive turbo engines use two (or more) turbos.

Twin turbos can be arranged so they operate either simultaneously or sequentially. A simultaneous system uses two turbos with identical characteristics. Both turbos operate all the time. Each turbo is generally fed from a

Many high-performance factory turbo cars use twin turbos to improve transient response and low rpm boost. Putting a single large turbo on a road engine can give impressive power figures but very poor driveability.

An incomplete installation using twin turbos on a Mazda rotary engine. While much more expensive than using a single turbo, twin turbos allow the use on the road of a small engine that develops a lot of power.

separate set of cylinders — in a 'V' engine, each bank powers a separate turbo, while in an in-line six, the front three cylinders feed one turbo and the rear three cylinders the other. The compressors of the two turbos, however, normally feed one inlet plenum. Intercooling on a twin turbo car can comprise separate intercoolers (one for each turbo) or a single intercooler into which the compressed intake air is combined.

Sequential twin turbos operate a single turbo at lower loads and bring on the second turbo at higher loads. The turbos can be of the same size or a large and a small turbo can be used. As the name suggests, they operate in sequence, with most sequential twin-turbo systems running the turbos in parallel. At low loads the second turbo is prevented from operating by the presence of a butterfly in the exhaust before or after its turbine. A butterfly valve in the compressor intake prevents the primary compressor blowing back through the secondary compressor when only one turbo is operating.

One of the two standard sequential turbos fitted to a Subaru EJ20 flat four. The wastegate actuator can be seen on the left, with the exhaust manifold valve prior to the turbo operated by the large actuator on the right.

This twin turbo Nissan RB26DETT engine (taken out to three litres) developed 526kW (705hp) at 25psi boost, using Autronic programmable management. The engine was fitted to a road car...

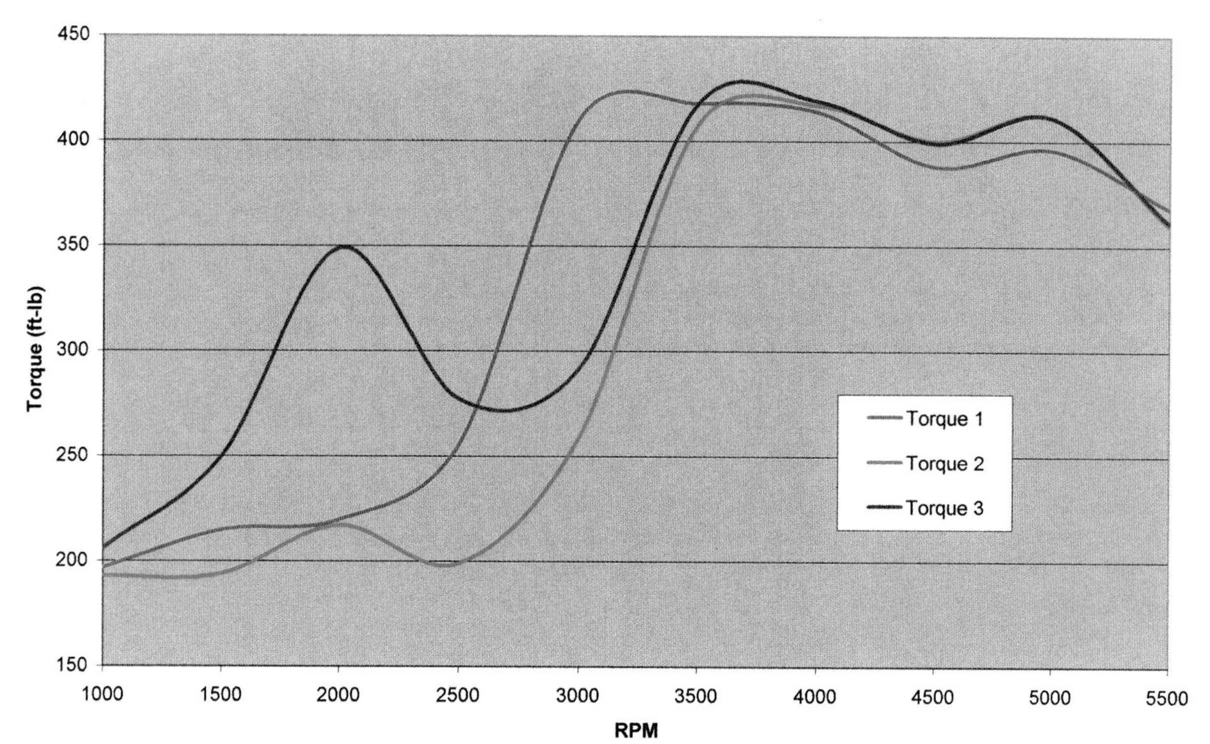

Figure 7.27
The torque output of a Toyota 2JZ-GTE sequential twin-turbo engine with the turbos operated in various configurations. Torque #2 was achieved running both turbos all the time; Torque #3 was measured with the turbos operated in sequence but the second turbo brought on abruptly; Torque #1 shows the results of bringing the second turbo on more gently.

Sequential turbo systems require very tricky handling of the changeover from one turbo to two. If the second turbo is brought onstream by the simple opening of the exhaust flap, the primary turbo will immediately slow, dropping boost pressure and so engine torque. In other words, when the exhaust flow suddenly needs to be shared, the first turbo slows somewhat. This means the second turbo needs to be brought up to speed relatively slowly, perhaps by pulse width modulating the exhaust flap opening actuator. However, depending on where the compressor butterfly is located, if the compressor valve is not opened simultaneously, compressor surge may occur. This makes the control of twin turbos by programmable management systems quite a task to set up in such a way that the on-road changeover from one turbo to two is seamless. Figure 7.27 shows the torque output of a twin turbo Toyota 2JZ-GTE engine, with the twin turbos operated in different configurations. The difference in the shape of the torque curves is staggering! The complexity of the intake and exhaust manifold valving used by manufacturers building sequential twin-turbo engines should not be underestimated — all those valves really are needed.

ANTI-LAG SYSTEMS

A turbocharger can develop boost only when there is sufficient exhaust gas flow to spin its shaft (and so the compressor wheel) quickly enough to flow the required amount of induction air. So when a driver first presses hard on the accelerator, there is a pause before the turbo spins fast enough to develop boost. This delay is known as turbo lag. Once boost is occurring, the volume of exhaust gas being produced quickly gets greater and greater, in turn rotating the turbo faster and faster! However, initially there is not the immediate response desired by some.

Anti-lag systems overcome this by allowing the development of boost at very low engine loads. This is achieved by allowing some combustion of the fuel externally to the engine — but still before the turbine. Fuel is allowed to pass through the engine in an unburnt state. This occurs because the ignition timing is retarded (or the ignition is selectively cut). However, before this fuel can burn in the exhaust manifold, air needs to be added. This air passes through either a separate pipe connected to the exhaust manifold, or through the standard throttle body or idle air bypass, either of which is kept open to a greater degree than normal during anti-lag operation.

This Mitsubishi Lancer GSR Evolution IV is one of the very few cars fitted as standard with a turbo anti-lag system. Unfortunately, it's turned off on the road versions! The valve that allows the passage of air from the intake manifold into the exhaust manifold can be seen on the right.

When the next internal combustion occurs, the 'anti-lag' fuel/air mix explodes in the hot exhaust manifold. These gases expand through the turbine and so spin the turbo up to speed. Flames belching in a series of explosions from the tip of the exhaust show the anti-lag function to be working — even when the boost gauge cannot be seen! During the period boost is actually being developed, the throttle body is either nearly or completely shut. Thus the intake manifold does not see boost until the throttle is opened — and when the throttle **is** opened, there is instant boost and very good bottom-end torque developed! In systems where the throttle body is partially open during anti-lag operation, rev-limiting tactics are used to provide idle speed control.

Anti-lag systems require the use of programmable management systems to coordinate the fuel, timing and throttle position variations required. MoTeC programmable management systems have provision for anti-lag; the parameters able to be set in MoTeC systems include:

- **Throttle Low** — Anti-lag is enabled when the throttle is below typically 2 per cent (and engine rpm is above the Minimum RPM parameter).
- **Throttle High** — The anti-lag function is disabled when the throttle is above typically 15 per cent.
- **Maximum Cut Level** — The percentage of ignition cut, typically set to 50 per cent.
- **Fuel Enrichment** — The percentage of extra fuel, typically set at 0 per cent unless extra air is added to the exhaust manifold.
- **Ignition Retard** — The amount the ignition timing is retarded during anti-lag functioning. Good anti-lag is achieved with ignition timing at 20-30 degrees ATDC, an ignition timing not possible with a distributor-type ignition.
- **Minimum RPM** — The engine speed above which anti-lag is allowed, typically 2000rpm.
- **RPM Limit** — This function is used when the throttle is being kept open to allow the flow of anti-lag air. It is typically set at 1500rpm and becomes the de facto idle speed control. As throttle position moves from Throttle Low to Throttle High, the RPM Limit is gradually blended out.
- **Maximum Time** — The longest time anti-lag can be continuously functioning, typically set at five seconds.

However, in its current form, anti-lag remains suitable only for competition use. The loud explosions, very high exhaust gas temperatures (950-1000 degrees C) and potential damage to the cat converter and mufflers make such systems inappropriate for road use.

TURBO TIMERS

Many people enthusiastically fit a turbo timer soon after buying a turbocharged car. A turbo timer is simply a device that allows the engine to run for a short period after the ignition is switched off. This allows you to lock the car and leave it while the engine continues running for a few minutes. Turbo timers are fitted to allow the oil to continue circulating through the turbo bearings, keeping them lubricated and cool while the turbo's temperature decreases. Switching off a hot turbo can result in the oil coking, damage to the bearings and even, in extreme cases, the blocking of the oil feed.

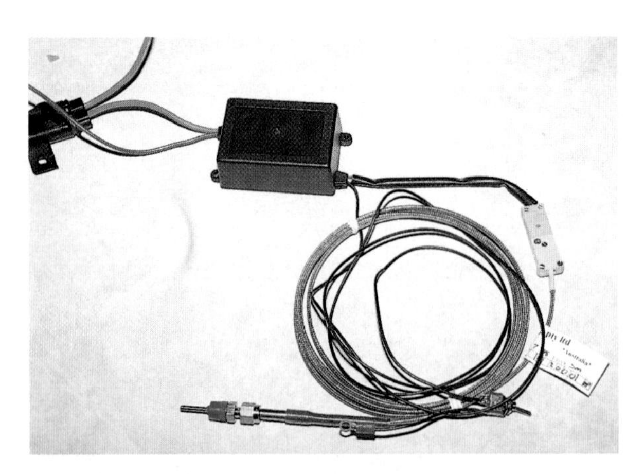

The best turbo timer is the person behind the wheel — you. However, if you must buy a turbo timer, an exhaust gas temperature sensing timer such as this one is preferable.

I believe turbo timers are a total waste of money; furthermore, they make your car easier to steal. In fact, in some jurisdictions, fitting a turbo timer is illegal. The most intelligent turbo timer is the person in the driver's seat — you. You can take into account the temperature of the day, how high the engine loads have been just before switch-off, and even the quality of oil being used.

If you have been driving the car very hard, you should drive more gently for a period before reaching your destination. For example, full-throttle, very high-speed, long-distance driving in high ambient temperatures will bring both exhaust gas temperature (EGT) and oil temperature up very quickly. In this situation, it's wise to start slowing down 10 or more kilometres before your destination. If you have EGT and oil temperature gauges fitted, you can actually watch both temperatures coming back to normal during these slow-down kilometres. When you reach your destination, you can then simply switch off the car. Note that the oil and exhaust temperatures drop far more quickly with gentle driving than with stationary idling! In a more normal urban conditions blast, idling the car for two or three minutes is sufficient. Sitting while the car is idling for this length of time is hardly an onerous task; if you proceed to fit a steering lock or arrange a windscreen shade, it's barely noticeable.

Remember that a prominent turbo timer is also a prominent ignition key bypass. Even when the engine is off, a thief can bridge the nicely accessible wires leading to the turbo timer and then be able to jump-start the car. The money you spend on a turbo timer is better spent on something that actually aids performance.

TURBO BLOW-OFF VALVES

Almost (but not quite) as useless as turbo timers are blow-off valves. Blow-off valves (sometimes also called bypass or dump valves) are valves that are plumbed between the turbo compressor outlet and the throttle body. They are triggered by plenum chamber vacuum, opening

Turbo blow-off valves are fitted by manufacturers to reduce noise (they're always of the recirculating type) and improve fuel metering.

when a certain amount of vacuum is sensed. They are installed on many high-performance turbo cars as standard equipment, and a great many people fit a large and expensive valve as one of the first modifications they carry out on a turbocharged car.

A blow-off valve vents the pressure buildup that occurs in front of the throttle when the throttle is snapped shut. This is said to be a good thing because:

- It allows the turbo to freewheel, rather than blowing air into a blocked tube. This means that when the throttle is again opened there is less lag.
- It reduces the possibility of compressor surge.

Fitting a large atmosphere-venting blow-off valve does nothing for performance and has little effect on turbo longevity, at least in the vast majority of road cars.

Blow-off valves can be plumbed so they either vent to the atmosphere or recirculate the air to the turbo compressor inlet. All OEM blow-off valves take the latter approach, while nearly all aftermarket installations vent the valve to the atmosphere. The reason the latter is done is that a *pssshht!* noise is created whenever the valve operates. Thus, on a charge up through the gears, the progress of the car will be punctuated by a *pssshht!* at each gear change. Some people like this.

The performance implications of fitting a blow-off valve are minimal — most large valves (expensive or otherwise) do nothing to make a normal turbo car either faster or more reliable. To explain why this is the case, a number of points need to be made. First, manufacturers fit recirculating blow-off valves to their turbo cars to **reduce** noise. If the throttle is shut quickly on a car not equipped with a blow-off valve, a high-pitched resonating noise will be created. This detracts from the aura of an expensive car, so manufacturers fit a blow-off valve to reduce this.

Second, airflow meters are very susceptible to measuring errors if the air does anything but flow through smoothly in one direction. If the throttle is shut suddenly (and there is no blow-off valve present) the turbo compressor will cease to flow air. The rush of the column of air through the airflow meter and the induction pipe to the turbo will suddenly stop, causing some reversion pulses to pass back through the airflow meter in the opposite direction to normal flow. The airflow meter will read this as a greater airflow than is actually occurring. If a recirculating blow-off valve is fitted that redirects the air into the compressor inlet after the airflow meter, this problem will not occur. This is the other reason manufacturers fit blow-off valves.

The 'keep the turbo spinning' school of thought overlooks the fact that the turbo needs exhaust gas flow through it to maintain its speed. When the throttle is shut, there's no exhaust gas flow, so the turbo will slow abruptly, anyway! This is the case with a plain bearing turbo; a ball bearing turbo will keep on spinning much longer — but it will do so, even with the throttle shut!

I'm sure that some **very** large turbos being used on relatively small engines will benefit from the fitting of a large blow-off valve. And perhaps these turbos — when running very high boost figures — will also benefit in longevity from the use of a blow-off valve. However, in the vast majority of performance turbo applications, aftermarket blow-off valves make little or no difference to how the car or turbo behaves. It can be quite illuminating to disconnect the vacuum line leading to a blow-off valve and drive the car in this configuration. In most cars, the performance difference will not even be noticeable. A standard-sized blow-off valve that does not leak under boost pressure does not usually need to be changed.

POWERING UP A FACTORY TURBO CAR

Modifying a car already turbocharged by the manufacturer has very major advantages overtaking almost any other approach to going fast. Not only is it very simple and cheap to get a major power gain, but the manufacturer has already done most of the hard work for you!

When they attach a turbo to an engine, original equipment manufacturers are very careful. Not only do they develop a new exhaust manifold on which the turbo sits, they also often develop a new intake manifold as well. Large injectors, knock sensing, a new ECU, a larger oil pump, new pistons or rods, sometimes a revised block — all are provided so the engine can be durable while developing typically 30 per cent more power than the naturally aspirated model. Elsewhere in the car, there will often be larger brakes and a revised suspension package. For you to make these modifications to a naturally aspirated car would cost a very large amount of money, and even then the testing and development of the aftermarket turbo engine will not remotely approach that which a major manufacturer will have carried out.

Powering up a factory turbo car is easy, cheap and effective. If done carefully, the engine will also last for many kilometres.

So, if you're after a performance package where it's possible to get substantial increases in power without spending much money, starting out with a factory turbocharged car is a very good approach.

The first step is to make sure the car is running perfectly. Its injectors must be clean, the compressions on each cylinder even and high, the turbo in good condition (no whines or oil burning) and the vehicle performance as-new. Never attempt to power-up a turbo car that's not in the pink of condition — at best you will be disappointed by how the mods don't appear to work and at worst, you will break something.

The next step is to fit a large exhaust. Turbo cars respond very well to exhaust upgrades, with the new exhaust allowing the turbo to come on boost earlier and stronger, and reducing combustion chamber contamination by waste gases. As detailed in Chapter 6, the pipe should be large and the mufflers straight-through. While a cat-back exhaust can be used, the section of exhaust directly after the turbine is very important to turbo performance, so where possible, the new exhaust should start from the turbine. Note that the cast-iron bend used off the turbo in some cars is quite restrictive, so careful examination should be made of this area before the pipe is simply attached at the first flange. This sort of inspection is often most easily carried out when the engine is out of the car, so if you can find the engine at a wrecker, inspect it carefully! If the elbow has poor flow, a flange will need to be cut that matches the turbine outlet and the new pipe started from there. Unless the manufacturer has used an overly small exhaust as a means of restricting boost pressure (rare, but not unknown!), the peak boost should not rise with a change of exhaust.

The intake system should be modified next. Chapter 5 covers this subject in detail, but basically you want as little restriction as possible to impede the airflow reaching the turbo. Note that the tuning of the length and diameter of the pipe that connects the airfilter to the turbo is not as important as the pipe that connects the filter to the throttle body in a naturally aspirated car. This means that in a transverse-engined FWD car where the turbo may be located just behind the grille without a radiator or other heat exchanger in front of it, a cylindrical airfilter can be mounted directly on the turbo intake if wished. You should always have a filter fitted — don't be tempted to remove it (even temporarily), as damage to the compressor wheel can easily occur through the ingestion of foreign matter. Also, cold-air induction on turbo cars is

Fitting a larger intercooler, lifting boost pressure, fitting a large exhaust and revising the intake system will give excellent, low-cost performance improvements to factory turbo cars. Using an interceptor to tweak the ignition and fuel maps can give further, subtle improvements.

even more important than on naturally aspirated cars, so make sure only air from outside the engine bay can be inhaled.

The engine-management system in pretty well all factory turbo cars will cope with a modified intake and a high-flow exhaust without problems. However, as with all turbo car modifications, you should always use fuel of the highest readily available octane and keep an ear out for the distinctive sound of detonation. A good exhaust and some intake mods will see peak power gains of around 15 per cent in most turbo cars. Taking this route gives an extremely cheap power gain.

The next step is to raise turbo boost pressure. However, before this is done, it's wise to consider the temperature of the intake air. If the car already has an intercooler, examine its size and position carefully. If there are any doubts about its effectiveness, measure the actual intake

Turbo cars are even more susceptible to the negative effects of breathing hot air than naturally aspirated cars. This airfilter has been placed directly on the turbo and when the bonnet is shut, hot air will be inhaled.

If a manifold pressure relief valve is fitted to the car as standard and the boost is then lifted, the valve needs to be disabled. In some cars it's simplest to remove it and plug the hole. This water pipe fitting (available from hardware stores) has the same diameter and thread as many pressure relief valves.

air temperatures of the car on the road. Most factory intercoolers are very small, so in nearly all cases, the standard intercooler should be replaced with a larger unit. If the car hasn't a standard intercooler, one should be installed or a water injection system fitted. Maintaining good power and reliability in a turbo car comes down to three things: low intake air temps, rich mixtures under load and an absence of detonation. Low intake air temp is a surprisingly important factor.

When turbo boost pressure is being increased, it's vital that careful testing is carried out. The reason is simple — random increases in boost blows up engines, usually because the engine-management system can't cope with the new power requirement or detonation occurs. The best approach is to increase the boost while the

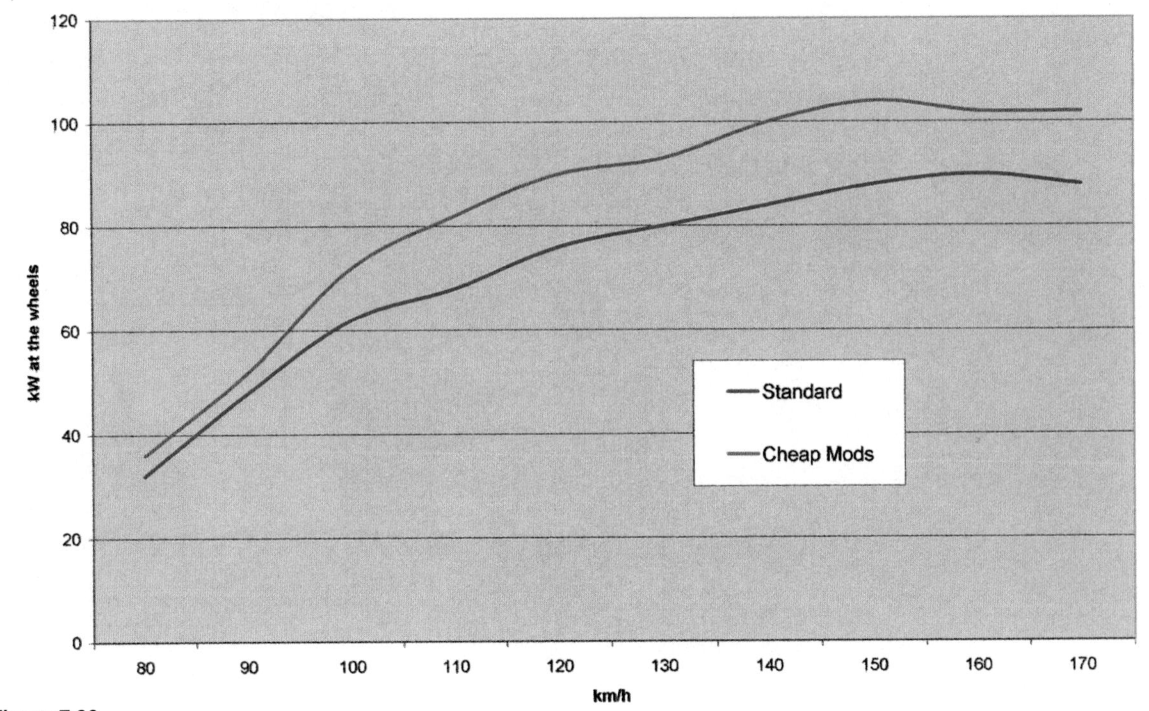

Figure 7.28
A dyno comparison between a standard MY98 Subaru Impreza WRX and one fitted with an intercooler water spray, a slightly modified exhaust (the standard resonators deleted and a new hi-flow rear muffler added) and a boost increase of 3psi. Peak power is up by 13 per cent, with a mid-range improvement of 21 per cent!

air/fuel ratio of the engine is being measured with good quality equipment on a chassis dyno. However, if the budget doesn't extend to that, during on-road acceleration runs, a passenger should carefully monitor the output voltage of the exhaust gas oxygen sensor using an LED Mixture Meter of the type covered in Chapter 3. A multimeter measuring the duty cycle of the injectors can be used to indicate how closely they are to running flat out (ie 100 per cent duty cycle), and an accurate boost gauge should also be used. Finally, detonation should be avidly listened for. The occurrence of detonation or the onset of leaning-out of the air/fuel ratio normally dictates the peak boost that can be used.

"How much boost are you running?" People with modified turbo cars swap boost figures as if they directly represent power. This is simply not so! The amount of power developed will depend on the mass of the intake airflow; boost pressure is only one of the factors that dictates this. Remember, getting the greatest amount of power on the lowest possible boost is always the best way — your car may be developing more power than another car of the same sort, even though you are using less boost pressure! For example, if the exhaust fitted to your car is far better than that fitted to another car of the same type, you would expect similar performance when your car is running less boost. Further, running very high boost is likely to substantially reduce turbo efficiency, giving higher intake temperatures. "He's running 20 pounds" doesn't mean a thing — it's the actual intake and exhaust airflows that are critical.

In the same way, high boost pressures don't kill engines. High combustion pressures do, and high combustion pressures accompanied by detonation certainly do! The only part of the engine likely to experience problems with high boost per se is the inlet plumbing, which may separate at clamps or blow off hoses. Note also that some factory blow-off valves are designed to vent high boost pressures as part of a safety system, so if your car has a blow-off valve of this type, it should be replaced with a non-leaking design.

Someone who tells you "You can run 17psi but after that watch out for the pistons" isn't really telling you much at all. What happens at 18psi? Does the air/fuel ratio start to lean out, does detonation occur, do the pistons suddenly fall in a heap because of the combustion pressures? The last is extremely unlikely.

The key to powering-up a factory turbo car is to be careful. You should always listen to the sound and feel of the engine. If you detect detonation, back off **instantly**. If the day is very hot, don't drive the car like a lunatic — save that for cool days! If you know you are running the car near the threshold of detonation, make sure you use only good fuel and periodically remove the spark plugs and examine them for degradation. The burning away of the electrode or the deposition of small globules of metal on the insulator tells you something is wrong. The fitting of the on-dash LED Mixture Meter will give an instant indication of a catastrophic lean-out condition. If you're careful, there's no reason why a turbo car modified in this way cannot have a very long and powerful life.

Even quite minor power-up mods performed on factory turbo cars can be successful. Figure 7.28 is a dyno comparison between a standard MY98 Subaru Impreza WRX and one fitted with an intercooler water spray, a slightly modified exhaust (the standard resonators deleted and a new hi-flow rear muffler fitted), and a boost increase of just 3psi. As can be seen, peak power is up by 13 per cent, with a mid-range improvement of 21 per cent! In this case, not only has the peak power been improved, but the **average** power is also substantially lifted. This spells a major performance increase.

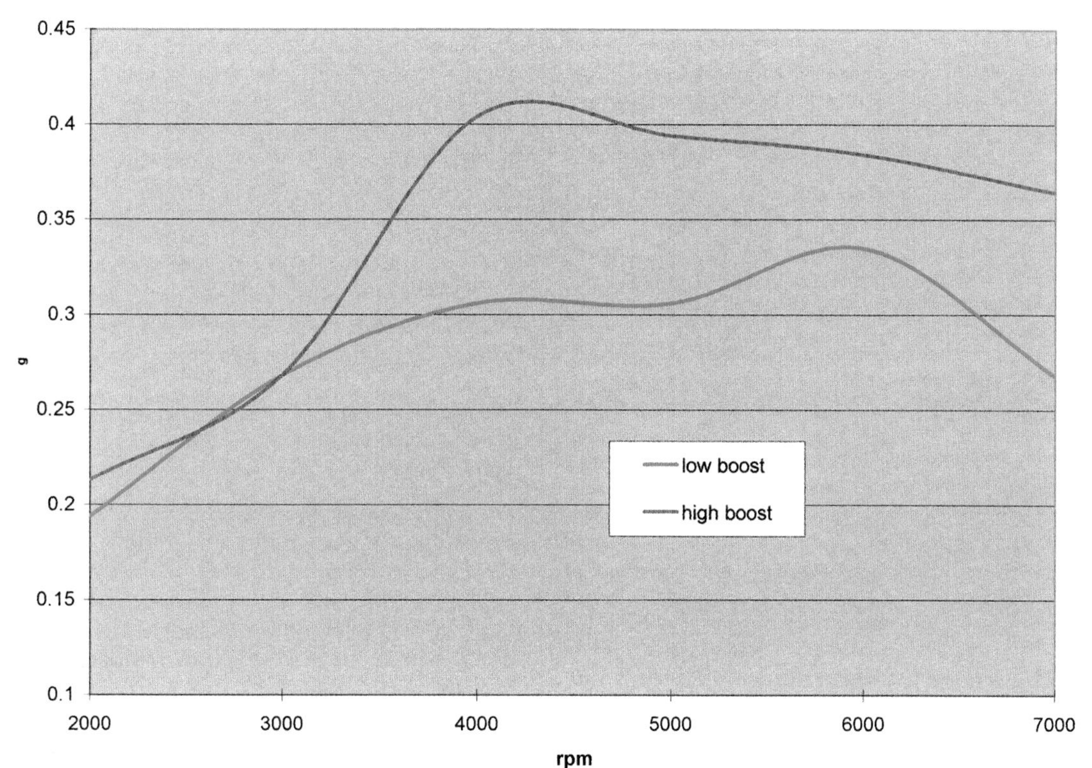

Figure 7.29
The on-road measured acceleration performance of a Subaru Liberty (Legacy) RS running slightly increased boost together with a three-inch exhaust and free-flow intake. Boost Control #2 system was used to bring on boost hard and fast. With the modifications, the on-road acceleration increased by as much as 35 per cent — and when your car is accelerating 35 per cent harder, you certainly notice it!

If the cast exhaust elbow off the turbo is restrictive, the new exhaust should start from the turbo itself, rather than from where the original exhaust began.

Figure 7.29 shows the on-road measured acceleration performance of a Subaru Liberty (Legacy) RS running increased boost together with a three-inch exhaust and free-flow intake. The Boost Control #2 system described above was used to lift boost from 11psi to 15psi. It can be seen how effective this control is at bringing on boost fast. With the modifications, the on-road acceleration increased by as much as 35 per cent — and when your car is accelerating 35 per cent harder, you certainly notice it!

The better flowing the exhaust on a turbo car can be made, the better will be performance. This Porsche has no exhaust flow restriction, on either the main or wastegate exhausts!

Factory turbo engines treated to a modified exhaust, intake, intercooler and 40-80 per cent more boost normally develop power outputs up to 30-40 per cent higher than standard, while retaining completely stock engine and turbo mechanicals. When you consider that most factory turbo cars are respectable performers in their own right, a genuine increase of 30-40 per cent in power normally provides a fun car!

A thermal blanket on the exhaust housing will reduce the radiation of heat, decrease underbonnet temperatures and so improve power.

The 30-40 per cent power increase is also the limit without either dramatically increasing the amount of money that needs to be spent, or decreasing reliability and driveability substantially. This is because going further requires major modifications to the engine-management system and turbo. In many cases, it's a good idea to stop at this point and enjoy your car, knowing you haven't spent a great deal of money and reliability and driveability are both likely to be very good. However, if you want more power, that's certainly also possible — see the section in this chapter on swapping turbos and Chapters 3 & 4 on engine management.

SUPERCHARGERS

Superchargers have been used for many years to improve the performance of road cars. However, in recent times, a variety of new supercharger types has appeared — compact and efficient designs that fit under the bonnet and are easy to mount and drive.

SUPERCHARGER TYPES

There are three main designs of superchargers used on cars. These can be divided into two categories: positive displacement and centrifugal. In a positive displacement design, every revolution of the blower pumps out a fixed volume of air. Centrifugal types have an airflow that rises as the square of their rotational speed — like turbos, they're more like fans than pumps. As you would then expect, the airflow versus rpm curve of a centrifugal supercharger is very similar to that of a turbo, although the centrifugal blower rotates at a slower speed so needs a larger impeller to generate the same airflow.

An example of the positive displacement supercharger is the Roots design, first commercially developed in 1866. More modern Roots blowers use twin helically-formed rotors, whose special rotor shape reduces pressure pulsations on the output side. Eaton is one of the largest commercial producers of this type of supercharger and their units have been used on recent Buick, DaimlerChrysler and Jaguar engines.

Another type of positive displacement blower is the screw supercharger which uses two rotors turning at different speeds. Because of the relative movement of the rotors, the volume of air trapped between the rotors reduces along their length, compressing the air through the outlet. Mazda has used an IHI Lysholm screw-type positive displacement blower on the Miller Cycle Eunos 800M.

Positive displacement blowers have airflow directly proportional to their speed. This means they can be matched to an engine to quickly provide peak boost and so give a very strong bottom-end response. Figure 7.30 shows the boost and power curves of the Eunos 800M. As can be seen, there is a near-constant boost available at all normally used engine speeds.

Centrifugal blowers use compressor wheels spun quickly through the use of step-up gears. These gears can be planetary or conventional in nature. Depending on the way they are geared, centrifugal blowers generally develop maximum boost at the redline, with lower boost at slower engine speeds. On cars with traction problems or breathing that falls away at higher revs, this boost curve can work well. Some centrifugal blowers are geared so they develop boost early in the engine rev range; the excess boost that would otherwise be developed at high rpm is bled off. Figure 7.31 (opposite) shows the boost curve of a Vortech centrifugal supercharger being used on a 5.7-litre V8.

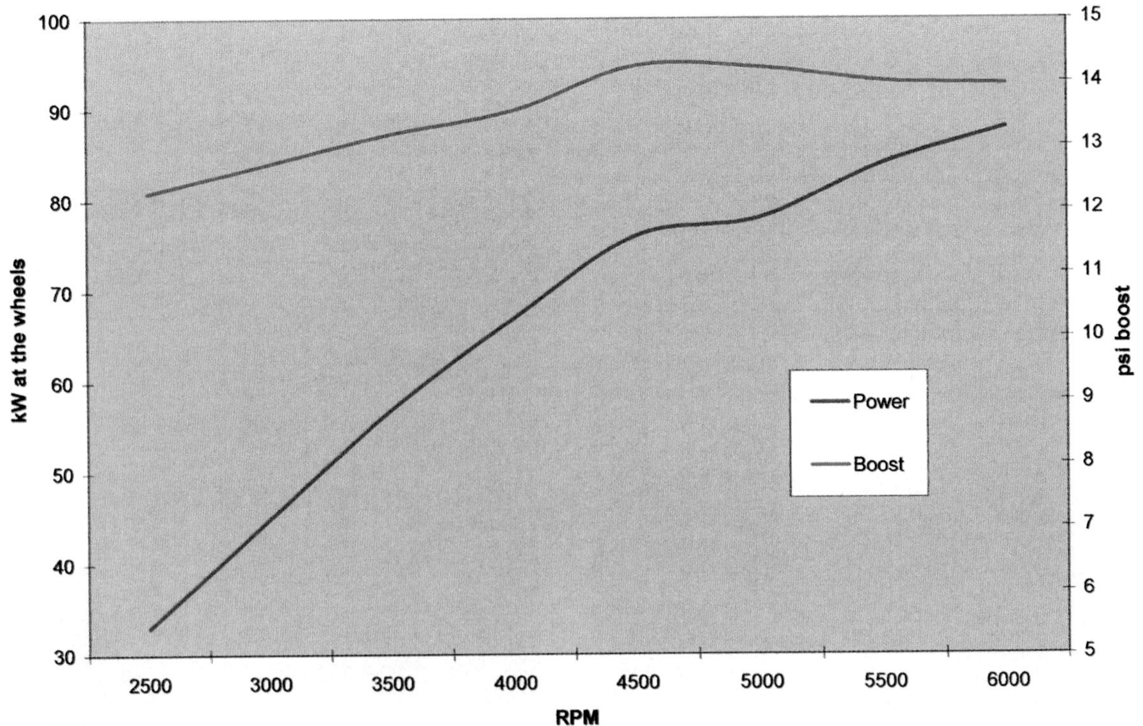

Figure 7.30
The boost and power curves of the Eunos 800M, which uses an IHI positive displacement screw-type blower. There is a near-constant boost available at all normally used engine speeds.

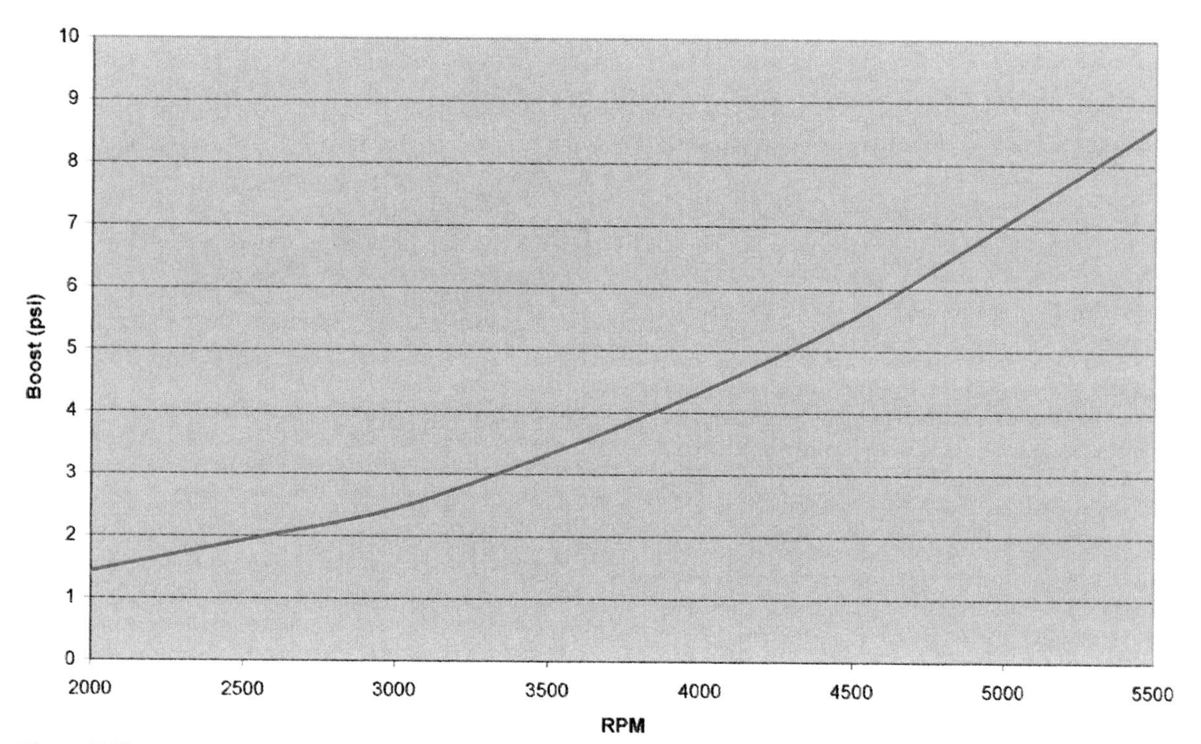

Figure 7.31
The boost curve of a Vortech centrifugal blower fitted to a 5.7-litre V8. With this type of supercharger design, boost increases with engine rpm.

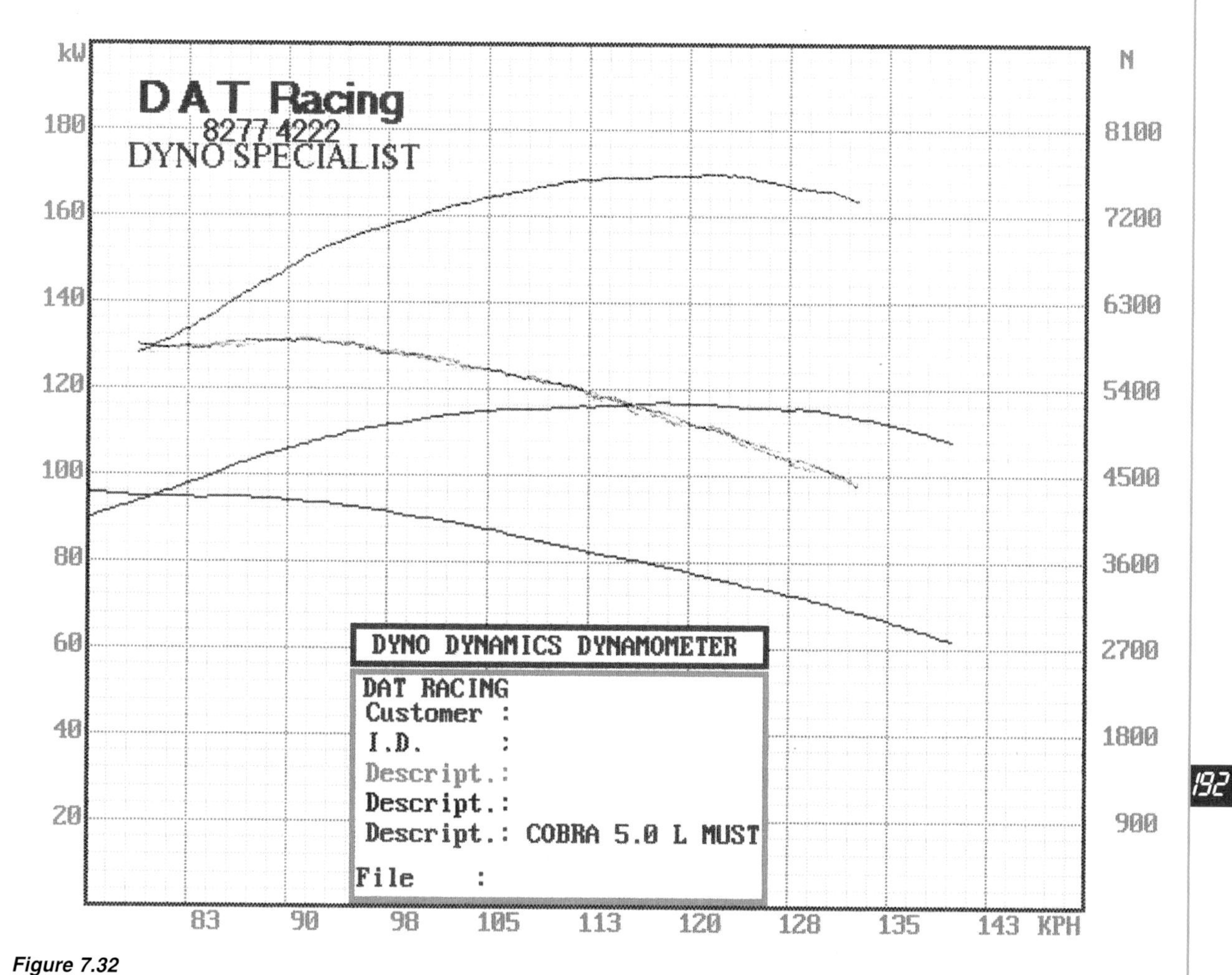

Figure 7.32
The power and tractive effort gains from a Vortech centrifugal blower mounted on a Ford 5-litre Windsor V8 running 7psi boost. Peak power has improved by 47 per cent!

Figure 7.32 (previous page) shows the power increase gained from the fitting of a Vortech centrifugal blower to another V8, this time a Ford 5-litre engine. Even though with this design of supercharger boost rises relatively slowly, on this engine there is a major power gain throughout the rev range.

The boost pressure developed by the supercharger depends on the amount of airflow being generated by the blower versus the breathing capacity of the engine. A large blower on a small engine will give a higher boost pressure, because the air is not being consumed as rapidly as it is being compressed. A standard engine with a blower may well have higher boost than a combination of the same blower and an engine modified for better breathing. Also, an engine whose breathing capacity sharply declines at higher rpm will experience a boost increase at these revs, even if a positive displacement blower is being used.

The supercharger input shaft is normally driven at a greater speed than that at which the engine crankshaft is turning. This step-up is achieved by using a smaller belt-drive pulley on the supercharger than on the crankshaft. The airflow output of the supercharger can be trimmed by changing the amount of overdrive, and this approach is normally used to set the maximum boost pressure that is achieved.

SUPERCHARGER EFFICIENCIES

A key factor in the design and appropriate applications of a blower is its efficiency. A supercharger with higher efficiency requires less power to drive it and has a lower outlet air temperature.

This positive displacement Eaton Roots-type blower has been mounted on the lower half of the modified factory inlet manifold. Note the use of a larger-than-standard throttle body, positioned prior to the supercharger.

All Roots superchargers have a relatively low efficiency when compared with Lysholm and centrifugal blowers. As an example, take a similar size Lysholm and Roots blower, each flowing 265cfm at a blower speed of 9000rpm and a boost of 11.5psi. Typically, the Roots supercharger will have an air temp **rise** of 94 degrees C, versus 78 degrees C for the Lysholm unit. The Roots blower will require 16.5kW (22hp) from the engine where the Lysholm unit will take 14.5kW (19hp), while the volumetric efficiency of the Lysholm unit will be about 85 per cent versus the Roots' 75 per cent.

Centrifugal blowers have an efficiency similar to Lysholm compressors, but their disadvantage comes from a boost curve that rises relatively slowly. As already indicated, without either throttling or venting excess air to the atmosphere (and both approaches reduce efficiency), maximum boost will be achieved only at max engine revs.

Two Sprintex screw-type blowers being used on a Nissan FJ20 engine, mounted in a tiny drag rail. The screw-type blower is another constant displacement design.

A cutaway view of a Sprintex supercharger shows the intermeshing screws. These superchargers have internal compression, the trapped air being squeezed into a smaller and smaller space as it is transported along the screws.

However, these specifications do not tell the whole story. As discussed in the next chapter, the initial airflow when the car first comes on boost will be relatively cool — the heat being absorbed by the supercharger housing, the intake plenum chamber and so on. Also, in actual vehicle applications, the designs that on paper are more efficient may, in fact, have quite high measured outlet air temperatures. The Eunos 800M with its IHI screw-type blower running about 14psi boost has a measured engine intake temperature of 75 degrees C after 10 seconds of full load running on the dyno on a 20 degrees C day. And that temperature was measured **after** one of the two small parallel intercoolers!

On the basis of their specifications, Roots blowers are the cheapest but have the lowest efficiency, Lysholm screw-type blowers have high efficiency but are expensive, and centrifugal blowers have high efficiency but do not develop maximum boost until the redline.

Positive displacement superchargers often use a bypass valve that, when open, equalises the pressure upstream and downstream of the supercharger. This significantly reduces the power required to drive the supercharger, allowing improved off-load fuel economy. GM on their Eaton (Roots-type) supercharged 3.8-litre V6 recorded a 1.5mpg improvement in US EPA highway and city fuel economy tests with the bypass installed. On that engine, GM used the bypass valve to achieve a number of outcomes.

A close-up of the lobes of a Roots-type supercharger. The intermeshing rotors pump air into the engine in direct proportion to the supercharger input shaft speed. This characteristic allows a constant boost level to be effected on the engine.

Figure 7.33 shows the valve status in a variety of operating conditions. Note the throttle is placed before the supercharger in this system.

MATCHING SUPERCHARGERS

Matching a supercharger to an engine has similar complexities to matching a turbo to an engine. However, a supercharger is a little easier in that only the intake airflow needs to be considered. One supercharger kit manufacturer takes the following approach to estimating the size of the supercharger required:

GM

Condition	Supercharger Outlet Pressure (kPa)	Supercharger Inlet Pressure (kPa)	Bypass Valve Position	Reason
Idle	Below –10 kPa	Below –10 kPa	Open	Fuel economy and idle quality
Deceleration	Below –10 kPa	Below –10 kPa	Intermediate	System diagnostics
Reverse	Below –10 kPa	Below –10 kPa	Intermediate	Transmission over-torque protection
Light Acceleration	From +10 to –10 kPa	Above –10 kPa	Closed	Boost response
Heavy Acceleration	Above 10 kPa	Above –10 kPa	Closed	Power

Figure 7.33
The opening status of the bypass valve used by GM on their Eaton supercharged 3.8-litre V6.

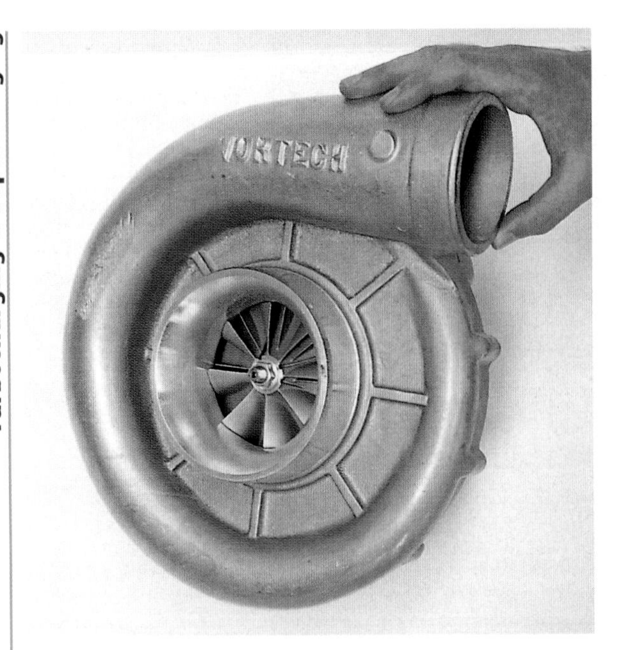

A Vortech centrifugal supercharger uses a compressor wheel similar to that of a large turbo. The output of air increases as the square of the rotational speed of the supercharger input shaft.

- Ascertain the peak power of the standard engine.
- From this figure calculate the mass airflow of the engine at peak power.
- Estimate the power increase required and so the additional mass of air needing to be flowed.
- Look at the supercharger specifications relating the blower speed and mass airflow output.

The supercharger used on the pictured Daewoo Lanos is an Eaton M62, whose specifications are shown in Figure 7.34 (opposite). At a supercharger speed of 12,000rpm, the M62 takes about 28hp (21kW) of power from the engine's crankshaft. At this speed (and when running a continuous 10psi boost) the **increase** in temperature will be about 90 degrees C, and the inlet flow to the supercharger will be about 380cfm.

As indicated earlier, altering the drive ratio of the supercharger makes minor trimming changes easy. Note also that a supercharger working well outside its zone of maximum efficiency will heat the intake air to a greater degree than a blower working within its best efficiency range, so measuring the engine intake air temperature can be used as an additional indicator of how well the supercharger is matched to its application.

Simple centrifugal superchargers use straight blades (right), while the curved blade designs (left) are more efficient. Where the option is available, select the supercharger that uses a curved blade compressor.

A well thought-out Eaton supercharger installation on a Daewoo Lanos, carried out by CAPA. A new air intake, airbox and throttle have been used, with the blower installed on the upper half of the modified stock intake manifold via an alloy adaptor plate. Drive is via a serpentine belt that also powers the alternator.

Eaton

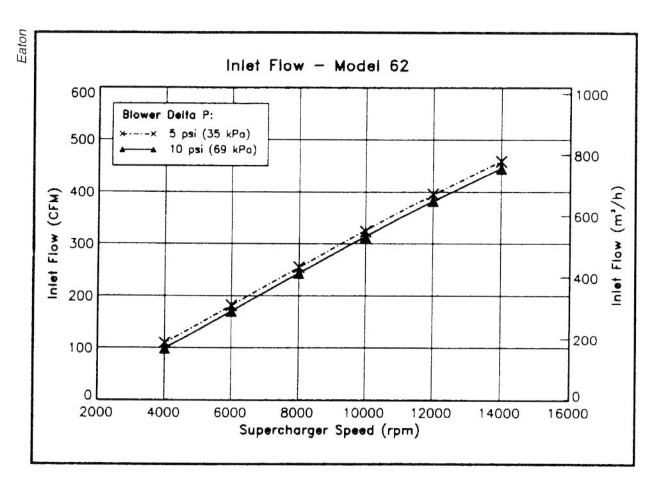

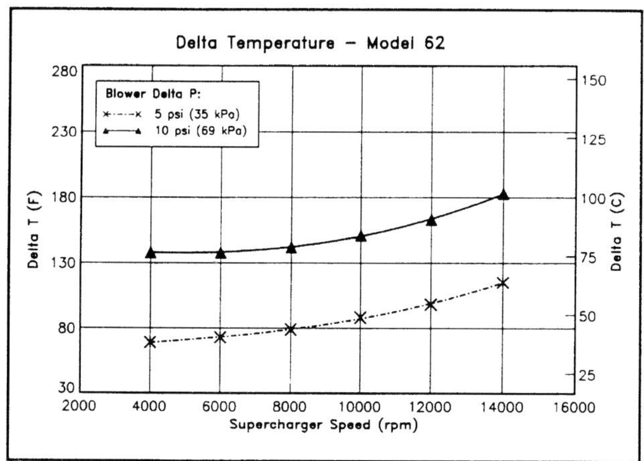

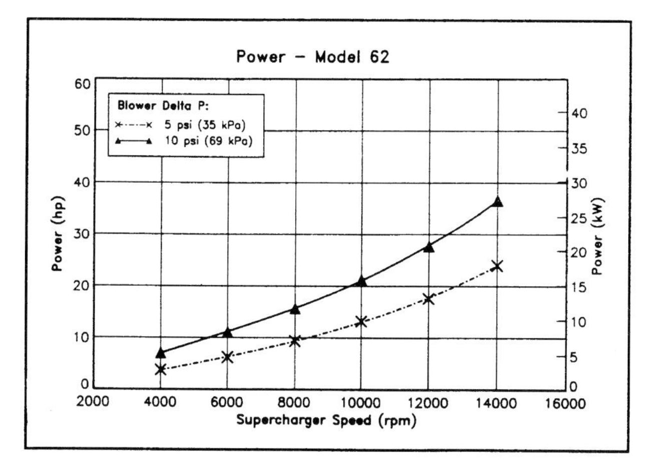

Figure 7.34
Major supercharger manufacturers have full specifications available for their blowers. This is part of the specs for the Eaton M62 Roots-type positive displacement supercharger. Delta temperature refers to the change in temperature across the supercharger, with the power consumed by the super-charger able to be read off the bottom graph. The inlet flow figures shown in the top graph are self-explanatory.

SUPERCHARGER DRIVE SYSTEMS

Modern blowers are usually driven by a ribbed belt. The belt may be additional to the normal engine drive belts or may be a replacement (longer) serpentine belt that also drives the water pump, alternator, power steering pump, etc. Some supercharger manufacturers recommend the use of cogged (Gilmer) belts with appropriate

A centrifugal supercharger design that uses the compressor wheel and housing from a turbo. Note the dual intakes to the compressor.

cogged pulleys. However, cogged belts should only be used on superchargers designed for this type of drive, as some superchargers rely on belt slippage to absorb shock loadings.

Serpentine belt systems inherently slip. This slippage normally occurs on deceleration, but can also occur under acceleration if the belt is too loose or the belt being used is not sufficiently wide for the power requiring to be transmitted. Excessive belt tension can lead to supercharger or crankshaft bearing failure — the latter more likely if a new, wider crankshaft pulley has been installed. The supercharger drive belt should be both appropriately wide and also have as much wrap-around as possible on the supercharger and crankshaft drive pulleys. Figure 7.35 (overleaf) shows the serpentine belt layout for the Jackson Racing Mazda Miata (MX5) kit. Note how the idler pulleys have been used to increase the supercharger pulley belt wrap around. A large accumulation of dust around the belt drive indicates the belt is slipping.

Multi-rib supercharger drive belts will wear more quickly than the normal belts on the engine. This is because they are transmitting more power and are usually at a higher tension. Jackson Racing suggests belt replacement

The screw-type blower uses gears to time the rotation of the rotors. These gears actually transmit little torque, the aerodynamic forces between the rotors transferring the majority of the power from one screw to the other. This Sprintex blower is pictured during an overhaul.

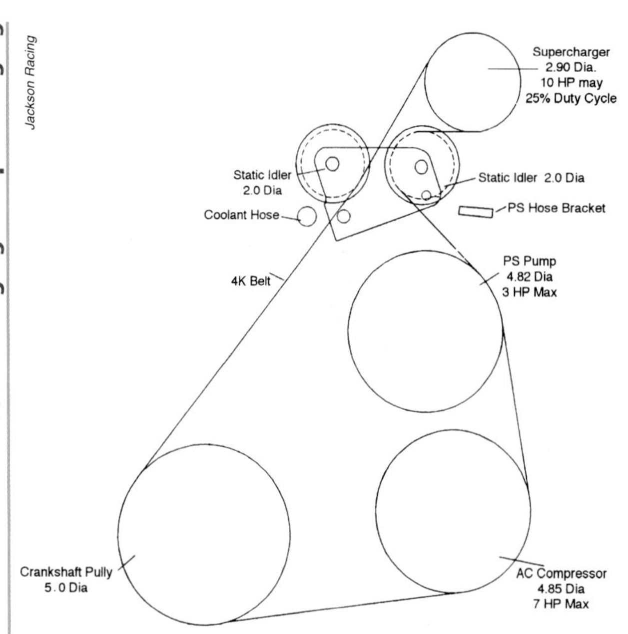

Figure 7.35
The serpentine belt layout for the Jackson Racing Mazda Miata (MX5) kit. Note how the idler pulleys increase the supercharger pulley belt wrap-around.

should occur when there are six or more cracks within a 25mm length of the belt, when looking at the inside of the belt and counting all the ribs combined. Multi-ribbed belts that either jump a groove or come off altogether do so because of a misalignment in the pulleys. This can be static (the pulleys are misaligned because of poor workmanship or kit design) or dynamic (the brackets distort during actual driving).

SUPERCHARGER KITS

Most people who install a supercharger buy a complete aftermarket blower kit for their car. When selecting a kit, there's a number of elements in the kit to consider carefully:

- The make, model and type of supercharger used
- The mounting system for the supercharger
- The drive system for the supercharger
- The type and quality of plumbing used
- The way the engine is modified
- The way the engine management is modified

Of these I would suggest it's the **last two** factors that are most important. Supercharger manufacturers and dealers usually have the first factors covered pretty well, but often there are glaring inadequacies in the way the engine and its electronic management system are modified.

Unlike a factory turbo engine that's being fitted with a larger turbo or is having its boost increased, most engines fitted with aftermarket superchargers were never intended by the manufacturer to be forced-aspirated engines. This means the block strength, piston type and strength, compression ratio, lubrication system, cooling system and driveline were all designed for perhaps 160kW (~215hp) — not the 300kW (~400hp) that can be relatively easily achieved when a blower is bolted to the side of the engine!

A centrifugal blower works very well on an engine with strong bottom-end torque but with top-end breathing that falls away. The shape of the supercharger boost curve (maximum boost at maximum engine rpm) then gives a strong torque curve throughout the rev range.

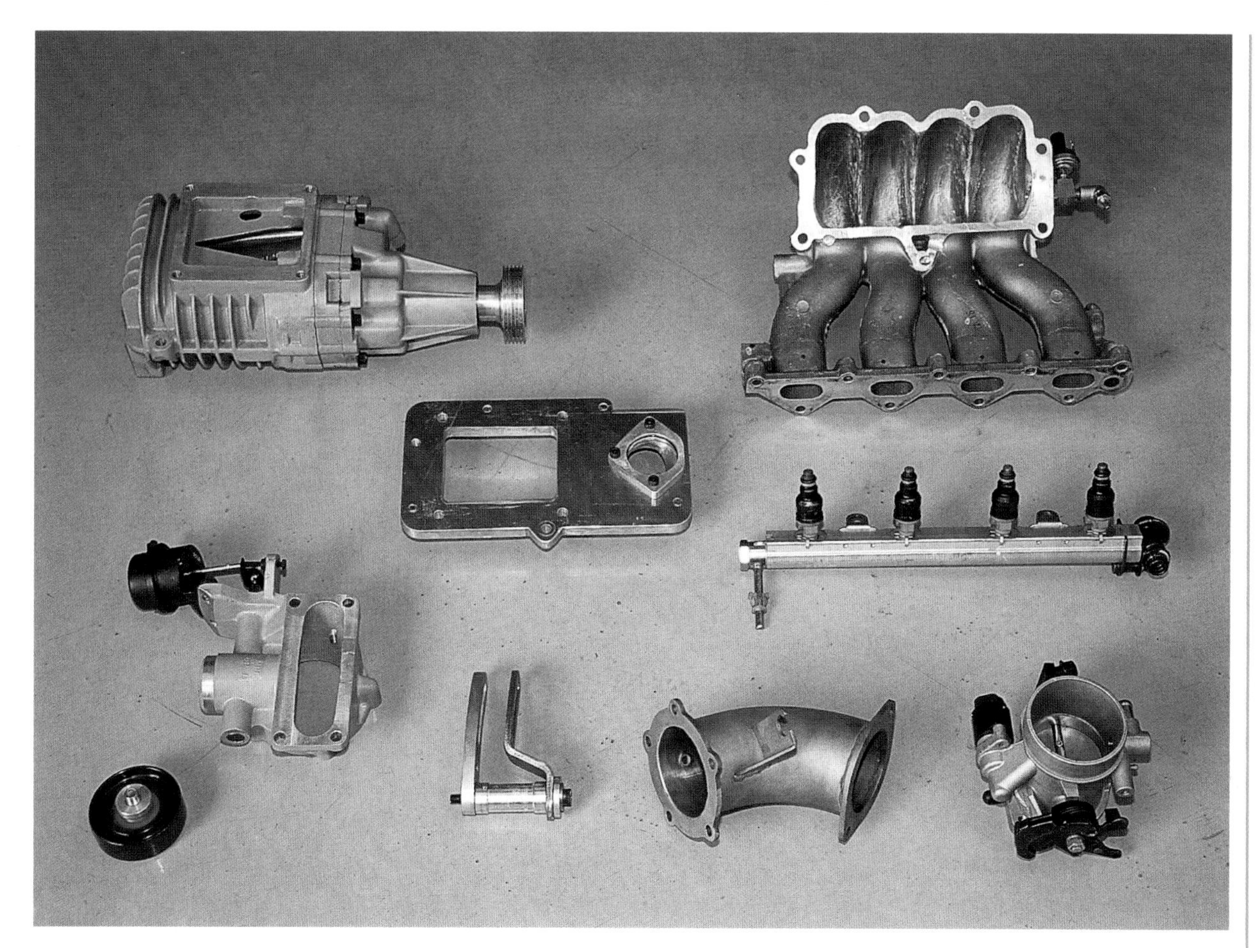

Spread out, the CAPA Daewoo Lanos package consists of (from upper left then clockwise) the Eaton supercharger; the lower half of the modified intake manifold; new injectors in the standard fuel rail; new throttle body; connecting pipe; new idler bracket and pulley; Eaton bypass valve; alloy adaptor plate.

The engine management system will have a fuel pump, ignition system, fuel injectors, fuel pressure regulator and engine management functionality all designed for a naturally aspirated engine. By the latter, I mean the standard management system will not recognise boost pressure and will retard ignition timing only gently when knock or increased intake air temperatures are experienced.

With power massively increased by the addition of a supercharger, the engine management system will require very major modification if OEM reliability and driveability are to be retained. Some blower kit manufacturers specify only a rising rate fuel pressure regulator and an on-boost ignition retard with their kits. This is very bodgy indeed. When assessing the approach that has been taken to engine management modification, you must look at it in the context of the percentage power increase claimed. Any gain over about 30 per cent on a previously naturally aspirated engine requires a larger fuel pump, larger injectors, engine management remapping or programmable management. A gain of 60 or 70 per cent or more requires new pistons and other appropriate engine mechanical upgrades.

One supercharger kit manufacturer states in its literature, "Optional bigger injectors should be fitted if the car is to be driven under high rpm and load conditions for long periods." From the latter, it can be assumed that the standard injectors (probably working at a much higher than normal fuel pressure) have too high a duty cycle at maximum power to be able to maintain that flow without

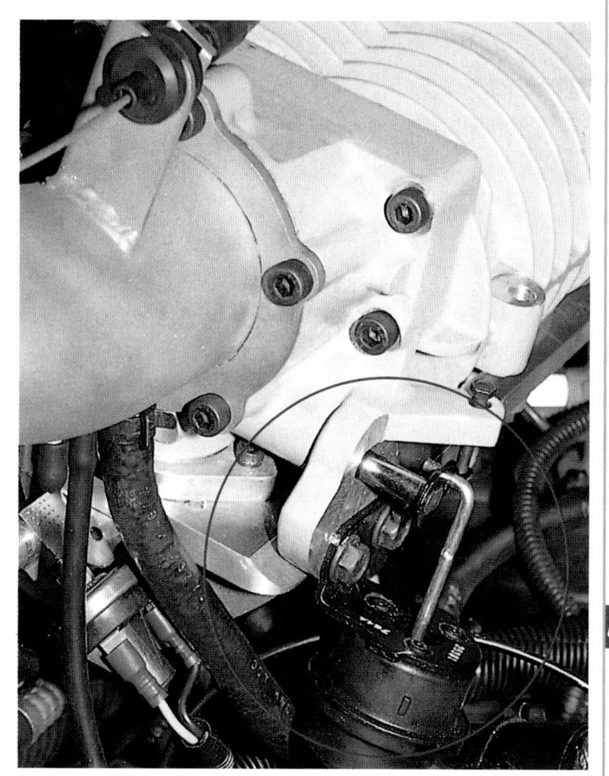

The Eaton bypass valve in its installed position. Opened at times of low load, or when boost needs to be limited, the valve can be used to improve fuel consumption by reducing drive loads during cruise and idle conditions.

A Vortech centrifugal blower installation on a Ford 5-litre Windsor V8, installed in a Cobra kit car. The power and torque gains from this installation are shown in Figure 7.32. Note that all centrifugal blowers have the throttle body located after the blower.

Another Vortech centrifugal blower installation. The red hose is an engine oil feed for the step-up gearing within the blower. On the left can be seen one of two Bosch throttle-closed blow-off valves that are fitted. In a centrifugal blower installation, compressor surge is possible without such a bypass valve used.

overheating. The difficulty that blower kit manufacturers face is that providing new injectors, a new ignition system and new ECU is likely to double the price of the kit over some of those available from their competitors. Accordingly, if you make major increases in power by converting a naturally aspirated engine to one boasting forced induction, and if you pay little for the conversion, you are simply going to have an engine with a short life!

Many kit manufacturers include water injection as part of their kit. This is because it's usually much simpler to fit water injection than to intercool an aftermarket supercharged engine. You can be sure of one thing: at least in some conditions, if the water injection system should fail, the engine is likely to detonate and die. Thus the water injection system should be of extremely high quality — the life of your engine is likely to depend on it! Chapter 9 covers water injection systems.

A kit provided with either a Roots or centrifugal blower is perfectly acceptable (even though the screw-type supercharger has higher efficiency) **if the kit is well engineered**. One way this can be assessed is to ask the manufacturer for the kit fitting instructions. Some manufacturers produce a few handwritten pages with rough notes and diagrams, while others have 30 or 40 pages of professionally produced, step-by-step instructions. If the company producing the kit cannot even afford to provide good instructions, what shortcuts will have been taken with the kit itself?

Finally, the manufacturer of the supercharger (not the kit as a whole) should be assessed for the quality of its product. Unlike with turbochargers — which are made in the hundreds of thousands by major companies — some supercharger manufacturers are almost backyard operators. A centrifugal compressor that uses straight blades, for example, is cheaper and easier to produce than one with curved blades, but the efficiency of the 'paddlewheel' straight blade compressor is poor. Companies such as Eaton, Vortech and IHI make very good products.

When selecting a kit, look for:

- Sophisticated engine-management improvements commensurate with the increase in power
- Engine mechanical changes commensurate with the increase in power
- Quality, heavy-duty brackets, drive system, fittings and plumbing
- A well-known supercharger sold with appropriate warranty conditions
- Intercooling or water injection of a quality that matches the power increase

Many supercharger installations are characterised by a whine from the supercharger. Even OEM blower installations, where engine quietness is a priority, have some supercharger noise. Some aftermarket installations are extremely noisy — a characteristic that may not be wanted. For this reason, it's best to at least go for a ride in a car supercharged with the same (or a similar) kit to the one being considered — before spending any money.

ELECTRIC SUPERCHARGERS

In any discussion of superchargers, the topic of electrically powered blowers is sure to come up. The concept has real attraction — imagine being able to flick a dashboard switch and have full boost just off idle! An electric supercharger could also be controlled very finely by electronic means, allowing potentially easy boost-governed traction control, different boost levels on different fuels and so on.

In fact, there are several types of electric superchargers available on the market. However, all suffer from a major — and unsolvable — problem. The power required to drive a supercharger capable of supplying the very large volumes of air needed by a performance engine cannot be adequately provided by a system powered from 12 volts.

To see why this is so, take the following example. To ascertain the effectiveness of a simple electric

In this Hyundai Excel installation, the first section of plenum and the original throttle body have been mounted directly on the intake to the supercharger.

supercharger, I attached a mains-powered electric leaf blower to the intake of a 2-litre, four-cylinder car and tested the vehicle on a chassis dyno. The leaf blower used had a 1.2kW motor and (according to the box) was capable of flowing 320cfm. This system did not produce any positive manifold pressure but still realised a 5 per cent gain over most of the power curve. Thus, a 1.2kW electric motor working with a fairly crude impeller design was capable of providing a noticeable power gain, even without developing boost.

At a mains power of 240 volts, a current flow of 5 amps is needed to provide a power of 1.2kW; however, at a nominal 13.8 volts, developing this much power requires a current flow of 87 amps! This very high current requirement of an automotive electric supercharger causes three problems:

- Finding a low-voltage motor whose windings won't melt at this current is difficult (and when such a motor is found, it has a life measured in hours).
- Controlling a motor drawing very high currents is difficult and expensive if proportional control is needed.
- The massive drain on the battery and alternator means the supercharger can be run only for very short periods.

The Mazda Miller Cycle Eunos 800 uses an IHI screw-type supercharger and two small air/air intercoolers, one for each cylinder bank. One of the intercoolers can be seen at the front, while the other is near the firewall, with cooling air fed from the plastic duct.

The crankshaft supercharger drive pulley in centrifugal installations generally needs to be large so the step-up ratio is sufficiently high.

A large Vortech throttle closed blow-off valve, recommended by Vortech for use in high-boost centrifugal blower applications.

This makes impossible the production of an electric supercharger that has a degree of effectiveness similar to that of a traditional supercharger.

However, on very small engines, an electric supercharger can work — there is an effective product available for the two-stroke Honda 250R engine of the type used in off-road four-wheel-drive motorcycles, for example. Also, currently being marketed is an electric supercharger claimed to be effective on cars. It uses a brushless motor spinning at 65,000rpm and is guaranteed for 80,000km. However, it has a price similar to that of a traditional supercharger and is capable of producing only low boost (less than 4psi) low in the rev range.

Another electric supercharger on the market is claimed to be as effective as a turbo — "increasing engine acceleration power up to 30 per cent, giving fuel savings of up to 30 per cent" and so on. When it is disassembled, the 'electric supercharger' turns out to be two tiny PC cooling fans mounted in a plastic housing...

If you are interested in supercharging a car, use a device driven directly from the crankshaft.

CARE OF A FORCED INDUCTION ENGINE

If you have fitted a turbo or supercharger to a naturally aspirated engine, or have increased boost in one already equipped with forced aspiration, there are some sensible precautions you should take if you and your car are to have a long and enjoyable life together!

First, the engine should never be operated at full throttle before it's sufficiently warmed. Allow time for the engine oil to reach operating temperature — something that usually takes much longer than indicated by the water temp gauge. Some OE forced-induction engines, in fact, prevent full boost being used until the engine has reached operating temperature.

Always use fuel that has the highest readily available octane. Fuels vary in quality from company to company, so it's probably not just your imagination when you think your car is happier on one particular brand. Fuel also becomes stale over time, so always buy fuel from a service station that has a high turnover of your chosen brew. If you buy fuel from a source different from the one you use regularly, listen for detonation even more carefully than usual.

The Jaguar XKR Coupe boasts an Eaton supercharged 4.OL V8 developing no less than 276kW and 525Nm.

Most manufacturers building blown engines currently are using Eaton Roots-type superchargers.

Detonation that becomes audible in a car that was previously running faultlessly may not, however, be the fault of the fuel quality. Other causes for the sudden occurrence of detonation include:

- Faulty fuel pump(s)
- Dirty injector(s)
- Dirty fuel filter
- Faulty spark plug(s) or aged spark plug leads
- Changed ignition timing
- Faulty ignition coil
- Improper functioning of the cooling system

So foreign materials don't damage the turbo or supercharger (either from within or without), make sure the air-filter is effective and the oil is clean. Some oiled foam aircleaners can allow the passage of dirt (sometimes it can even be seen as an oily smear on the intake to the compressor) and in ram air intake systems, the pickup of small pebbles can lead to the puncturing of paper element filters. To avoid internal damage, change the oil filter and oil at frequent intervals. Some aftermarket supercharger manufacturers suggest an appropriate change interval is 5000km and this matches well with OE intervals in forced-aspirated cars. Good-quality oil should be used (synthetic or semi-synthetic) and it's best to use OE oil filters. If you drive the car very hard in high ambient temperatures, fit an engine oil cooler.

Interestingly, supercharger manufacturer Vortech states: "Insufficient fuel delivery (pump or calibration) is the most common reason for performance and/or engine durability problems."

A toothed drive belt system can be used when supercharger belt slip is a problem. However, the supercharger must have been designed with this type of drive system in mind.

In a forced-aspirated engine, the spark plugs are subjected to higher combustion pressures and, if working with a CDI system, high voltages. Accordingly, conventional spark plugs should be inspected every 10,000-15,000km. The use of plugs one or two ranges colder than standard will help protect against detonation and will usually not foul, even in street-driven applications. Inspection of the plugs at frequent intervals is added security against engine damage occurring without your being aware of it.

When a turbo or supercharger compresses air, the air is heated. While this hot air can be fed straight into the intake of the engine (and often is), there are two disadvantages in taking this approach.

First, warm air has less density than cool air — this means it weighs less. It's important to know it's the **mass** of air breathed by the engine that determines power, not the volume. So if the engine is being fed warm, high-pressure air, the maximum power possible is significantly lower than if it's inhaling cold, high-pressure air. The second problem with an engine breathing warm air is that the likelihood of detonation is increased. Detonation is a process of unstable combustion, where the flame front does not move progressively through the combustion chamber. Instead, the air/fuel mixture explodes into action. When this occurs, damage to the pistons, rings or head can very quickly happen.

If the temperature of the air can be reduced following the turbo or supercharger, the engine will have the potential to safely develop a higher power output. Intercoolers are used to produce this temperature drop.

INTAKE AIR TEMPERATURES

How much hotter the air gets as it's being compressed depends on the pressure ratio (how much it is being compressed) and the efficiency of the compressor. This means the theoretical outlet temperature can be calculated if three factors are known: the inlet air temperature, the compressor efficiency and the pressure ratio.

Before this can be done, the temperatures and pressures need to be expressed in the right units. First, temperatures need to be converted to Kelvin (K), a measurement of absolute temperature.

$$\text{K} = \text{degrees C} + 273.15$$

A temperature of 35 degrees C is therefore the same as 308.15K (or 308K for our purposes).

Boost pressures also need to be converted to pressure ratios. Note that 1 Bar = 14.5psi

$$\text{Pressure ratio} = \frac{\text{Boost Pressure in Bar} + 1}{1}$$

A boost of 1.5 Bar therefore becomes a pressure ratio of 2.5.

Let's have a look at an example.

If the inlet air temperature to a turbo is 20 degrees C (293K) and the boost pressure is 1.1 Bar (pressure ratio = 2.1), the theoretical outlet temperature will be:

$$\begin{aligned}\text{Theoretical outlet temp} &= 293 \times (2.1)^{0.286} \\ &= 293 \times 1.236 \\ &= 362\text{K (89 degrees C)}\end{aligned}$$

This means there's a temperature **rise** of 69 degrees C (89 degrees - 20 degrees = 69 degrees).

However, this doesn't take into account that the compressor efficiency will be less than 100 per cent. If we assume a compressor efficiency of 70 per cent (typical for a good turbo):

$$\text{Actual temp increase} = \frac{69 \text{ degrees}}{0.7}$$

$$= 98.6 \text{ degrees C}$$

This is a temperature increase of 98.6 degrees which, added to the ambient temp of 20 degrees, means the actual outlet temp will be:

20 + 98.6 = 118.6 degrees
(119 degrees C when rounded off)

While the theory is fine, there's a number of factors that affect the accuracy of the calculated figure. First, it's difficult to accurately estimate the efficiency of the compressor. And even if such a figure is available, it doesn't necessarily apply to all the different airflows that the compressor is capable of producing. In other words, there will be some combinations of airflow and boost pressure where the compressor is working at peak efficiency — and other areas where it isn't. While a well-matched compressor should be at peak efficiency most of the time, in some situations it will be working at less than optimum efficiency. This will change the outlet air temperature, usually for the worse.

Second, the turbo- or supercharged car engine is not working under steady-state conditions. A typical forced induction road car might be on boost for only 5 per cent of the time, and even when it **is** on boost, it is perhaps for only 20 seconds at a stretch. Any decent

An air/air radiator consists of a large heat exchanger. Forced-induction air passes through the core, shedding heat to the airflow caused by the car's movement. This is an A'PEXi intercooler for a Skyline GT-R.

forced-induction road car will be travelling at well over 160km/h if given 20 seconds of full boost from a standstill, meaning longer periods of high boost occur only when hill-climbing, towing or driving at maximum speed. While all the engine systems should be designed with the full load capability in mind, in reality very few cars will ever experience this. This factor means the heatsink capability of the intake system must be considered.

If the inlet air temperature of the engine in cruise condition is 20 degrees C above ambient, then on a 25-degree day the inlet air temp will be 45 degrees C. After 30 minutes or so of running, all the different components of the intake system will have stabilised at around this

David Bryant

A non-intercooled engine like this Nissan RB30ET is severely limited in the power it can safely develop. Fitting an intercooler should be amongst the first steps in gaining more power from a turbo car.

Some manufacturers mount the air/air intercooler on top of the engine. This severely affects efficiency, especially when the car is not moving.

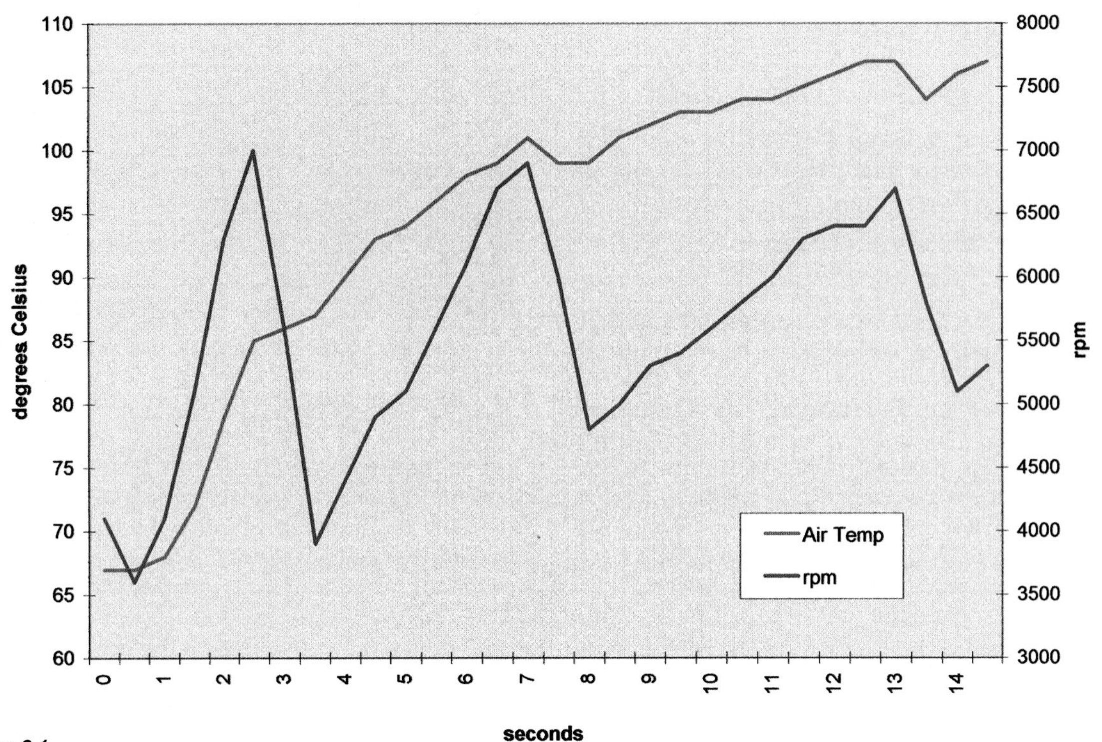

Figure 8.1
The increase in intake air temperature recorded during a quarter-mile dragstrip of a non-intercooled Nissan FJ20DET turbo four fitted to a Nissan Bluebird. The engine was using 14.5psi (1 Bar) boost and Haltech engine management. As can be seen, in the 14.7 seconds the car took to travel down the standing quarter-mile, the intake air temp rose from 67 degrees C to 108 degrees C.

temperature. If the engine then comes on boost and there is a sudden rise in the temp of the air being introduced to this system, the temperature of the turbo compressor cover (or blower housing), inlet duct, throttle body, plenum chamber and inlet runners will all increase. These components increase in temp because they are removing heat from the intake air and so limit the magnitude of the initial rise in the actual intake air temperature. As a result, the infrequent short bursts of boost used in a typical road-driven forced-induction car often produce a lower initial intake air temperature than expected.

This **doesn't** mean intercooling is not worthwhile — it certainly is — but that the theory of the temperature increase doesn't always match reality. However, taking a theoretical, calculated approach at least gives a general idea of the temperature increase likely to be experienced.

An example of an actual intake air temperature increase during a quarter-mile dragstrip run is shown in Figure 8.1. It shows the logged intake air temperature and engine rpm on a non-intercooled Nissan FJ20DET turbo four fitted to a Nissan Bluebird. The engine was using 14.5psi (1 Bar) boost and Haltech engine management. As can be seen, in the 14.7 seconds the car took to travel down the standing quarter-mile, the intake air temp rose from 67 degrees C to 108 degrees C.

AIR DENSITY CHANGES

In the example above there was a temperature increase from 20 degrees to 119 degrees. But how does this temperature change affect the all-important density of the air? The density of air depends on two factors — its temperature

The Subaru Impreza WRX normally has an intercooler mounted in the engine bay, but here a front-mounted core has replaced it.

This 5.7-litre V8 has been supercharged with a STA centrifugal blower. It uses a large air/air intercooler located in front of the radiator.

and pressure. The drop in density due to increased temperature is directly proportional to the ratio of the temperatures, when they are expressed in Kelvin. So the drop in density of the air at 119 degrees C (392K) versus 20 degrees C (293K) is found by:

$$\textbf{Density difference} = \frac{293}{392}$$

$$= 0.75$$

In other words, the temperature increase would have caused a drop in density by 25 per cent had the air still been at the same pressure. But, of course, its pressure is now higher because the turbo or supercharger has compressed it! If the air pressure is doubled without a change in temperature, the density is doubled. To work out the effect of **both** the loss of density because of the temp rise **and** the increase in density because of boost pressure, the two factors are multiplied.

$$\textbf{Increase in Air density} = \frac{\textbf{Inlet Air Temp (K)}}{\textbf{Outlet Air Temp (K)}} \times \textbf{pressure ratio}$$

$$= \frac{293}{392} \times 2 = 1.49$$

In other words, the increase in density has been 49 per cent at 1 Bar boost with a compressor outlet air temperature of 119 degrees C. This means that, all things being equal, the engine can develop 49 per cent more power.

INTERCOOLER EFFICIENCY

An intercooler will do two things: it will lower the temperature of the intake air and at the same time cause a slight drop in boost pressure. The latter comes from the restriction to flow caused by the intercooler.

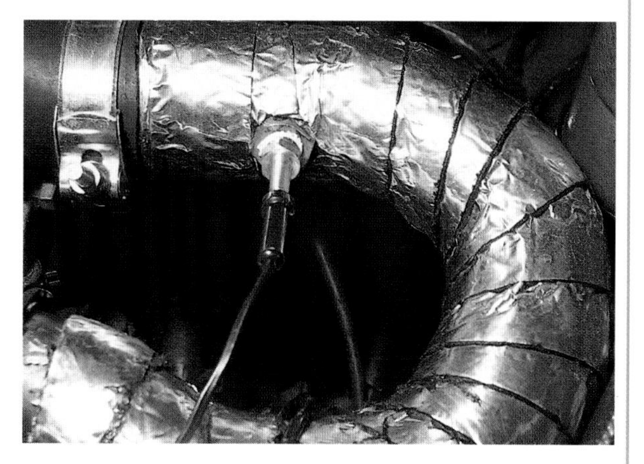

The return pipe from this intercooler has been insulated with aluminium tape applied over the top of ceramic fibre mat. This prevents the induction air being re-warmed as it passes through the engine bay. Note the thermocouple probe to measure intake air temps.

A back-fire can split open an intercooler with ease...

This twin turbo Honda (Acura) NSX has its intercooler cores mounted over the engine. The cores are cooled by ducts rising from the bottom of the engine bay, with air extraction occurring through louvres in the (not shown) clear plastic rear window.

Some restriction is unavoidable because the flow through an efficient intercooler core needs to be turbulent if a lot of the air is to come in contact with the heat exchanger surfaces. However, if the pressure drop is too high, power will suffer. A pressure drop of 1-2psi can be considered acceptable if it's accompanied by good intercooler efficiency.

Intercooler efficiency is a measurement of how effective the intercooler is at reducing the inlet air temperature. If the intercooler reduces the temperature of the air exiting the compressor to ambient, the intercooler will be 100 per cent efficient. It will also be a bloody marvel, because no conventional intercooler can actually achieve this!

Intercooler efficiency is given by:

$$\textbf{Intercooler Efficiency} = \frac{\textbf{Actual Temperature Drop}}{\textbf{Maximum Possible Temp. Drop}}$$

For example, if on a 20-degree C day the outlet air temp of the turbo is 110 degrees C and the temperature of the air after it has passed through the intercooler is 45 degrees C, then:

$$\textbf{Intercooler Efficiency} = \frac{\textbf{110 degrees - 45 degrees}}{\textbf{110 degrees - 20 degrees}}$$

$$= \frac{\textbf{65 degrees}}{\textbf{90 degrees}}$$

$$= \textbf{0.72 (or 72 per cent)}$$

FINAL CALCULATIONS

Let's take an example that puts together all the figures. You fit an intercooler that's 72 per cent efficient and has a pressure drop of 0.14 Bar (2psi) to an engine running 1.5 Bar (22psi) boost developed by a turbo compressor that's 70 per cent efficient.

It's a 25-degree Celsius day.

First, the theoretical compressor outlet air temp:

$$\textbf{298K x (2.5 pressure ratio)}^{0.286} \textbf{ = 387K = 114 degrees C}$$

114 degrees C - 25 degrees C = a temperature increase of 89 degrees C

$$\frac{\textbf{89 degrees C}}{\textbf{0.7 compressor efficiency}} \textbf{ = 128 degrees C temp. increase}$$

128 degrees C temp increase + 25 degrees C ambient = 153 degrees C air temperature coming out of the turbo!

The maximum possible temperature decrease through an intercooler:

153 degrees C - 25 degrees C = 128 degrees C

An intercooler that is 72 per cent efficient will give an actual temperature drop of:

128 degrees C x 0.72 = 92 degrees C

A turbo outlet temp of 153 degrees C - 92 degrees C intercooler drop = actual engine inlet air temp of 61 degrees C.

Boost pressure was 1.5 Bar but with the pressure drop through the intercooler it becomes:

1.5 - 0.14 = 1.36 Bar.

$$\frac{\textbf{1.36 + 1}}{\textbf{1}} \textbf{ = a pressure ratio of 2.36}$$

So with an intercooler we have a temp of 61 degrees C at a boost pressure ratio of 2.36, compared with a non-intercooled 153 degrees C at a pressure ratio of 2.5.

$$\text{Intercooled: } \frac{\textbf{298K}}{\textbf{334K}} \textbf{ x 2.36 pressure ratio}$$

= 2.1 times non-turbo air density

$$\text{Non-intercooled: } \frac{\textbf{298K}}{\textbf{426K}} \textbf{ x 2.5 pressure ratio}$$

= 1.75 times non-turbo air density

If the engine developed 100kW (134hp) in naturally aspirated form, with 1.5 Bar (22psi) boost it could develop a theoretical 175kW (235hp) without an intercooler and 210kW (281hp) with an intercooler! On this high-boost example, fitting the very big intercooler is good for a 20 per cent power gain. However, again reality intrudes — the picture is often even better than this! Any forced-induction car seems to go disproportionately harder when the ambient temp is low — much more so than the physics suggest. The same applies to a car equipped with a very good intercooler — throttle response is superior and power markedly so.

If you're using a boost pressure level of more than about 7psi (~0.5 Bar), an intercooler is generally a worthwhile investment.

INTERCOOLER TYPES

Most intercoolers fall into two categories: air/air and air/water. There are also those special designs that cool the intake air to below ambient temperature, using ice, the air-conditioning system or direct nitrous oxide sprays, but they will not be covered here.

Air/Air Intercoolers

Air/Air intercoolers are the most common type, both in factory forced-induction cars and aftermarket. They are technically simple, rugged and reliable. An air/air intercooler consists of a tube and fin radiator. The induction air passes through thin rectangular cross-section tubes that are stacked on top of each other. Often inside the tubes are fins that are designed to create turbulence and so improve heat exchange. Between the tubes are more fins, usually bent in a zig-zag formation. Invariably, air/air intercoolers are constructed from aluminium. The induction air flows through the many tubes. The air is then exposed to a very large surface area of conductive aluminium that absorbs and transfers the heat through the thickness of metal. Outside air — driven through the core by the forward motion of the car — takes this heat away, transferring it from the intake air to the atmosphere.

Described above is what is normally called the intercooler 'core' — the part of the intercooler that actually effects the heat transfer. However, there also needs to be

Here poor quality plumbing has been squeezed past the headlight of a Mitsubishi Cordia Turbo. Where there is absolutely no room for well-made ducts, it's better to use a water/air intercooler rather than an air/air design.

an efficient way of carrying the intake air to each of the tiny tubes that pass through the core. End-tanks are used for this, being welded at each end of the core. While some cores are 'double-pass' (the inlet and outlet tanks are at one end separated by a divider, while at the other end the air does a U-turn), most cores are single-pass, with the inlet at one end of the core and the outlet at the other.

A bonnet scoop normally feeds this Mazda 13B rotary turbo intercooler, but here a ducted enclosed approach has been taken. This is certainly better than nothing — but a bonnet intake is better!

Good intercooler manufacturers have two specifications available: the pressure drop at a rated airflow (with the airflow often expressed as engine power) and the cooling effect (normally expressed as a temperature drop at that rated flow).

However, many intercooler manufacturers have no data available on either of these factors! To some extent this doesn't matter greatly — the design of the intercooler is normally limited by factors other than heat transfer ability and pressure drop. Because an air/air intercooler uses ambient air as the cooling medium, an air/air intercooler cannot be too efficient — simply, the bigger the intercooler, the better. In fact, the maximum size of an air/air intercooler is normally dictated by the amount of space available at the front of the car and the size of your wallet, rather than any other factors!

It's easy to see how cost is a vital factor — those forced induction cars produced by major car companies as homologation specials (either for rallying or circuit racing) have quite huge intercoolers that dwarf the ones fitted by the same companies to their humdrum cars. Nissan used an air/air core no less than 60x30x6cm on their R32 Nissan Skyline GT-R and the Mitsubishi Lancer Evolution vehicles also use huge intercoolers. The 'bigger is better' philosophy can be clearly seen at work in these cars.

Many factory-fitted intercoolers are undersized. Air/air cores no larger than a paperback book can be found in turbo cars with a nominal maximum output of 150kW (~200hp). Cars equipped with this type of intercooler can be held at peak power for only a very short time before the increasing inlet air temperature causes the ECU to retard timing or decrease boost. A car fitted with this type of tiny factory intercooler is almost impossible to dyno test — the intake air temp rises so fast that rarely can more than two consecutive dyno runs be made before the

Truck intercoolers come in two sizes — huge and huger! A lot of modification of this core will be needed before it can be fitted into the Mazda MX6 Turbo. However, using a discarded truck intercooler is a cheap way of building a very large 'cooler.

intake air temp is so high that the engine detonates... On the other hand, the aforesaid Skyline GT-R has a measured intake temp of 45 degrees C on a 35-degree C day at 14.5psi (1 Bar) boost and a sustained full-throttle 250km/h!

Intercooler Mounting

When either increasing the size of a factory intercooler or installing a new one for a custom forced-aspiration car, care needs to be given to the location that is chosen. The first point to consider is the amount of ambient heat present. An intercooler core **absorbs** heat just as well as it sheds it. This means an underbonnet intercooler core can easily become an intake air **pre-heater** if care isn't taken with its location. Turbo cars run especially high underbonnet temps, so a bonnet vent designed for intercooler cooling while the car is underway can easily become a 'chimney' ducting out hot air while the car is stationary — hot air that passes straight through the intercooler core. In fact, the behaviour of the intercooler while the vehicle is stopped is very important if you're in the habit of caning the car in traffic-light grands prix!

By far the best location for an intercooler is in front of the engine radiator. The car manufacturer will have aerodynamically tested the vehicle to ensure that large volumes of air pass through the engine cooling radiator, so an intercooler placed in front of that is sure to receive a great amount of cooling air. Note the intercooler should be in front of any air-conditioning condenser as well.

The air/air core should be ducted with the cold air if at all possible. Many people simply place the intercooler at the front of the car, hoping the air being forced through the front grille will all pass through the intercooler. However, if there is an easier path for the air to take, that's the way it will go. Sheetmetal guides can be used to channel the air coming in the grille through the intercooler, and foam rubber strips can be used to seal the escape routes the air might otherwise take. Taking this approach also improves the car's coefficient of drag.

The plumbing leading to and from the intercooler should produce only a minimal pressure drop. Factory turbo cars often use intake ducts that smoothly increase in size from the diameter of the turbo compressor outlet (often only 50mm or so) to the inlet diameter of the throttle body (perhaps 80mm) and, if this can be done, it's an approach that should be followed. Intercooler plumbing should have gentle curves and be as short as possible. Don't forget when you are planning the plumbing that the engine (and so also the blower or turbo!) moves around, while the body-mounted intercooler core does not. This means some rubber or silicone hose connections must to be incorporated in the plumbing to absorb the

Care should be taken that the intercooler is positioned so it can get enough airflow. There's not much of a problem in that regard with this Mitsubishi Lancer GSR!

This Skyline GT-R is using the standard air/air intercooler that has been stripped of paint. The measured average intake air temperatures rose by a few degrees Celsius after this occurred.

movement. The use of rubber mounts between the intercooler plumbing and the bodywork will also help absorb some of this movement.

The return duct from the intercooler should be insulated to avoid it picking up heat from within the engine bay. Lagging the pipe with fibreglass or ceramic fibre matting works effectively without being too bulky. The pipework can be finished off with a wrapping of aluminium adhesive tape of the type sometimes used to seal roofs. Also note that when planning the intercooler pipework, the compressor cover of a turbo can be easily rotated to allow the outlet to come out at a different angle. This can reduce the number and tightness of the bends required.

Some people believe that if they fit a very big intercooler with large ducts, the volume of charge air within it will unduly slow throttle response. Their concern is unjustified, however; throttle response problems (for example, turbo lag) are largely the result of other factors within the forced induction system, not the volume of air within it.

Sourcing the Core

There's a number of ways of getting together a very good air/air intercooler. Those companies specialising in the production of intercoolers (Spearco in the US is one of the largest) have a huge variety of cores and end-tanks available. However, as an aluminium item of fairly intricate construction, they're not cheap. For a really big air/air intercooler complete with end tanks, expect to pay about as much as you would for a turbo.

An alternative are Japanese importing wreckers. While few factory turbo cars have really large intercoolers (and even fewer factory supercharged cars have them!), there's at least a couple of large ones available. As mentioned previously, the Nissan Skyline GT-R and Mitsubishi Evolution model Lancers have very good intercoolers. The Nissan Pulsar GTiR also has a large intercooler, while the Mazda RX7 single turbo Series 4 has an engine-mounted intercooler that has a good flow, despite its appearance. Welding two of the RX7 intercoolers in series

Several manufacturers use water/air intercooling in their turbocharged vehicles. This is a GT4 Celica, here in rally guise.

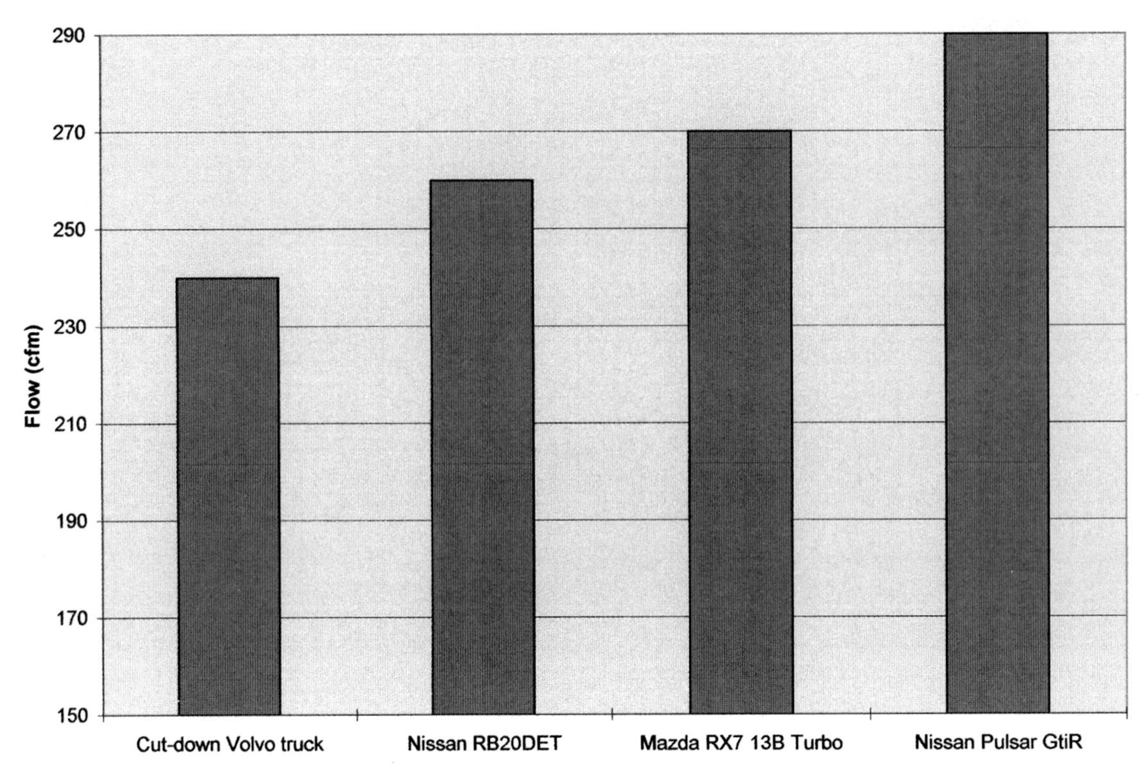

Figure 8.2
The flowbench figures for three Japanese factory intercoolers and one modified truck intercooler are shown here. The flow testing was carried out at a pressure differential of 28 inches of water. As can be seen, modified truck intercoolers aren't always free-flowing.

has been shown to work very well. The flowbench figures for three Japanese factory intercoolers are shown in Figure 8.2.

You can also produce your own intercooler by modifying heat exchanger cores designed for other duties. However, having personally done so, I can advise that it involves a great deal of work! One source of efficient heat exchangers are old air-conditioners. Domestic and industrial refrigerative air-conditioners use copper tube and aluminium fin heat exchangers for both their evaporators and condensers. When the air-conditioner is discarded (perhaps because of a faulty compressor), these components are sold off at scrap value — less than the price of a few sparkplugs! If you're patient and handy, you can cut off each end of the core and make plates that fit over the multiple copper tubes. Making end tanks that attach to these baseplates is then straightforward. The resulting copper-cored air/air intercooler is efficient and very, very cheap.

The Subaru Liberty [Legacy] RS also uses water/air intercooling, with the heat exchanger positioned near the firewall.

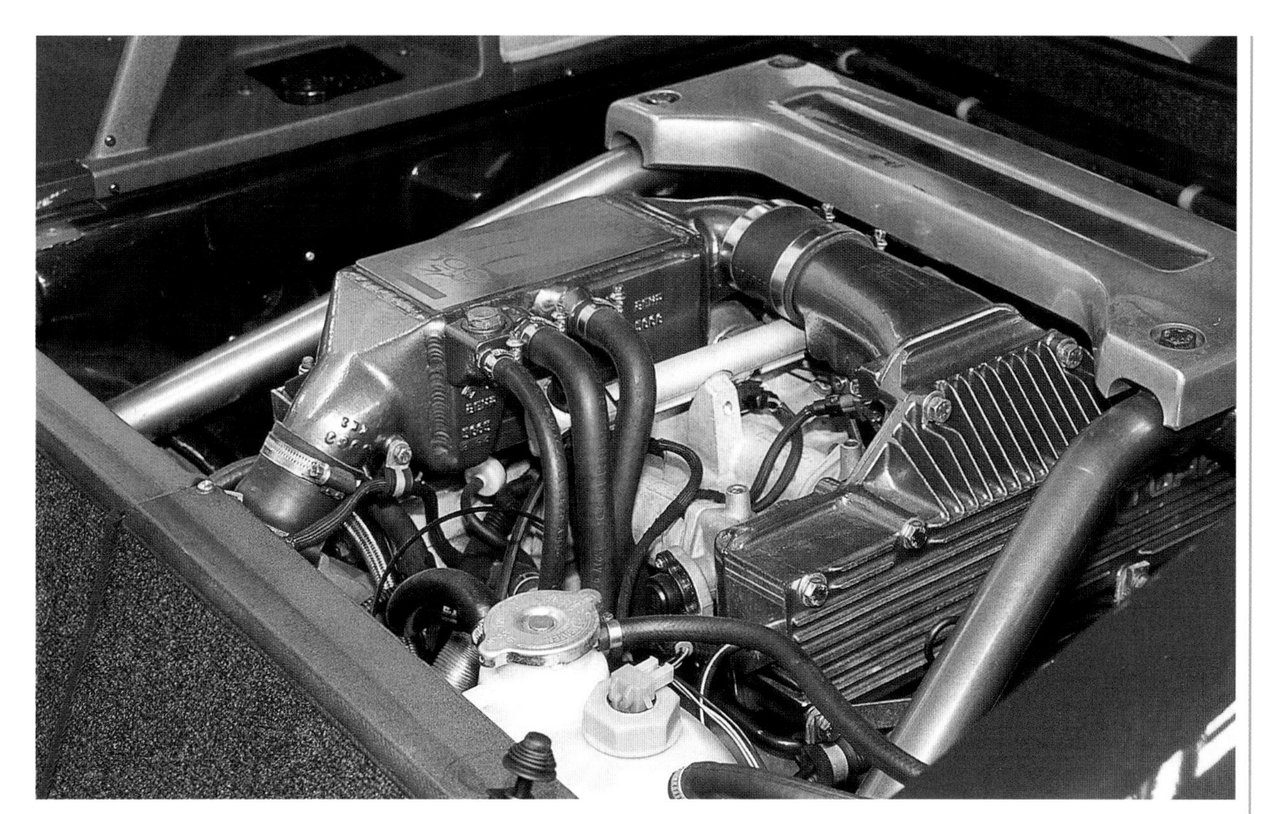

Lotus uses a compact heat exchanger in their water/air system. Note that each of the factory heat exchangers has positioned on it air-bleed fittings or the filler cap.

Another alternative is to visit truck wreckers. Diesel turbo truck intercoolers are absolutely huge. They can also often be picked up very cheaply from insurance repair jobs, where the core has been twisted slightly or one end tank is damaged, perhaps. If you choose with an eye to modification, the core will be able to be shortened without new end tanks being required — which substantially reduces the amount of work! This way you need only make new blanking plates for the ends of the shortened tanks. However, be aware that reducing the number of tubes of a truck intercooler in this manner can also reduce its flow by an unacceptable amount.

Water/Air Intercooling

Water/air intercooling is used less frequently than the air/air approach. However, it has several benefits, especially in cramped engine bays. A water/air intercooler uses a compact heat exchanger located under the bonnet and normally placed in-line with the compressor-to-throttlebody path. The heat is transferred to water, which is then pumped through a dedicated front-mounted radiator cooled by the airflow generated by the car's movement. A water/air intercooler system consists of these major parts: the heat exchanger, radiator, pump, control system and plumbing.

Technically, a water/air intercooler has some distinct cooling advantages. Water has a much higher specific heat than air. The water's specific heat value shows how much energy a given mass of the substance can absorb for each degree temp it rises by. A substance good at absorbing energy has a high specific heat, while one that gets hot quickly has a low specific heat. Something with a high specific heat can obviously absorb (and then later get rid of) lots of energy — good for cooling the air.

Air has a specific heat of 1.01 units (at a constant pressure), while the figure for water is 4.18 units. In other words, for each increase in temp by one degree, the same mass of water can absorb some four times more energy than air. Or, there can be vastly less flow of water than air to get the same job done. Incidentally, note that pure water is best — its specific heat is actually degraded by 6 per cent when 23 per cent anti-freeze is added! Other commonly available fluids don't even come close to water's specific heat.

The high specific heat of water has a real advantage in its heatsinking effect. An air/water heat exchanger designed so it has a reasonable volume of water within it can absorb a great deal of heat during a boost spike. Even before the water pump has a chance to feed in cool water, the heat exchanger will have absorbed considerable heat from the intake airstream. It's this characteristic that makes a water/air intercooling system as efficient in normal urban driving **with the pump stopped**

This is an Isuzu Piazza air/air core that's being turned into a water/air heat exchanger. Note the strong cast alloy end tanks that make enclosing the core much easier.

An air/air Piazza core being enclosed in sheet aluminium to form a water/air heat exchanger. After this welding is carried out, the heat exchanger needs to be pressure-tested for water leaks.

as it is with it running! To explain: the water in the heat exchanger absorbs the heat from the boosted air, feeding it back into the airstream once the car is off boost and the intake air is cooler. I am not suggesting you don't worry about fitting a water pump, but it is a reminder that, in normal driving, the intercooler works in a quite different way from the manner in which it needs to perform during sustained full throttle.

The Heat Exchanger

Off-the-shelf water/air heat exchangers are much rarer than air/air types. Water/air intercooling has been used in turbo cars produced by Lotus, Subaru and Toyota, and a few aftermarket manufacturers also produce them. If you want to make your own, the easiest way to go about it is to jacket an air/air core. Pick an air/air intercooler that uses a fairly compact core that still flows well. If it uses cast alloy end tanks (as opposed to pressed sheet aluminium), so much the better. (Plastic end tank types need not apply!) The core is then enclosed in 3mm aluminium sheet, TIG welded into place. Water attachment points can be made by welding alloy blocks to the sheetmetal, with these blocks then drilled and tapped to take barbed hose fittings. Pressure-test the water jacket to make sure it does actually seal, and make sure the water flow from one hose fitting to the other can't bypass the core. Small baffles can be used to ensure the water does fully circulate before exiting.

The finished modified Piazza intercooler installed on a Nissan RB20DET engine. Note the air-bleed screw on top and the very direct path the air has from the turbo compressor to the throttle body.

A water/air heat exchanger made by enclosing a Mazda 13B Turbo air/air core. Note the use of a rotary engine coolant filling cap and fitting.

Another type of water/air heat exchanger can be made using a copper tube stack. These small heat exchangers are normally used to cool boat engine oil, exchanging the heat with engine coolant or river or sea water. While the complete unit uses a cast-iron enclosure so is too heavy and large for car applications, the core piece itself can be enclosed to make a very efficient heat exchanger. Comprising a whole series of small-bore copper tubes joining two endplates, the core is cylindrical in shape and relatively easy to package. The induction air flows through the tubes while a water-tight sheetmetal jacket can be soldered around the cylinder. The resulting heat exchanger is a little like a steam engine boiler, with induction air instead of fire passing down the boiler tubes!

Another type of water/air heat exchanger can be made from a boat oil cooling stack. This copper-cored heat exchanger was photographed while fuel flow testing was being carried out.

As with air/air designs, the more efficient you can make the heat exchanger, the better the potential system performance. If you plan to use an off-the-shelf heat exchanger that has specifications available for it, you'll be interested to know the 150kW (~200hp) turbo Subaru Liberty (Legacy) RS uses a factory-fitted water/air exchanger that has a 4kW capacity. This heat exchanger also works quite effectively when engine power is increased to about 210kW (~280hp). Remember in your design considerations that you want a reasonable store of water in the actual heat exchanger (two or three litres at least) to help absorb the temperature spikes.

Radiator and Pump

The front-mounted radiator for the water/air intercooler should be completely separate from the engine cooling radiator. Some turbo trucks use the engine coolant to cool the water/air intercooler, but their efficiency is much reduced by taking this approach. Suitable radiators that can be used include large oil coolers, car air-conditioning condenser cores and scrap domestic air-conditioning condensers. If you use a car air-conditioning condenser, there's likely to be available a small dedicated electric fan that attaches to the core easily. This fan can be triggered to aid cooling when the vehicle is stationary. The radiator should at least match (and preferably exceed) the cooling capacity of the heat exchanger, but again finding proper specifications is often difficult. The Subaru Liberty (Legacy) RS with the 4kW heat exchanger uses quite a small radiator, only 45x35x3cm.

Diaphragm pumps like the one shown here can be used to circulate water in a water/air intercooling system. While durable and widely available, these pumps are a little noisy. Incidentally, note the engine's custom-made moulded PVC cool air intake mounted next to it.

An electric pump is the simplest way of circulating the water, with the type of pump chosen influenced by how the pump is to be operated. Some factory systems have the pump running at low speed continuously, switching to high speed at certain combinations of throttle position and engine airflow. If you follow a similar approach, the pump chosen must be capable of continuous operation. Another approach is to trigger the pump only when on boost, or to trigger a timing circuit that keeps the pump running for another, say, 30 seconds after the engine is off-boost. The latter type of operation will mean the pump operating time is drastically reduced over continuous running.

Twelve-volt water pumps fall into two basic types: impeller and diaphragm. An impeller pump is of the low-pressure, high-flow type. In operation it's quiet with low vibration levels. A diaphragm pump can develop much higher pressures but generally with lower flows. A diaphragm pump is noisy and must be rubber-mounted in a car.

Suitable impeller-type pumps are used in boats as bilge pumps and for deck washing. They're relatively cheap and have very high flows — 30 litres a minute is common. However, they are **not** designed for continuous operation and generally don't have service kits available for the repair of any worn-out parts. Diaphragm pumps are used to spray agricultural chemicals and to supply the pressurised water for use in boat and caravan showers and sinks. They're available in very durable designs suitable for continuous running and have repair kits available. Flows of up to 20 litres a minute are common and they

Enclosing a car air-conditioner evaporator core can make an extremely efficient water/air heat exchanger. This one is working hard on a big-turbo Nissan RB25DET six mounted on an engine dyno.

develop enough pressure (~45psi) to push the water through the front-mounted radiator and heat exchanger without any problems.

The factory water/air intercooler system in the Subaru Liberty (Legacy) RS uses an impeller-type pump rated at 15 litres a minute (all flow figures are open-flow). It is automatically switched from low to high speed as required. This — and other factory water/air intercooler pumps — are ideal because they were designed to circulate the water in water/air intercooling systems! However, they are generally very expensive to buy new, but if one can be sourced secondhand it is ideal.

A cheap and simple impeller pump is the Whale GP99 electric pump. It's so small that the in-line pump can be supported by its hose connections. It flows 11 litres a minute and has 12mm hose fittings. It's 136mm long and 36mm in diameter and is suitable only for intermittent operation. This pump is available from marine and caravan suppliers.

The Flojet 4100-143 4000 is a diaphragm pump suitable for water/air intercooler use. The US-manufactured pump uses a permanent magnet brush-type fan-cooled motor with ball bearings and is fully rebuildable. The pumping head uses four diaphragms, which are flexed by a wobble plate attached to the motor's shaft. The 19-litre/minute pump uses 0.75-inch fittings and is 230mm long and 86mm in diameter. It's available from companies supplying agricultural spray equipment.

The Flojet pump needs to be mounted either vertically with the pump head at the bottom, or horizontally with the vent slots in the head facing downwards. This is to stop any fluid draining into the motor if there are any sealing problems in the pump head. At its peak pressure of 40psi, the pump can draw up to 14 amps; however, in intercooler operation the pressure is vastly less, so the pump draws only about 5.5 amps at 12 volts. The pump is noisy (as all diaphragm pumps are), but mounting it on a rubber gearbox crossmember mount effectively quietens it. Note these pumps are much louder when mounted to the car's bodywork than they are when sitting on the bench!

Control Systems

As already indicated, there's a number of ways of controlling the pump operation. The simplest is to switch the pump on and off with a boost pressure switch. This means that whenever there's positive manifold pressure, the pump circulates the water from the heat exchanger through the radiator and back to the heat exchanger. If boost is used frequently and for only short periods, this approach works well. However, it's better if a timer circuit is used so the pump continues to operate for a short period after boost is finished. Such a circuit is shown in Figure 8.3. Figure 8.4 (opposite) shows how to lay out the components on a piece of tag strip.

If the circuit is assembled as shown, the automotive relay will be held on for about 40 seconds after the on/off switch is released. Fitting a 470uF capacitor in place of the 1000uF component will halve this 'on' time, while using an even lower value capacitor will reduce this time still further. In a pump control application, the on/off switch can be replaced with a boost pressure switch.

A suitable pressure switch is an adjustable Hobbs unit, available from auto instrument suppliers. However, this

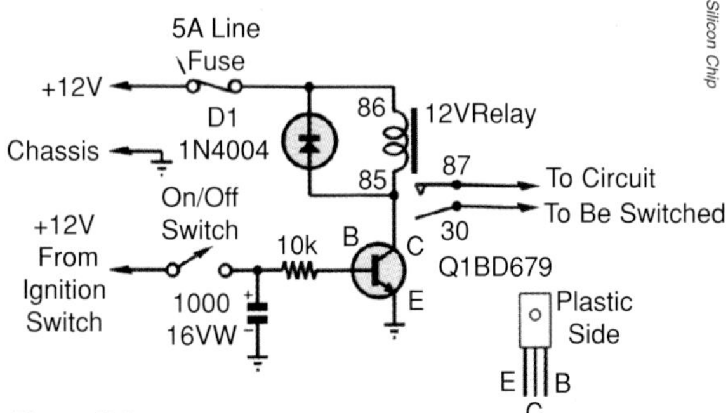

Figure 8.3
This electronic timer is easily built to provide an extended 'on' time for a water/air intercooler pump. If the circuit is assembled as shown, the relay will be held on for about 40 seconds after the on/off switch is released. Fitting a 470uF capacitor in place of the 1000uF component will about halve this 'on' time, while using a lower-value capacitor will reduce this time still further. In a pump control application, the on/off switch can be replaced with a boost pressure switch.

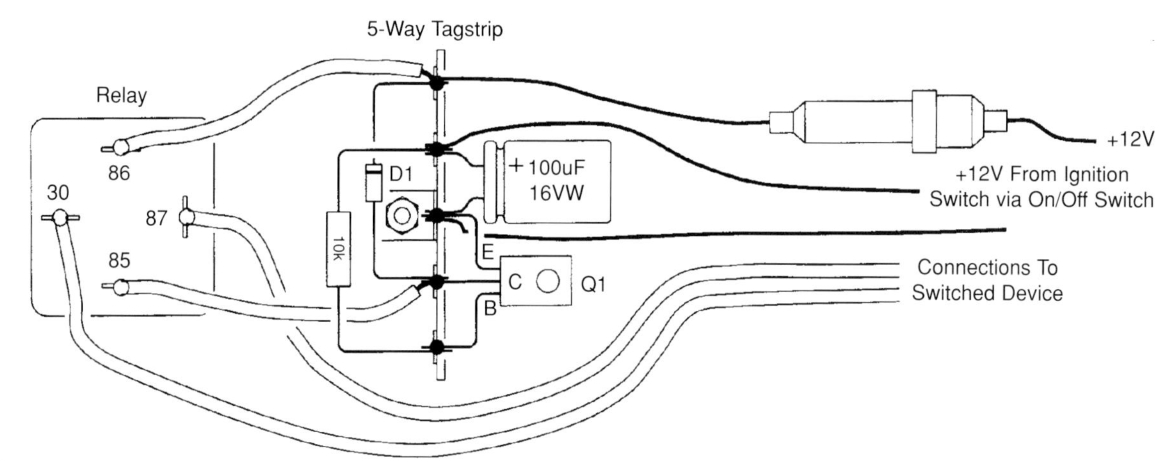

Figure 8.4
The timer circuit shown in Figure 8.3 is easily assembled on a piece of tag strip.

switch is relatively expensive and a cheaper unit is easily found. Spa bath suppliers stock a pressure-operated micro switch that's ideal for forced-aspirated car use. The pressure switch is designed to work as part of the air-actuated switching system that's used in a spa bath so bathers don't have to directly operate high-voltage switches. The switch triggers at around 1psi and costs about half that of a traditional automotive pressure switch. If a switching pressure above 1psi is required, simply tee a variable bleed into the pressure line leading to the switch. Adjusting the amount of bleed will change the switch-on point.

Another approach to triggering pump operation is to use a throttle switch. A micro switch (available cheaply from electronics stores) can be used to turn on the pump whenever a throttle position over, say, half is reached. A cam can be cut from aluminium sheet and attached to the end of the throttle shaft. If shaped with care, it will turn on the switch gently and then keep it switched on at throttle positions greater than the switch-on opening throttle angle.

A large transmission oil cooler can be used as the radiator for a water/air intercooler system. Other alternatives include the engine oil cooler from a Mazda rotary engine and the condenser core from a domestic air-conditioner.

If a two-speed pump operation is required, the pump can be fed current through a dropping resistor to provide the slow speed. When full speed is required, the dropping resistor can be bypassed. Suitable dropping resistors are the ballast resistors used in older ignition systems or the resistor pack used in series with some injectors. The value of the resistor used will depend on the pump current and

The standard water/air intercooler pump from the Subaru Liberty (Legacy) RS uses an impeller-type pump rated at 15 litres/minute.

its other operating characteristics. In all cases, the resistor will need to dissipate quite a lot of power, so will need to be of the high-wattage, ceramic type. The resistor will get very hot and can be placed on a transistor-type heatsink mounted within the airstream, perhaps behind the grille. When experimenting with resistors and a pump, you should know that placing the multiple resistors in parallel will increase pump speed while wiring the resistors in series will slow the pump.

Another approach is to use a temperature switch, so the pump doesn't run when the intake air is not actually hot. This situation can occur on boost if the intake air temperature is very low because the day is cold. Overly cold intake air can cause atomisation problems, although this is not normally a problem in a high-performance car being driven hard! However, running the pump when the intake air is perhaps only 5 degrees C is pointless and it can be avoided by placing a normally open temperature switch in series with the boost pressure or throttle position switches. If the switch closes at temperatures above, say, 30 degrees C, the pump will operate only when it actually needs to. A range of suitable, cheap temperature switches is available from RS Components (stores in most states and countries).

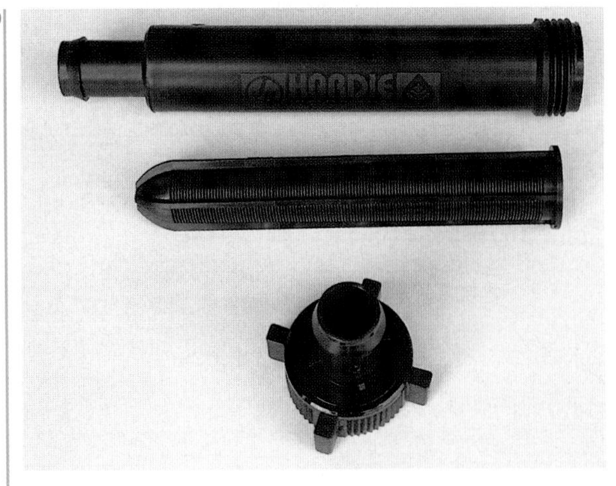

A cheap garden irrigation filter can be used to protect the pump in a water/air system. These filters are available from hardware stores.

Finally, automotive electronics company Fueltronics makes a module that can switch on a pump once the output of two engine management sensors pass two adjustable thresholds. One of the module's inputs is designed to detect a falling voltage, and the other input a voltage that's increasing. The 'falling voltage' input can be used with the standard management's inlet air sensor, while the 'rising voltage' input can be used with an airflow meter or throttle position sensor. In this way, the pump can be switched on (or its speed increased) when both high throttle openings **and** high intake air temperatures are present.

Note that in all pump control systems a relay should be used to operate the pump. A circuit with two pump speeds, temperature or pressure switch control, and radiator fan control is shown in Figure 8.5. Note the intercooler radiator fan is triggered only when the pump is working at full power. A dashboard override switch has also been included to be used when you want to manually operate the system in its high-load configuration.

The Water Plumbing

The most obvious place for the pump within the system is immediately after the radiator so it's then subjected only to relatively cool water temperatures. However, this can't always be done because some designs of pump are reluctant to suck through the restriction posed by the radiator. Depending on the design of the radiator, its flow restriction may be substantial. During the assembly of the system it's therefore wise to set it all up on the bench. Check water flows with the pump running (at different speeds, if this is the approach to be taken) and with the pump in different positions within the system. The pump position that yields the greatest water flow should be the one adopted — even if that places the pump immediately after the heat exchanger. In practice, the

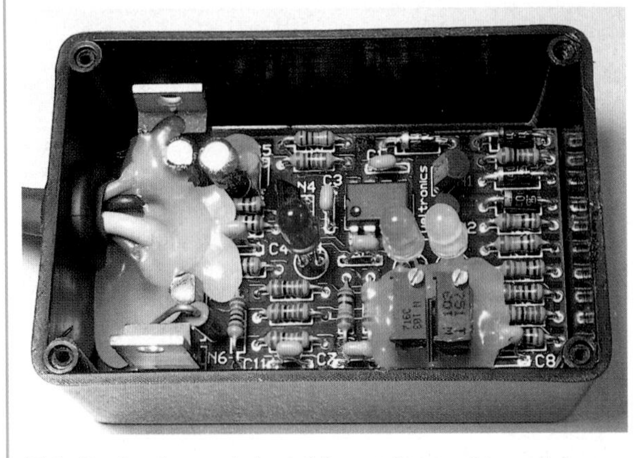

*This Fueltronics control module can be used to switch on a water/air intercooler pump when intake air temperatures **and** engine load are high. The module monitors the outputs of the standard engine management system sensors.*

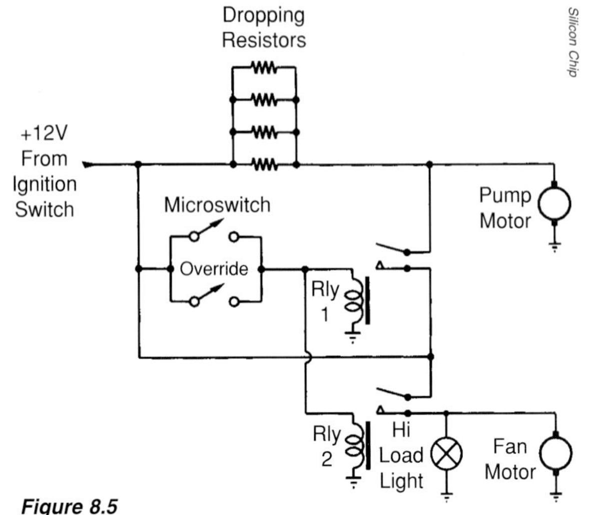

Figure 8.5
This water/air intercooler circuit provides two pump speeds, temperature or pressure switch control and radiator fan control. Note the intercooler radiator fan is triggered only when the pump is working at full power. A dashboard override switch has also been included to be used when you want the pump to operate at high speed with the fan running.

temperature of the water exiting the heat exchanger will not be extremely high if the water volume circulating through the system is adequate.

A header tank should be positioned at the highest point of the system. This should incorporate a filler cap and can actually be part of the heat exchanger if required. Note that a water/air system can be pressurised if required by the use of a radiator-type sealing cap. Ensure that the system design allows air to be bled from any spots where it will become trapped. Air in the system degrades performance and can cause pump problems. A filter placed in front of the pump is a good idea and very cheap water filters can be found in the garden irrigation section of hardware stores. These filters use a fine plastic mesh design and can be easily placed in-line.

Figure 8.6 (opposite) shows an example of the plumbing in a water/air intercooling system.

SELECTING AN INTERCOOLING SYSTEM

Both air/air and water/air systems have their own benefits and disadvantages. Air/air systems are generally lighter than water/air, especially when the mass of the water (1kg a litre!) is taken into account. An air/air system is less complex and if something does go wrong (the intercooler develops a leak, for example), the engine behaviour will normally change noticeably. This is not the case with water/air, where, if a water hose springs a leak or the pump ceases to work, it will not be immediately obvious. However, an air/air intercooler uses much longer ducting and it can be very difficult to package a bulky air/air core at the front of the car — **and** get the ducts to it! Finally, an air/air intercooler is normally cheaper than a water/air system.

A water/air intercooler is very suitable where the engine bay is tight. Getting a couple of flexible water hoses to a front radiator is easy and the heat exchanger core can be made quite compact. A water/air system is very suitable for a road car, with the thermal mass of the water allowing temperature spikes to be absorbed with ease. However, note that if driven hard and then parked, the water within the system will normally become quite warm through underbonnet heat soak. This results in high intake air temperatures after the car is re-started, as the hot water takes some time to cool down. Figure 8.7 shows a comparison of the two intercooling techniques.

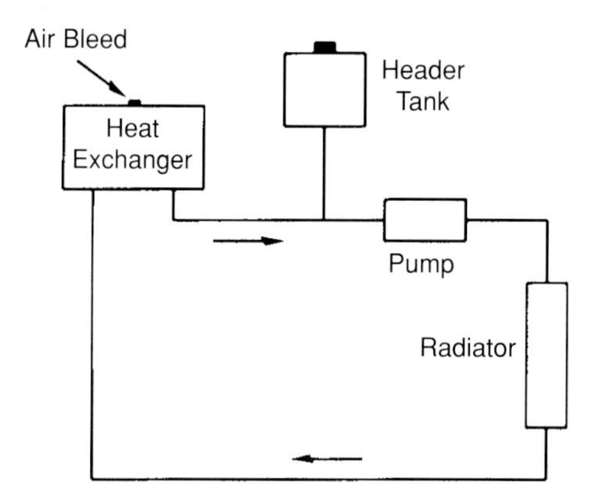

Figure 8.6
The layout of the plumbing in a typical water/air system. Note the pump is drawing from the heat exchanger and pushing water through the radiator because it's normally the radiator that's the greatest restriction to flow.

MEASURING INTERCOOLER PERFORMANCE

The intake air temperature can be measured easily and cheaply. Tandy Electronics/Radio Shack sells a Micronta-branded 'bath thermometer' that uses a digital LCD display and remote probe, and measures temps up to 110 degrees C. Other commonly available LCD digital thermometers can measure up to 70 degrees C, which is high

This Nissan Bluebird demonstrates the importance of intercooling. Running an FJ20DET turbo engine and Haltech engine management, the data-logged intake temperature graph is shown in Figure 8.1 (page 205).

enough once an intercooler is installed. Even faster in reaction time is a bead K-type thermocouple, read off on a dedicated instrument. Some multimeters have inputs for this type of thermocouple.

I strongly advise that when you install an inlet air temperature gauge, you monitor the temperature over a period of a few weeks rather than on just a quick drive around the block. This is because you'll often find that the highest intake air temps are recorded at idle in city conditions, rather than on full boost as you might first think. Increasing the size of the intercooler won't help lower this, although better heat shielding (especially around the turbo or exhaust) and insulating the return pipe from the intercooler will. Depending on the application, any

This Carel K-Type thermocouple display meter is a fast-response way of monitoring inlet air temps. The meter can also switch on devices once a pre-set temp has been reached. The 12-volt meter can be permanently mounted in the car. It's available from industrial air-conditioning wholesalers.

Type of Intercooling	Advantages	Disadvantages
Air/Air eg Nissan 200SX	• Efficient • Cheap • Used Japanese-import cores readily available	• Longer induction air path • Packaging of large intercoolers difficult • Large pipes to and from intercooler required
Water/Air eg Subaru Liberty RS	• Short induction air path • Easy to package • Excellent for short power bursts (ie typical road use)	• Heavier • More complex • More expensive • Heat exchangers harder to source

Figure 8.7
The advantages and disadvantages of air/air and water/air intercooling systems.

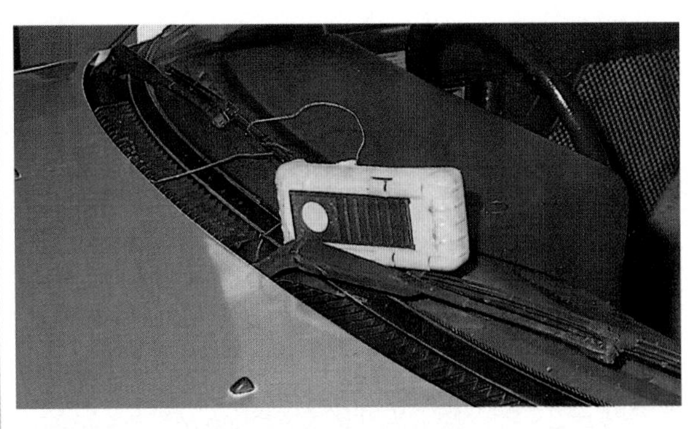

Perching a temperature-sensing multimeter on the windscreen and inserting its probe into the inlet manifold allows the quick and easy monitoring of inlet air temp. However, an incomplete picture of what goes on is then often realised.

on-boost temps that are about 30 degrees C or more higher than ambient are cause for concern. Obviously, the lower the inlet air temperature the better, but the law of diminishing returns applies very much to intercoolers!

The other parameter of importance is the pressure drop across the intercooler. To measure this, all you need is a normal boost gauge. First, plumb the boost gauge in as close to the turbo or supercharger outlet as possible. On many turbo cars this can be done by inserting a T-piece into the wastegate hose. Drive the car and note the peak boost in one gear. Next, move the pressure gauge sensing hose to the engine side of the intercooler. Normally, this is easily done by T-ing into a hose attached to the plenum chamber. Measure the peak boost in the same gear as used for the previous measurement. If the difference in the readings is more than 1-2psi, the pressure drop is too high and a bigger intercooler should be fitted. (This technique also measures at the same time throttlebody and plumbing restriction, but most of the pressure drop is usually through the intercooler.)

Measuring the actual intake air temperature and the actual pressure drop is by far the best way of assessing the effectiveness of an intercooler. Doing it in this way is better than using flowbench figures or calculating probable results from manufacturer's data because it takes into account the actual intercooler installation and engine operating parameters. While this statement seems pretty obvious, many people make statements like, "I heard the standard intercooler is pretty restrictive, so I changed it", although measuring the actual performance of the standard 'cooler is only a five-minute job.

An intercooler water spray can improve the efficiency of (especially) air/air intercooler cores.

INTERCOOLER WATER SPRAYS

An intercooler water spray is a simple addition to an intercooler that improves its efficiency. On air/air intercoolers the water is sprayed over the core, while on water/air intercoolers the spray is onto the intercooler radiator. A fine mist of water is used, with the evaporating water absorbing heat and so reducing the temp of the intercooler. It works best on air/air intercoolers.

Most intercooler water sprays are let down by the use of poor pumps and even poorer-quality spray nozzles. Windscreen washer pumps and garden irrigation plastic nozzles do not make for the best system! While such a spray is better than nothing, a spraying system that uses a high-quality pump and brass nozzles is both more reliable and more efficient. The diaphragm pumps mentioned above as suitable for use with water/air intercoolers are also good for intercooler water sprays. For a spray application, less volume is required but at the highest possible pressure. One commonly available pump is the Shurflow 2088-423-244 Multifixture Pressure Pump. This pump has a maximum pressure of 45psi and an open-flow rate of 10.6 litres/minute.

US company Spraying Systems makes a range of nozzles and fittings that are ideal for intercooler water sprays.

When looking for a pump, keep in mind that it should have the following characteristics:

- High pressure (45psi or above)
- Flow of at least 2 litres/minute
- Quiet operation
- Small size
- Low price!

In addition to agricultural spray suppliers, caravan and boating shops often carry this type of pump.

Many of these pumps are designed with a built-in pressure switch. This is fitted so that the pump will automatically turn on and off when it is being used to supply a tap. The pump starts when the tap is opened and then stops when the tap is closed — imagine a sink tap in a boat and you get the idea. However, when the pump is used for an intercooler spray, this feature can cause problems. This is because when the pump (rated at maybe 10 litres/minute) is flowing through only a couple of 0.5-litre/minute nozzles, the pressure in the supply line soon rises high enough to switch off the pump. With the pump off but the nozzles still flowing, the pressure then drops and the pump starts again.

If you don't want the pump constantly cycling on and off like this, T into place a bypass line from the pump outlet

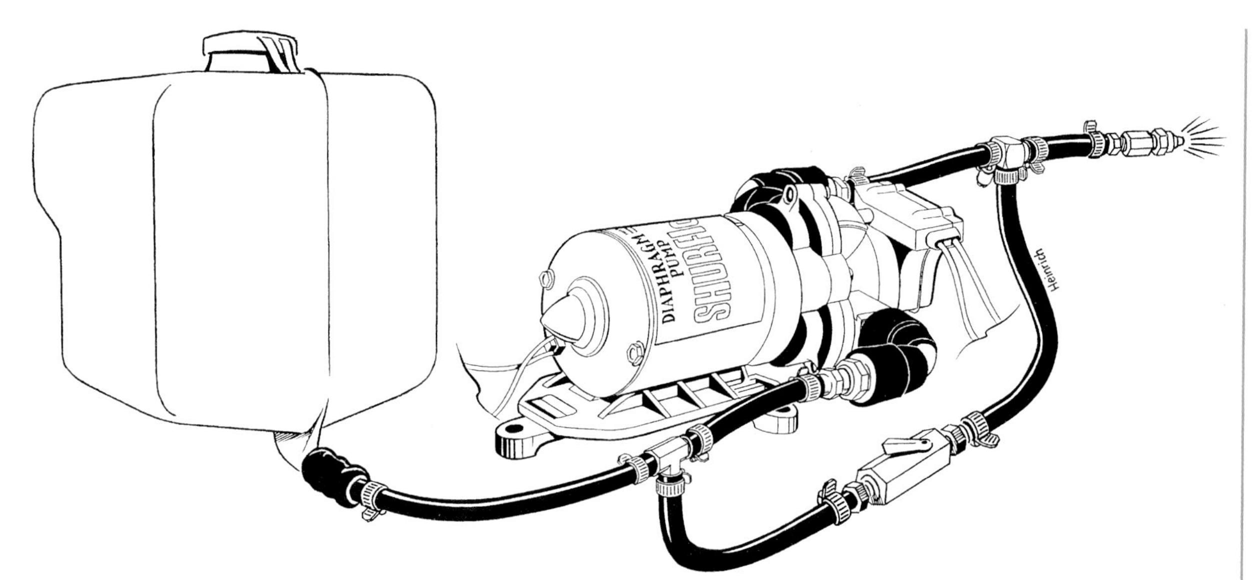

Figure 8.8
An intercooler water spray using a pump with a pressure switch requires a bypass if the pump isn't to constantly cycle on and off. Changing the bypass flow by adjusting the ball-valve will also vary how much water passes through the nozzle.

Spraying Systems also makes this flat spray nozzle and pre-nozzle filter. Check valves that stop the spray nozzle dripping are also available.

to the inlet. Placing a ball valve (or similar in-line tap) in this hose will enable you to adjust the bypass flow until the pump is able to run continuously at full pressure. Because you may need to later adjust it, make sure that this bypass tap remains accessible even with the pump mounted. Figure 8.8 shows the approach taken.

The spray nozzles should be picked specifically for your application. While miniature plastic nozzles are available very cheaply from hardware stores, their durability (especially if exposed to engine heat) isn't very high. At agricultural supply shops you'll find a range of nozzles. The US company Spraying Systems is one that has extensive catalogues listing different brass spray nozzles that are available — most agricultural spray shops have these catalogues.

A nozzle that's very suitable for intercooler use is the Spraying Systems TX-4 ConeJet spray tip. This nozzle develops a hollow cone, finely atomised spray, with a water flow of about 200ml/min. Because all the Spraying Systems nozzles are part of a professional system, there are also some very good extras available for these nozzles. One useful addition is the 4193A combination filter/check valve, which prevents the nozzle from becoming blocked and also stops it dripping. Different mounting fittings for the nozzles are also available, including bulkhead fittings and adjustable-angle swivel fittings.

One good source of water tanks is boating supply shops — there you'll find a stock of outboard motor fuel tanks in a variety of sizes. The tanks have several benefits for this application: built-in pickups with strainers, splash-proof venting, sturdy design, large fillers and good tie-down brackets. A tank of around 10-15 litres is a good compromise of size and practicality. If you want an indication of a low water level, an alarm can be triggered by a float switch (from RS Components, for example) that can be easily installed through the tank wall.

The control techniques discussed above for the water/air intercooler pump can also be used for an intercooler water spray pump. When considering the most appropriate control system, remember that if the spray works only when the car is on high boost, the cooling effect of the spray will be occurring too late. It's therefore best in a road car if the control system causes the spray to operate frequently and for short periods, so the intercooler is already cool when it's required to work hard. In this way, its heatsinking effect will be magnified.

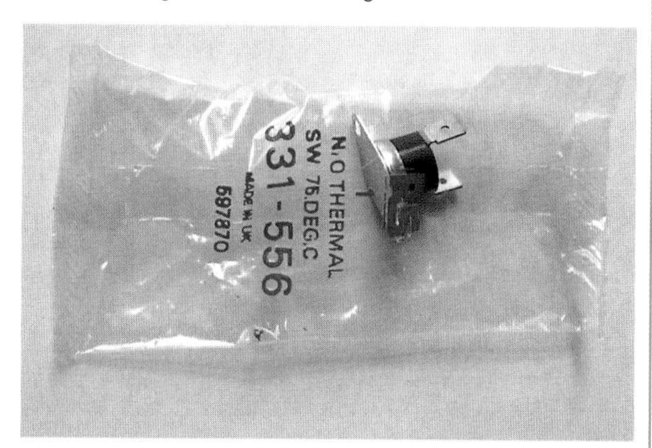

A temperature switch like this can be used in series with a pressure switch to turn on the intercooler water spray when the engine is on boost ***and*** *the water spray is actually needed.*

Preventing Detonation

Detonation occurs when the air/fuel mixture burns in an uncontrolled manner, causing cylinder pressures to suddenly spike. (While there are some differences between detonation, knock and pre-ignition, the terms are used interchangeably here.) Detonation can cause the breakage of rings, pistons and heads. The importance of preventing detonation should not be underestimated — high-load detonation can destroy an engine in a few seconds. When audible, detonation is usually a faint 'tink tink' sound, a little like a hammer hitting the crown of a piston. However, detonation doesn't always sound like this — any odd noise emanating from the engine under peak loads should be suspect!

Detonation can be heard when someone selects the wrong gear — second gear from a standstill, for example.

The results of detonation in a Nissan RB26DETT 2.6-litre twin-turbo six. The sparkplug has been shattered and the sharp fragments have caused numerous small dents in the squish zone as they have been sandwiched between the piston and the head.

All engines can tolerate some of this type of low-speed detonation but high-speed, high-load detonation must be avoided if an engine is to survive.

The likelihood of an engine detonating depends on a variety of factors, most of which influence the peak cylinder pressures that occur:

- The compression ratio
- The ignition timing advance map and how accurately it can be maintained
- The octane of the fuel
- The air/fuel ratio map and how accurately it can be maintained
- The combustion chamber design
- The engine load
- The camshaft specifications, especially the amount of cam overlap
- The volumetric (breathing) efficiency
- The temperature of induction air and the engine
- (In a forced-induction car) boost pressure

Most builders of performance engines juggle compression ratio, ignition timing (and sometimes boost) to achieve the greatest amount of power possible without having detonation occur. Detonation is most likely to be a major problem in a car that's been fitted with an aftermarket supercharger without having its compression lowered, or in a turbo car that has had the boost increased.

An older turbo car like this Mitsubishi Cordia may not warrant the expense of intercooling — water injection can be a relatively cheap and simple fix for detonation.

COMPRESSION RATIO

The compression ratio is one of the most important factors determining whether or not detonation occurs. The compression ratio is the ratio of the volume in the cylinder when the piston is at the bottom of its travel (Bottom Dead Centre) to the volume remaining when the piston is at the top of its travel (Top Dead Centre). In other words, it's how tightly the air/fuel mix is squeezed on the compression stroke. Figure 9.1 shows how the static compression ratio is calculated. A 3-litre engine with an 11:1 compression ratio will have 550cc volume in the cylinder (including the combustion chamber) at BDC and only 50cc when the piston is at TDC. Calculating the static compression ratio is a straightforward exercise in measurement, and with standard engines the figure is usually available in manufacturers' literature. However, the actual pressure that occurs in the cylinder at the end of the compression stroke is dependent on both the compression ratio and the mass of charge inhaled during the intake stroke. The ratio of this pressure to that which would occur if the cylinder were initially filled with air at atmospheric pressure (and then the valves closed) is known as the effective compression ratio. In a forced-induction engine, the effective compression ratio is greater than the geometrically computed ratio; in a naturally aspirated engine with a great deal of valve overlap it's less.

Bosch

Figure 9.1
An engine's static compression ratio is calculated by $V_h + V_c$ divided by V_c. TDC is Top Dead Centre and BDC is bottom Dead Centre.

A supercharged engine uses a low geometrical compression ratio so the effective compression ratio — and consequently the peak pressure — will not be excessive. For example, a turbo engine might have a geometric compression ratio of 8:1 while the naturally aspirated version of the same engine has a compression ratio of 10.5:1. Engines using cams with lots of valve overlap can have increased geometric compression ratios without the risk of detonation that would otherwise occur.

Using a higher compression ratio allows more energy to be extracted from the air/fuel charge when it's burnt. Engines with higher compression ratios thus give increased power and/or lower fuel consumption. Increasing the compression ratio from 8:1 to 10:1 improves theoretical efficiency by 7 per cent, although these gains decrease as the compression ratio is increased still further. In a turbo car, the use of a relatively high static compression allows the engine to have good off-boost responsiveness and fuel economy.

An aftermarket supercharged car is an ideal candidate for water injection — by the time the supercharger is fitted under the bonnet, adding the pipework for an intercooler is often difficult.

If the maximum efficiency is to be realised, it would then seem that all engines should run very high compression ratios. However, compression ratios of typical engines vary widely, with some much lower than others. So why would anyone build an engine with a low compression ratio? The limiting factor on the compression ratio is the occurrence of detonation. As a result, engines are specified to have the highest compression ratio possible before detonation, while not retarding ignition timing excessively or using special fuels.

IGNITION TIMING AND AIR/FUEL RATIO

Both the amount of ignition advance and air/fuel ratios that are employed for different combinations of load and rpm have a dramatic effect on the occurrence of detonation. These topics are covered in Chapter 4.

FUEL OCTANE

The octane rating of fuel describes its resistance to knocking, as measured on a special variable compression ratio test engine. Two different octane ratings are given to fuel: the Research Octane Number (RON) and the Motor Octane Number (MON). For a given fuel, the two ratings are usually slightly different, with the MON the lower. The RON describes the fuel's ability to resist detonation during acceleration, while the MON indicates the fuel's propensity to knock at high engine speeds. Most of the time when the octane number of a fuel is being referred to in general conversation, it's the RON that's used. However, in some countries, the figure gained by adding the MON and RON, and then dividing this number by two, is used.

The table below shows the octane of some VP racing fuels, and including this 'R+M/2' approach.

As can be seen from the table, there are fuels available with considerably higher octanes than normal pump fuels. So, how relevant are these fuels for street use? The answer is — not very! First, if your road car requires a special fuel, its running costs and range will be severely curtailed. Second, all the highest-octane fuels shown in the table contain lead — and leaded fuel cannot be used when catalytic converters or oxygen sensors are fitted. As with octane boosting additives (covered below), these special fuels are therefore best saved for special occasions.

Fuel Name	C-10	C-12	C-14	C-14+	C-15	C-16	C-18	CSP	Performance Unleaded
Motor Octane Number	96	108	114	115	115	117	116	96.6	
Research Octane Number	104	110	116	117	117	118	116	107.9	
R+M / 2	100	109	115	116	116	117.5	116	102.3	100

One of the pistons from the damaged Nissan RB26DETT. Inspection of sparkplugs for degradation should always be undertaken when detonation is suspected.

One arena in which special fuels were very successful was Formula 1 racing. In the 1980s, Formula 1 cars used 1.5-litre turbocharged engines to develop astronomical specific power figures. Peak powers of over 750kW (~1000hp) were documented, with rumours of flash dyno readings of 970kW (1300hp)! These engines used relatively high compression ratios (eg 9.4:1), boost pressures of up to 45psi and 14,000rpm redlines. To prevent detonation, large air/air intercoolers were used along with very special fuels. Regulations required that the fuel used could not exceed 102 RON; however, the detonation limits found with fuels of different compositions (each not exceeding 102 RON) varied substantially. The fuel that allowed the greatest ignition advance comprised 84 per cent toluene and 16 per cent n-heptane to achieve a RON of 101.8. This fuel had a stoichiometric ratio of 13.7:1 and poorer atomisation than traditional fuels. An air/fuel ratio of 11.9:1 was used to achieve maximum power. Honda also tested fuels containing 30 and 60 per cent toluene. When compared with the fuel containing 30 per cent toluene, the 84 per cent toluene fuel allowed another 12 degrees of ignition timing advance, presumably at full load!

This perfume atomiser develops a very fine spray using only low pressure, with a jet of air used to blast the fluid into very small droplets. The same approach works very well with boost-driven water injection systems.

An air atomising nozzle uses both boost pressure and water connections. This SUE18A air atomising nozzle from US company Spraying Systems is available from industrial and agricultural nozzle suppliers. A system using this nozzle is described on page 228.

OCTANE BOOSTERS

One of the most popular ways of preventing detonation is to increase the octane of the fuel. This is especially relevant to two types of car: those with engines that use knock detection systems to automatically vary the ignition timing or boost being used, and turbo cars where the boost level is easily adjusted. But how much extra performance is actually gained when an octane booster is added to a forced-induction engine — one that's able to vary its ignition timing in response to the input of a knock sensor?

Testing was carried out on an aftermarket supercharged 5-litre V8, electronically managed by the standard ECU working under the direction of Kalmaker programmable management software. The management system used knock sensing, allowing it to adjust ignition timing to suit the fuel octane being used. Seven different lead-free octane boosters were individually mixed with fuel, and dyno power runs were then made. Six of the octane boosters were concentrates commercially available in small bottles, while the seventh was toluene (also know as toluol or methyl benzene).

The blown engine used the standard compression ratio but was developing well over twice the standard power output, so combustion pressures were very high indeed. During the power runs, the knock sensor's output could be studied on the PC laptop screen — and it was always active! The supercharged car used a simple water-injection system, which was operating during the testing.

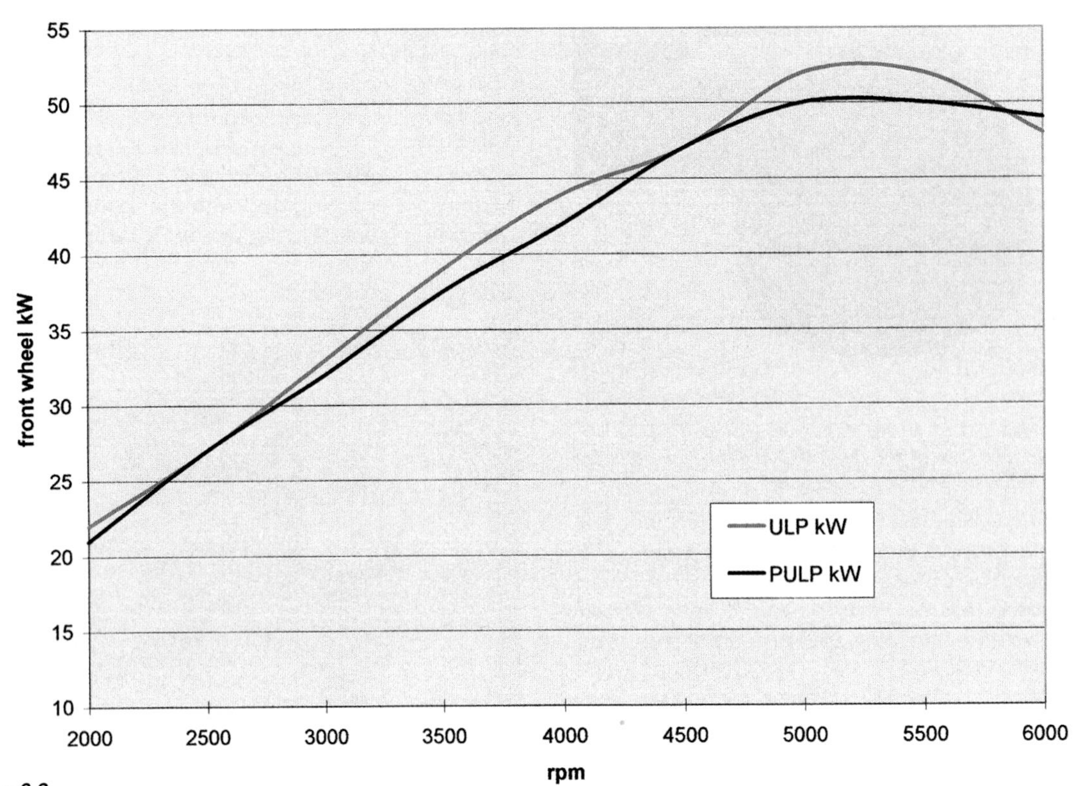

Figure 9.2
Increasing fuel octane while making no other tuning changes does nothing for performance in many cars. This Nissan Micra markedly declined in power when filled with Premium Unleaded Petrol (PULP) rather than its normal Unleaded Petrol (ULP).

Of the seven commercial octane boosters tested in this manner, only one gave a modest power increase through the complete rev range. One additive **decreased** measured power by 8 per cent through most of the rev range, while many of the others had a clear-cut power advantage only over the last few hundred rpm of the engine's rev range. The test results indicate that simply adding an octane booster to such a supercharged engine without extensive retuning does not make a major difference to power.

A similar situation also occurred with a very different car — a 1.3-litre four. The engine developed a peak power at the wheels of 52kW (70hp) on the standard unleaded fuel that it was designed to use. Filling the tank with higher-octane premium fuel resulted in an immediate **decline** in peak power of 4 per cent, as shown in Figure 9.2. Advancing the whole ignition timing map (by rotating the distributor) resulted in a peak gain of just 1kW — not worth the expense of running the higher-octane fuel.

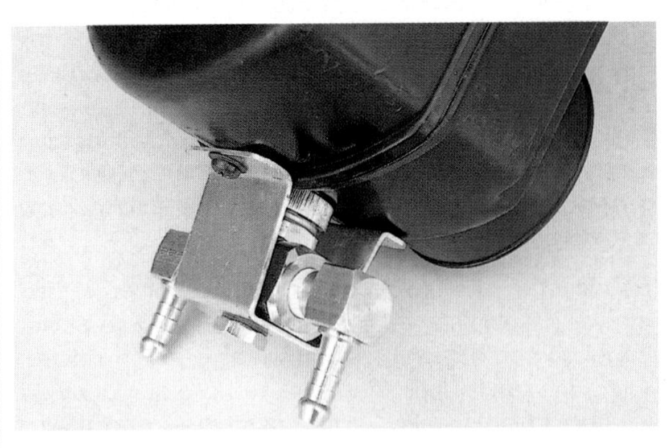

A water injection air atomising nozzle, which needs to be placed prior to the compressor. Here it has been attached to a standard resonating box that's positioned before the turbo in a Subaru.

In a turbocharged engine, the addition of a good octane booster will allow more boost before detonation, therefore usually allowing the development of more power. However, the management system must provide appropriate mixtures and timing at the increased boost level if this approach is to succeed.

Toluene is widely used as an octane booster. Typically, it's added at concentrations of about 10 per cent. Acetone is also sometimes used in the same way. However, both of these hydrocarbons can cause fuel system problems. Acetone is more widely used as a paint stripper and both it and toluene are not very kind to rubber components such as fuel hoses! Note that toluene has a high carbon content, so may lead to sooty sparkplugs — even when the mixture isn't overly rich.

The commercial octane booster that performed best in the test discussed above (and has performed well in other testing) has the following ingredients listed on its container: heavy aliphatic solvent naptha, medium aliphatic solvent naptha, methylcyclopentadienyl manganese tricarbonyl, heavy aromatic naptha and xylene. Another off-the-shelf octane booster with good results also lists xylene as an ingredient, along with toluene. Incidentally, it's false economy to buy a cheaper, lower grade fuel and then add an octane booster!

Note that an anti-detonant such as toluene need not be mixed with the fuel. Instead, it can be added to the intake air through a separate injector, proportionally controlled by either its own ECU or simply switched on and off. Taking this approach requires the development of a completely separate fuel system for the anti-detonant. As this handles an inflammable fluid, it should be constructed with the same regard to safety as the original fuel system.

WATER INJECTION

Water injection is a technology that's nearly as old as the car itself. However, like many automotive technologies, it has waxed and waned as fashion has dictated. Water injection has the ability to suppress detonation, allowing the use of higher cylinder pressures. It's easy to control and relatively simple to install. In times of tight emission controls, decreasing fuel octane and rising petrol costs, water injection is one of the best ways of controlling detonation. And it has another major advantage over other approaches — the 'fuel' is available at almost zero cost!

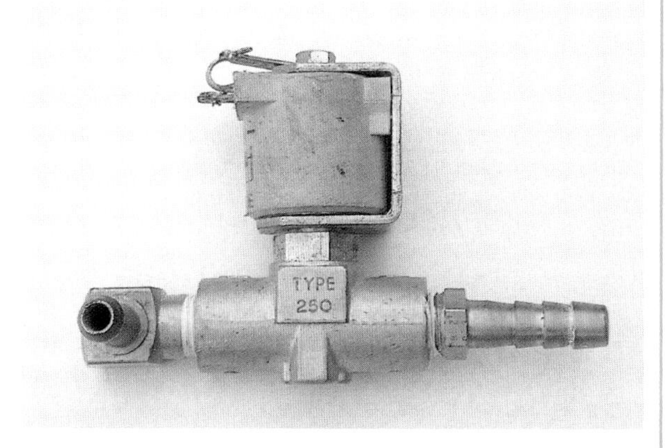

An in-line solenoid should be fitted to allow definite switch-on and switch-off of the water spray. This solenoid is a brass unit normally used as a petrol lock-off in an LPG system.

How it Works

Water injection suppresses detonation in three ways. First, when the water is injected into the intake system prior to the cylinder head, the small droplets absorb heat from the intake air. Water has a very high specific heat rating (it can absorb lots of energy for little temperature increase), so the intake air is initially cooled. Next, the small drops of water start to evaporate. Water has a very high latent heat of evaporation (its change of state absorbs a lot of heat), so the intake air charge is cooled still further. Finally, when the remaining water droplets reach the combustion chamber, steam is produced. This acts as an anti-detonant and also keeps the interior of the engine very clean, so preventing the buildup of carbon 'hot spots'.

Water injection was first experimented with in the 1930s. At the time it was discovered that detonation could initially be prevented by enriching the air/fuel ratio. As cylinder pressures rose still further and that approach ceased to be effective, the injection of water into the intake air stream was found to prevent detonation. Interestingly, the detonation remained suppressed even if the air/fuel ratio was then leaned out. This occurred because previously the excess fuel was being used to cool the combustion process. When water replaced fuel in performing this function, less fuel was then required.

This has major implications for both emissions and fuel economy at high engine loads. In fact, Saab, on some of their recent turbocharged cars, has used water injection at high loads in conjunction with leaner air/fuel ratios to reduce emissions output and improve fuel consumption. To put this another way, at high engine loads it's possible to reduce the amount of fuel being used, replacing it with water without sustaining any loss of power!

While I have referred to 'water' injection, many systems add a 50/50 mix of water and methanol, or water and methylated spirits (which comprises ethanol and a small amount of methanol). Research carried out during World War II indicated that pure water is best at suppressing detonation, while a 50/50 mixture of water and methanol permits the greatest power output before detonation occurs. One reason for this may be that the alcohol burns more slowly than petrol, causing peak cylinder pressures to occur at a later crankshaft rotation, so increasing torque.

The question of whether a water injection system can increase engine power is a contentious one. While the intake air will be lower in temperature (and so denser) when a water injection system is operating, the presence

Here is a Spraying Systems SF3 Fogging Nozzle in operation. As can be seen, the droplet size is very small indeed.

of an increased amount of water vapour in the air means there's less room for oxygen. It's for this reason that dry air (ie air with a low relative humidity) can allow an engine to develop more power. So when the air is cooler but its water vapour content is higher, will more power be developed? If no changes are made to air/fuel mixtures, theoretically the two factors almost exactly cancel each other out.

This means if water injection is used without any changes being made to the tuning of the engine, improvements in power are possible but not probable. However, if the engine air/fuel ratio is leaned out, or boost is increased, or the ignition timing is advanced, more power is very likely. Supercharged aircraft engines using water injection had mechanisms that leaned out the air/fuel ratio simultaneously with the operation of the water injection. However, it's very important to note that making changes to the air/fuel ratio and ignition timing at high engine loads can be very dangerous to the health of the engine. Such changes should be made with care — **it's very easy to blow up a forced-induction engine with random leaning of the mixtures and/or ignition timing changes!**

Both methanol and methylated spirits combine well with water when it's required that a mix be injected. However, it's important to realise that both of these mixtures are inflammable and so the anti-detonant injection system's storage container, pump and lines should all be designed and installed with the carriage of an inflammable liquid in mind — even though it's ostensibly a water injection system!

It has been suggested in some circles that the water can be directly added to the petrol by using a solvent such as acetone. However, I have not heard of anyone actually doing this!

Water Injection Systems

A water injection system should:

- Distribute the water equally to each cylinder
- Automatically start the water flow before it's required
- Have positive shut-off (eg via a solenoid valve) when water injection is not required
- Either warn the driver or decrease engine power (eg by dropping boost) should the water supply be exhausted
- Be very reliable

Many aftermarket water injection systems do not satisfy any (let alone all!) of these criteria. It's common to find a windscreen washer bottle feeding a pump that sprays water into the intake airstream through a coarse nozzle — or even through a bare tube!

To be most effective, a water injection system should add water in proportion to the changing airflow. In other words, the flow of water should match the flow of air, with small amounts of water being added at low loads and high amounts at high loads. If very accurate control of the water injection quantity is available, maximum water flow per cfm of induction air should occur at peak torque when cylinder pressures are at their highest.

The water should be injected in as fine a spray as possible. This results in each drop being smaller, increasing the surface area to volume ratio and so promoting evaporation. The smaller drops are also less inclined to fall out of the air, wetting the intake manifold walls and perhaps then being distributed unevenly from cylinder to cylinder. A small droplet size requires a high-pressure pump and a well-designed spray nozzle. Note that a normal fuel injector and high-pressure roller cell fuel pump **cannot** be used in a water injection system. Both components will corrode rapidly if flowing a mixture containing large amounts of water.

The UK company URL produces some very sophisticated water injection systems. The company has developed their own pumps, which work at high pressures and low flows. The pumps use an approach a little like a bicycle pump. Water is drawn in during the induction stroke of the solenoid-like pump, then pushed out past a valve by internal spring pressure. The stainless-steel armature pulses in this way 50 times a second, delivering up to 160ml a minute at over 70psi. The pump has built-in electronics to control this pulsing, with a control signal able to vary the flow. While URL use a sophisticated ECU to control some versions of the system, the availability of the control signal input means the output of the airflow meter or MAP sensor may be able to be used for the same purpose.

An alternative to a pump is to use boost pressure to force the water through a nozzle. If this approach is adopted, the spray can be used only in a forced-induction car, and with the water introduced prior to the compressor. A very special nozzle is also needed if the spray is to be sufficiently fine to pass through the compressor without long-term damage occurring. People using coarse droplet water injection in front of turbos have reported that over a period of time the edges of the compressor blades develop a serrated edge — presumably from the impact of the water droplets.

Water can be injected at a number of different points within the intake system. In a naturally aspirated car, the nozzle is usually situated prior to the throttle body. In a forced-induction car, the nozzle can be situated:

- Before the compressor
- After the compressor
- After the compressor but before an intercooler
- After an intercooler

URL suggests a nozzle position just prior to the throttle body for road cars, while the supercharged aircraft of many years ago used up to 18 nozzles positioned around the supercharger exit diffuser. Testing of the two systems discussed below indicated the best nozzle location is found through experimentation.

The amount of water that will evaporate into the air versus the amount that will be carried into the engine in the form of small droplets will be dependent on the air's relative humidity. For the same mass of water vapour in a given volume of air, the relative humidity will fall as the air temperature increases.

Thus, the amount of water added to the air to achieve saturation (ie a relative humidity of 100 per cent) will be dependent on the temperature of the intake air. This makes an intake air temperature sensor very important in mapped water injection systems.

Water Flow (ml/minute)	Water Pressure (psi)	Air Pressure (psi)
75	3	5
90	4.5	5
140	10	10
200	22	20

Figure 9.3
The flow performance of the SUE18A air atomising nozzle from US company Spraying Systems shown on page 224. Unlike many nozzles of this type, it works effectively at pressures below 7psi.

If the flow of water is initially high and is then slowly reduced, practical testing can be used to show the flow that gives best results. Testing on aircraft engines indicates the mass of water required to suppress detonation is 20-30 per cent of the weight of the total liquid charge (that is, the water plus the petrol) being consumed. Note that all water injection systems should be configured to inject water only when there are high intake airflows.

Boost Pressure Water Injection

A good-quality boost-pressure-controlled water injection system can be built using off-the-shelf components. The system gives an extremely fine spray and can be used with water/methanol as well as pure water. While the injection of water is not proportional to load (it is proportional to boost pressure), the variation in the supply of water is still better than that which many systems provide. The greatest benefit of the system is that it's maintenance-free, other than requiring the refilling of the water tank and the occasional cleaning of the filter.

The nozzle used is the single most expensive part of the system. It's specifically designed to evaporate water into air and at high boost pressures can produce a completely atomised spray. In fact, during testing, the nozzle was placed inside a moving car so its spray behaviour could be monitored. The spray was so fine that nothing inside the cabin got wet — the tiny droplets drifted out of the open window before having a chance to settle on the floor and seats!

The nozzle is an air atomising design produced by the US company Spraying Systems. It has two connections — one for compressed air and the other for water. The compressed air is directed out of two orifices so it collides with the water stream, scattering it into the tiny droplets. The pressure to supply both the water and the compressed air comes from the turbo or supercharger. The part number for the nozzle is SUE18A and it's available from agricultural irrigation and spray suppliers. Unlike many air atomising nozzles, the SUE18A works effectively at pressures below 7psi. Figure 9.3 summarises the actual flow performance of the nozzle at different water and air pressures.

As can be seen, 200ml/min flows through the nozzle when it's supplied with water and air at around 20psi (1.4 Bar) boost. If this flow is too great, a ball valve can be placed in the water supply hose to allow easy adjustment of the flow. If the ball valve is partially closed to restrict the water supply, the remaining water will then be even better atomised! If more than 200ml/minute of water is required, I suggest you use two or more nozzles.

The water supply for the nozzle should pass through a small water filter to avoid filter blockages. An appropriate

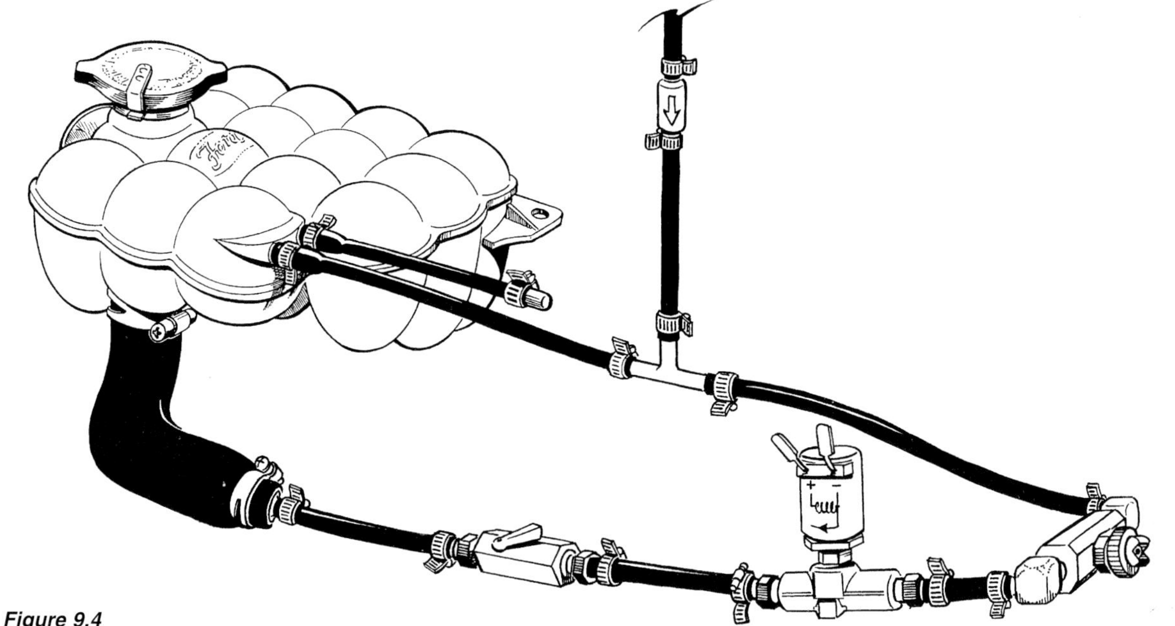

Figure 9.4
This system uses boost pressure to force water and air through an air atomising nozzle, giving a very fine spray. Here a Ford radiator header tank has been used as the pressure vessel, being fitted with a sealed cap. The ball valve can be used to adjust the flow of the water.

Water ball valve position	Rolling 60-90 km/h in second gear (averaged seconds)	Detonation
Fully closed	2.2	Yes
Two-thirds open	2.1	No
Half open	2.0	No
Quarter open	2.1	No
Fully open	2.1	No

Figure 9.5
The results of testing the system shown in Figure 9.4, when it was fitted to a Subaru Liberty [Legacy] RS turbo. Tuning the amount of water being added proved critical in stopping detonation while at the same time improving performance.

filter is available from the suppliers of the nozzle; alternatively, a small garden irrigation filter can be used. The nozzle must be mounted so it flows into the intake system prior to the compressor and the water should be injected after the airflow meter (if present) to prevent the water droplets upsetting the air metering.

The fluid storage container must be pressurised if the water is to be forced through the nozzle. A custom tank can be made or a large pressurised radiator header tank pressed into service. Preferably the tank should be at least five litres in volume for each nozzle used. Note the tank must be capable of handling the constant cycling of internal pressures up to the peak boost level. A low fluid warning buzzer should be fitted.

Plumb the system as shown in Figure 9.4 (previous page). The ball valve to the right of the pump is used to adjust the supply of water, while the boost-pressure triggered solenoid has been found to be a necessary inclusion if the spray is to switch on and off sharply. An alternative to the use of the solenoid is the installation of a vacuum-operated valve (such as a PCV valve) to vent the tank back to the inlet system, causing the tank pressure to more closely follow boost when the throttle is lifted. However, the solenoid valve is the safer of the two approaches: if the water ever flows into the intake when the engine is stopped, very major engine damage can be caused when an attempt is next made to re-start it! A brake booster valve is a cheap one-way valve that can be used on the boost pressure feed.

The system as described here was built and tested on a Subaru Liberty (Legacy) RS turbo. The car was equipped with the standard water/air intercooler and a three-inch cat-back exhaust, K&N cone filter in a revised location and up to 20psi (1.4 Bar) boost. The nozzle was mounted on a custom-fabricated mixing chamber placed between the airflow meter and the turbo. Testing was carried out on a 30-degree C day using a 50/50 mix of methylated spirits and water and a water switch-on point of 1psi. The air/fuel ratio, ignition timing and turbo boost were not altered during this testing.

Figure 9.5 (above) shows the results. It can be seen that tuning the amount of water being added proved critical in stopping detonation while at the same time improving performance. On a 19-degree C day, the performance gain was much smaller, but in all situations where water was flowing, audible detonation was completely prevented.

Pumped Water Injection

If you require that the water be injected after the turbo or supercharger, or if a large pressurised tank is unwieldy, a pumped system is a better option. This type of system is also suitable for a naturally aspirated car.

The pumped water injection system again uses Spraying Systems nozzles. The SF2 and SF3 Fogging Nozzles are designed for humidifying the air in chook sheds, among other agricultural applications! The nozzles require only pressurised water to produce a fine spray, although note that the droplets are not as fine as those produced by the air atomising nozzle described above. The specifications of the two nozzles are shown in Figure 9.6. The nozzles can be equipped with in-built filters and check valves that prevent their dripping.

The electric pumps discussed in Chapter 8 for intercooler water sprays and water/air intercooler circulation pumps are also suitable for a pumped water injection system. A US-made Shurflow Multifixture Pressure Pump was used in the tested system. This pump is suitable only for water, so if a water/methanol mix is to be used, a different pump

Spray Tip Number	Orifice Diameter	Water Flow (ml/min)				
		20 psi	30 psi	40 psi	60 psi	80 psi
SF-2	0.020"	70	80	100	120	140
SF-3	0.030"	100	120	140	170	200

Figure 9.6
The flow specifications of the Spraying Systems SF2 and SF3 Fogging Nozzles. These are suitable for use with high-pressure water pumps.

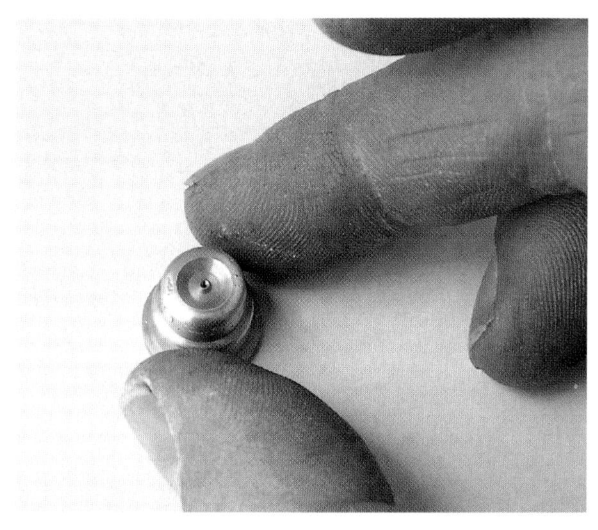

Obtaining a fine droplet spray is more difficult in pumped water injection systems, so nozzle selection is critical.

should be sourced. Such a pump is the Shurflow 8000-541-236. This pump develops a higher pressure of 60psi and is equipped with chemically resistant Santoprene diaphragms and Viton valves. However, it's more expensive than the 2088-423-244 pump. If the spray nozzle is located after the throttle butterfly, a solenoid valve should be inserted in the water supply line so there's no possibility of water being sucked through the pump during periods of high manifold vacuum.

The easiest activation technique for the pump is the use of a manifold pressure switch that simply turns on the water injection when the engine is on boost. If proportional control of the amount of water injected is required, a number of approaches can be taken. A two-stage control can be adopted using a dropping resistor to slow pump speed, as is covered in the section in Chapter 8 in relation to water/air intercooling. Finer control than that can be gained if the pump speed is electronically controlled. Jaycar Electronics produces a kit that can control pumps drawing up to 20 amps. The kit (cat no KC-5225) is called the 12-volt Motor Speed Control and it uses a pot to adjust pump speed. This pot could be dash-mounted so flow adjustment could be made for various boost pressures, weather conditions or driving styles!

The Spraying Systems Fogging Nozzles can be fitted with the strainer/check valve shown in the foreground. This prevents foreign matter blocking the nozzle orifice and also stops the nozzle from dripping.

Another entirely different way of varying the water injection rate is to drive the pump with a trailer electric brake controller. These devices are available from caravan and similar suppliers and are designed to energise the electro-magnets that are located in the brake assemblies of some caravans and trailers. One class of controllers does this by measuring the braking force the car is undergoing and increasing its output voltage proportionally. I envisage the controller reversed in orientation so it measures acceleration. If it were then connected to the water injection pump, the harder the car accelerated, the higher would be the pump speed and so the greater would be the addition of water into the intake air!

A boat outboard motor fuel tank is a suitable water storage container that can be used in pumped water injection systems. These have the required fittings and tie-down brackets, and are cheap and durable.

The pumped water injection system covered above was trialled on the Liberty (Legacy) RS. The water injection nozzle was installed just before the throttlebody and the pump was operated with a simple manifold pressure switch. During testing, various combinations of pure water and methanol were tried, at different flow rates and switch-on boost pressures. No consistent performance gains were realised with this system but, when the water injection was in operation, audible detonation was always eliminated.

Water Injection versus Intercooling

In Chapter 8, I cover intercooling in some detail. So which is better when running a forced-aspirated car — water injection or intercooling? Each has its own advantages and disadvantages. Intercooling is a reliable means of reducing intake air temperatures and so decreasing the likelihood of detonation and, depending on the

A high compression engine is very reliant on knock sensing to stop detonation. Fitting a water injection system can prevent the excessive retarding of ignition timing that otherwise occurs on poor fuels or with increased combustion pressures.

approach chosen, can comprise a very simple system. However, it should be noted that while air/air intercoolers have few component parts, water/air intercooling is more complex than water injection. Intercooling systems require little or no maintenance, and a good intercooling system will provide an engine power increase in addition to preventing detonation.

However, intercoolers are much larger than water injection systems and are generally harder to package. Finally, all intercoolers cause a restriction to intake flow.

Water injection is very effective at preventing detonation. It is not subject to efficiency drop-offs through heat

Water injection can be activated by the use of a throttle microswitch. Here a cam made from alloy plate is used to turn on the switch.

Another approach is to use a boost pressure switch, such as this from a spa bath.

soak and causes no restriction to intake flow. It is easy to fit as an add-on to an existing system and, because its components can be spread around the car, it's generally very easy to package. Unlike intercooling, water injection will not necessarily give a power increase. However, the biggest disadvantage is the requirement to carry a relatively large water tank — and to keep filling it!

A water injection tank can be placed within the inner guard and filled through an extension snout. OEM windscreen washer reservoirs can be used when it's not required that the tank be pressurised.

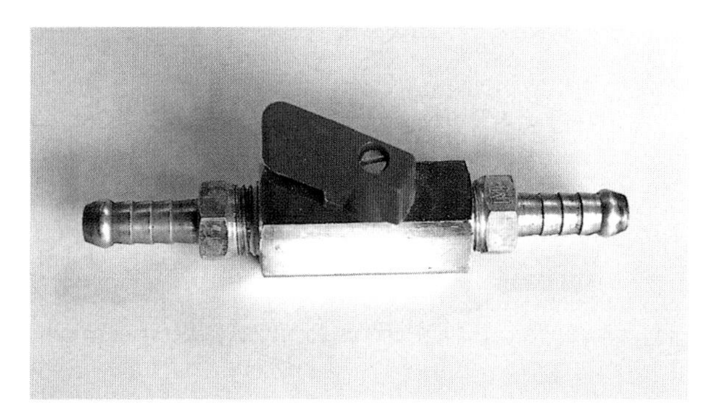

A brass ball valve can be used to adjust the water flow in any water injection system. These valves are available quite cheaply from companies specialising in pneumatics and hydraulics.

The most common reason for failure of a water injection system is that the water tank is empty! To avoid this occurring inadvertently, a low water level alarm such as this should be fitted.

Figure 9.7 summarises the good and bad points of taking each approach. It should be noted that these points apply only when good-quality systems of either type are used.

	Advantages	**Disadvantages**
Intercooling	• Reliable	• Weight
	• System can be very simple	• Bulk
	• Always improves power	• Major underbonnet changes
	• System break-down usually immediately recognisable	• Usually poses a flow restriction
	• No ongoing maintenance	
Water Injection	• Very effective in preventing detonation	• Requires regular filling of water tank
	• System components can be spread around car	• System breakdown can be difficult to recognise
	• Can be used to inject octane booster	• Large filled water tank is heavy
	• Generally low cost	• Variable flow systems are complex
	• No intake flow restriction	• Effectiveness will vary depending on weather
	• Reduces emissions	

Figure 9.7
*The advantages and disadvantages of intercooling and water injection approaches. Note that these points refer only to **good** systems of each type!*

The fuel delivery system in an EFI car is required to deliver a continuous, high-pressure supply of cool fuel to the injector rail(s). The fuel supply system consists of a number of parts, each of which is important if the system is to work reliably.

TANK AND PICKUP

The fuel tank itself seldom needs modification, unless the increased performance of the engine means it drinks fuel so quickly the tank needs to be enlarged! However, the in-tank fuel pickup is often a source of concern in modified cars. If the power of the engine is doubled, then at the same air/fuel ratio, the amount of fuel consumed each second is also doubled. Where there is just gravity and a little suction from the pump to move fuel through the pump pickup, this can cause problems.

A typical standard fuel pickup consists of a plastic mesh strainer positioned either on the end of the fuel pump itself (for in-tank pumps) or on the end of the pipe leading to the pump. The strainer is designed to protect the pump from ingesting materials that could damage it and uses quite a fine mesh. If the strainer closes up and so reduces its exposed area, the supply of fuel to the pump will be decreased. Where the strainer is easily removed, one solution is to insert a coarsely wound spiral spring within the strainer. This holds it in a more dilated state, preventing it from closing.

The pickup should be positioned such that it has access to fuel at all times. This includes times of cornering, acceleration, braking and when the fuel level is very low. If in any of these situations the pickup draws in air rather than fuel, disaster can strike. This is especially so on a forced aspirated engine where sudden fuel starvation can cause the melting of pistons or valves. Even on a less highly strung engine, sudden loss of power can cause major handling difficulties if it occurs mid-corner! Most standard cars will fuel-starve if cornered hard enough with a low fuel level, so once the cornering levels have been increased, this can become a major problem in itself.

INTERNAL SWIRL POTS

There are two ways of overcoming fuel pickup problems. One is to use an external swirl pot (covered shortly) and the other is to build an internal swirl pot into the standard tank. An internal swirl pot is an open-ended container placed at the base of the tank — a little like a plastic

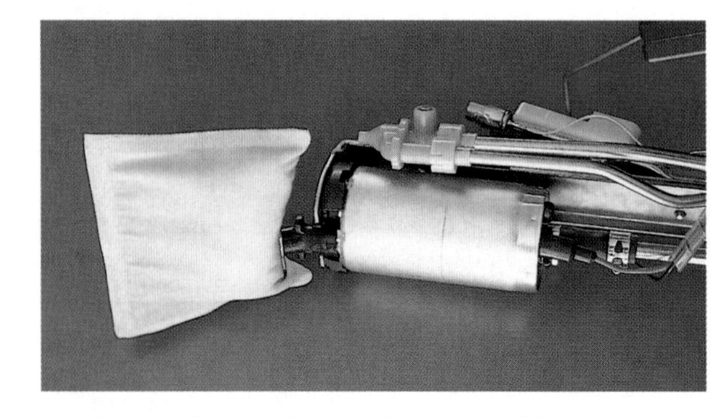

Most standard pumps have a plastic mesh filter on the pump pickup. On some cars, this can close up, restricting flow. Dilating the filter with a coarsely wound spiral spring can fix this problem.

Cars retaining the standard fuel tank and pickup that are driven hard on the road or track usually starve during cornering, even when there is a reasonable amount of fuel left in the tank. Modifying the fuel pickup can solve this.

bucket sitting on the floor of a rainwater tank. The feed for the fuel pump is taken from the base of the swirl pot, with the small dimensions of the container stopping fuel sloshing away from the pickup too readily. The return pipe from the injector rail passes fuel back into the swirl pot, while its open top and holes (or an opening) around its base allow fuel from the rest of the tank to enter it. Incidentally, the reason the return line is brought back into the swirl pot rather than returning fuel to another point in the tank is so that in times of **very** low fuel level, the car will keep running until the last drop has gone!

Before an internal swirl pot can be installed, the tank needs to be cut open. **Note that before welding, cutting or grinding can occur, a petrol tank must be professionally cleaned and then checked for the presence of residual petrol vapour. Not doing this may result in an explosion and probable injury or death!**

Most cars run simple internal swirl pots like this open-top plastic container. After the fuel tank is professionally cleaned, it can be cut open and the swirl pot and fuel pump supply fittings upgraded for better flow.

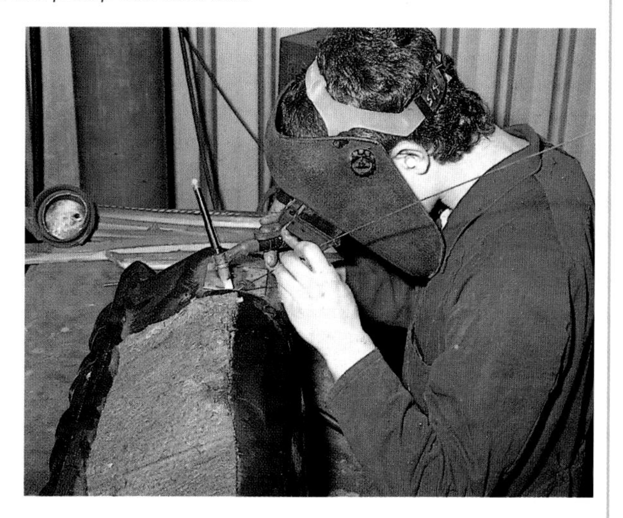

Having the standard fuel tank modified is a cheap and effective way of improving the fuel pickup and supply lines. It also means space does not need to be found for an external swirl pot and hoses.

Once the base of the tank has been cut open, a swirl pot can be installed around the fuel pickup. A simple swirl pot can be made from a short length of steel tube, 120mm long and about 100mm in diameter. The dimensions are not vital — any piece of tube that holds a third to half a litre can be used. The tube should be spaced a little off the floor of the tank so fuel can enter at its base, in addition to flowing through the open top. This is important because otherwise fuel will not be able to reach the pickup when the level is lower than the wall height of the swirl pot. An easy way of attaching the pot to the tank floor is to weld short lugs to the base of the swirl pot. Matching slots can then be cut in the floor of the tank and the swirl pot lugs pushed through them, allowing the welding of the pot to occur externally of the tank. Figure 10.1(overleaf) shows the layout of a typical in-tank swirl pot.

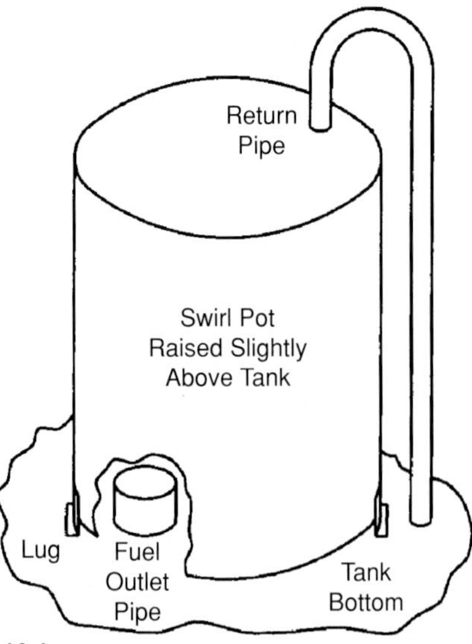

Figure 10.1
A swirl pot can be built into the base of a steel petrol tank, removing the need for external swirl pots. The in-tank swirl pot is simply an open-ended container placed slightly above the floor of the tank, with the fuel pickup drawing from within it. The return pipe from the injector rail passes fuel back into the swirl pot, while its open top and the opening around its base allow fuel from the rest of the tank to flow into the pot.

The term 'swirl pot' originally arose because the return line was frequently aimed along the inside wall of the pot, causing the fuel to be swirled around within the receptacle. This was done to prevent aeration of the fuel, which can cause problems. While the swirling approach can still be followed, many OE and aftermarket manufacturers don't bother. In fact, the term 'swirl pot' is often replaced with 'surge tank'.

The fuel pickup tube should be quite large in diameter. Many in-line EFI pumps are designed to take half-inch (12.7mm) lines on their suction side and such a tube size can be used with advantage right into the tank pickup. A strainer will then need to be either adapted from another fuel tank or an alternative type of pre-filter fitted. Don't be tempted to run without a filter in front of the EFI pump — damage to the pump will result from it drawing in even the tiniest piece of grit or rust. In-tank pumps will need to be mounted so their fuel pickup is deep within the swirl pot.

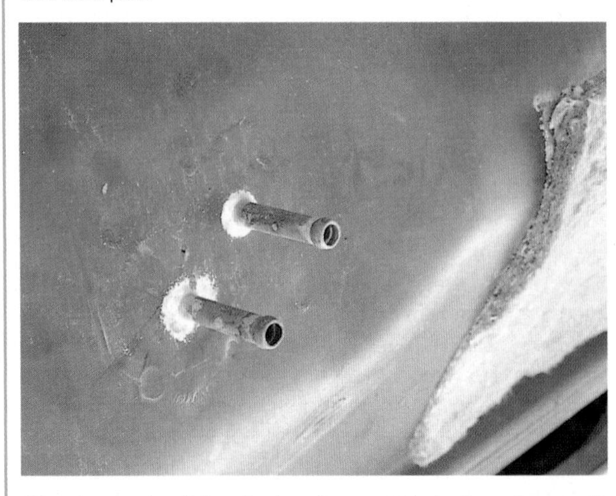

The standard outlet and return lines on this tank are too small. A high-flow EFI pump fed from one of these fittings will overheat, in addition to flowing little fuel. Fuel pump supply lines are normally 13mm or so in diameter.

This external swirl pot system is not very pretty but it does show the layout of one such system. The low-pressure fuel pump (gold cylinder) supplies fuel to the swirl pot (vertical cylinder on the left) through a low-pressure filter. The high-pressure fuel pump draws from the swirl pot, pushing the fuel through a high-pressure filter to the fuel line that runs to the injector rail(s). Note the poor mounting of the swirl pot.

EXTERNAL SWIRL POTS

An external swirl pot is simply a small reservoir filled with fuel pumped by a low-pressure pump from the main tank. A high-pressure EFI pump draws from this reservoir. Even if the low-pressure pump momentarily starves, the secondary pump will keep on supplying fuel without any fuel delivery problems occurring.

Swirl pots can be made of mild steel, stainless steel or aluminium (as here). Note the large-diameter tapped boss at the base of the swirl pot for the pump feed.

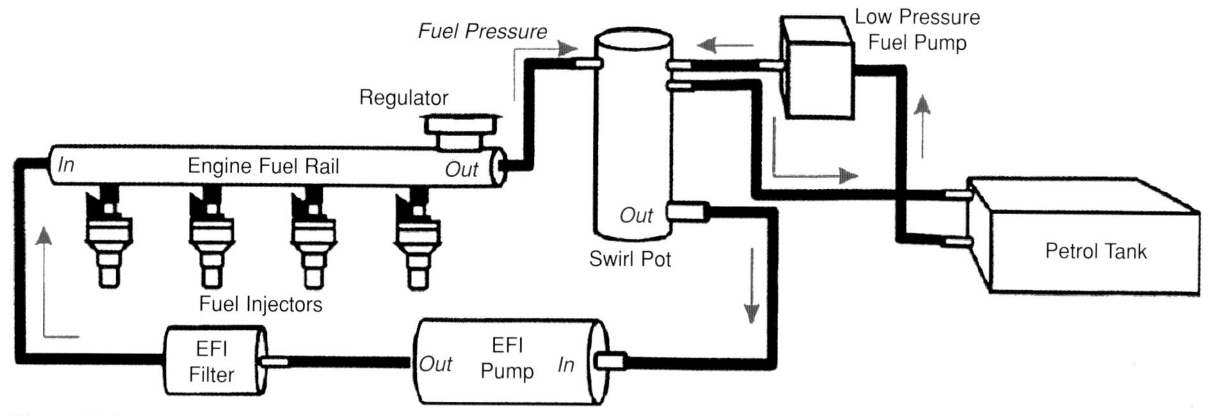

Figure 10.2
An external swirl pot system with the injector rail returning fuel to the swirl pot. Taking this approach means the low-pressure pump need supply only enough fuel for the engine — it doesn't have to match the flow of the high-pressure pump. However, hot fuel is rapidly recirculated through the injector rail.

There are two quite distinct ways in which the swirl pot can be incorporated into the fuel delivery system. The first approach is where the return line from the injector rail goes back into the external swirl pot (surge tank), while the second directs the fuel line back to the main tank. Figures 10.2 and 10.3 show these different layouts.

Taking the Figure 10.2 approach has several advantages — but also important disadvantages as well. On the pro side, the capacity of the low-pressure pump can be reduced. If the return line from the injectors goes back into the main tank (instead of to the external swirl pot), the low-pressure pump must be able to supply **all** the fuel which the high-pressure pump can draw. It's important to realise this is a much higher flow than is actually being used by the engine, especially under normal cruise conditions. This is because a great deal of the fuel pumped by the high-pressure pump is recirculated — pumped back in the return line rather than squirted through the injectors. Therefore, if the system is set up as shown in Figure 10.3, the low-pressure pump must be rated at a higher flow than the high-pressure pump.

The other advantage of recirculating the injector return line to the external swirl pot is the car can run lower on fuel before coughing and spluttering. If the level in the main tank has dropped to the extent that the low-pressure pump cannot pick up the fuel, the swirl pot level will fall and fall until the high-pressure pump ceases working and the car stops. If that same fuel were being put back into the (much smaller) external swirl pot, it would be picked up more easily and fed down the fuel line to the injectors.

However, there's one major disadvantage in recirculating the injector feed fuel back to the external swirl pot. Depending on the size of the low-pressure pump, much of this fuel will go around and around — to the fuel rail and then back to the external swirl pot, to the fuel rail and then back to the swirl pot. As it does, it will get hotter and hotter. On a highly stressed engine, this hot fuel can cause detonation, as well as reduce power because of its lower density. The speed of heat buildup in the fuel will depend on the size of the swirl pot — the smaller it is, the more quickly the fuel will get hot.

It's important to realise that when even a standard high-performance car is being dyno'd, the fuel in the tank can become hot enough to reduce power — although there might still be 20 litres in the tank! Here's another example of this hot fuel problem. A highly boosted turbo rotary speedway car (being run on methanol) had a change made to the fuel system. Instead of fuel being recirculated back to the main tank, the unused methanol went back to a small swirl pot. After the change was made, the engine detonated to death...

If you do use a system where the fuel is returned to a swirl pot, the pot needs to be at least two litres in volume, and

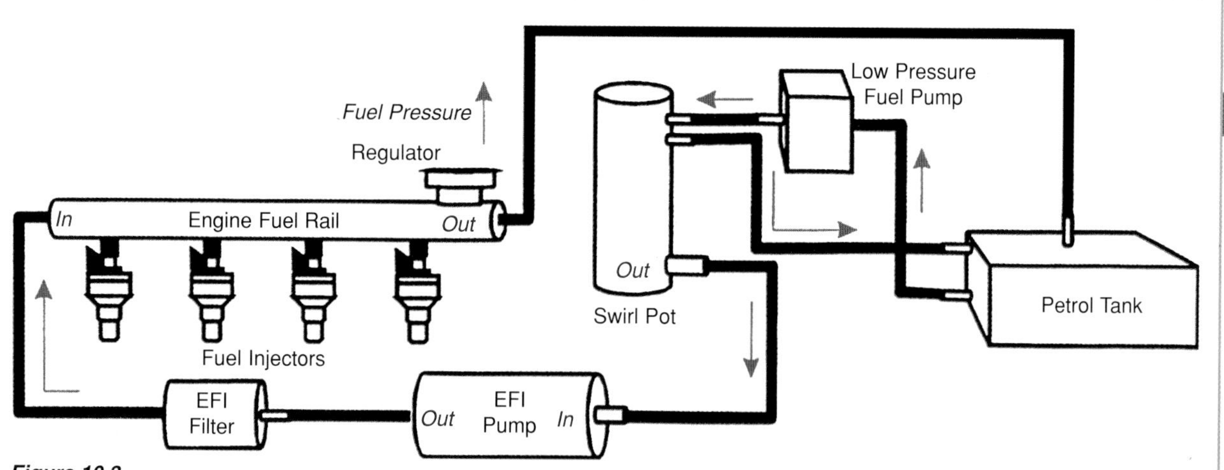

Figure 10.3
An external swirl pot system with the injector rail returning fuel to the main fuel tank, not the swirl pot. With this approach, the low-pressure pump must have a higher flow than the high-pressure pump, but the fuel will stay cooler.

On this car, the high-pressure pump, low-pressure pump, swirl pot and high-pressure filter are all tucked under the floor. The swirl pot is made from mild steel and is the black object on the right.

during dyno or track work the swirl pot temperature should be monitored. If high fuel temperatures become a problem, a heat exchanger can be mounted on the fuel supply line. An oil cooler or a small condenser from a domestic air-conditioner can be used for this task. However, take care that the cooler is protected against stone damage that could puncture it.

A swirl pot can be made from mild steel, stainless steel or aluminium. Cheapest and easiest is to use mild steel, with a short length of three- or four-inch exhaust tube low in cost and readily available. The end caps can be cut from 3mm steel sheet with an electric jigsaw and gas-welded into place. Very cheap barbed hose fittings can be sourced by cutting off the ends of steel fuel lines on cars about to head to the crushers. These can be brazed into place and the swirl pot very thoroughly cleaned inside before being used. Taking this route can result in a swirl pot that costs nearly nothing! However, care should be taken that water does not enter the fuel, or rust inside the pot can occur. A stainless-steel swirl pot will avoid the potential rust problems.

A more common approach is to make the swirl pot from aluminium. Again, a section of tube can be used, this time with the end caps TIG'd on. On conventional fuels there won't be any problems with corrosion and, if required, the swirl pot can be highly polished. Sections of small-diameter aluminium pipe can be welded to the swirl pot to provide the fluid connections or, alternatively, alloy spuds can be welded into place and then drilled and tapped to take barbed brass fittings.

Note that the number of plumbing points needed will depend on how the swirl pot is to be integrated into the fuel system. This means you should plan the fuel system before making the swirl pot! Swirl pots are often mounted in the boot. However, this makes any fuel leakage potentially very dangerous, so an external swirl pot is much better mounted under the car, if there is room. Note the swirl pot should be a vertically oriented container — some people have been known to turn swirl pots on their side, which creates a whole new fuel surge problem!

PUMPS

Because in an EFI system the fuel is pumped in a circle from the tank to the injector rail and then back to the tank, it's impossible to over-fuel the engine by using a fuel pump that's too big. However, as noted above, the fuel may become overly hot if it's circulated through the fuel rail(s) too frequently or if the return line flows into an overly small swirl pot. This means the pump used shouldn't far exceed the required capacity — but, then again, it most certainly shouldn't be too small either! (Alternatively, the pump speed can be changed so it becomes variable in flow — this is covered shortly.)

When sizing the primary high-pressure fuel pump, a conservative rule of thumb is to use the following equation: kW x 8 = cc of fuel per minute required. So for example, a 200kW engine will need about 1600cc (1.6 litres) per minute **at the rated fuel pressure**. It's vital that the fuel flow is assessed at the peak pressure that will be required. This is because all fuel pumps decrease in flow when they're required to work into a higher pressure. If you're using a rising-rate fuel pressure regulator (covered later), the pump flow may drop to only two-thirds of its rated flow at the highest pressure.

If selected carefully, an off-the-shelf pump from a standard fuel-injected car can be used in most performance applications. Selecting an OEM pump means it will be widely available, relatively cheap and the pump can sometimes be sourced from a wrecker. If required, dual pumps can be used in series to increase the total fuel flow. The pumps shown in Figure 10.4 (opposite) are suitable for many performance applications. Note that the figures shown for the R32 Nissan Skyline GT-R pump are not misprints — it's a very capable pump indeed!

The aftermarket isn't flooded by high-volume, high-pressure EFI pumps, but there's still a number available. Among

Many standard Bosch roller-cell fuel injection pumps (including those for early mechanical injection systems) flow very well. In all but the highest power or high boost pressure applications, one (or two) of these pumps will be sufficient.

Pump	Zero Restriction Flow (litres/minute)	Flow at 36-40 psi (litres/minute)	Rated Power At 36-40 psi Fuel pressure (kW)
Ford XE Falcon/ZK Fairlane/FD LTD 1984-88	-	1.8	225
Nissan R31 Skyline 1986-90	-	2.0	250
Holden VK Commodore 1984-86	-	2.2	275
Holden VL Commodore, barbed fittings, 0580 464 033	2.8	2.3	290
Mercedes Benz 6 cyl K-Jetronic, screw fittings on outlet, 0580 254 942	2.8	2.3	290
Mercedes Benz 6.9 litre K-Jetronic, screw fittings on outlet, 0580 254 975	3.2	2.7	340
Porsche K-Jetronic, screw fitting on outlet, 0580 254 984	3.6	3.1	390
Bosch Motorsport, screw fitting on both ends, B580 152 460	4.1	3.6	450
Nissan R32 Skyline GT-R in-tank	-	4.5	560

Figure 10.4
An off-the-shelf pump from a standard fuel-injected car can be used in most performance applications. If required, dual pumps can be used in series to increase the total fuel flow.

the best is the T-Rex pump from US blower manufacturer Vortech. This pump flows 3.4 litres/minute (over 420kW) at a huge 80psi. No problems here with lifted fuel pressure slowing down fuel flow! Other even more highly rated pumps from US aftermarket manufacturers include the Paxton Kamikaze (4.6 litres/minute at 45psi) and the SX Performance 18201 (five litres/minute at 45psi).

While low-pressure electric fuel pumps can be salvaged from carby cars, often the flow of these pumps is quite poor. It's generally simplest to buy an aftermarket electric fuel pump designed for use with a powerful carby-fed V8. A Holley 'Red' flows over six litres/minute and a Mallory 110 nearly seven litres/minute. Both of these pumps can then be used to directly feed a high-pressure pump, or to change the fuel in a swirl pot at a high rate. Note again that if the fuel return line isn't directed back into the swirl pot, the low-pressure pump must have sufficient flow to match the high-pressure pump, rather than match the engine's power requirements.

The SX Performance 18201 pump flows 5 litres/minute at 45psi. It's made from anodised aircraft quality alloy and uses 'aerospace composite material' for the rotor.

FUEL LINES

The higher the fuel pressure, the more fuel a line of a given diameter can carry. That's why it's almost always unnecessary to fit half-inch fuel lines to an EFI car, while such a size fuel line is common on big carby engines. The fitting used on the output of the fuel pump provides a good guide to the size of fuel line required. Generally, you won't need to go over a 1/4-inch or 5/16 inch (6.4mm or 8mm) internal diameter. Sizing of the lines by looking at the pump fittings also applies to the intake side of the pump where, as noted above, half-inch (12.7mm) lines are commonly used. On both sides of the pump, sharp angled bends (for example 90-degree fittings) should be avoided.

Flow testing of the entire fuel system when it's installed in the car (or set up on the bench to accurately replicate the car configuration) is a worthwhile step to undertake.

FUEL INJECTOR RAIL

Fuel injector rails are generally large in diameter to minimise the pressure drop that occurs when each injector opens. In some designs, the rail also holds the injectors in place, while others have no structural function. Because they are subject to internal pressure waves caused by the opening and closing of the injectors and external vibration from their location on the engine, fuel rails should be ruggedly constructed. They are usually made from machined aluminium extrusions or fabricated plated steel.

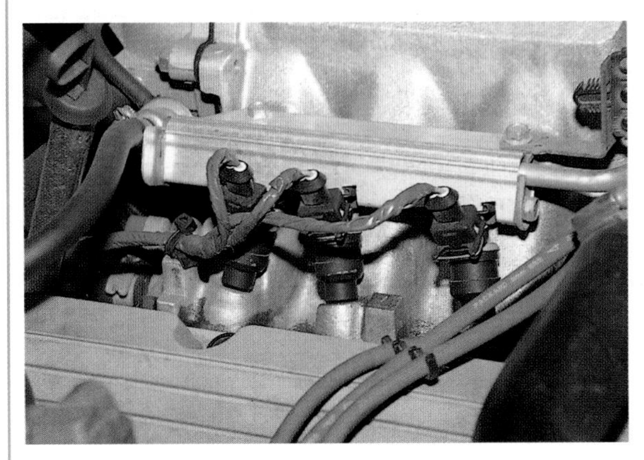

The fuel injector rail is normally of much larger diameter than the individual injector feeds, decreasing the pressure drop that occurs as each injector squirts.

FUEL PRESSURE REGULATOR

The function of the fuel pressure regulator is to maintain a constant headroom above manifold pressure. This means, for example, a 45psi fuel pressure regulator will provide a measured fuel pressure of 45psi only when the engine is not running — and in a naturally aspirated engine, when it is running at full power! This is because it is in only those two situations that manifold pressure is zero. In a forced aspirated engine, the fuel pressure will rise above the regulator's nominal value as boost increases. A rising-rate fuel pressure regulator increases fuel pressures at a rate that's not proportional to the increase in manifold pressure. An increase in manifold pressure of 1psi might mean a fuel pressure increase of 10psi. This technique is used to squeeze more fuel through the same injector pulse widths. This is useful on a modified engine running standard management or any engine where the injectors are a little small for the use being made of them.

This SX Performance 15401 fuel pressure regulator will maintain a fuel pressure of 45psi (plus or minus 2psi) with fuel flows from 0-7.5 litres per minute. It's made of anodised alloy.

While standard cars use pressure regulators designed to maintain the constant fuel pressure headroom above manifold pressure, many race cars use a pressure regulator that holds a fixed pressure relative to the atmosphere. However, because the management system is mapped with this (effectively) varying pressure differential across the injectors in place, the air/fuel ratios are still maintained appropriately.

Aftermarket fuel pressure regulators are required when:

- The fuel flow is substantially increased (power increases of 100 per cent or more, for example)
- Manifold boost pressure has been dramatically lifted
- A naturally aspirated engine has been converted to forced induction
- Adjustable fuel pressure is required

However, on the vast majority of modified street engines, the stock fuel pressure regulator is fine.

This Vortech rising-rate pressure regulator increases fuel pressure disproportionately to boost. This allows more fuel to pass through the injectors as turbo or blower boost rises.

FILTERS

Filters are a very important part of the fuel system. Both injectors and high-pressure EFI roller cell pumps are very susceptible to damage from the ingestion of foreign materials. High-pressure fuel filters are always moderately large and almost always use a metal body to cope with the bursting pressure. Other than the size and type of hose fittings used, most high-pressure filters are interchangeable. This means if your car uses a very expensive filter, it's possible to replace it with another from a cheaper car. However, make sure the new filter still flows

Normal off-the-shelf high-pressure fuel filters are interchangeable. A filter such as this one is suitable with systems with quite high fuel flows and, if even more fuel flow is required, two filters can be used in parallel.

as well as the old one did. One easy way to check this is to flow some fuel through each of the filters using a funnel and a metre or so of hose. Rig up the hose and funnel above the filter so gravity alone causes the flow. Then all you need do is compare the relative volumes of fuel that flow through each of the filters in the same amount of time.

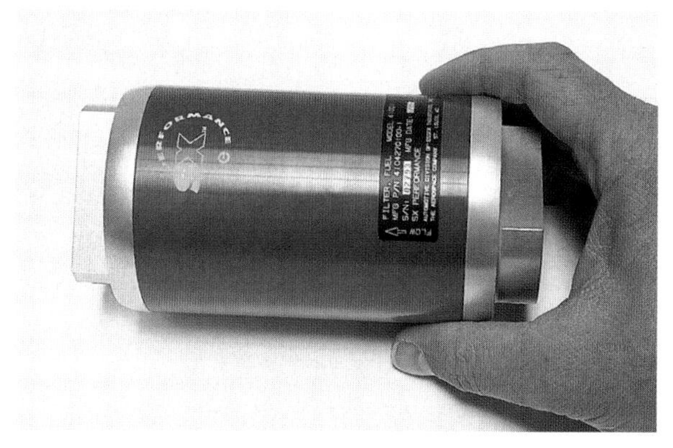

The SX Performance 41001 filter uses a red anodised alloy body and replaceable filter elements. Either 10-micron paper or 60-micron stainless-steel elements are available.

As indicated earlier, a mesh sleeve should be used on the in-tank pickup. However, there is an alternative low-pressure filter that can be used if the high-pressure pump is mounted externally to the tank. Trucks use a diesel fuel spin-on filter that screws to an attractive alloy filter head. Large hose diameters are used and the filter elements normally contain the same paper that's used in petrol filters. The large lines and filters mean the pressure drop through the filter is low, allowing it to be used in front of the high-pressure pump. If you're on a real budget, the filter heads can be bought from truck wreckers — the Mitsubishi Canter is one mid-sized diesel truck that has a suitable filter head. It uses a Ryco Z188 filter or equivalent.

The reusable stainless-steel filter element from the SX Performance filter can be cleaned in normal fuel system cleaner.

You should also be aware that the inlet to both the high-pressure pump and the injectors use small mesh cones inserted into them. If dirt has gone through the fuel system, these tiny filters should be pried out and cleaned. On a low-pressure pump, conventional filters can be used on either the inlet or outlet. However, the filters should be sized according to the flow capacity of the pump. This is not normally a problem if a filter with a suitable hose barb is used.

The Brain from Paxton Products is a fuel pump speed controller. It allows the fuel pump to be run more slowly at times of low load. However, any high current dropping resistor could probably be used to perform the same function.

PUMP WIRING

Because the output of pumps varies significantly when the voltage on which they are operating changes, care should be taken that very near to full battery voltage is available at the pump itself. Any voltage drop will be minimised if heavy wiring and a relay are used to control the pump operation. If the pump flow is well down over the results of bench testing (or down over the pump's specifications) the voltage at the pump itself should be checked. A voltage drop of greater than one volt between the running pump and the battery requires further investigation. Fuel pumps should be controlled by the ECU so that in the event of an accident and a stalled motor, the pump(s) switch off after a short time.

Some cars use a pump speed control module that varies pump speed under the direction of the ECU. The pump may have two or more speeds, with the faster speed used at high loads and the lower speed used at idle and low loads. Taking this approach has major benefits — pump wear, noise and current consumption are all decreased; and heating of the fuel is minimised. A car using this type of system may have a measured voltage at the pump that is well down on battery voltage. Note also that a pump controlled in this manner cannot easily have its flow measured when the car is just idling.

BENCH TESTING THE FUEL SYSTEM

While performing any fuel system testing, remember that petrol in both vapour and liquid forms is extremely inflammable. Great pains should be taken to ensure that flames (including hidden pilot flames inside gas water heaters and the like) are nowhere nearby. Sparks should also not occur during testing. For this reason, the testing is best done outdoors or in a very well ventilated area.

The testing technique that's used for fuel pumps depends on whether it is a low-pressure (ie swirl pot feed) or high-pressure (ie injector feed) pump that's being tested. A low-pressure pump is most easily tested. Simply connect the hoses to the pump, feed **clean** fuel to it and measure the flow output. If running the pump for 30 seconds results in the pumping of one litre, the pump flows two litres a minute. While this simple test will allow a quick and easy comparison of pump flow rates, a number of other factors should be introduced for more accurate testing. First, the pump should be fed a regulated voltage. Pump outputs vary substantially at different voltages, so a power supply voltage of 13 volts (ie typical of that available to a pump in a running car) should be used. Second, the flow of some pumps is dramatically changed when a filter and/or fittings are introduced in front of the pump. For example, a low-pressure pump that's required to suck through a 90-degree brass fitting may drop substantially in flow. For this reason, testing with the best accuracy occurs with the complete system bench-mounted.

Testing a high-pressure pump requires the use of a pressure gauge and an adjustable restriction (eg a needle valve) in addition to a measuring container and regulated power supply. The test rig is set up as shown in Figure 10.5. The pump draws from a container of **clean** petrol, pushing the fuel past the variable tap. A pressure gauge is tee'd into this line. The fuel pump is started and the needle valve closed until the pressure gauge shows the maximum fuel pressure that will be required. If this is unknown, typical EFI factory equipment pressure regulators have a zero-vacuum pressure of 45psi. With such a regulator, this is the maximum fuel pressure that will be required in a naturally aspirated car, but with a forced aspiration car, it's important you add to this the maximum boost figure. Thus, a typical forced aspirated car using a peak boost pressure of 10psi will have a maximum required pressure of 55psi.

This Inter-Ject fuel system tester can measure fuel flow and pressures on the installed fuel system. When used in conjunction with a chassis dyno, this type of workshop equipment allows the quick, safe diagnosis of fuel system problems.

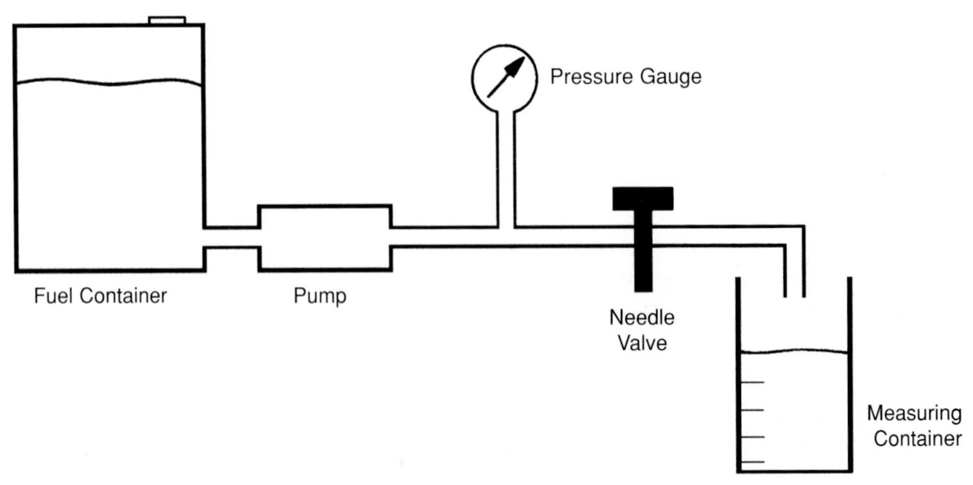

Figure 10.5
High-pressure pumps should be flow tested only at the pressure required in the application, never under free-flow conditions. A simple way to do this is to set up the test rig shown. The pump draws from a container of clean petrol, pushing the fuel past an adjustable needle valve. A pressure gauge is tee'd into this line. The fuel pump is started and the needle valve slowly closed until the pressure gauge shows the maximum fuel pressure that will be required. The flow is then measured at this pressure.

This Microtech Fuel Flow Meter allows the accurate measurement of fuel being returned to the car's tank. Using the meter is safer and more convenient than flowing the return fuel into a measuring container.

With the needle valve closed until the pressure gauge reads the maximum pressure that will be required, the amount of fuel flowing past the valve and into a container can be measured. If the pressure is set to 60psi and 1.5 litres flows past the needle valve in a minute, the pump is capable of flowing 1.5 litres/minute at a fuel rail pressure of 60psi. You'll find that the flow of the pump drops off rapidly as you increase the fuel pressure, so it's wise to do a complete test at different fuel pressures while you have the test rig set up. That way, if an increase in fuel pressure later proves necessary, you'll know whether the fuel pump can keep up.

Again, for maximum accuracy, set up the complete fuel system on the bench. In an external swirl pot system, this can include the low- and high-pressure pumps, low- and high-pressure filters and all the fittings and lines. Taking this approach removes many of the assumed factors from the test.

ON-CAR TESTING OF THE FUEL SYSTEM

Checking the pumps in a car can be done in a couple of ways. First, when the pumps are running, they should have an even, continuous low-pitched buzz rather than a high-pitched scream or a noise that varies suddenly in pitch. Switching on the ignition will start the pumps, which should then run for five or so seconds. If continuous running of the pumps is required, the two heavy wires at the pump relay can be bridged. The fuel pumps should then run for as long as you have the wire link in place.

The flow of the fuel delivery system can also be checked. One way to do this is to locate the return line from the fuel pressure regulator to the tank. Detach this line at a

This competition car is using aviation fuel, which is compatible with normal pumps, injectors and lines.

Even with the power increased, many cars do not need a higher rated fuel pump.

convenient fitting and feed it into a measuring container. On a forced aspirated engine you should also pressurise the vacuum line of the fuel pressure regulator — using a small syringe is the easiest way to do this. Running the pump and measuring the volume of fuel that flows into your container will show how well the pump pickup, pump(s), fuel lines and filters are flowing. This technique can also be carried out while the car is under load on the dyno. However, to reduce the danger inherent in having raw fuel fed into a container while the car is running, a fuel flow meter can be used instead. This uses a rotating paddle flow transducer that's inserted temporarily into the fuel return line, allowing the flow to be read off on a digital meter.

The volume of fuel flowing in the return line even under full load should be at least 0.5 litres per minute. If no fuel flows from this line under these conditions, it means all the pumped fuel is being used by the injectors — and that's too close to danger for comfort!

An external pump is easily upgraded for a much higher capacity pump such as this aftermarket unit.

A quick and easy approach to measuring fuel system flow is to plumb a fuel pressure gauge to the fuel rail. Generally, if the pressure starts to drop under full load, not

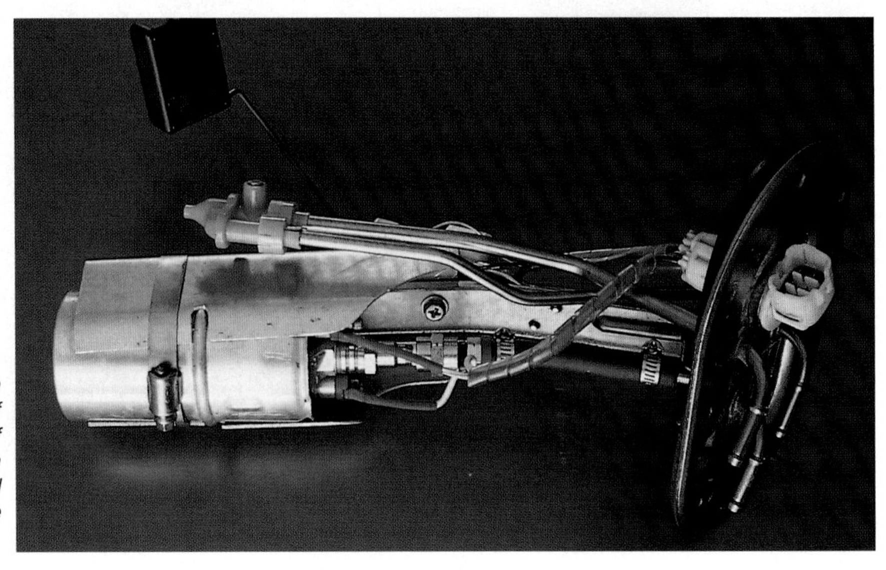

The upgrade of an in-tank pump often requires the manufacture of new brackets and the changing of the pump wiring. In this case, an aluminium bracket has been used to mount the new pump lower in the tank.

enough fuel is being supplied. However, this procedure shouldn't be used as a final diagnosis of the fuel system — just as the first indication that something is wrong. Likewise, if programmable management is being used and the injector pulse widths are being increased but the mixtures stay lean, it's a good indication that something is wrong. Some workshops have commercially available fuel system diagnostic machines that allow the quick and easy measurement of on-car fuel pressure and flow. These are accurate and very safe to use.

Injectors vary in coil resistance, fuel flow, mounting system and fuel system connection. When swapping injectors, each of these needs to be taken into consideration.

UPGRADING THE STANDARD PUMP

Where the engine power has been increased it's often the case that just the fuel pump needs to be upgraded. The fuel lines, tank pickup and filtering system are often quite satisfactory. If this is the case, there's a number of ways in which the upgrade can be carried out. The easiest technique is to place a new pump in-line after the standard pump. When the standard pump isn't required to pump into a pressure, its flow will be increased. This means that if a second pump is installed to do the hard work, the standard pump may well flow enough to be an effective feed pump in the upgraded system. Before the second pump is selected, the flow of the standard pump should be measured in a zero-pressure situation and then the new pump sized appropriately.

Another approach is to simply swap the standard pump for a better one. Where an external pump is used, this approach is straightforward. However changing an in-tank pump can be more involved, because the new pump is unlikely to be physically identical to the old. This means you should pick a new pump that has the same type of pickup system and be of a similar size. In-tank pumps usually use either a short exposed strainer on a tube, or a wire mesh pickup built into the pump body. Even then, it's likely that new brackets will need to be made — brackets that are quite likely to be required to support the fuel level sender unit as well!

The other way around fitting a new in-tank pump is to boost the voltage going to the standard pump. If you feed a constant 15 volts or so to the pump, its flow will be much

To fit these new injectors into this manifold the fuel rail has been spaced a little higher by the use of washers.

higher than if it's being powered by only 12 volts. Kenne Bell sells an electrical device called Boost-a-Pump, which lifts and then regulates the pump voltage. Pump output can be increased anywhere from 1 to 50 per cent over stock. However, the cost of this product makes it viable only if upgrading the factory pump is nearly impossible.

FUEL INJECTORS

If the power of a car is upgraded considerably, the standard-sized injectors are likely to be too small. Whether this is so can be assessed by examining the duty cycle required of the injectors to maintain correct air/fuel ratios. The duty cycle is a means of expressing the 'on' time of the injectors and is measured as a percentage. Maximum injector duty cycle is measured at peak power, normally achieved at high rpm. While commentators express various opinions on the maximum permissible duty cycle, anything over 95 per cent is pushing it! Duty cycle can be measured using a good-quality multimeter or can be read off from the control screen in most types of programmable engine management systems.

If the injector sizing needs to be undertaken before the engine actually runs, the probable power required per injector should be assessed. How much power an injector is good for depends on the air/fuel ratio and fuel pressure used. Also, the relationship between power and fuel flow for injectors is a little different from that used above for fuel pumps. This is because the fuel pump needs to be sized so that — even at maximum power — there is still a circulation of fuel through the system to keep it cool and so pressure regulation occurs, whereas this is not a requirement for injectors. Unlike a fuel pump, an injector **can** be too large for the application. The control accuracy of large injectors is not good at the very small opening times required when the engine is idling or at light loads. This means that using injectors that are much too large for the application can cause driveability problems.

So how much power requires how much injector flow? A good rule of thumb is kW x 7 = cc per minute per injector. So if a six-cylinder engine develops 250kW, each injector is required to deliver enough fuel for (250 divided by 6 = 41.6kW) about 42kW. 42 x 7 = 294cc per minute or, in round figures, 300cc per minute. From Figure 10.6 you can see there are several Nippon Denso and Bosch injectors that have this level of flow.

Injector Specifications

Courtesy *Advanced Engine Management*

Manufacturer	Part Number	cc per minute	Test Press. kPa	Ohms	Colour	Vehicles	Engine
Bosch	0 280 150 208	133	300			BMW	323
Bosch	0 280 150 716	134	300				
Nippon Denso		145	255	2.4	light green	Toyota	4KE
Nippon Denso		145	255	2.4	green	Toyota	1GE
Bosch	0 280 150 211	146	300				
Lucas	5207007	147	270			Ford	1.6L
Lucas	5206003	147	300				Starlet
Bosch	0 280 150 715	149	300				
Nippon Denso		155	290	13.8	red/ dark blue	Toyota	3EE
Nippon Denso		155	290	13.8	violet	Toyota	3EE, 2EE
Nippon Denso		155	290	13.8	sky-blue	Toyota	1GFE
Nippon Denso		155	290	13.8	violet	Toyota	4AFE
Lucas	5207003	164	300			Buick	3.0
Lucas	5208006	164	250			Renault	
Bosch	0 280 150 704	170	300				
Bosch	0 280 150 209	176	300			Volvo	B200,B230
Nippon Denso		176	290	13.8	light green	Toyota	4AFE
Nippon Denso		176	290	13.8	grey	Toyota	4AFE
Bosch	0 280 150 121	178					
Nippon Denso		182	255	2.0	dark grey	Toyota	4AGE
Nippon Denso		182	255	2.4	grey	Toyota	4ME,5ME, 5MGE
Bosch	0 280 150 203	185					
Bosch	0 280 150 100	185	300				
Bosch	0 280 150 114	185					
Bosch	0 280 150 116	185					
Bosch	0 280 150 125	188					

Manufacturer	Part Number	cc per minute	Test Press. kPa	Ohms	Colour	Vehicles	Engine
Lucas	5208003	188	250			Alfa	
Lucas	5206002	188	250			Toyota	
Lucas	5208007	188	250			BMW	325E
Lucas	5202001	188	250			914	1.8L
Lucas	5208001	188	250			Nissan	280ZX
Bosch	0 280 150 614	189	300				
Lucas	5207013	201	270			Jeep	4.0L
Nippon Denso		200	290	1.7	dark grey	Toyota	3SFE
Nippon Denso		200	290	1.7	beige	Toyota	4YE
Nippon Denso		200	290	1.7	orange	Toyota	22RE
Nippon Denso		200	290	1.7	brown	Toyota	3VZE
Nippon Denso		200	290	2.7	pink	Toyota	4AGE
Nippon Denso		200	290	13.8	dark blue	Toyota	3SFE
Nippon Denso		200	290	13.8	orange/ blue	Toyota	22RE
Nippon Denso		200	290	13.8	brown	Toyota	3VZFE
Nippon Denso		200	290	13.8	red	Toyota	2VZFE
Nippon Denso		210	255	2.4	blue	Toyota	4AGE
Nippon Denso		213	290	13.8	sky blue	Toyota	3FE
Nippon Denso		213	290	13.8	beige	Toyota	4AGE
Nippon Denso		213	290	13.8	yellow	Toyota	5SFE
Bosch	0 280 150 706	214	250				
Bosch	0 280 150 712	214	250			Saab	2.3L Turbo
Bosch	0 280 150 762	214	300			Volvo	B230F
Bosch	0 280 150 157	214	250			Jaguar	4.2L
Lucas	5207011	218	300			Chev	5.7L
Bosch	0 280 150152	230	?			Alfa	Turbo
Lucas	5207002	188	250			Chev	5.0L
Lucas	5204001	188	250			Fiat	

Manufacturer	Part Number	cc per minute	Test Press. kPa	Ohms	Colour	Vehicles	Engine
Bosch	0 280 150 201	236	300				
Lucas	5208005	237	250			Chrysler, BMW	
Lucas	5208004	237	250			Ford	98CID
Bosch	0 280 150 151	240	?			BMW	633
Nippon Denso		250	255	1.7	yellow/ orange	Toyota	22RTE
Nippon Denso		250	290	13.8	green	Toyota	4AGE
Nippon Denso		250	290	13.8	violet	Toyota	4AGE
Nippon Denso		250	255	13.8	brown	Toyota	3SGE
Nippon Denso		251	290	13.8	violet	Toyota	1UZFE
Bosch	0 280 150 001	265	300				
Bosch	0 280 150 002	265	300				
Bosch	0 280 150 009	265	300				
Nippon Denso		282	290	13.8	light green	Toyota	2RZE
Nippon Denso		282	290	13.8	violet	Toyota	2TZFE
Bosch	0 280 150 802	284	300			Volvo, Renault	B200Turbo, J7R Turbo
Nippon Denso		295	255	2.7	yellow	Toyota	7MGE
Nippon Denso		295	255	1.6	pink	Toyota	22RTE
Nippon Denso		295	255	13.8	green	Toyota	3SGE
Bosch	0 280 150 811	298	350			Porsche	944 Turbo
Bosch	0 280 150 200	300	300			BMW	
Bosch	0 280 150 335	300	300			Volvo	B230 Turbo
Bosch	0 280 150 945	300			red/ brown	Ford	MotorSport
Nippon Denso		315	290	13.8	pink	Toyota	3SGE
Nippon Denso		315	290	13.8	light green	Toyota	7MGE
Bosch	0 280 150 804	337	300			Peugeot	505 Turbo
Bosch	0 280 150 402	338	300			Ford	
Bosch	0 280 155 009	346	300			Saab Turbo	
Bosch	0 280 150 951	346	300			Porsche	

Manufacturer	Part Number	cc per minute	Test Press. kPa	Ohms	Colour	Vehicles	Engine
Bosch	0 280 150 015	380	300				
Bosch	0 280 150 024	380	300			Volvo	B30E
Bosch	0 280 150 026	380	300				
Bosch	0 280 150 036	380	300			MB	4.51
Bosch	0 280 150 043	380	300			BMW	
Bosch	0 280 150 814	384	300				
Bosch	0 280 150 834	397	300				
Bosch	0 280 150 835	397	300			Chrysler	
Nippon Denso		430	255	2.9	black	Toyota	7MGTE, 3SGTE
Bosch	R 280 410 144	434	300			Bosch R Sport	
Bosch	0 280 150 400	437	300			Ford	4.5L
Bosch	0 280 150 401	437	300			Ford	
Bosch	0 280 150 041	480	300			MB	6.9L
Bosch	0 280 150 403	503	300	0.5	blue	Ford	
Nippon Denso		365	255	2.9	red/ orange	Toyota	4AGZE

Figure 10.6
The specifications of nearly 100 injectors.

In addition to the variations in fuel flow from injector to injector, you will also notice in the tables that the injector resistance (ohms) varies. Injectors fall into two broad classes in terms of their coil resistance: 'low' and 'high'. Low-resistance injectors have 2-3 ohm coils, while high-resistance injectors use coils of around 14-16 ohms. High-resistance injectors require only a simple on/off ECU switching circuit to operate. When a 16-ohm injector has 12 volts applied to it, the current builds up over a very short period (around half a millisecond) until it stabilises at 0.75 amps. However, low-resistance injectors would draw an excessive current if switched in a similar manner.

Instead, an ECU peak-and-hold circuit is used, where the injectors are opened with a current of typically 2 amps, and are then held on with a 0.5 amp current. (Other peak-and-hold injectors use a 4-amp peak and a 1-amp hold current.) If you are upgrading a car to larger injectors, the new injectors should be of the same resistance class as those being replaced. If you are fitting programmable management, you must make sure the ECU will be happy with the coil resistance of the injectors you have selected.

Note also that injectors vary substantially in their size and shape. Some cars run side-feed injectors, while others are end-feed O-ring types, where the fuel rail holds the injectors in place. Still others use barbed hose fittings. Unless you're prepared to do custom machining and fabrication work, you need to be aware of the required injector size and shape before you make the selection. Finally, injector wiring plugs also vary; however, it's usually not much drama to wire in new plugs — assuming you can get them!

A final note: the data in Figure 10.6 is provided only as a guide. When installing new injectors, you should always proceed with care, testing flow rates and measuring actual air/fuel ratios if you are at all unsure of the injectors you have selected.

The tables are provided courtesy of Advanced Engine Management, the makers of Wolf programmable engine management systems. You'll find lots more interesting information on fuel injection and the Wolf range at www.wolfems.com.au

Performance Testing

It's extremely easy to fall into the trap of assuming that because you have just spent time and/or money modifying your car, it will be faster. However, this very human trait should be strongly resisted! Otherwise, you may well be heading in the completely wrong direction with your mods, not realising that the car is, in fact, getting slower. Don't rely on just 'feel'; this is notoriously inaccurate, especially if the modification has also made the car louder or lower. Performance modifications that are **not** effective are quite common — a point soon noticed if you attend a street meeting at a drag strip. Often, modified cars post slower times than the times the same cars achieved in new car road tests! And that's on a sticky track ... You should **always** test how well a modification has worked after you have performed it. Even if the news is bad, it will at least guide you in making the right steps next!

VEHICLE PERFORMANCE

Vehicle performance consists of many different parameters. Some of these are:

- Acceleration
- Top speed
- Handling
- Braking
- Fuel consumption
- Emissions
- Aerodynamics

But before you can work out how to test each of these parameters, you need to know what each is all about.

1. Acceleration

Acceleration is one of the most important factors in assessing car performance. For straight-line travel, it's a measurement of how quickly a change occurs in vehicle speed. It's measured in metres per second **per second** or in 'g'. g represents the acceleration caused by gravity, the force that draws objects towards the centre of the earth. 1g is equivalent to an acceleration of 9.8 metres per second, per second. That is, every second the speed of a falling object **increases** by 9.8 metres per second, or 35.3km/h. This is the unit used when for, example, a car is said to be accelerating at a constant 0.4g.

More commonly, vehicle acceleration is measured indirectly. The length of time it takes a car to accelerate from one speed to another, or from standstill to a certain speed is often used. An example of this type of acceleration measurement is when a 0-100km/h sprint is completed in 8.4 seconds, or when a 60-90km/h split takes 2.3 seconds. Standing-start acceleration times are made from a dead stop, while rolling times are made when the car is already moving.

Rolling times can be measured in one of two ways: from one steady speed to the second speed, or when the car accelerates past the first speed until it attains the second speed. The difference between steady-state rolling times and accelerating rolling times can be great, so it should be clearly specified what sort of rolling times are being measured. Another rolling acceleration time is one that's conducted through the gears. From 20km/h in first gear through to 130km/h in third gear is an example of a through-the-gears test. This type of measurement can be used to show how quickly a turbo spools up after each gear change, for example.

If the car is to be drag raced, the peak acceleration achieved on launch may be an important factor to consider. This can be expressed by stating the car pulled (for example) 1.2g on launch. At the drag strip, acceleration is indirectly measured. The time taken to travel from a standstill to a point one quarter of a mile (402.25 metres) down the strip is recorded, and the speed the vehicle has attained by the end of the quarter-mile is also indicated. A car may have a 14.3-second quarter-mile at 101.3mph, for example. (Imperial measurements are always used in drag racing.)

The method of measuring acceleration that's best for you depends on the type of use you make of your car. For a car that's to perform well at a drag racing meeting, the quarter-mile time and speed are obviously all-important. For a mostly urban-driven car, the 0-100km/h time will be of great importance. And for a car where you want blistering overtaking performance, an in-gears time from perhaps 90-120km/h will be the best parameter to measure.

The best general-purpose acceleration measurement is the quarter-mile time.

2. Top Speed

Once upon a time, a common question of the owner of a good car was, "How fast will she go, mate?" These days, with strictly enforced speed limits and an expectation that you must be a maniac if you drive fast, top speed has dwindled in importance. The concept is simple — it's the maximum speed attainable by the car in flat, windless conditions. For example, a car may have a top speed of 200km/h.

3. Handling

Acceleration can involve not only change of speed but also change of direction, so that when a car is cornering (even at a constant speed) it's said to be accelerating. The unit universally used for this lateral acceleration is 'g'. Cornering acceleration can be recorded in either steady-state conditions or can be of the peak acceleration type. Steady-state cornering is normally carried out on a skid-pan, while peak cornering accelerations can be recorded wherever there is the room. An example is of a car with a steady-state cornering acceleration of 0.8g and a peak cornering acceleration of 1.0g.

While the measurement of cornering acceleration is straightforward, measuring overall vehicle handling is much more difficult. This is because the concept 'car handling' is a little nebulous — a car that handles well corners quickly but also does not bite back if the driver slightly oversteps the mark. Handling is **not** simply about lateral acceleration — a car may have relatively low levels of grip but still handle very well. This means that while many measurements can be made that attempt to show vehicle handling, in the final analysis, determining whether handling improvements have occurred is up to you.

4. Braking

Braking can be measured in terms of either the stopping distance from a specified speed, or peak or average deceleration. In the former, the distance it takes the car to come to a complete stop from the initial speed is recorded. An example might be 53.1 metres from 100km/h. Peak deceleration is the maximum deceleration recorded, even if that measurement occurs only for a very short period. A car might pull a peak g reading of 1.01g, with an average deceleration for the stop of 0.74g. Of the different measurements, the stopping distance is the most useful.

5. Fuel Consumption

Fuel consumption is a measurement of the volume of fuel used per unit distance. It is most frequently expressed as litres per 100 kilometres (l/100km) or miles per gallon (mpg). Most countries have official fuel consumption tests that are carried out on all new vehicles. The tests that arrive at these figures use a special chassis dyno to replicate a drive cycle that's indicative of city or highway driving. Other official figures that may be available are constant speed consumption at 90km/h or 120km/h. While these tests provide good means of comparing the fuel consumption of different cars, the figures seldom match what normal people driving on real roads actually achieve!

6. Emissions

Emissions refer to the pollutants that are released by the car. All new cars are required to pass laws that specify the maximum allowable emissions and it is increasingly common that modified cars must also be emissions-tested. Depending on the laws in the country in which the car is sold, as many as one in 40 new cars being prduced may need to be individually emissions-tested before sale. Emissions tests are carried out in very sophisticated laboratories using a special chassis dyno and extensive gas analysis equipment. Legal requirements for emissions vary from country to country and sometimes from state to state within a country. An example of emissions requirements are maximum outputs for carbon monoxide of 9.3 grams/km, hydrocarbons 0.9 grams/km and oxides of nitrogen 1.9 grams/km. Standards regarding evaporative emissions are also required, with an example being 2 grams in the specified test.

7. Aerodynamics

Aerodynamic performance relates to how the flow of air past the vehicle affects its behaviour. The aerodynamic performance of cars is covered extensively in Chapter 13.

TESTING VEHICLE PERFORMANCE

There are lots of ways of testing the effectiveness of your modifications. Some tests cost nothing and can be completed quickly, while others cost thousands of dollars and take days to complete.

Acceleration

Stopwatch Testing

The humble stopwatch built into most digital watches is vastly underrated as a performance measuring device. Today when we have data-logging, infra-red timing beams and laptop PC lap-timing that can measure 0.001-second changes, a hand-held stopwatch seems too simple. But it is impossible to humanly feel a performance improvement **fifty times** bigger than 0.001 seconds (ie 0.05 seconds) in the sprint to 100km/h — so is it necessary to have such accurate timing? Certainly, if such small time increments **can** be easily measured that's great, but it is simply not a requirement that timing systems have this degree of accuracy to be useful on road cars. With care, stopping and starting a hand-held stopwatch can be carried out with a consistency of less than 1/10th of a second. And it should be noted that any performance mod that results in less than a 10th of a second gain isn't very successful!

Practise starting and stopping a stopwatch while you view the second hand on an analog clock or the digital seconds display on another watch. Time a number of 10 second increments on your stopwatch and see how close you actually get to 10.0 seconds. I have just stepped away from the keyboard to do this and have got the following results: 9.94, 10.10, 9.98, 9.92, 9.94 and 9.96 seconds. You can see that when the single figure furthest from the median is excluded, the timing is only 0.6 per cent inconsistent! Obviously, with the g-forces pushing you back in your seat and the speedo needle whipping around while the road races towards you, the in-car timing won't be this accurate, but using a digital stopwatch is far better than most people think.

Using a stopwatch to measure acceleration performance in this way does not require that the speedo be accurate. If you wish to compare your results with figures gained in other cars, of course the speedo must be correct, but in the vast majority of cases you will be simply comparing the performance before and after making a modification. Unless you change the tyre diameter, final drive ratio or some other gearing aspect, the speedo accuracy won't change during this period.

The greatest variation in standing-start acceleration times is that caused by differing launches. A high-powered manual RWD car and any manual FWD or constant AWD car is very difficult to launch consistently. In both types of two-wheel-drive cars, wheelspin will occur if too many revs are used on launch, while a constant four-wheel drive car will very easily bog down. For this reason, accurately measuring performance changes with these cars is better done from a rolling start. All cars with automatic transmissions lend themselves very well to consistent 0-100km/h timing.

In any type of testing where a hand-held stopwatch is to be used, pick a flat, preferably deserted road. Certainly, don't attempt to perform any timing runs in urban areas — it's just too easy for a child to run out onto the road from a house, or a car to suddenly appear in front of you. Set the watch to zero (it helps if it beeps when started and stopped) and then rev the engine to the launch rpm. In an automatic car, load the engine with the brakes while applying some accelerator. Release the clutch (or brakes in an automatic car) as you press the stopwatch and then accelerate hard through the gears. Prepare to press the 'stop' button as the speedo needle sweeps around, hitting the button a fraction before you see the needle actually pass over the mark. It is safer if a passenger riding directly behind the driver operates the watch, but I have often timed cars in this way while alone.

If you are after the most accurate figures possible, make three or four runs in each direction. You will quickly see if the figures form a pattern or are simply all over the place. Don't confuse timing inaccuracies with the performance of the car changing during the testing; as the engine heats up, power (especially in a forced-induction car) will often decline.

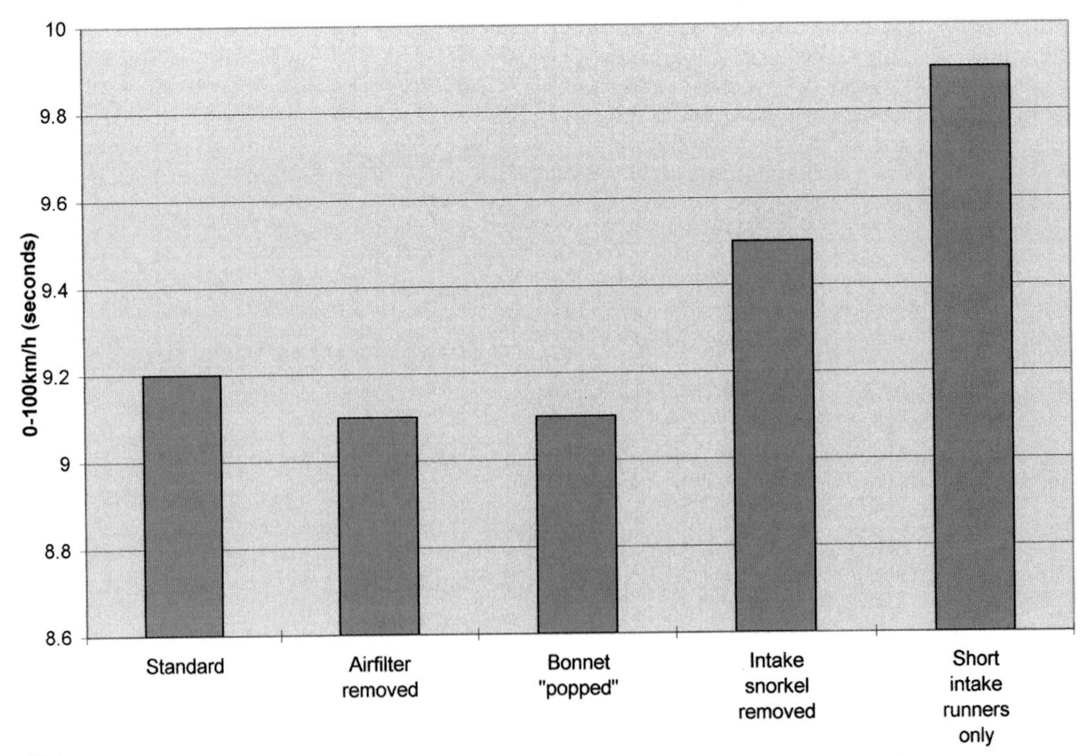

Figure 11.1
Hand timing using a digital stopwatch is a cheap, easy and effective testing technique. Here, the differing 0-100km/h performance of a car has been assessed with various minor intake mods.

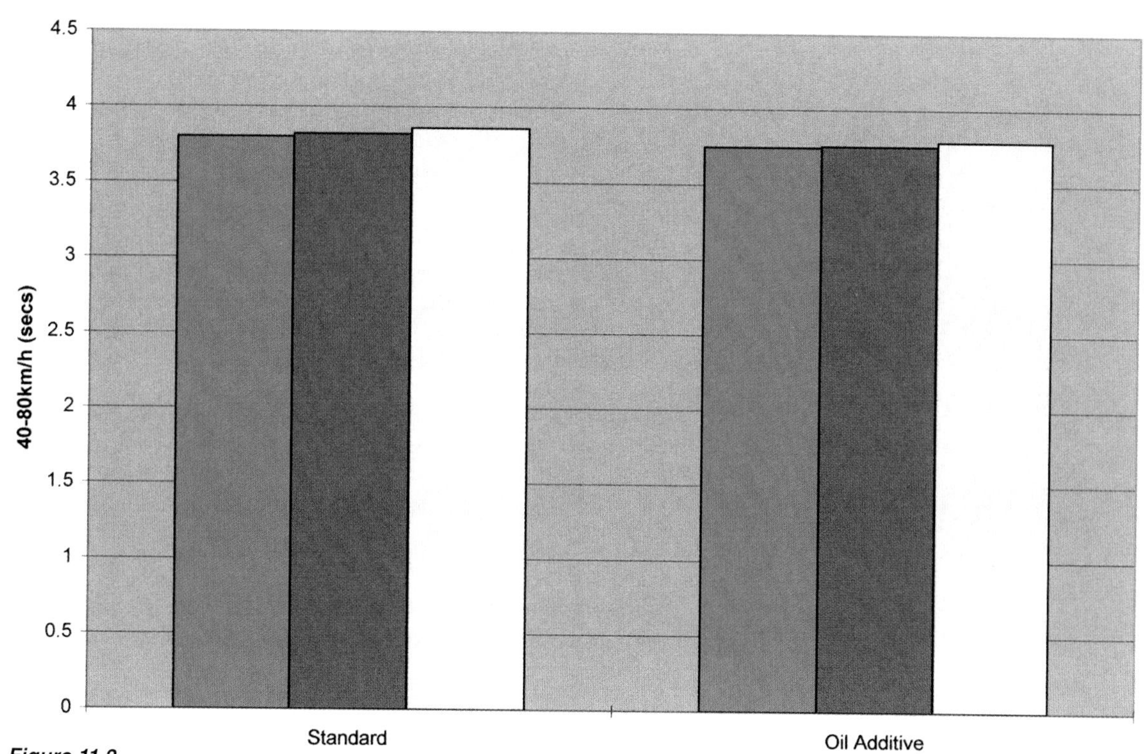

Figure 11.2
Stopwatch testing of the performance claims made by the distributors of an oil treatment showed there was no marked difference after the engine had been treated. The consistency of the timing that's possible using this technique can be seen.

An example of stopwatch testing is that carried out on a six-cylinder family car. The automatic car recorded a 0-100km/h time of 9.2 seconds in standard trim. With the air-filter removed, this improved to 9.1 seconds. The airfilter was then replaced and the bonnet 'popped' to the safety catch. This let more air get to the factory intake snorkel and the time stayed at 9.1 seconds even with the filter back in its box. The bonnet was then returned to its closed position and the intake snorkel to the airbox removed, allowing the engine to breathe hot air. The 0-100 time then lengthened considerably to 9.5 seconds. The engine is fitted with a dual-length intake manifold that changes from long to short runners at a certain engine speed. For the final test it was permanently held in its short runner position, resulting in a slow 9.9 second 0-100. These results are shown in Figure 11.1 (opposite).

Another example of stopwatch testing was carried out on a car using an oil additive that was claimed to give immediate, major power gains. The acceleration increment that was used was a rolling time from 40-80km/h in second gear. With the standard oil in the car, the acceleration runs gave times of 3.80, 3.82 and 3.86 seconds. While showing the times to the hundredth of a second isn't normally done with stopwatch testing, I have done so here to show how consistent this type of testing can be. The super-dupa oil additive was poured in and the timing runs repeated. This time, the recorded times were 3.75, 3.76 and 3.79 seconds. When averaged, this means the 'after' time was 0.07 seconds quicker than the 'before' times. Seven one-hundredths of a second is totally insignificant, so there was no real performance gain made when using this product. Figure 11.2 clearly shows this.

Accelerometer Testing

Performance measuring accelerometers are available in two types — electronic and mechanical.

Electronic accelerometers are most often integrated into full digital performance computers. They use an internal electronic g-sensor, which can detect how much acceleration the car is undergoing at any time. To work out the car performance, the instrument splits up an acceleration run into many tiny time 'slices'. It looks at how hard the car is accelerating during each time slice, and therefore knows how far the car must have travelled during that time period. From the acceleration figures it also knows how fast the car must be going, so the time taken to reach a certain speed (eg 100km/h) or cover a certain distance (eg a quarter-mile) can be calculated.

One example of an electronic accelerometer is the G-Tech Pro. A tiny device (only 70x55x15mm), it simply sticks to the inside of the windscreen with a suction cap, is fed power from the cigarette lighter socket and reads off performance times. In addition to 0-100km/h times, it can also record standing quarter-miles, estimate the peak power available at the wheels and measure positive and negative accelerations. In calculating how much power is available at the wheels, the G-Tech Pro uses the formula that power is speed x acceleration x car mass. This ignores aerodynamic drag, so the testing is best carried out at lower speeds.

Before commencing a performance run, the G-Tech Pro is attached to the inside of the windscreen and the device is levelled. Just before the timing run starts, the display flashes 'Go'. The car doesn't have to be launched at the exact moment the display indicates — the device starts timing only when it senses that the car is moving. For a 100km/h time, the car is braked after it has exceeded

Small electronic accelerometers like this G-Tech Pro are useful devices for measuring acceleration, braking and cornering.

100. The display then shows the 0-100 time in seconds and also the 100-0 braking distance in metres. If you keep on accelerating after 100 has been reached, the G-Tech Pro automatically switches to quarter-mile timing. It then shows both the quarter-mile time and the speed reached at the end of the quarter-mile. The instrument is accurate enough to assess the worth or otherwise of performance modifications and is easy and convenient to use.

The other type of accelerometer is mechanical in nature. Compared with an electronic accelerometer performance computer, it requires much more work from the user but is also much cheaper! A mechanical accelerometer measures acceleration only in g units, not 0-100km/h or quarter-mile times.

A mechanical accelerometer can use either a vertical pendulum that's deflected across a scale by acceleration, or a tube shaped in a semi-circle in which a small ball bearing is moved. A US company called Analytical Performance some years ago produced one of the best of the latter type of accelerometers, called the G-Curve. Their accelerometer consisted of an engraved alloy plate into which was let a long curved glass tube. The tube was filled with a damping fluid and a small ball bearing was sealed inside. A very good handbook was also provided with the instrument.

Unfortunately, it appears that Analytical Performance is no longer in business, but a substitute accelerometer can easily be assembled. Boat and yachting supply companies sell clinometers that are designed to measure the angle of boat heel. One such clinometer is the Lev-O-Gage, which in construction is very similar to the G-Curve. However, because it's designed to measure heel angles, the scale is calibrated in degrees rather than g units. Like the G-Curve, the glass tube is filled with a damping fluid to prevent the ball overshooting.

Both accelerometers are attached to the car in the same way. To measure longitudinal acceleration, the instrument is mounted level and parallel with the direction the car is moving. This means the accelerometer is often mounted on the passenger-side window. The G-Curve came with suction caps to allow this to be easily done, but the Lev-O-Gage is designed to be fixed in place with double-sided tape, so a suitable bracket should be made and then equipped with suction caps (available from rubber supply shops).

So how does the accelerometer work? When the car accelerates, the ball climbs up one arm of the curved tube, showing how hard the car is accelerating. To convert the degrees reading of the clinometer to g readings, simply use a scientific calculator to find the tangent ('tan') of the number of degrees indicated. This means if the car is accelerating hard enough to move the ball to the 20-degree marking, the acceleration is about 0.36g (tan 20 = 0.3639). However, note that converting the degree readings into g's isn't necessary for most testing, where you are only trying to see changes rather than measure absolute values.

Like the G-Tech Pro, the mechanical accelerometer must be accurately levelled before performance measurement can take place. The accelerometer will clearly show any road gradient, so levelling is best done on a flat road. If the road is **not** level (and many aren't), the instrument needs to be adjusted so the error when the car is parked facing in opposite directions on the same spot is of the same amount but in different directions. For example, the clinometer might show +2 degrees with the car facing in one direction and -2 degrees with it facing in the other. If the road was level, the instrument would therefore show 0 degrees. Using a mechanical accelerometer requires a sharp-eyed assistant armed with a paper and pencil to record the data.

One example of the use that can be made of a mechanical accelerometer is the measurement of vehicle acceleration in a single gear. A gear is selected and the car driven at as low a speed as is possible in that gear. After warning your assistant you're about to start the run, yell "now!" and quickly push the accelerator to the floor. Every 1000rpm after the initial rpm figure, yell "now!" Each time you yell, your assistant records the accelerometer reading. In many cars the acceleration will be too quick for the assistant to keep up, so on the first run do, for example, 2000rpm, 4000rpm and 6000rpm, and in the second run do 3000rpm, 5000rpm and 7000rpm.

The G-Curve is a simple mechanical accelerometer, where the amount of acceleration is shown by the movement of the ball in the fluid-filled tube. (In this pic, the ball is at the far right of the tube.)

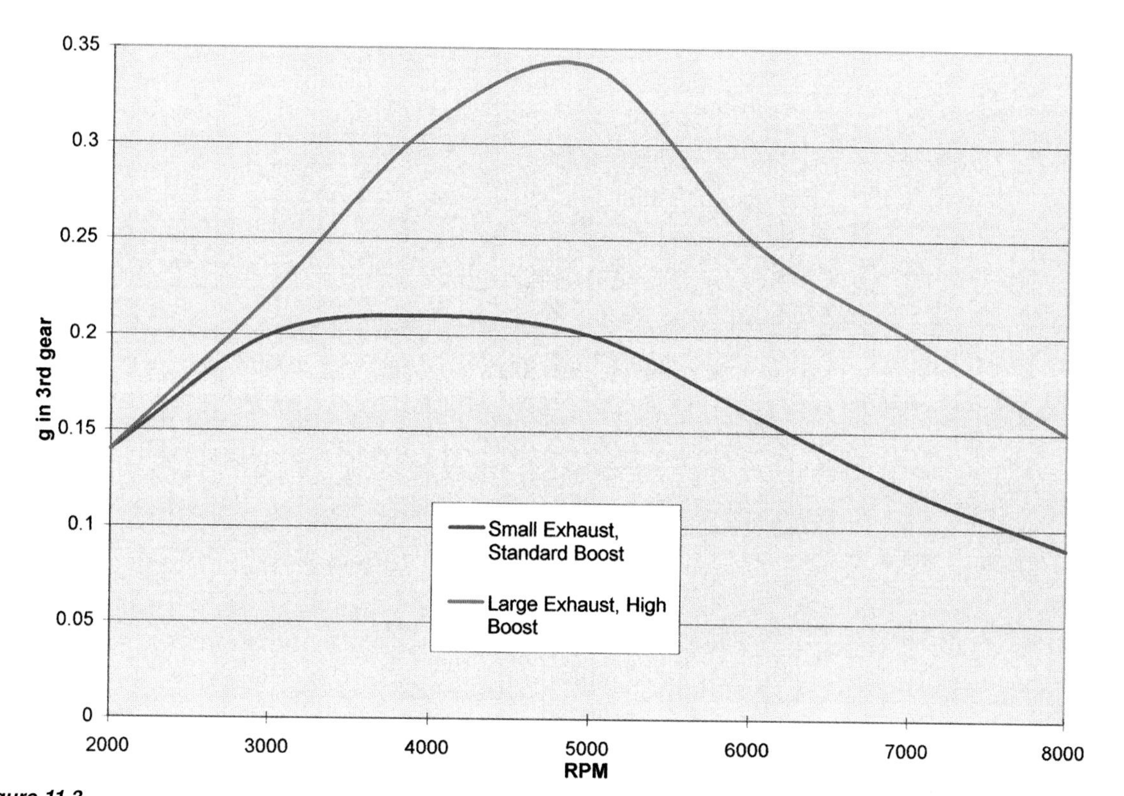

Figure 11.3
Using an accelerometer allows the change in vehicle performance to be cheaply and accurately measured. After a lift in boost pressure and the fitting of a large exhaust, this turbo car has increased in peak acceleration by 62 per cent!

Figure 11.3 shows an example of some of 3rd gear acceleration runs done in this way on a turbo car. In the first run the car was tested with the standard exhaust and standard boost pressure. The vehicle was then modified with a large exhaust and increased boost. As can be seen, the improvement was spectacular! The low rpm performance remained the same in both configurations as the turbo was yet to begin boosting, but once the turbo started developing boost — wow! The peak acceleration in 3rd gear went from 0.21g to 0.34g, a 62 per cent

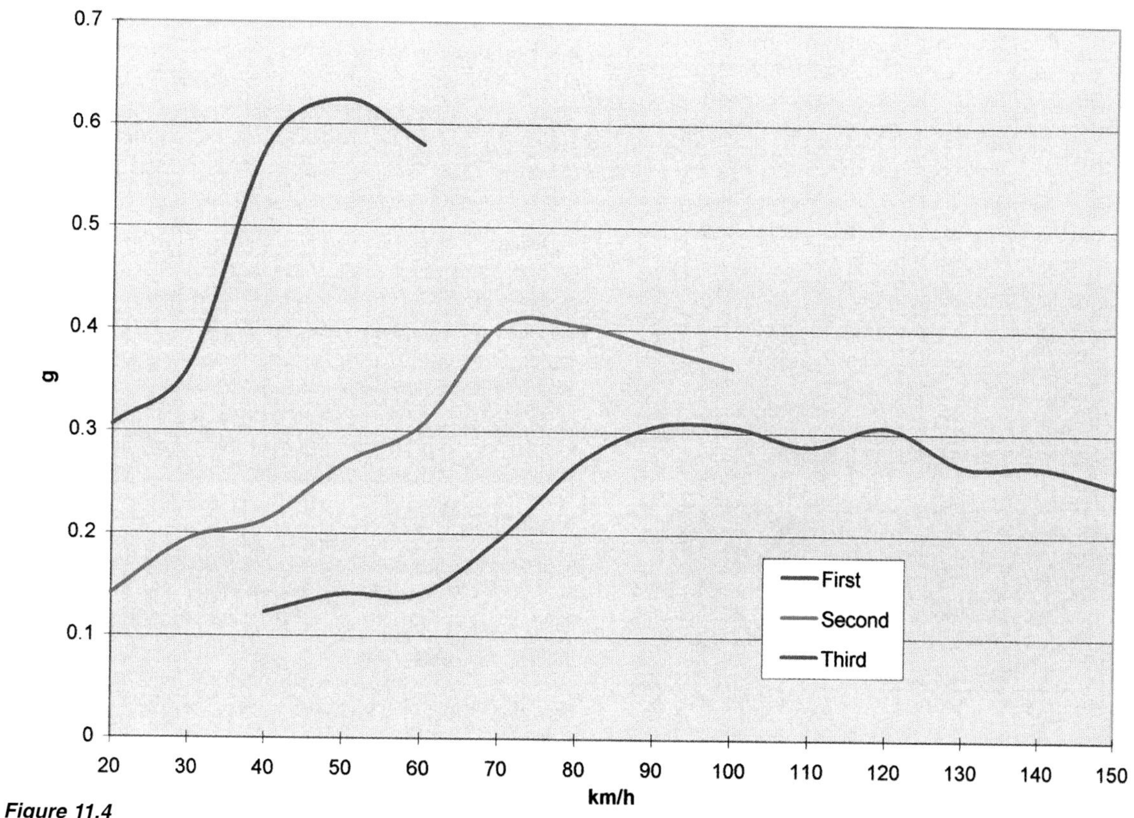

Figure 11.4
An accelerometer can be used to plot the acceleration in each gear. This allows the accurate pinpointing of the best rpm at which to change gear. In this car, the maximum possible acceleration occurs if all gears are held to the redline.

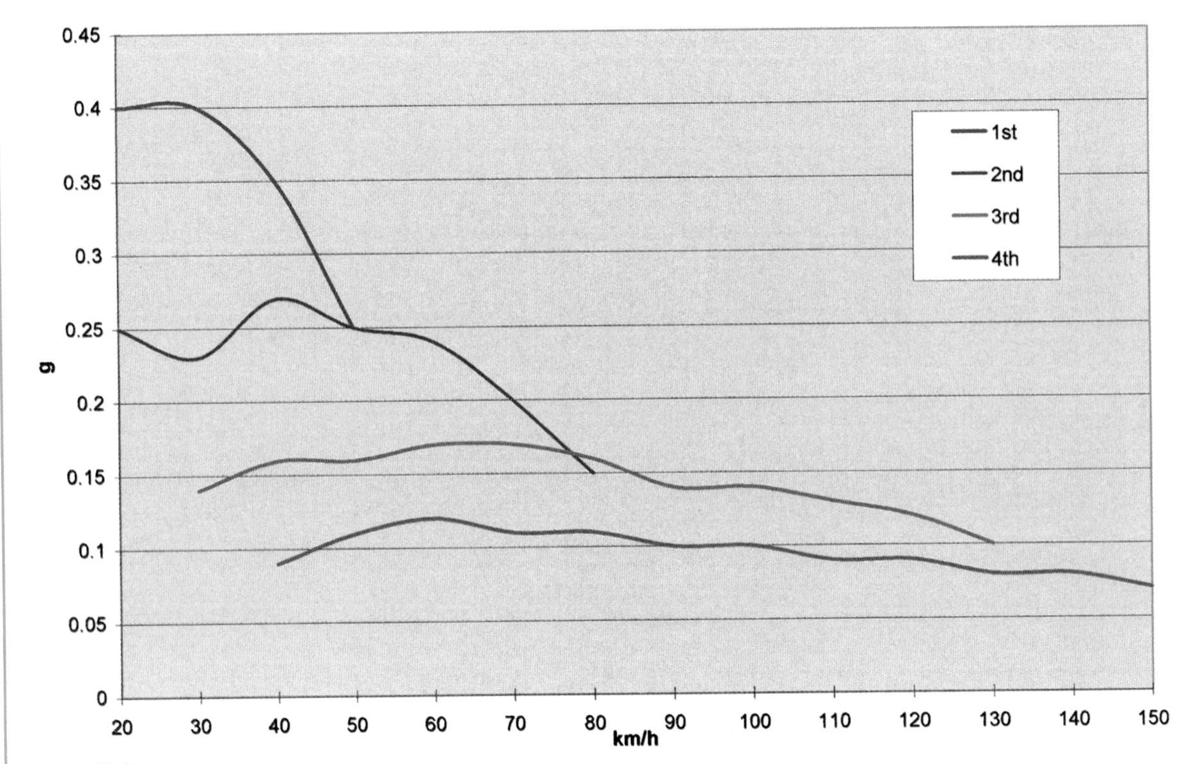

Figure 11.5
With this car, a slightly early change from second to third gear (at about 77km/h) will improve acceleration. This is because holding second gear to the redline results in the car being slower than if the early change to third gear is made. Accelerometer testing can easily be used to collect this data.

increase! The acceleration also remained higher for much longer, with the car accelerating an amazing 67 per cent harder at the redline in modified form.

With this technique you can quickly see both how much any mods have changed acceleration and where in the rev range the power increases (or decreases!) have come. The plotted acceleration line will be exactly proportional to vehicle tractive effort — how hard the road is being pushed back by the tyres. Any improvement in acceleration means more power is getting to the driving wheels.

Another very effective use of the accelerometer is to plot the best gear change points. Instead of acceleration being plotted against engine speed, as was described above, it's plotted against road speed. The acceleration in each gear is measured, with the numbers jotted down each 5km/h or 10km/h. Each gear is tested from as low a speed as possible to the maximum speed possible before the engine redline. Doing this results in the sort of diagram shown in Figure 11.4 (previous page). As can be seen, in this particular car the maximum possible acceleration occurs if all gears are held to the redline. This is because each time a gear change is made, acceleration drops markedly. The first gear change drops acceleration by a massive 47 per cent — it appears this modified car could use either a taller first gear or a higher engine redline.

If a professional accelerometer is not available, a boat clinometer can be used. This works in the same way, with a ball moving in a fluid-filled tube. (In this pic, the ball is out of sight at one end!)

However, as you can see in Figure 11.5, short-shifting will give quicker acceleration in some cars. Here, a slightly early change from second to third gear (at about 77km/h) will improve acceleration. This is because holding second gear to the redline results in the car being slower than if the early change to third gear is made. Note that using an accelerometer in this way is **far** better than trying to calculate the best shift points using torque curves, gear ratios and so on.

Dragstrip Testing

Many drag strips allow normal road cars to use their facilities. Such meetings are often called 'Off Street Legal Drag Racing', or 'Street Meets'. While they allow you to line up your best mate and see whose car is faster, they also give you very accurate performance times. Note, however, the times achieved on a drag strip are often not repeatable on the road and may, in fact, be considerably optimistic. This is the case because the launching area of a drag strip is normally coated with both tyre rubber and special chemicals that promote grip. Also, if there is a strong tailwind, the times will be quicker than

Both the speed and elapsed time are important information — the speed a strong indication of the car's power/weight ratio and the ET showing both this and how quickly your car can get away from a standstill.

can be achieved in still conditions. However, these disadvantages are more than outweighed by the accuracy and detail of the timing.

Before attending a street drag meet, you should make sure your engine's fluids are at the correct levels and the tyre pressures are at whatever pressures you run on the street. If you want to establish a really quick time and you have a powerful car, all but constant four-wheel-drive cars will require sticky slick or semi-slick tyres. However, if you are using the drag strip to measure the performance of the car in street trim, obviously leave the standard tyres on the car.

Before you are allowed to compete, your car and your attire will be scrutineered — checked for safety aspects, in other words. The inspection of the car is normally fairly basic — the security of the battery tie-downs will be checked, as will the steering, seatbelts and bodywork. You will need to be covered from neck to toe, so long sleeves, trousers and closed shoes must be worn. A helmet is a necessity and you should make sure it complies with local design rules and regulations. Normally, there's a sticker to this effect inside the helmet.

Many people at street drag meets perform large burnouts. A section of the track well in front of the start line is sprayed with water, allowing the easy loss of traction. Competitors move their cars forward until the drive wheels are within this area. They then spin the wheels, generating smoke and at the same time heating up their tyres, which makes them stickier. Well, that's the theory, anyway. However, road car tyres don't respond in the same way as drag slicks to a burnout, and the gains made by performing a burnout may not be worth the financial cost of wearing out your tyres! Also, most RWD cars require that the brakes are used to hold the car from accelerating off down the strip during the burnout, meaning the rear brakes become very hot. Most FWD and many RWD cars do not have limited-slip differentials

A dragstrip is a very accurate testing ground. However, it can be hard on a car and too many burnouts like this one will cost plenty in tyres and driveline wear...

Another advantage of a dragstrip is that if something goes wrong, there are plenty of professional helping hands available.

and the very rapid spinning of one wheel can cause damage. Very, very few four-wheel-drive cars have traction problems, so there's normally no necessity for them to perform burnouts. In summary, in a road car, burnouts are usually better for the driver's ego than for dragstrip performance.

At a drag meet, the car must be positioned very accurately on the starting line. This process is called 'staging'. At the top of the Christmas Tree (the vertical stack of starting lights), two clear lights are positioned. These show your location as you are staging. You need to slowly drive the car forward until the first light shines. Once you have done this, edge forward very carefully until the second light illuminates. You are now in the correct position for a start. Note that each of the two lanes has its own complete set of lights, so you can also easily monitor when the other car has staged. When both cars have staged, a start is usually not far away.

The Christmas Tree has (from top to bottom) the two clear staging lights, either three or five amber lights, a green light and a red light. At some street meets, all the amber lights come on at once, followed almost immediately by the green. At others, the amber lights flash down the tree in sequence. In either case, green means 'go'!

If you are attending a street drag meet simply to measure your car's performance, it's important that you realise the timing does not start until you drive off and so break a timing beam. In 'real' drag racing, your reaction time (the time between the illumination of the last amber and your driving off) is very important. However, in a street meet you can stay on the line for a whole second after the light turns green if you wish. You probably won't hang around for that length of time, but there is certainly enough time to collect your thoughts and so perfect your launch.

At the other end of the drag strip, you will be travelling quite quickly, so expect to have to brake firmly once you have passed the quarter-mile point. The end-point of the timing is normally indicated by timing equipment or painted markings on the track. In some track timing systems, your terminal speed is measured over both the last 60 feet of the quarter-mile and an additional 60 feet, so keep your foot down for a moment after you pass the line.

Once you have completed your run you can pick up the time card that shows your measured performance. It also shows the performance of the car you ran against, so make sure in your excitement that you read **your** car's performance, not that of the other car! Quite a lot of information is shown on a typical street drag time slip. First, there will be your reaction time. As explained, this isn't important if you are simply measuring the performance of your car — a long or short reaction time doesn't change the measured car performance. The time taken for your car to reach a distance from the start line of 60, 330, 660 and 1000 feet will be shown. The 1000-foot time will also be accompanied by your measured speed at that distance. Finally, there will be the quarter-mile time and the terminal speed.

In general terms, the terminal speed is indicative of the power/weight ratio of your car while the elapsed time shows both this and launch acceleration. If wheelspin occurs on the starting line (as with most street cars), the elapsed time may easily be 0.5 seconds slower but the terminal speed will be only 1mph or 2mph down.

Handling

The terms used to describe vehicle handling are unique to the dynamics of vehicles. Understeer occurs when the front of the car does not follow the cornering line but instead runs wide, while oversteer is where the rear of the car slides wide. However, these descriptions are only the broad brush strokes — as you will see below there are lots more specific terms used as well.

Describing Handling

One of the keys to improving handling is identifying which factors are deficient. At its most general, a description of the handling of a car might be "it doesn't go around corners very well". A more informed person might say of the same car "it understeers", while someone really trying to identify the problem might state that "the roll linearity changes on turn-in, promoting some initial understeer". None of the descriptions is wrong, but the last will make it a lot easier to identify and then fix the problem.

Mitsubishi Motors uses the following subjective assessment criteria when evaluating a model's ride and handling. Each factor is ranked by test drivers on a scale of 1-10.

Racetrack testing is safe and allows the same test to be done repeatedly. However, a racetrack is an alien environment to a road car, so tests should be carefully structured if information relevant for a road car is to be gained.

- Straight-line stability
- Predictability
- Steering response on-centre
- Steering response off-centre
- Steering linearity
- Steering effort buildup
- Steering on-centre feel
- Roll rate
- Roll linearity
- Transient understeer/oversteer balance
- Steady-state understeer/oversteer balance
- Tuck-in
- Roll angle
- Ride float
- Ride pitch
- Smooth surface isolation
- Coarse surface isolation
- Patched surface isolation
- Broken surface isolation
- Impact harshness

While your first reaction when you look at the list might be "what's this got to do with me?", many of these criteria are relevant to testing normal road car handling. For example, a car that leans over quickly in a corner but then doesn't lean further even with higher g-forces (it has poor 'roll linearity') will be quite unsettling to drive, tending to lurch on turn-in. A car with good steady-state understeer/oversteer balance that loses its composure when the change of direction is very sudden (such as occurs through a tight S-bend) has a particular handling characteristic that can be named — poor transient understeer/oversteer balance. Looking carefully at the list allows you to split the handling behaviour into separate sections, letting you concentrate on just that one specific aspect when you test drive the car.

Slalom Testing

The ability of a car to change direction quickly at relatively low speeds can be measured by a slalom. This is a test where a number of plastic cones ('witch's hats') are placed in a straight line a predetermined distance (eg 20 metres) apart from each other. The car approaches the slalom at a constant speed (eg 80km/h) and is driven in and out of the cones as shown in Figure 11.6. The time taken for the car to travel from one end of the course to the other end is recorded. Obviously, the quicker the car can be driven in and out of the cones, the better! This test records how throttle-responsive the handling of the car is as well as indicating the quality of lateral grip, turn-in and so on. As with all handling tests, it also measures how good the driver is! A really good driver will use lift-off oversteer to turn the car in and have the tail of the car penduluming in and out of the cones while the front of the car remains very close to the cones.

An example of slalom testing was carried out on a sedan equipped with a factory traction control system (TCS). On a dry course, the car recorded a time of 14.6 seconds with the TCS switched on and 14.7 seconds with the TCS off, meaning the car was marginally faster with the traction system working. However, using a shorter track

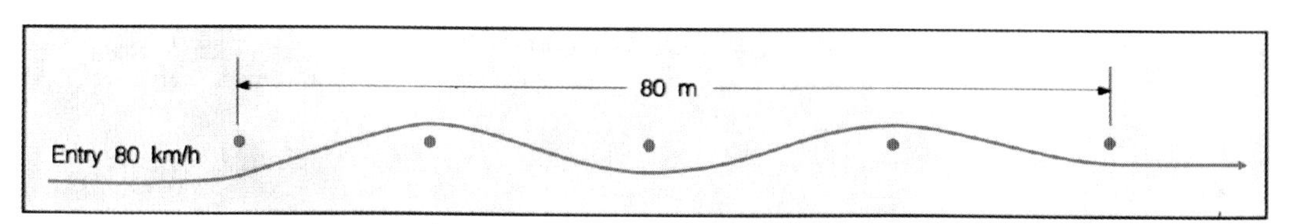

Figure 11.6
Slalom testing can be used to find changes in low-speed handling. Here, the cones are 20 metres apart, but they can be placed closer together if there's insufficient space to perform the testing with this cone layout.

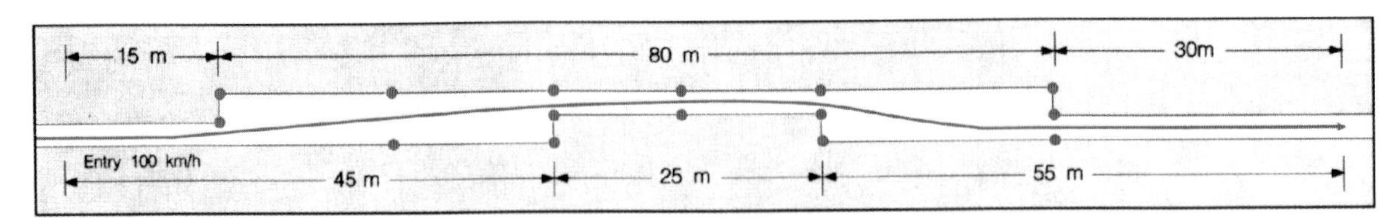

Figure 11.7
The ISO Double Lane Change Test can be used to assess higher-speed handling stability. The 100km/h entry speed is the minimum recommended, thus extensive run-off areas are required.

that had been wetted with a hose, the time with TCS off averaged 7.3 seconds, versus 7.7 with the system switched on. That means that in wet conditions the car was faster through the slalom with the system switched off! This was the case because whenever it sensed wheelspin, the Traction Control System shut down power to an excessive degree.

If required, the slalom cones can be more tightly spaced and lower speeds (eg 40-60km/h) used, reducing the amount of run-off room needed. However, tyre squeal and engine noise will be prominent, so even though a deserted carpark looks ideal, other problems often intrude...

Lane Change Testing

A test that provides a very good indication of vehicle stability and control is the double lane change test. This requires that the car be swerved from one lane to another and then back onto the original path. Cones (as shown in Figure 11.7) indicate the edges of the lanes. The entrance speed into the lane change test is at least 100km/h, which certainly makes the manoeuvre look much more difficult when you're behind the steering wheel.

The lane change test is a very good indicator of transient stability — how quickly the car settles after a violent change of direction. For example, if the car is inclined to oversteer during the lefthand turn-in, the driver will find it very difficult to balance the car before suddenly swerving right! In fact, in this situation the tail will whip back the other way and the car will probably spin. This means the lane change test requires a reasonable amount of run-off area if it is to be safely carried out. The driver should wear a helmet during this test. The maximum speed with which it is possible to negotiate the test without hitting any cones can be used to quantify performance.

Skid Pan Testing

Skid pan testing is a deceptively simple test. The skid pan is simply an area of flat concrete or bitumen that has a large circle painted on it. Typically, the circle has a diameter of 60 metres. The car is driven around the outside of the circle with the car kept accurately on the cornering line. The speed is then gradually increased until the car starts to slide. The car will either understeer or oversteer once the limit of cornering grip has been reached. Because the speed is only gradually increased, the lateral acceleration is of the steady-state type. An accelerometer (of either the electronic or mechanical types described above) placed laterally (sideways) in the car will gradually increase in reading, until it reaches its maximum. This figure shows how much cornering grip can be developed.

However, if the driver either turns the steering wheel further or changes the throttle angle, the behaviour of the car will also change. In a FWD car, lifting off the throttle may induce the front to tuck-in as understeer is reduced, or alternatively the back of the car may suddenly move out as throttle-off oversteer occurs. A RWD car may initially understeer, with this characteristic gradually changing into power oversteer as more throttle is used. A four-wheel-drive car may understeer, followed in some cars by a type of 'four-wheel-drive hop' as all wheels slide together.

While it would first appear that driving a car on a skidpan must be very boring (just driving around and around in circles!), it is, in fact, much more akin to driving around a corner that never ends. A skidpan is an extremely effective way of measuring maximum cornering grip and also the car's behaviour on the limits. Because of the higher speeds used, a skidpan needs a much larger area of bitumen than a slalom. However, again the actual requirement is fairly simple — just a large flat area with a

A slalom tests vehicle turn-in, throttle control and balance. However, a good driver is needed if the runs are to be consistent.

A skidpan is an excellent handling test venue. The understeer/oversteer balance can be explored and roll angles and lateral acceleration easily measured.

sealed surface. If an area is to be temporarily pressed into surface, witch's hats can be used instead of a painted line. The circle can be accurately drawn using just a length of rope, some chalk and two people. The people hold opposite ends of the rope, keeping it taut. One person stays stationary in what will become the centre of the circle, while the other describes the circle around him, marking the ground with the chalk as he or she moves.

The best equivalent of a skidpan available to normal road users is a roundabout of about the same diameter. This can be used if traffic is light and fast driving around it is not likely to embarrass other road users (or yourself). However, note that you can only use the roundabout 'skidpan' in one direction — clockwise or anticlockwise, depending on the side of the road on which cars are driven where you live! Also, the roundabout is not as easy to drive over as a painted line if you lose control...

Track Testing

Testing on a purpose-built racetrack allows you to assess handling, braking and engine performance. Most racetracks allow road cars to use their facilities in the form of Club Sprinting, practice days or when you hire the whole track for just your personal use. Most racetracks have a combination of different types of corners, from very tight low-speed hairpins to open high-speed sweepers.

The simplest way of assessing changes in performance is to measure lap times. Lap times can be kept by an assistant using a stopwatch. To be as accurate as possible, the assistant needs to line up an object (for example, a sign or pole) on the other side of the section of track directly in front of him or her. The stopwatch is started when the front of the car breaks an imaginary line between the assistant and the object and then is stopped when the front of the car again passes through the imaginary line.

Another approach is to have the driver note the speed with which the car is travelling on different parts of the circuit. The car can then be driven relatively gently until this section is entered. This helps to remove one of the problems of driving road cars on a race circuit — brakes. Even high-performance road cars will start to lose braking effectiveness after only a few laps of a race circuit, a

The lane-change test marked out by the cones behind these two cars is an excellent 'real world' test that can be carried out safely at high speed on a track.

Circuit laps are lots of fun but often don't prove much in a road car. Tyre and brake wear can be phenomenal!

process that will inevitably slow lap times. If you're testing the performance changes of a new swaybar or springs and take a few laps to get into a groove, the brakes might start going 'off' before you can even start to explore the performance gains. And, of course, any really major acceleration or cornering benefits will result in the brakes having to work even harder than before! Of course, if the test is of brakes, all this will be to the good.

There is another very important point to keep in mind. A road car whose suspension is optimised for the track will generally **lose** handling performance on the road. This is because the conditions are really quite different. Racetracks are smooth and little suspension travel is needed. On a racetrack you know where the corners go, so the suspension can be set up to be much less forgiving of a mistake. You can enter even the tightest corner on the track relatively quickly because there are no surprises in store. The real benefit of track testing is that there are run-off areas and no traffic ever comes the other way! This means testing the handling on the edge of control is much safer in a track environment than it can ever be on the road.

Tyres

It can be quite useful to monitor the behaviour of the tyres during handling tests. First, the temperature of the tread of each of the four tyres will vary, with the hottest tyre being the one working the hardest. Each tyre can have its temperature measured in three places — in the middle of the tread, towards the inner edge and towards the outer edge. This will show if any part of the tyre is experiencing an unfair share of the punishment. For example, a car with a lot of bodyroll when understeering through a righthand corner will have an extremely hot outside edge to the lefthand front tyre. Inappropriate wheel alignment specs will also show through tyre temperature testing.

Tyre temperatures are most easily measured with a probe-type thermocouple and dedicated meter. While these are available from motorsport shops, it's generally much cheaper to buy from an electronics shop either a dedicated thermocouple meter or a multimeter with a thermocouple input. A sharp thermocouple probe should be pushed into the tyre tread a millimetre or two when tyre temps are being measured.

Tyres should also be inspected very closely for wear during handling testing. If the handling is poorly balanced, tyre wear will very quickly be visible. An hour or so of track testing can be enough to completely wear out a brand-new tyre! The same can also occur on a skidpan. But even when the conditions are less extreme, the pattern of tyre wear will show some of what's going on. An overworked tyre can develop a wrinkled tread appearance, lift tread blocks, develop a saw-tooth pattern in the surface of the tread, or even have wear marks on the sidewalls rather than on the tread. During the slalom test mentioned above, the front tyres lost sections of tread 10-15cm long from their outside shoulders!

Tyre wear should always be monitored closely during any handling tests. Even when not as blatant as this, wear patterns will show which tyres (and which parts of the tyres) are doing the most work.

Braking Tests

As indicated above, you can test the brakes very much harder on a racetrack than is possible on the road. One of the reasons for this is that at the track there are run-off areas to be taken advantage of if the brakes fade all the way to nothing. On the road, the brakes starting to go mushy is a very strong pointer to slow down! Serious testing of brakes at a track will involve measuring brakepad thicknesses at regular intervals of 10 or 15 laps. Heat-sensitive paint can also be applied to the outside edge of the disc to show how hot the discs are getting. However, unless you are comparing the performance of

Serious brake testing will require the replacement of disc rotors as well as pads at frequent intervals. This type of testing is best carried out on the track.

different brakepads, a test of this magnitude is generally not done on a road car. Testing a number of sets of pads in this way would also normally require that the discs then be replaced as well!

But doing just three or four quick racetrack laps is a very effective test for the brakes in a road car. Brake ducts, new pads or the effectiveness of grooved or drilled discs can be quickly tested in this manner. Criteria to examine include how late you can commence braking, the initial bite of the pedal and fade.

Deceleration brake testing involves carrying out repeated stops from a fixed speed. An electronic or mechanical accelerometer is used to measure the deceleration, with the electronic device also capable of reading off stopping distances and/or times. A brake pedal force transducer is sometimes sandwiched between the driver's foot and the brake pedal.

The extent to which brakepads can be tortured is easy to see in this photo of a pad after track testing.

An example of some results of 100-0km/h brake testing can be seen in Figure 11.8. At the same time, 160-0km/h tests were carried out, along with laps of a race track. However, this type of result needs to be treated with great care. While on paper Set #5 appears to have performed without distinction, they were in fact given the top rating by the driver because in track testing they had minimal fade and allowed much later braking.

Heat-sensitive paint that changes colour when subjected to predetermined temperatures makes monitoring disc temperatures straightforward.

Fuel Consumption Tests

The easiest way of testing fuel consumption is to fill the petrol tank, zero the trip meter (or record the odometer reading) and then drive the car. When you need fuel, again fill the tank to the brim and record how much fuel was used and how far you have travelled. For example, if you use 48.6 litres to travel 402km, the rate of fuel consumption is 48.6 litres divided by 4.02 hundred kilometres = 12.089 litres/100km or, in round figures, 12.1 l/100km. If you wish to convert litres/100km to miles per Imperial gallon (mpg), divide 282.48 by the litres/100km figure. To convert litres/100km to miles per US gallon, divide 235.21 by the litres/100km figure.

In day-to-day operation it's often easiest to zero the trip meter each time you fully fill the tank. The distance you travel before the gauge shows empty is an easy figure to remember, giving you a frequent indication of the sort of fuel economy you're getting. A sudden increase in fuel consumption could point to a blocked airfilter, blocked catalytic converter or dragging brake.

Greg Brindley

Pads	Maximum Pedal Force (N)	Deceleration (g)	Stopping Time (s)	Peak Average
Set #1	272	0.96	0.74	3.3
Set #2	267	0.92	0.76	2.9
Set #3	285	1.01	0.74	2.9
Set #4	171	0.91	0.75	3.1
Set #5	223	0.93	0.66	3.3

Figure 11.8
Deceleration brake testing involves carrying out repeated stops from a fixed speed. An accelerometer is used together with a brake pedal force transducer. However, this type of result needs to be treated with great care.

Emissions testing uses a specified drive cycle and is carried out on a special chassis dynamometer. All the tailpipe output is collected and a proportion of it sampled.

In those cars equipped with trip computers, watching the instantaneous fuel consumption readout can be surprisingly informative. For example, the display quickly shows that selecting too high a gear for the driving situation can increase fuel consumption.

While some people who enjoy performance cars care little about fuel economy, I believe it's a very important parameter of car performance. Not only does a car with poor fuel consumption cost more to run, it also produces more emissions. Well-modified cars should have good fuel consumption if the engine management mapping is carried out well and the volumetric efficiency of the engine is high.

Emissions Tests

Since emissions tests are made to fulfil legal requirements, the type of test that's regarded as acceptable will depend on what your local lawmakers have to say. If a complete and detailed emissions compliance test is required, it will have to be carried out at a major emissions testing laboratory.

A typical emissions test (carried out in a lab both temperature and humidity controlled) is described here. Before being tested, the car is filled with tightly specified 96 RON fuel and the carbon canister in the car's emission system is filled with a mixture of butane in nitrogen. The car is then placed on a single-roller 100kW chassis dynamometer that's software-driven to provide the correct variation in loads for the conditions being simulated. A constant-speed fan placed at the front of the vehicle provides engine cooling during the test.

During the test cycle, a driver sits in the car watching a display on a large monitor. On the monitor is a moving trace that indicates the speed the car must be 'driven' at, with a cursor showing the actual vehicle speed. The driver must try to keep the cursor on top of the specified speed line at all times. This means the car's speed follows the appropriate cycle, with the driver applying the amount of throttle needed to maintain the required speeds. Importantly, full throttle is not used at any point in most drive cycle emissions tests.

All exhaust emissions from the tailpipe of the car are collected through a large hose that leads to special separators and venturis. Horiba analysis equipment is used to measure oxides of nitrogen, hydrocarbons and carbon monoxide. In some tests, methane concentrations are also recorded. In addition to tailpipe emissions, a SHED test is also undertaken. The SHED (Sealed Housing for Evaporative Determination) test is carried out both before and after the drive cycle test. An external pad heats the fuel within the car's tank and the hydrocarbon emissions from the whole car are then measured.

The engineers working in emissions testing facilities talk about the fine line they tread between emissions and fuel economy. Contrary to popular belief, full power rarely suffers due to emissions legislation because the car does not ever get measured at full power during the test! However, advanced ignition timing at low loads (which is good for fuel economy) can lead to excessive production of oxides of nitrogen, to give just one example of the compromises that must be reached.

If less than a full legal test is required, it may be able to be completed in a workshop equipped with a chassis dynamometer and five-gas analysis equipment. Specified by the authorities will be the loads, engine and road speeds, and the permitted maximum emission outputs. I expect this area of emissions testing to tighten up considerably in years to come — today's relaxed approach is soon likely to disappear. In some circumstances where an engine conversion is taking place, the legal authorities might simply specify that the emissions of the new engine must be at least as good as the engine being replaced. Remember, this means some testing must be carried out before the old engine is removed from the car!

Good emissions are the result of very tight engine parameter control, good programming of the engine management system, an effective catalytic converter and a well-maintained car. None of these factors spells doom for performance-modified cars. However, cheap and cheerful tricks like fooling the coolant temperature sensor so the ECU supplies more fuel, and fitting extra injectors switched on by manifold pressure, are very unlikely to satisfy legal emissions requirements.

DYNAMOMETERS

Dynamometers are instruments that measure torque and speed. From those two factors, power is calculated. Usually abbreviated to dynos, they're available in two different categories: engine and chassis dynos. An engine dyno measures power at the crankshaft and requires that the engine be removed from the car and mounted on the dyno itself. A chassis dyno has the car driven onto it and measures power at the tyres.

I stated above that dynos measure speed and torque — not power. So what's torque? Torque is the measurement of how much twisting force is being applied to a shaft. When you tighten a nut, you apply a torque to it via a lever called a spanner. Imagine you're tightening a wheelnut using a very large spanner that's one foot long. You place the spanner on the nut so it's parallel to the ground and then hang a one-pound weight from the end of it. You are then applying a torque of one foot-pound (abbreviated to 1ft-lb) to the nut. In the metric system you'd need to use a one-metre-long spanner and hang a 102-gram weight on it to apply a torque of one Newton metre (1Nm).

A dyno measures torque using a brake and a lever. The brake tries to slow down the engine, while the engine at the same time tries to spin the brake around. A lever attached to the brake resists this rotation, pushing (or pulling) on a special sensor. With the length of the lever

One of the engine dyno cells at a Mitsubishi engine manufacturing plant. Manufacturers use engine dynos almost exclusively for the development and durability testing of engines.

known and the force on the sensor able to be easily measured, the engine's torque output can be calculated. Engine speed is measured electronically. With speed and torque available, the power output can be calculated.

In the Imperial system:

$$\textbf{Horsepower} = \frac{\textbf{Ft-lb x rpm}}{\textbf{5252}}$$

In the metric system:

$$\textbf{Kilowatts} = \frac{\textbf{Nm x rpm}}{\textbf{9550}}$$

The data from engine dynos is couched in terms of power, torque and engine speed, but chassis dynos show power and tractive effort. Tractive effort is the force pushing back the road at the tread of the driving tyres. It can be considered as engine torque multiplied by the ratios of the gearbox and final drive and divided by tyre radius. The shape of the tractive effort curve will reflect the engine torque curve, but with much higher values. Tractive effort is measured in pounds or newtons.

The amount of power developed by an engine will vary depending on the day's weather. Changes in temperature, relative humidity and barometric pressure will all cause variations in the power developed. Some dynos have sensors to measure these parameters, with in-built correction software taking them into account. The most common correction is for intake air temperature, which is normally measured with a probe on the end of a long cable. While dynos are frequently used as test and measurement instruments, most of their upsurge in popularity in recent times is because a dyno makes the mapping of programmable engine management systems much easier to carry out. This topic is covered in more detail in Chapter 4.

Dyno Brakes

A dyno is equipped with a brake that absorbs the power being developed by the engine. Many high-performance engines have a very large amount of power, so if it isn't to melt, the dyno brake must be equipped with a very effective cooling system.

Water brakes are fitted to many engine dynos and also some chassis dynos. This type of dyno head uses a rotor that spins within a housing through which water is pumped. The rotor is equipped with blades that fling the water outward against cups cast into the outer housing. The power is absorbed by the water, which also acts to cool the dyno head. This water is then reduced in temperature by being circulated through either a cooling tower or a very large tank of water. Pumping differing volumes of water through the dyno brake controls the amount of load applied. Good dyno control software and a fast-response water valve are required to gain accurate control, especially over small loads. A water brake dyno can hold a load continuously for long periods.

A water brake dyno head. This type of brake can absorb a continuous load with the heat both absorbed and dissipated by the circulating water.

An Eddy current dyno head uses the magnetic field generated by current-carrying coils to slow cast-iron discs whizzing past. This engine dyno sprays cooling water onto the discs.

Another common dyno brake is the eddy current type. Widely used in chassis and engine dynos, this design uses two or more iron discs that rotate in close proximity to electrical coils. Electric current is fed through the coils, which has the effect of magnetically retarding the discs. This type of dyno can be very finely controlled and can also hold a continuous load. Cooling is by either air or water circulation.

The final type of common dyno brake is the inertial design. This uses the idea that a heavy flywheel requires the application of power to increase its rotational speed. The dyno control software is programmed with the flywheel's moment of inertia and it monitors the acceleration and speed of the flywheel. From this, the power being applied to the flywheel can be measured. Inertial dynos cannot hold a continuous load, so do not require cooling systems.

Engine Dynos

Because of the need to remove the engine from the car before it can be attached to the dyno, engine dynos are used only for major testing and development. The engine is bolted to the dyno head using a suitable gearbox bellhousing; new engine mounts are also made to hold the engine in place. The fuel system is plumbed, the engine management ECU is wired and the throttle cable connected. An uncommon engine may well take a whole day to be set up on an engine dyno.

In addition to being able to measure power and torque outputs very accurately, an engine dyno workshop is normally equipped with a very wide range of instrumentation. This means the behaviour of the engine at all loads

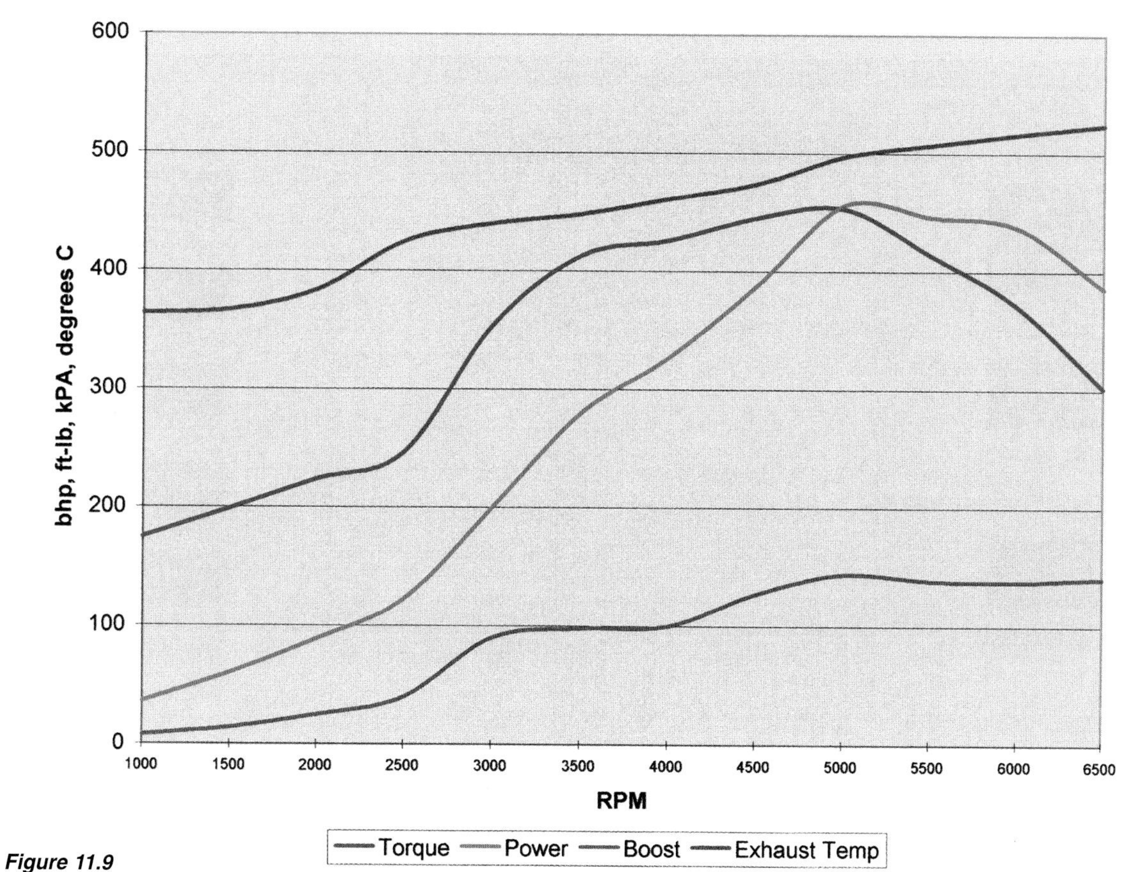

Figure 11.9
The data from an engine dyno run of a turbo 3-litre Nissan RB30ET six-cylinder engine shows the sudden lift in the torque curve when boost arrives. The continuous increase in exhaust gas temperature during the power run can also be seen. Engine dynos allow the logging of many engine parameters.

This view inside a water brake dyno head shows the cups and vanes the water impacts against.

can be accurately monitored — something impossible on a chassis dyno or in normal driving. Engine dyno workshops can measure all or some of these parameters:

- Air/fuel ratio
- Severity of detonation
- Individual cylinder exhaust temperatures
- Manifold pressure
- Exhaust back-pressure
- Oil pressure
- Oil temperature
- Intake air temperature
- Intake airflow
- And, of course, power, torque and engine speed

Most engine dynos have their own permanently mounted fuel supply systems. This one uses a plastic fuel cell and two Bosch pumps.

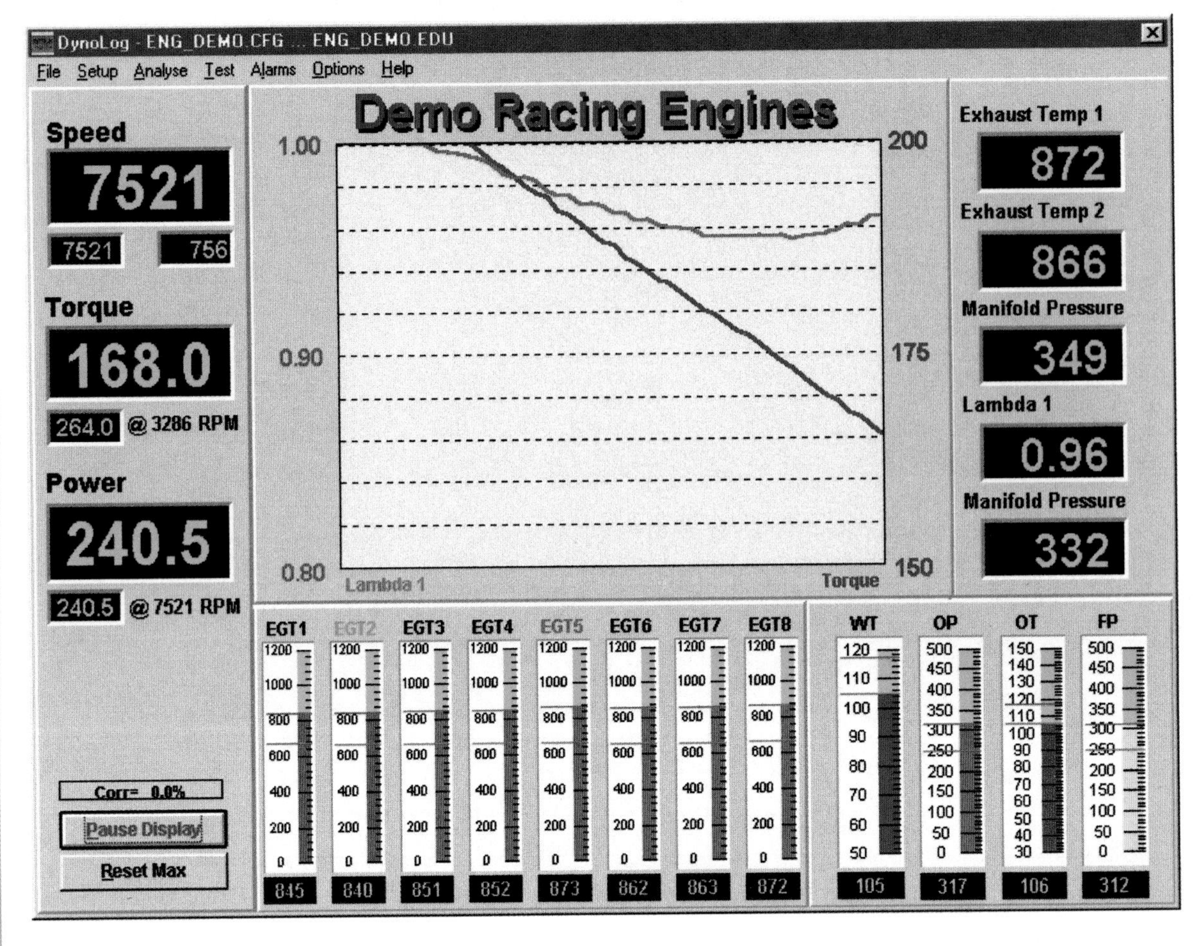

Figure 11.10
This Dynolog display is an example of the sophisticated PC control and logging software that can be used with an engine dyno.

When selecting an engine dyno workshop, investigate which engine parameters can be measured. These should also be able to be data-logged, allowing the careful analysis of the data after the power run has finished. At a minimum, the workshop should be able to provide (for every 500rpm of engine speed) printouts of power and torque; oil, inlet air and coolant temps; exhaust gas temperature(s); and manifold pressure.

An oxygen sensor being used to monitor the air/fuel ratio while the engine is dyno'd. Adjacent is a K-Type thermocouple that senses exhaust gas temperature.

The engine air/air intercooler has been submerged in a large container of water to cool the induction air while the engine is working on the engine dyno.

Figure 11.9 (previous page) shows an example of some of the data that can be gathered during an engine dyno test. In this case, the engine was a modified Nissan RB30ET 3-litre turbo six. Figure 11.10 shows an example of an engine dyno control system's PC screen display.

If you have an intercooled forced induction engine, investigate the intake air cooling technique used by the workshop. In some dyno shops, your engine's intercooler is used, perhaps submerged in a tub of water if it's an

An engine dyno control room. The engine and dynamometer are in the next room, behind the double-glazing. Very extensive instrumentation allows the monitoring of many engine parameters.

air/air intercooler. Other workshops have their own water/air system within the dyno room. This is quite an important issue if the engine management is being mapped because the ignition timing used will be strongly affected by the intake air temperature. The intake air temperature on a forced-induction engine will also have a large bearing on the developed power and torque.

Because engine dynos allow the engine's performance to be very carefully analysed, all intensive engine development uses engine dynos. For example, car manufacturers use engine dynos almost exclusively. In addition to measuring power and torque outputs, the manufacturers run engines for very long periods to test for reliability and wear.

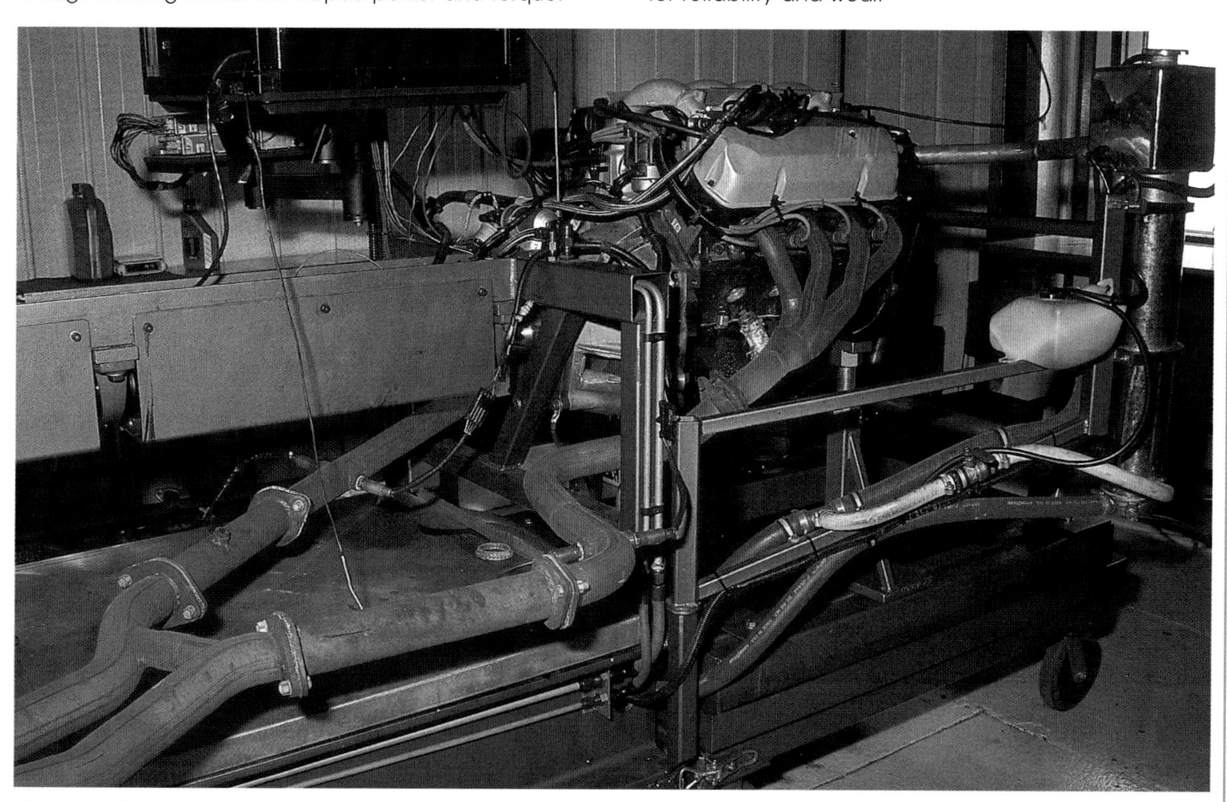

Automatically controlled durability testing of engines can be carried out on engine dynos. This V8 is being put through its paces at a race team's headquarters.

A Nissan SR20DET turbo four being mapped on an eddy current engine dyno. Note the water/air intercooler being used, the huge dyno exhaust and the airflow meter that's monitoring air mass consumption.

An example of a dyno test program is that undertaken by a major car manufacturer. Their standard engine durability test is an eye-opener — and one that very few modified motors could pass! An engine is removed from the assembly line and run in for the first four hours. This consists of gradually increasing the load in 30-minute increments until at the end of the four hours the engine is working with full load at 5000rpm. Following this are three hours of running at full power. The engine is then checked before it operates continuously at full throttle and full load for more than **eight days**! Then for a further eight days it is cycled between idle, 4500rpm and 5500rpm while under load.

A gas analyser being used in conjunction with a diesel engine chassis dyno. Five-gas analysis (oxygen, carbon monoxide, hydrocarbons, oxides of nitrogen and carbon dioxide) is often used.

Other tests carried out include one 53-hour torture test where the coolant is held at 123 degrees C and the oil temperature is an incredible 155 degrees C! To test for cylinder head sealing, another test involves the engine being run up straight from dead-cold to 5500rpm at full throttle for two minutes, after which it is shut down and flushed through with coolant at 20 degrees C for another two minutes. Immediately following that, it's straight up to full load at 5500rpm again. The test follows this sequence for 2500 cycles!

Chassis Dynos

A chassis dyno is much simpler to use than an engine dyno. A chassis dyno is equipped with large rollers, usually sunk into a pit. The car is driven onto the dyno so the drive wheels rest on the rollers. On a four-wheel-drive dyno, the distance between the front and rear sets of rollers can be changed to cater for different-wheelbase cars. The car is strapped down so it cannot move laterally or longitudinally, with the non-driving wheels chocked on two-wheel-drive dynos. Front-wheel-drive cars tend to be more unstable on the dyno than RWD cars, so when FWD cars are tested, small 'trainer wheels' are locked into place either side of the front wheels. A large fan is placed at the front of the car to cool the radiator and any other heat exchangers located there. The dyno operator sits in the driver's seat and operates the

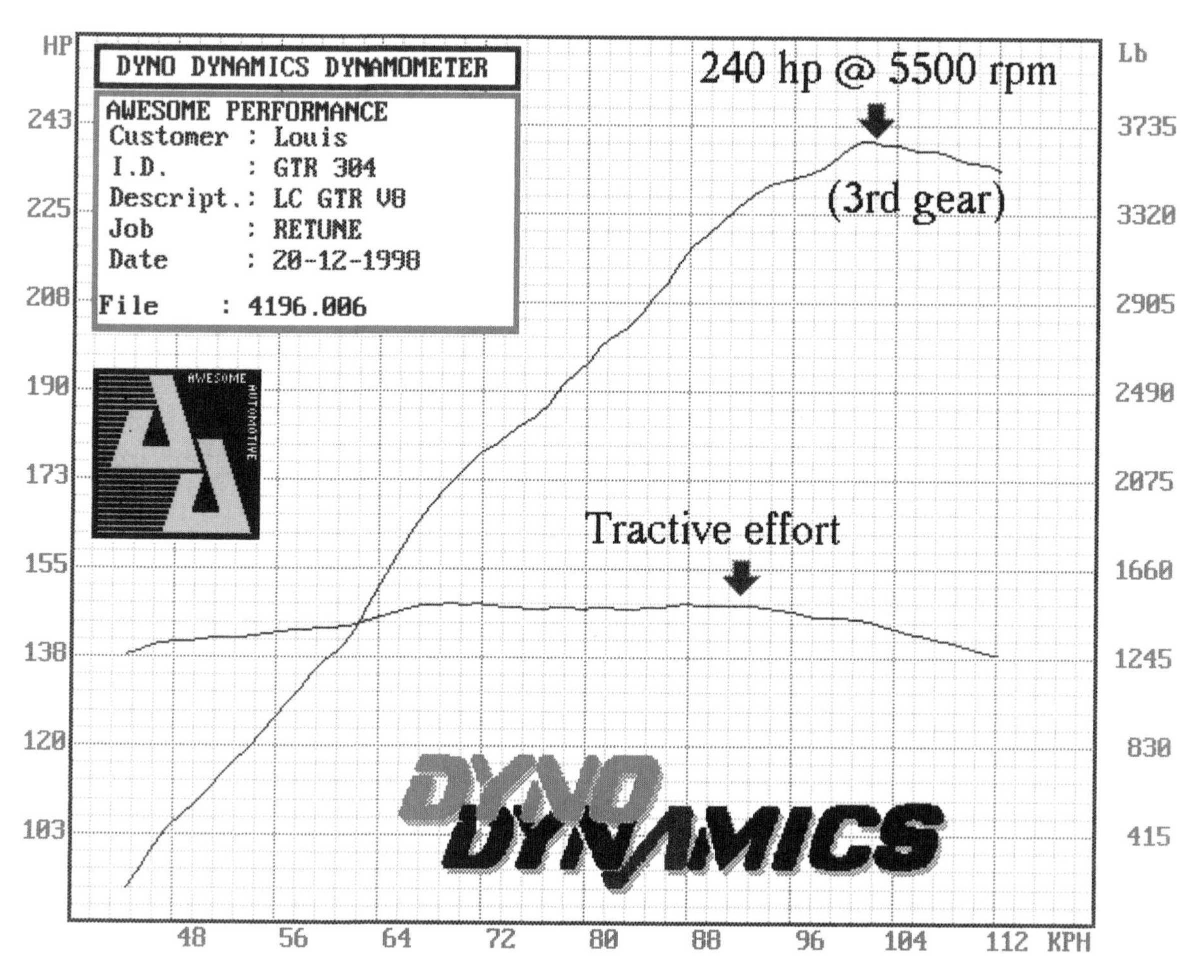

Figure 11.11
A ramped power pull carried out on a Dyno Dynamics chassis dyno shows both the developed power and tractive effort curves. This power run was completed in third gear.

accelerator, clutch and brake. The steering wheel does not need to be touched. The dyno is controlled via a hand-held pendant that contains controls for load, speed and other factors.

The simplest test is a power pull. The 'ramp speed' (rate at which the dyno lets the engine speed increase) is set and the car is run at very low rpm in the selected gear. Any gear can be used, but second, third or fourth gears are usually selected. The higher the gear, the lower the tractive effort trying to pull the car off the dyno. Very powerful cars are therefore sometimes tested in the higher gears, although second gear is most commonly used. The driver instructs the dyno as to the type of test being undertaken, holds the low rpm in the selected gear and then gives the engine maximum throttle. Flicking a switch allows the dyno to ramp up in speed. As the engine rpm increases, the dyno software draws a trace on a PC screen, showing the power being developed. When the engine redline is reached, the test may be completed, or a switch on the controller may be flicked, causing the dyno to then ramp the engine back down. Averaging ramp-up and ramp-down power pulls removes from the reading the inertia of the dyno rollers. Figure 11.11 shows a typical ramped power pull which was conducted on a 5-litre V8.

Another dyno approach is to 'drive' the car through the gears. The car is driven along at very low speed in first gear, the throttle floored and a switch on the dyno control pendant moved. The car accelerates smoothly in first gear until the redline is reached. The gear change occurs and the car then accelerates in second gear. The process continues through each gear until the rollers are turning very fast indeed. Power curves for each gear can be drawn in this way over a speed range of typically 20-180km/h. This type of test shows transient responses, such as how quickly turbo boost rises after each gear change. Note the car has to be tied down very well, especially if a car with a manual transmission is being used! Figure 11.12 (overleaf) shows a power run of this type, conducted with the engine in the automatic transmission car standard, and then with part of the engine airfilter blocked off.

As with any test or measuring device, a dyno can be quite **inaccurate** if used incorrectly. Certainly, modifications resulting in very small changes in power (1-2 per cent) are almost impossible to accurately measure. This is because other factors affecting power also change during the dyno test. The engine oil changes temperature, as does the final drive and gearbox fluids. The inlet manifold becomes hotter, as does the block and cylinder head. Likewise, the tyres change in their behaviour as well! Even if an inlet air temperature correction probe is used, it's impossible to cancel out all other factors that cause variations in power.

The best way of using a chassis dyno to find power changes is to follow this procedure. First, warm the car

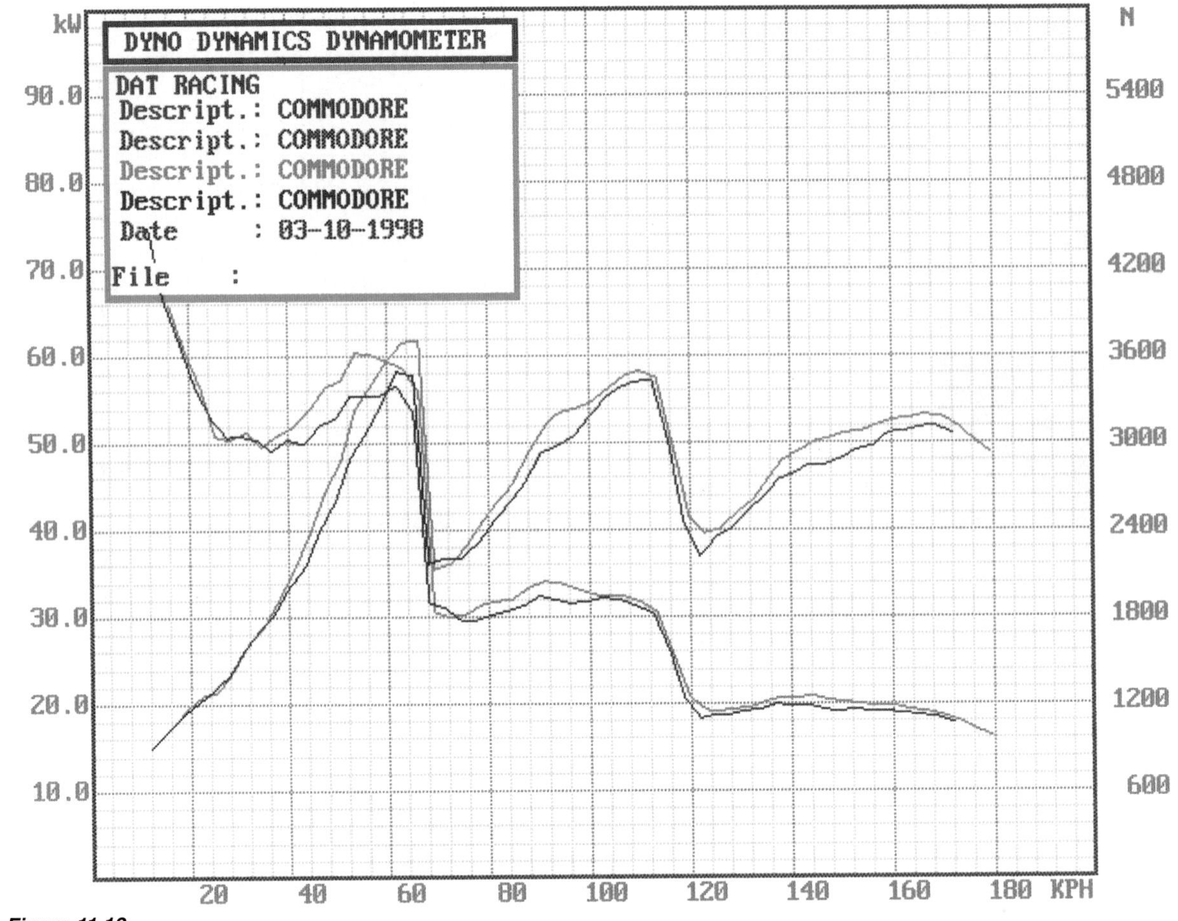

Figure 11.12
A through-the-gears power run completed on a chassis dyno. The green lines show the power and tractive effort outputs in standard form and the pink lines show the outputs with the filter partially blocked. Tractive effort decreases with each upward gear change, while power peaks at high revs in each gear.

to operating temperature. Note the reading of any temperature gauges (coolant, engine oil, and transmission oil) the car is equipped with and monitor these during the power runs that follow. If they change, either let the car cool down again or re-start the testing at the higher temperatures. If an air temperature correction probe is available, it should be mounted in the intake duct to the engine. A ramp-up and ramp-down power pull should then be completed and averaged. Another run should be done, followed by a third. By the third averaged run, the power curve should have stabilised and a fourth should produce a result very close to the third run. If the power continues to change (for example, dropping each time), the car will be very difficult to test on a chassis dyno. This can happen, for example, with turbocharged cars using small intercoolers.

Even when doing simple 'before' and 'after' testing of a modification, a high-speed air/fuel ratio meter should always be used to indicate mixtures. In some cars, even a new exhaust is sufficient to cause a lean mixture that's both potentially dangerous to the engine and also may dramatically affect the power being developed. In this situation, completing a dyno test of a new exhaust could well show a power decrease, the loss being caused by a lean air/fuel ratio rather than the exhaust itself.

To highlight how a dyno can easily produce variations in measured power, a stock six-cylinder family car was placed on a chassis dyno. The bonnet was open and power runs were averages of ramp-up and ramp-down runs. The first power pull resulted in a peak power of 98kW. Next, the temperature correction probe was removed from the car's airbox, lifting the measured power by a couple of kilowatts and giving a noticeable 'improvement' through much of the rev range. The dyno ramp speed was then doubled, which gave a major power 'gain' at both ends of the rev range. Other tests included shutting the bonnet and altering the dyno correction factors. Six quite different dyno graphs were generated in the one session — with the car mechanically untouched! It's for this reason that dubious performance products that have their power gains supported by 'before' and 'after' dyno graphs should still be mistrusted.

When using a chassis dyno it's important to realise that the gearbox, final drive and even tyres are not perfect transmitters of torque. As a result, some of the power of the engine is absorbed by the drivetrain. When measuring power at the wheels it's therefore normal to assume that the power being developed at the crankshaft is, in fact, higher. However, the exact proportion of power lost in the drivetrain is a contentious issue.

One way of assessing the power lost in the drivetrain is to test brand-new cars on a chassis dyno. Doing this allows comparison of the manufacturers' engine dyno power output figures with those achieved at the wheels. An example of such a test was carried out on a five-speed manual Ford V8. The claimed power for the engine in this

In this Eddy current chassis dyno, the torque-sensing long arm and load-sensing cell can be seen in the foreground. Dynos measure torque and speed, and from this calculate power.

RWD car was 185kW (248hp), while a peak figure of 143kW (192hp) was recorded on the chassis dyno. This equates to a drivetrain loss of 42kW (56hp), or 23 per cent. Another 5-litre V8 car, from a different manufacturer, was tested at the same time. This RWD car had a claimed power output of 195kW (261hp) and recorded 154kW (206hp) at the wheels — a 41kW (55hp) or 21 per cent loss.

However, in separate testing, a standard FWD 1.3-litre four with a claimed 55kW (74hp) at the flywheel had a measured power of 52kW (70hp) at the wheels — a power loss of just 3kW (4hp), or 5 per cent! And to make it even more confusing, any constant four-wheel-drive car tested on a chassis dyno appears to have a huge driveline loss. For example, a standard Subaru Impreza WRX with 155kW (208hp) at the flywheel has a measured power of only 95kW (127hp) at the wheels. That's a massive loss of 60kW (80hp), or 39 per cent! And the Subaru seems to go awfully well (it has a standing quarter in the mid 13s) for a 1250kg car with only 95kW at the wheels...

Another test carried out was where a highly modified Nissan FJ20DET engine was mapped on an engine dyno and later tested on a chassis dyno. The engine was installed in a RWD Nissan Gazelle, a car that uses an independent rear suspension design that makes it a little harder to avoid wheelspin on a chassis dyno. On the engine dyno, a peak power of 358kW (480hp) was developed, while the chassis dyno showed a peak of 235kW (315hp) at the wheels. This is a 123kW (165hp), or 34 per cent driveline loss, but the car did have some wheelspin on the chassis dyno.

Some dynos use coast-down tests that claim to be able to calculate driveline losses. However, by definition, the coast-down occurs with no torque transmission occurring — and the losses experienced in a transmission will depend at least in part on the loads it is undergoing with high torque inputs. In short, it's impossible to **accurately** estimate engine power from that measured at the wheels — a precise engine power figure can be gained only by using an engine dyno. However, a chassis dyno comes into its own when quick measurements of changes in engine performance need to be made.

While rear-wheel-drive cars tend to be very stable on a chassis dyno, front-wheel-drive cars require the use of 'trainer wheels' to prevent the cars moving laterally.

A four-wheel-drive dyno allows the testing of constant four-wheel-drive cars. The dyno is able to adjust the torque being transmitted through the front and rear axles by varying the load placed on each.

Tyres, Suspension & Brakes

Audi

Everyone at some stage in his or her driving life has misjudged the entry speed to a corner. Whether the car then goes on to hold the line through the corner or leaves the road backwards depends on both the driver's skill and the car's handling. A car with good handling is not only rewarding and fun to drive, it's also safer.

Don't confuse handling with grip. Many modified cars grip very well because they have been equipped with wide, sticky, low-profile tyres. However, handling becomes most relevant when the car **ceases** to grip. Does the car grip and grip — and then slide so suddenly you cannot catch it? Or does the car signal its intentions early, allowing you to feel you're running out of adhesion and giving you time to prepare for the slide? If the car behaves in a predictable manner it will always be safer and more fun to drive quickly.

This is also a good argument for uprating the suspension **before** you outlay a large amount of money on the tyres. If the rubber on the car is relatively slippery, you will be able to modify the car's suspension while still being able to test the car safely. For example, if the car understeers (the front of the car runs wide when cornering hard), you will be able to assess the effect of changing a swaybar. If you have very sticky tyres fitted, reaching the understeer limit may occur only when you are going so fast that you understeer off the road before you can react...

TYRES

The only part of the car touching the ground is the tyres. This means all cornering, braking and acceleration forces must be transmitted to the ground through just these four rubber footprints. This fact puts the tyres among the most important components in the whole car; for a performance driver, the tyres are an absolutely vital element in making the car fast and safe.

Sidewall Information

While to the casual observer it might appear that all tyres are the same, tyres vary substantially. The writing on the wall of a tyre describes in detail its size, date of manufacture, speed and temperature rating, and sometimes its rubber hardness. It's important that all this information is understood if the best tyre selection is to be made.

The following is an example of the information found on the sidewall of one tyre:

Kumho Ecsta 711

This indicates the make, model and tread pattern of this tyre. Just as there are good and bad makes and models of cars, there are tyres that are good and tyres that (for a performance driver) are bad. Looking at the make and model of tyres fitted as standard to good performance cars can be quite an education!

225/50R16 92V

This is one of the most important parts of the information shown on the sidewall. It refers to the size, and the load and speed rating of the tyre. In this case it shows that the width of the tyre is 225mm from one sidewall to the other; the height of the sidewall is 50 per cent of the tyre's width; the tyre fits onto a 16-inch rim; it has a load rating of 92 (630kg), and it has a V (240+km/h) speed rating. Figure 12.1 (opposite) shows all the available tyre speed ratings. 'R' in this context means it's a radial tyre.

Rotation (→)

Most performance tyres are directional, meaning that when they're fitted to the car they should rotate in only one direction when the car is moving forwards. This prohibits the swapping from side to side of wheels fitted with directional tyres — something people have been known to do while rotating the tyres! Fitting the tyre so it turns in the wrong direction will increase noise and decrease handling performance.

Speed Rating	N	P	Q	R	S	T	U	H	V	W	Y	ZR
Speed (km/h)	140	150	160	170	180	190	200	210	240	270	300	240+

Figure 12.1
The speed ratings commonly used on road car tyres. 'ZR' is being replaced with the more specific ratings of 'W' and 'Y'.

Maximum load 630kg, 300kPa (44psi) maximum pressure

The maximum static load the tyre should be subjected to is shown by the load figure, which can be expressed either as a load rating or in kilograms. It's possible to get a tyre correct in every specification except load rating, so don't forget to check this when buying new tyres. Figure 12.2 shows the available load ratings. The maximum pressure refers to the maximum inflation pressure when the tyre is cold.

Tread steel 2 + Rayon 2 + Nylon 2, Sidewall Rayon 2 ply

This information refers to the internal construction of the tyre. Figure 12.3 (overleaf) shows the internal construction of one high-performance tyre.

Treadwear 320, Traction AA, Temperature A

This important triumvirate — known as the UTQG data — is not marked on all tyres, but it's certainly worth checking out if it is. It applies to summer car tyres with rim diameters of 13 inches and larger, and is usually marked only on tyres that have the US as one of their markets. The treadwear number is a comparative rating based on the wear rate of the tyre during field testing carried out by the manufacturer. The higher the number, the longer the wear life of the tyre — a tyre with a marking of 160 gives 60 per cent more mileage than a tyre marked with 100. The traction rating refers to wet braking performance in a straight line. It's tested with locked wheels on wet concrete and wet asphalt. The grades AA, A, B or C are used to indicate this, with AA the highest. The temperature resistance of the tyre is graded A, B or C, with C the minimum required to meet US standards.

Load Index	kg	Load Index	kg	Load Index	kg	Load Index	kg	Load Index	Kg	Load Index	kg
0	45	20	80	40	140	60	250	80	450	100	800
1	46.2	21	82.5	41	145	61	257	81	462	101	825
2	47.5	22	85	42	150	62	265	82	475	102	850
3	48.7	23	87.5	43	155	63	272	83	487	103	875
4	50	24	90	44	160	64	280	84	500	104	900
5	51.5	25	92.5	45	165	65	290	85	515	105	925
6	53	26	95	46	170	66	300	86	530	106	950
7	54.5	27	97.5	47	175	67	307	87	545	107	975
8	56	28	100	48	180	68	315	88	560	108	1000
9	58	29	103	49	185	69	325	89	580	109	1030
10	60	30	106	50	190	70	335	90	600	110	1060
11	61.5	31	109	51	195	71	345	91	615	111	1090
12	63	32	112	52	200	72	355	92	630	112	1120
13	65	33	115	53	206	73	365	93	650	113	1150
14	67	34	118	54	212	74	375	94	670	114	1180
15	69	35	121	55	218	75	387	95	690	115	1215
16	71	36	124	56	224	76	400	96	710	116	1250
17	73	37	127	57	230	77	412	97	730	117	1285
18	75	38	130	58	236	78	425	98	750	118	1320
19	77.5	39	133	59	243	79	437	99	775	119	1360

Figure 12.2
The Load Indices of tyres and their equivalent mass ratings.

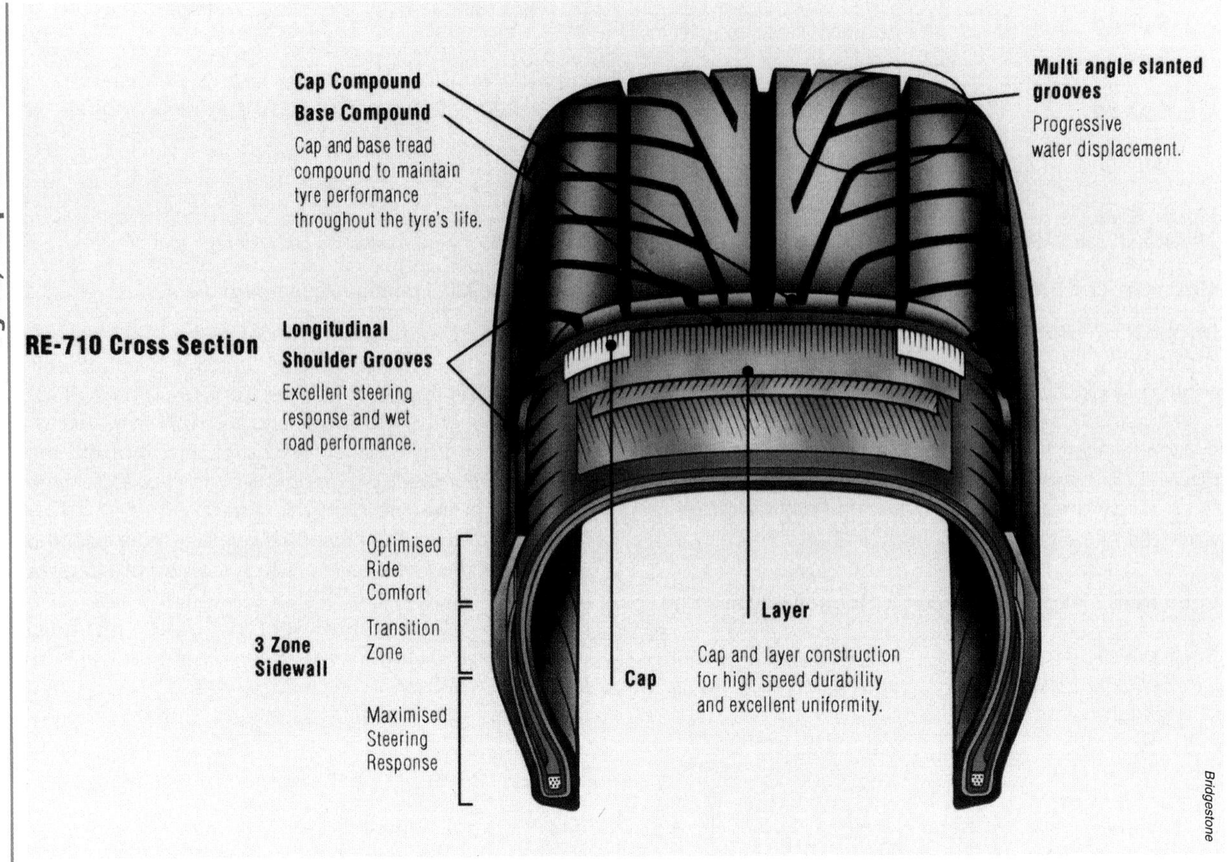

Figure 12.3
The internal construction of the Bridgestone RE-710.

DOT YOF8 YA3F 408

DOT shows that tyre meets or exceeds US Department of Transportation safety standards, YOF8 indicates the manufacturer and plant code number, YA3F is the tyre size code number and 408 is the date of the tyre's manufacture, where 40 indicates Week 40 (ie in August) and 8 refers to 1998. The latter information is important if you want to make sure the tyres you are buying are 'fresh' — tyres deteriorate with age, especially if stored in hot environments.

In addition to this sidewall information, there are often warnings about the need to have the tyre correctly fitted and not to overload or under- or over-inflate the tyre.

The width, profile, speed rating, rubber compound and other factors vary from tyre to tyre. Making a good tyre choice is crucial to gaining good handling.

Selecting Tyres

Good-quality tyres are very expensive. If four new tyres are being bought (perhaps together with new wheels on which to fit them), the financial outlay can be very large indeed. The scary part of this is you can't be sure of exactly what you're getting until you have paid for the tyres and are driving around on them.

Every person who owns a car is a tyre expert: you need only ask someone their opinion of the tyres fitted to their car!

"Fantastic," they might say. "They grip incredibly well. You can't go wrong with these."

Or, "No, these are terrible — I only got 15,000 kilometres out of them!"

Neither response is relevant unless you know the person's driving style, the type of car they drive and what they actually want from a tyre.

However, one reason you might be asking for opinions is that it's simply impossible to select a tyre by inspection or by reading a sales brochure. Even the **worst** tyre produced by a company will still be glowingly praised in the brochure! "New high grip compound ensures exceptional car control. Advanced computer-designed tread with numerous lateral sipes ensures excellent traction." Does that mean it's the company's best tyre or worst tyre? Who knows!

Unfortunately, most tyre dealers have very little knowledge of the tyres they sell. This applies especially to high-performance tyres. No tyre dealer in my experience buys

*A tyre that is **just** road legal but is designed primarily for circuits will provide extraordinary levels of adhesion on dry roads. In a mainly dry climate, such a tyre can be an excellent choice for a road car.*

a range of tyres, hires a racing track or section of closed road and tests each type of tyre. To do so would simply cost too much money. This means dealers' recommendations are based on one of a few perspectives:

- How the tyres have performed on their own car
- How the tyres performed at a racetrack day organised and paid for by the tyre manufacturer to showcase their tyres to dealers
- Customer responses after they have had the tyres fitted

You can see that each of these is a very doubtful rationale on which to tell you Tyre A has better wet-weather handling than Tyre B but it wears noticeably quicker; or that Tyre C needs some heat in it before it really starts to perform; or that Tyre D has stiffer sidewalls and so a better turn-in response than the others.

Tyres should be inspected frequently. In addition to obvious problems such as this, check both the inner and outer sidewalls for damage often indicated by bulges.

Over the years, so-called 'specialists' have given me some totally ridiculous tyre advice. One man was surprised that I criticised the performance of some tyres he had sold me. "They're Japanese and you've got a Japanese car," he said. I still don't understand the relevance of his remark, years later! Another glowingly praised the tyres fitted to his road car, a vehicle that also saw trackwork in club sprints. His car was a small RWD Alfa Romeo with little power — I wanted to fit the tyres to a four-wheel-drive turbo car weighing 1500kg and with about 260kW available! He was taken aback when I suggested his advice was irrelevant in the circumstances.

So after all that, where do you start in your tyre selection?

Your first step when selecting new tyres should be to study the Tyre Placard or Owner's Manual. This will tell you what tyres the manufacturer fitted as standard and what are the legal requirements of the tyres fitted to the car. Don't simply read the information from the tyres already on the car — if the car has had other owners, it's quite likely the wrong tyres may have been fitted at some time in the car's life! One car I heard of even had tyres with an incorrect load rating fitted from the factory.

The placard might read:

16x8JJ, 225/50ZR16 The tyres fitted to this vehicle shall have a maximum load rating not less than 630kg or a load index of 92 and a speed category symbol of not less than 'V'.

This means the rims fitted by the factory are 16 inches in diameter and eight inches wide, with a rim flange configuration of JJ. The original 225/50 tyres were 'Z' speed-rated and were (of course!) radials. However, the placard says it's permissible to fit tyres of a 'V' speed rating but the load index must be at least 92. These are the legal minima and so must be observed. Don't forget that after an accident the first thing an insurance assessor checks is the tyres...

Once you have determined what legal requirements must be met, you need to decide whether tyre life or grip is more important. Some people state they have had a wonderful life from a tyre and it gripped really well. Perhaps it did grip well, but a tyre with a softer rubber would have gripped even better! The UTQG treadwear indicator gives a good indication of the rubber softness. As indicated earlier, the lower the number, the softer is the rubber. This allows a quick comparison to be made between different tyres from which you may be choosing. Where the tyres are equal in all other respects, I will always choose the tyre with the **lower** treadwear indicator. Why? Because soft, sticky tyres can make an awesome difference to vehicle grip — something I am prepared to trade off for a shorter tyre life.

Next, look at where the tyre fits within the company range — with the ranking based on price. Within a single company's tyre lineup, the better the tyre, the more expensive it is. However, comparing prices doesn't work when looking at tyres from different companies. At the time of writing, Korean tyre companies are trying to buy market share in a field dominated by Japanese, European and US firms. The prices of the Korean tyres are perhaps as much as 30 per cent lower than the price of similar tyres from the 'traditional' manufacturers. This currently makes these tyres very good value for money.

When you have narrowed your choice, select the tyre with the highest speed rating you can afford. The reason you should do this is that tyre manufacturers use better rubber compounds and construction in their tyres that are designed to travel at very high speed. A downside is that if the tyre is punctured, some repairs that would be allowed on lower-speed tyres are not permitted.

South Pacific Tyres

Cornering Force Generation

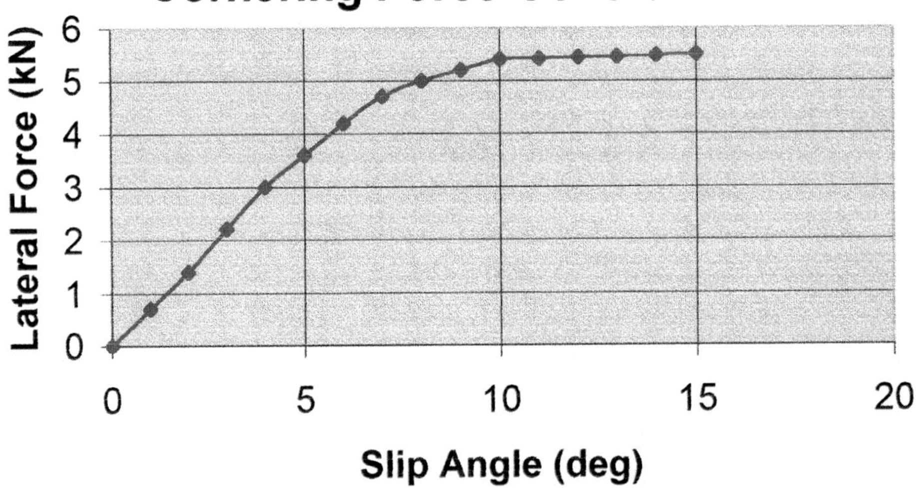

Figure 12.4
The actual contact patch and slip angle versus lateral force curve for a 205/65 15 steel-belted radial tyre mounted on a six-inch-wide rim, with a load of 570kg and an internal pressure of 27.5psi. The tyre has a contact area 183mm long and 148mm wide, giving a length/width ratio of 1.24. With a slip angle of 5 degrees (ie; the tyre is pointing in a direction 5 degrees divergent from the direction that the wheel is actually following), it generates a lateral cornering force of 3.7kN.

With a brand and type of tyre fixed in your mind, ask the tyre retailer whether he or she can suggest a previous customer with a similar car who has bought the tyres. Obviously, if you can speak to that person, you'll be able to find out what they think of the tyres. Make sure you quiz the person on their driving style, tyre expectations and previous tyres, though! Car clubs are also a good source of tyre experience. Selecting the wrong tyres can be a very expensive mistake, so research your choice thoroughly!

Changing Tyre Size

Changing the size of the tyre fitted to the car can improve vehicle roadholding and other handling aspects. Most frequently, the new tyres that are selected have an increased section width (they are wider) and a lower aspect ratio (they have a lower sidewall height). Increasing just the width of the tyre while retaining the same aspect ratio will increase the diameter of the tyre, also increasing its rolling circumference. An increase in rolling circumference will give higher vehicle gearing (less engine rpm per km/h) and will also introduce speedo error. Changing tyre circumference in this way may also affect the operation of the ABS, Traction Control System, Automatic Stability Control and four-wheel-drive control. These systems can be particularly badly affected if the front/rear tyre circumferences vary substantially from each other. So changing the overall tyre diameter from standard is not a good approach to take!

South Pacific Tyres

Cornering Force Generation

Lateral Force (kN)

6
5
4
3
2
1
0

0 5 10 15 20

Slip Angle (deg)

Figure 12.5
*The contact patch and slip angle curve for a 225/55 16 steel-belted radial tyre mounted on a seven-inch rim with the same vehicle load (570kg) as in Figure 12.4 and an internal pressure of 26psi. It has an actual footprint 173mm long and 185mm wide, giving a length/width ratio of 0.94. With 5 degrees of slip angle it generates 4.75kN of lateral force — 28 per cent more than the 205/65 tyre. Note how both tyres have the same contact patch **area**, despite their different sizes.*

When a tyre with an increased width but lower profile is selected, the shape of the contact patch of the tyre on the road will change. Changing to a wider tyre will give a contact patch that's shorter fore/aft but wider. The change in the shape of the contact patch is likely to decrease ride comfort and may decrease self-aligning torque (the propensity for the wheel to wish to track straight). It will, however, improve turn-in, as a shorter contact patch takes less time to 're-arrange' itself when establishing the cornering force.

Figures 12.4 and 12.5 show the contact patches and slip angle versus lateral force curves for two typical steel-belted radial tyres. Figure 12.4 is for a 205/65 15 tyre mounted on a six-inch-wide rim, with a load of 570kg and an internal pressure of 27.5psi. It has an actual contact area 183mm long and 148mm wide, giving a length/width ratio of 1.24. With a slip angle of 5 degrees (ie the tyre is pointing in a direction 5 degrees divergent from the direction the wheel is actually following), it generates a lateral cornering force of 3.7kN. Figure 12.5 is for a 225/55 16 tyre mounted on a seven-inch rim with the same vehicle load (570kg) and an internal pressure of 26psi. It has an actual footprint 173mm long and 185mm wide, giving a length/width ratio of 0.94. With 5 degrees of slip angle it generates 4.75kN of lateral force — 28 per cent more than the 205/65 tyre.

Note that a wider tyre has the **same contact patch area** as the narrower tyre — incredibly, a wider tyre does **not** put 'more rubber on the road'! The handling gains of wider, lower-profile tyres come from the change in shape of the contact patch, and the use of stiffer sidewalls and softer tread rubber. The shorter fore/aft contact patch of a wider tyre can actually decrease longitudinal grip — braking and accelerating. However, the softer rubber used in high-performance, wide tyres usually offsets this decrease in longitudinal performance.

Ascertaining the potential change in rolling diameter that occurs when a tyre of a different size is fitted requires some calculations. First,

$$\text{Sidewall height} = \frac{\text{Aspect ratio}}{100} \times \text{section width}$$

For example, let's assume you currently have a 225/50 tyre fitted to a 16-inch rim. As indicated earlier, the aspect ratio of 50 in the description indicates that the tyre's sidewall height is 50 per cent of its width.

$$\text{Sidewall height} = \frac{50}{100} \times 225 = 112.5\text{mm}$$

As there are two segments of sidewall making up the diameter of the tyre, the tyre walls in this case make up 225mm of the tyre's height. The other portion of the diameter is made up by the rim, which in this case is 16 inches in diameter. Multiplying the rim diameter in inches by 25.4 gives its diameter in millimetres.

$$\text{Rim diameter} = 16 \times 25.4 = 406.4\text{mm}$$

The two sidewall heights (225) + the rim diameter (406.4) = an overall height of 631.4, or 631mm.

A replacement tyre must retain an overall diameter within a maximum of 5 per cent of the standard tyre's diameter — but the smaller the variation from standard, the better. For example, fitting a 245/50 tyre to the same 16-inch rim gives a calculated diameter of 651mm, an increase of 3 per cent. Fitting a 245/40 16 tyre gives a diameter of 602mm, a decrease of nearly 5 per cent. Changing the rim diameter to 17 inches and using a 245 wide tyre requires that the tyre has a profile of 40 if the overall diameter is to be kept very close to standard.

However, while these calculations are useful for gaining an initial idea of the tyre diameter changes that may occur with swaps, tyres vary in their actual diameters on a model-to-model basis. Even those tyres having the same nominal size often have different actual rolling diameters. The tyre's specification sheet will show the overall tyre diameter, static loaded radius, rolling circumference and revolutions per kilometre. You can use any of these specifications to compare the diameter of the old and new tyres.

When increasing tyre width, it's very important that the tyre not foul any suspension components or the bodywork of the car. Wheels deflect laterally a little under cornering loads, and wheels that are able to be steered (either passively or actively) must be able to pivot through their extremes of movement without the tyres fouling. The arcs through which the wheel passes as suspension movements occur may also cause the tyre to rub. If the outside wall of the tyre touches the inner part of the guard, enough clearance may be able to be gained by having the guard lip rolled, so that it protrudes less into the wheel well. Note that the offset of the wheel will also have a major bearing on the clearances around the tyre.

To get the best performance from tyres, they should be expertly fitted and balanced. Rotating the tyre on the rim can be used to reduce the mass of balance weights that need to be attached.

The tyre clearances should be checked carefully when the tyre/wheel combination is first fitted. Then, after some initial driving, both the inner **and** outer surfaces of the tyre should again be checked for signs of rubbing.

Tyre Pressures

Tyre inflation pressure is easy to throw into the same basket as any other maintenance item on a car. You check the tyre pressures at the same time as making sure the coolant level hasn't dropped, and so on. However, it's important to realise the tyre pressures have a dramatic effect on handling, ride and tyre life.

First, tyre pressure is dependent on the temperature of the tyre — that's why 'cold' inflation pressures are specified. As the tyre warms, the internal pressure rises by about 1psi per 6 degrees C. Tyres increase in temperature dramatically when they're working hard or when they're in contact with a very hot road surface (or both!), so the actual working tyre pressure is likely to be significantly increased over its cold value. Because of this, tyre pressures should be checked only when the tyre is cold. Furthermore, changes in ambient temperatures will affect tyre pressures, so the pressures you set on a 35-degree C day will not be the same as those experienced when the day temperature is only 20 degrees C.

The tyre pressures recommended on the tyre placard or in the owner's manual should be considered the lowest tyre pressures to use. Manufacturers tend to recommend tyre pressures on the lower side because they're concerned about ride comfort.

As a performance driver, ride comfort is less of a priority to you, so generally a pressure 3-4psi above that recommended is more appropriate. Setting slightly higher tyre pressures will improve the car's handling — more about this in a moment.

A tyre that's grossly under-inflated is very dangerous. It's important to realise that a visual inspection of a tyre running perhaps only 20psi instead of its recommended 35psi will not always indicate that the tyre is near-flat. (However, as a matter of course you should visually inspect tyres as you get into and out of the car.) If you drive quickly on a tyre in this condition of under-inflation, you risk a tyre blowout, the tyre rolling off the rim or a sudden and massive slide while cornering.

None of these is a very pleasant outcome, so the importance of frequently checking tyre pressures simply cannot be underestimated. Whenever I start driving fast, I mentally review the last time I inspected the tyres and checked their inflation pressure. If it occurred more than a week ago, I change my mind about putting the hammer down!

Tyres that are permanently under-inflated tend to wear out both the outer and inner edges first, while tyres over-inflated wear the centre of the tread at a greater rate. Periodically inspecting the tread of your tyres will reveal if either of these is occurring. However, the wear of performance tyres is also greatly affected by the alignment angles used. Tyres with optimal pressures and alignment angles that are driven hard will also wear both inner and outer edges.

WHEELS

Wheels vary in a number of aspects, including:

- Styling
- Construction material
- Construction technique
- Width and diameter
- Offset
- Stud pattern

The majority of wheels are made from either pressed steel, or cast or forged alloy, with performance cars universally using alloy wheels. Most alloy wheels are made in one piece, although two- and three-piece wheels are also available. These latter rims allow the mixing and matching of the different parts so that unusual sizes and offsets can be produced more easily.

One-piece alloy wheels can be made in a number of ways. The simplest and cheapest is gravity casting, where the molten metal is poured directly into a mould and then allowed to harden. Counter pressure casting uses a vacuum to draw the molten metal up into a mould, keeping the metal a more constant temperature. This casting technique results in a stronger wheel with a more uniform metal density. The most advanced wheel-forming method is forging. Forging compresses a billet of alloy into a wheel using extreme pressure and heat. Forged wheels are up to three times stronger than conventionally cast wheels while also being substantially lighter.

Most people buy alloy wheels for looks alone, but there are also some technical advantages to using alloy rims that will justify your purchase! First, an alloy wheel is typically lighter than its steel equivalent. (However, it should be noted that most people don't bother fitting standard-sized alloy wheels, so the larger wheel and tyre that are actually used often have a significant increase in mass over the standard wheel/tyre combination!) Keeping the

Alloy wheel masses vary significantly from wheel to wheel. This BBS wheel is very light; a forged alloy wheel of the same design will be even lighter!

wheel/tyre mass as low as possible improves roadholding and acceleration and braking performance. (There's more on this topic in the section below.)

In addition to being lighter, the open design of most alloy wheels and the greater heat conduction of the alloy mean that brake cooling is substantially improved. An alloy wheel is also more rigid than one made of steel, reducing wheel/tire deflection when cornering. Finally, when new wheels are being fitted, they can be specified to have the appropriate offset and size to allow wider, lower-profile tyres to be fitted without body fouling or gearing changes.

The offset of the wheel refers to lateral distance between the mounting hub of the wheel and the centreline of the tyre. This and other wheel characteristics are shown in Figure 12.6. An offset is regarded as positive if the hub mounting surface is towards the outer edge of the wheel. Positive offset wheels are the most common. Negative offset wheels have the hub mounting surface toward the rear of the wheel, while zero offset wheels have the hub mounting surface and the centreline of the tyre on the same plane.

To measure the offset of a wheel, the backspace must first be determined. Wheel backspace is the distance from the rear edge of the rim to the hub mounting surface. To determine the wheel backspace:

- Position the wheel face down.
- Lay a straight-edge across the back of the wheel.
- Measure the distance from the straight-edge to the hub mounting surface.

To then determine the offset from this:

- Position the wheel on a flat surface and measure its overall width.
- Divide by two and subtract the backspace.

The offset is of critical importance if the tyre is not to foul the bodywork or suspension, and if the steering geometry of (especially) front-wheel-drive cars is not to be altered. In fact, in most situations, the front track of a front-wheel-drive car (the distance from centreline to centreline of the front tyres) should not be altered at all, in order that the negative scrub radius of the steering is not adversely affected (for more on scrub radius, see later in this chapter). Increasing offset can also cause bearing and steering component failure due to the increased leverage on the suspension components. Incidentally, the use of spacer plates to change the effective wheel offset should be avoided — a spacer plate also causes much greater loads to be applied to wheel studs.

An example of the different offsets that need to be used to cater for changes in rim and tyre width can be seen in the table below, which has been formulated for one particular make and model of car.

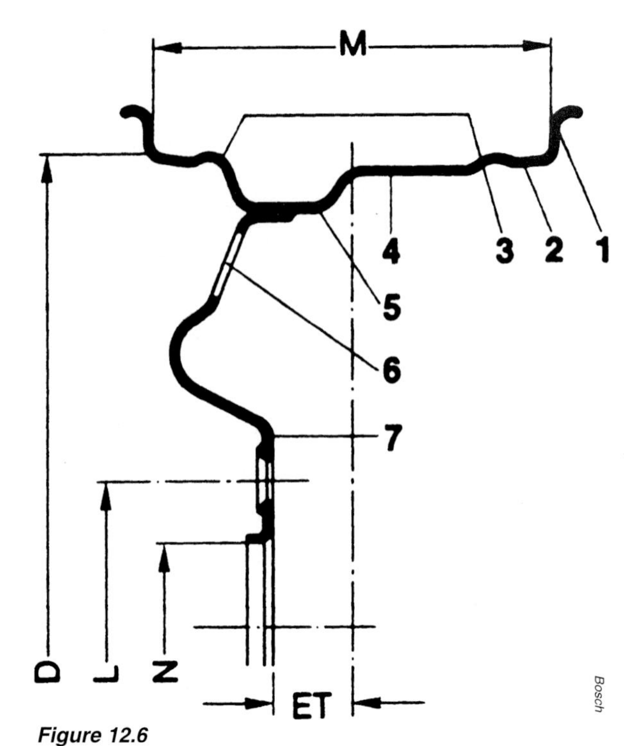

Figure 12.6
*The section through a 6J x 14H2 steel wheel. **1** – J rim flange, **2** – 2.5 degrees tapered bead seat, **3** – double hump H2, **4** – rim, **5** - drop centre, **6** – ventilation hole, **7** – wheel nave, **D** – 14-inch rim diameter, **L** – pitch circle diameter, **M** – six-inch rim width, **N** – centre hole, **ET** – rim offset.*

The stud pattern of a wheel is described in terms of the Pitch Circle Diameter (PCD) and the number of studs. A wheel description of '4x100' means there are four studs and the PCD is 100mm. The PCD is the diameter of an imaginary circle drawn through the centreline of the studs. The PCD (and number of studs!) of a wheel should exactly match the hub of the car. It's possible to fit wheels whose PCD is out by a few millimetres, but then the wheel does not sit properly on the hub and the studs can be highly stressed, sometimes even to the point of breaking off.

The width of the wheel should be a good match for the section width of the tyre. Using a rim width that's 90 per cent of the tyre's section width works well for a road car. However, this rule does not need to be observed precisely — Figure 12.7 (opposite) shows the range of rim widths that can be used with different tyres. Note that a tyre's actual section width depends to some extent on the rim on which it's mounted — changing about 5mm for every 0.5 inch change in rim width.

WHEEL AND TYRE MASS

As briefly mentioned above, the mass of the wheel/tyre combination has some significant performance implications. It's important that it be kept as low as possible for two reasons: unsprung mass and rotating inertia.

	Standard	Alternatives				
Tyre size	225/50	255/45	245/45	255/40	255/35	265/35
Rim size	16 x 8	16 x 9	17 x 10	17 x 9	18 x 10	18 x 9
Rim offset	+30	+15	+10	+15	+25	+20

Aspect Ratio	Metric Size Designations	Approved Rim Width Ranges	Aspect Ratio	Metric Size Designations	Approved Rim Width Ranges
35 Series	275/35R17	9.0-10.5	**60 Series**	185/60R13	5.0-6.5
	335/35R17	11.0-13.0		205/60R13	5.5-7.5
	345/35R15	11.0-13.5		185/60R14	5.0-6.5
40 Series	285/40R15	11.0-13.5		195/60R14	5.5-7.0
	215/40R15	7.0-8.5		205/60R14	
	245/40R17	8.0-9.5		195/60R15	5.5-7.0
	255/40R17	8.5-10.0		205/60R15	5.5-7.5
	265/40R17	8.5-10.0		215/60R15	6.0-7.5
	235/40R18	8.0-9.5		225/60R15	6.0-8.0
45 Series	195/40R15	6.5-7.5		235/60R15	6.5-8.5
	215/45R15	7.0-8.5	**65 Series**	185/65R15	5.0-6.5
	255/45R15	8.5-10.0		195/65R15	5.5-7.0
	205/45R16	7.0		205/65R15	5.5-7.5
	225/45R16	7.5-9.0		215/65R15	6.0-7.5
	245/45R16	8.0-9.5	**70 Series**	165/70R10	4.5-5.5
	215/45R17	7.0-8.5		175/70R12	5.0-6.0
	235/45R17	8.0-9.5		175/70R13	5.0-6.0
	255/45R17	8.5-10.5		185/70R13	5.0-6.5
50 Series	175/50R13	5.0-6.0		185/70R14	5.0-6.5
	195/50R14	5.5-7.0		195/70R14	5.5-7.0
	205/50R15	5.5-7.5		205/70R14	5.5-7.5
	215/50R15	6.0-7.5		185/70R15	5.0-6.5
	225/50R15	6.0-8.0		205/70R15	5.5-7.5
	195/50R16	5.5-7.0		215/70R15	6.0-7.5
	205/50R16	5.5-7.5	**82 Series**	155R12	4.0-5.0
	225/50R16	6.0-8.0		155R13	4.0-5.5
	235/50R16	6.5-8.5		165/70R13	4.0-5.5
	215/50R17	6.5-7.5		175R14	4.5-6.0
55 Series	205/55R14	5.5-7.5		185R14	4.5-6.0
	185/55R15	6.0-6.5		165R15	4.0-5.5
	195/55R15	5.5-7.0			
	205/55R15	5.5-7.5			
	205/55R16	5.5-7.5			
	225/55R16	6.0-8.0			

Figure 12.7
The recommended rim widths for tyres with different section widths and aspect ratios.

The unsprung mass refers to the mass of the tyres, wheels, brakes, part of the suspension linkages and the other components that move vertically with the wheels. Unsprung mass in a typical car can comprise as much as 15 per cent of the overall vehicle mass. Leaving aside the springiness of the tyres, these masses move in direct accordance with the road surface.

The ratio between unsprung and sprung mass is important because the vertical accelerative forces imparted to the unsprung mass by road bumps must be controlled by the sprung mass. When a tyre encounters a bump, the wheel is accelerated vertically at a speed dependent on the characteristics of the tyre, the size of the bump and the forward speed of the car. The greater the

If buying used alloys, you should be aware that this size marking is only one part of the story. The offset, stud number and PCD also need to be correct if the wheel is to fit without problems.

accelerated mass (the unsprung weight), the greater the kinetic energy that must be absorbed by the suspension if the sprung mass (the body of the car) isn't also to be vertically accelerated. If the sprung mass is very large in comparison to the unsprung mass, the inertia of the sprung mass will provide a greater resistance to vertical movement, giving a better ride and keeping the tyres firmly pushed against the road. Thus, keeping unsprung mass as low as possible will not only improve ride but also give better roadholding.

If you're on a tight budget, it's sometimes possible to get a larger-diameter steel rim, rather than having to buy more expensive alloy wheels. These 15-inch rims replace the standard 14-inch rims and were fitted when this car was used for police duties.

While changing the mass of the suspension arms, driveshafts and so on is difficult, reducing the mass of the wheels and tyres is straightforward. In addition to decreasing unsprung mass, another reason this is advantageous is in reducing rotational inertia.

Unlike most parts of the car, the wheels and tyres are rotationally accelerated when the car speeds up. The inertia of the wheels resists this change, meaning the input of a substantial amount of engine power is required just to spin the wheel/tyre combination faster. Like miniature flywheels, the heavier the wheels are, the greater the power required to change their speed — and the less power that's then available to accelerate the car linearly!

US magazine *Sport Compact Car* found some of these effects when their late-model Honda Civic was fitted with an upgraded wheel/tyre package. The starting point was a set of 185/65 Dunlops on 14-inch steel wheels. Each wheel/tyre weighed 15.5kg. They then changed these to 205/40 Nitto tyres on 17-inch alloys, increasing each corner's mass to 19.5kg. This 4kg (26 per cent) per wheel increase in mass was enough to drop the peak power measured at the wheels on a Dynojet chassis dyno by nearly 5 per cent!

Alloy wheels vary in mass depending on their size, styling and construction technique. A typical 15x6 cast alloy wheel weighs about 8kg, versus 9.5kg for a steel rim of the same size. Typically, 16x7 alloy wheels have masses from 7-9kg, while 17x9 alloy rims weigh 9-10kg each. Very large alloy wheels can weigh more than 14kg each. Likewise, tyres also have different masses, although tyres sized from 205/60 14 right through to 235/45 17 are usually all in the range of 10-12kg. From a rotational inertia perspective, it's better to have a heavier wheel and a lighter tyre rather than a heavier tyre on a lighter wheel. However, having both a light tyre and a light wheel is better again!

Although stronger than steel rims, an alloy wheel can still be damaged. While repairs can be made, it's better in a high-performance car to buy a replacement rim if damage like this occurs.

WHEEL ALIGNMENT TERMS

Wheel alignment is a critical aspect in modifying car suspension. Having performance-oriented wheel alignment geometry at the back and front can make a tremendous difference to stability, turn-in, grip and balance. In many cars, gaining sufficient suspension adjustment to allow the setting of these optimal angles will require the fitting of special adjustment mechanisms (eg camber kits), but these are almost always worth the expense.

Although suspension alignment settings are made while the car is stationary, it's the dynamic suspension angles that are actually important — that is, the wheel angles that occur when the car is under the influence of braking, acceleration, cornering and aerodynamic forces. Static wheel alignment is carried out on the premise of predicting the dynamic angles that will actually occur. Wheel alignment recommendations are therefore only a starting point, with fine-tuning changes subsequently made by examining tyre wear and/or tyre temperatures.

There's a number of wheel alignment specifications. Toe is the difference in the distance measurement between the front and rear of the left and right rims. Like many

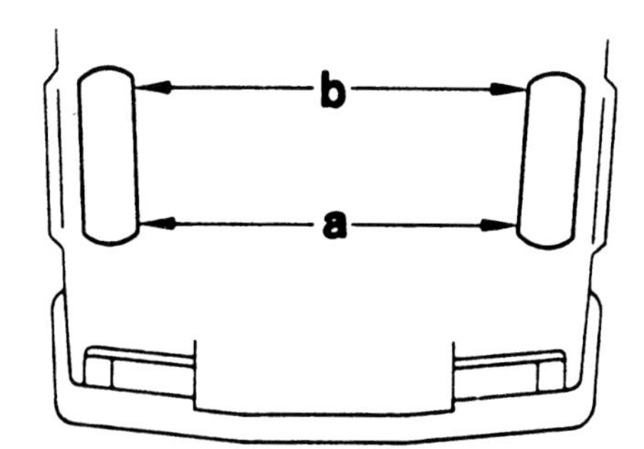

Figure 12.8
*Toe is the difference in the distance between **a** and **b**. If **a** is a smaller distance than **b** the car is said to have toe-in.*

wheel alignment terms, this is initially hard to picture. However, imagine you are standing, looking down at your feet. When you point your toes inwards, you are adopting a toe-in stance. With your toes pointing outwards, you have toe-out. Toe is expressed in millimetres or degrees. A car with zero toe has wheels that are parallel with the long axis of the car. Figure 12.8 shows how toe is measured.

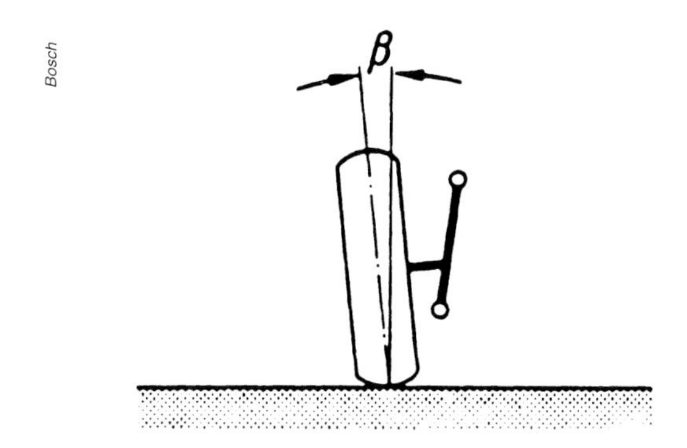

Figure 12.9
Camber is the inclination of the wheel when viewed from the front or rear of the car. Here β is the camber angle, which is measured in degrees.

Camber is the angle the wheel leans in or out from the vertical, when viewed end-on to the car. If you squat down and look at the front or back of a performance car, you may notice the wheels lean in at the top. This is termed negative camber. Positive camber is said to occur when the wheels lean outwards at the top. Camber is expressed in degrees. Negative camber is much more frequently used in performance applications because, when the body rolls during cornering, the heavily loaded outside tyre then stands upright, putting as much rubber as possible on the road. Figure 12.9 shows how camber is calculated.

Castor applies only to steered wheels. It is the angle between the steering axis and the tyres' vertical centreline, when viewed from the **side** of the car. In a MacPherson strut car, it's the angle the strut leans back from vertical. Castor is shown in Figure 12.10. The further ahead of the tyre contact path the imaginary extension line of the steering axis impinges on the road, the greater the amount of positive castor. In other words, if the wheel is moved forward relative to the body (or the top of the strut backwards), positive castor will increase. Castor is measured in degrees.

The greater the castor, the greater the camber that occurs during cornering. This camber will be negative on the outside wheel (good) and positive on the inner wheel

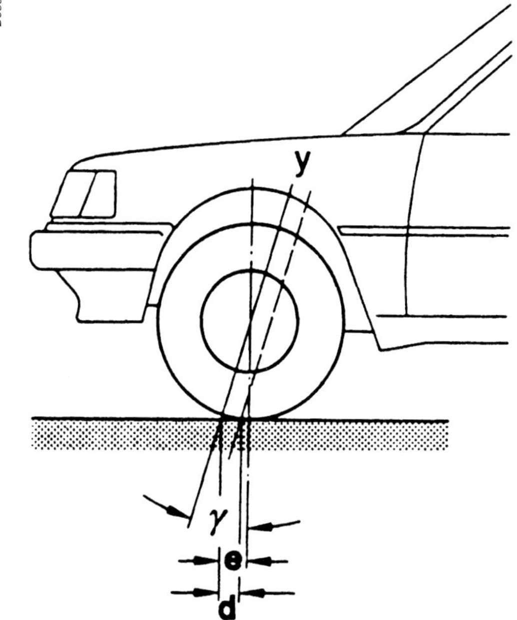

Figure 12.10
*Castor refers to the angle of the steering axis to the vertical. **y** – steering axis, γ – positive castor angle, **d** – castor offset, **e** – positive castor.*

(doesn't matter because it is under little load). However, too much castor can make the car slow to respond to initial steering inputs. Incidentally, if you can't see all these angles in your head, I suggest you make a model of a wheel and MacPherson strut from an Allen key and a round pencil eraser. Push the rubber onto the 'stub axle' formed by the short arm of the Allen key. Give your 'MacPherson strut' more positive castor by placing the contact patch of the rubber well ahead of the top of the Allen key, and then swivel the angled upright of the Allen key in your fingers. As you do so, you'll see that the outside wheel leans over (adopting negative camber) during the turn.

The angle to the vertical of the steering axis when viewed from the **front** (or rear) of the car is referred to as the kingpin inclination. If I again use a MacPherson strut car as the example, with the wheel set square to the road (ie no camber), the top pivot point of the strut will be further towards the centreline of the car than the bottom pivot point. (To achieve his with the Allen key model, you'll need to open its 90-degree bend out a little.) When this imaginary pivot line is extended down to the road, the

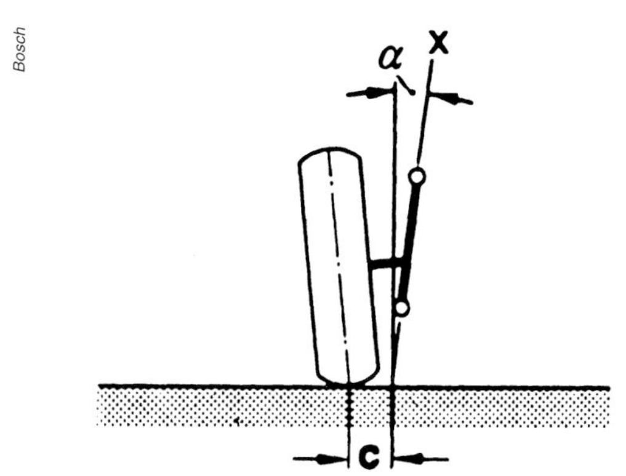

Figure 12.11
*The kingpin angle and steering scrub radius can be seen here. α – kingpin angle, **x** – steering axis, **c** – positive scrub radius.*

Fitting alloy wheels gives definite performance advantages ... even if most people do it just for looks!

distance between the point at which this line impinges on the road and the centreline of the tyre is called the scrub radius (or offset). Figure 12.11 (previous page) shows this concept. Negative scrub radius occurs when the line reaches the ground outwards of the tyre centreline, positive when the line reaches the ground inwards of the tyre centreline, and zero scrub radius occurs when the tyre centreline and the kingpin inclination line coincide. Many FWD cars use negative scrub radius geometry, which helps stability when the car is braked with each front wheel on a surface with a different level of grip. Changing the wheel offset (eg by fitting wider wheels) on cars that use negative scrub radius geometry can therefore substantially reduce vehicle stability.

SWAYBARS

A swaybar — sometimes called an anti-roll or stabiliser bar — is a torsion bar that resists bodyroll by linking the suspension movement of the wheels at the one end of the car. Swaybars are placed on the front, rear or both front and rear suspensions systems. A swaybar resists bump (suspension compression) on just one side of the car. However, because the swaybar is mounted on pivots, the suspension of both left and right wheels can compress together without additional spring resistance being imposed. Except over one-wheel bumps, this means a stiffer swaybar will not cause a stiffer ride, unlike the fitting of higher rate springs.

However, in an independent suspension car, to a greater or lesser degree, a swaybar creates an effective axle, linking the suspension movement of the two wheels. Hence, the size of the swaybar that's chosen affects the required shock absorber and spring rates. Large swaybars cause an increase in NVH (noise, vibration, harshness) and over-barred cars tend to cause the body to jerk sideways while passing over one-wheel bumps.

The stiffness of a swaybar is dependent on its shape and construction. The longer the lever arm acting on the torsion bar, the softer will be the effective rate. To reduce

If wider tyres and/or wheels are being fitted, care should be taken that the tyres don't rub on the suspension components or bodywork during suspension and steering movements. Note this car has four-wheel steering — a factor sometimes forgotten when checking a car for tyre fouling.

Having both the front and rear wheels aligned for maximum performance can make a staggering difference to handling. The specifications used in this type of alignment are often very different from the standard specs.

unsprung and sprung mass, some swaybars use a hollow rod, but most comprise a solid bar. For a given shape of constant-diameter, solid swaybar, the change in rate can be easily calculated for different diameters.

Gaining optimal wheel alignment often requires the fitting of suspension 'kits'. Here, adjustable upper A arms have been fitted, allowing the adjustment of camber.

Swaybar stiffness increases as the fourth power of the diameter. For example, a swaybar might have a diameter of 22mm and you are considering changing it for one which is 26mm in diameter. 22^4 (22x22x22x22) give a stiffness factor of 234,256 units. The second bar's stiffness is 26^4 which is 456,976. (Divide one by the other and you can see that the second bar's stiffness is almost twice (1.95 times) as high.) As you can see, small changes in diameter make for large changes in stiffness!

If your car has front and rear swaybars fitted, you can easily work out the ratio of front/rear swaybar stiffness. If you are happy with the car's balance — you just want less roll — you can maintain this ratio when increasing the diameter of the bars. For example, a car with a front 22mm bar and a rear 17mm bar has a front-to-rear roll bar stiffness ratio of 2.8:1. This is worked out by taking the stiffness of the front bar (22^4) and dividing it by the stiffness of the rear bar (17^4). If the front bar is increased to 24mm, retaining the same front/rear ratio means the rear bar needs to be about 18.5mm in diameter. This was worked out by 24^4 divided by 2.8 (the front/rear ratio in this example), with this figure then fourth rooted. A fourth root is the opposite of raising a number to the fourth power, and can be easily done on a scientific calculator.

Cars with MacPherson strut suspension can be fitted with new pillow ball upper mounts. These allow adjustment of camber and, because of the absence of a rubber bush, provide tighter handling.

Setting of the rear alignment should not be overlooked — it's of equal importance to the front alignment. This rear camber adjustment eccentric bolt is standard on the Subaru Impreza WRX.

The thicker a swaybar, the more it will resist roll at that end of the car. However, as it does this, it will also try to pick up the inside wheel. For example, an overly stiff rear bar will lift the inner rear wheel off the road. This can result in that end of the car being more prone to sliding. However, having no swaybar (or one that's too soft) can allow major weight transfer to occur laterally, overloading the outside tyre and also causing it to slide! So both over- or under-stiff swaybars can cause a similar end result, although the body attitude will certainly be different at either extreme of swaybar stiffness!

Changing the swaybars is normally carried out for one of two reasons: as a first, cost-effective suspension upgrade, or as a last-stage tweak to determine the final over/understeer attitude. If adjustable bars are fitted, both approaches can be accommodated.

SPRINGS

Coil springs consist of a torsion bar wound in a spiral. Describing the design in this way makes it easier to understand that the stiffness of a coil spring will be dependent on the wire thickness, the diameter of the spring and the number of free (active) coils.

The stiffness of a spring is expressed as 'spring rate'. This is a measurement of how far the free spring compresses when a weight is placed on top. Spring rate is measured in pounds per inch or newtons per millimetre. A 100 pound per inch spring compresses by 1 inch when 100 pounds load is placed on it. It will compress by 2 inches with a 200 pounds and so on. (To convert from pounds/inch to Newtons/mm, divide by 5.7.)

Where the bump and rebound of the spring are the same as the travel of the wheel, the spring rate and the wheel rate are the same. This is the case on MacPherson struts, for example. Where the spring is compressed and expanded through the action of a suspension lever, the spring rate and the wheel rate are **not** the same. An example of this occurs in semi-trailing arm rear suspension, where the spring rate is much higher than the wheel rate. The difference between spring and wheel rates makes meaningless the casual swapping of the spring rates that are being used in a variety of cars.

Spring specifications can be determined in four ways. One way is to simply look in the workshop manual! For the R32 Nissan Skyline GT-R, for example, the workshop manual lists the springs as having the specs in the table opposite.

The use of strut bars can help reduce body deflections, allowing the better retention of the static suspension angles. However, the change is usually very subtle.

Spring Specification	Front	Back
Spring rate	23.5 N/mm (134 lb/inch)	26.5 N/mm (151 lb/inch)
Free length	405mm (15.94 inches)	345mm (13.58 inches)
Diameter	Upper: 110mm (4.33 inches) Lower: 80mm (3.15)	Upper: 100mm (3.94 inches) Lower: 90mm (3.54)
Wire diameter	12.3mm (0.48 inches)	11.8 mm (0.465 inches)
Number of active coils	7.92	7.29

The spring can also be removed from the car and tested on a machine that applies a load and then measures the resulting spring deflection. However, it's easy enough to do that for yourself!

To measure the spring rate you need to:

- Use a tape measure to find the free length of the spring.
- Carefully stand on the spring and have an assistant measure the compressed length.
- Weigh yourself on some bathroom scales.

The figures might look something like: 15 inches uncompressed and 12.75 inches compressed, with a body weight of 196 pounds. To work out the rate, do the following calculation:

$$\text{spring rate} = \frac{\text{weight (eg your body)}}{\text{uncompressed length} - \text{compressed length}}$$

$$\text{spring rate} = \frac{196}{15 - 12.75}$$

spring rate = 87 pounds/inch

To calculate the spring rate purely on its physical specifications, you need the:

- Wire thickness in inches
- Number of active coils (ie those fully free to move)
- Coil diameter in inches (centre of wire to centre of wire)

The equation is:

$$\text{Spring rate} = \frac{11{,}250{,}000 \times \text{wire diameter}^4}{8 \times \text{number of active coils} \times \text{coil diameter}^3}$$

So if the spring has a wire thickness of 0.5 inches, 6 active coils and a diameter of 5.6 inches, the calculation looks like this:

$$\text{Spring rate} = \frac{11{,}250{,}000 \times 0.5 \times 0.5 \times 0.5 \times 0.5}{8 \times 6 \times 5.6 \times 5.6 \times 5.6}$$

$$\text{Spring rate} = \frac{703125}{8429.6}$$

Spring rate = 83 pounds/inch

Note the use of variable rate springs makes the application of this equation not quite as straightforward as it first appears. With a variable-rate spring, measuring the actual rate at a variety of deflections is the best approach to take.

The spring rate is one of the most important specifications of springs. A high-rate spring will be stiff, resisting bodyroll but also giving a harder ride. Upgraded suspensions always use stiffer springs as one of the modifications. One reason is that the ride height is usually lowered so the centre of gravity is closer to the road and the car does not have so much lateral weight transfer when cornering. A lowered car with the standard spring rates will tend to bottom out — that is, the suspension will hit the bump stops, having used up all of its travel. Fitting shorter springs that are also stiffer will prevent this happening as frequently, because the stiffer springs will resist compression to a greater degree.

The car's ride height will depend not just on the spring rate and the free length, but also on the spring pre-load, if any. When springs are mounted on struts, they're compressed and then held captive in that compressed state. Pre-load holds the spring in place when the strut reaches full extension (which occurs when the car is jacked up or the wheels leave the road), but it also means the car does not settle to the extent you might first expect when the weight is on the springs.

For example, a spring might have a rate of 100 pounds/inch and a free length of 15 inches. When it is mounted on the strut it is compressed to (say) 11.6 inches, so the pre-load is 3.4 inches. To compress the spring by each inch takes 100 pounds, so the compressive force acting on the spring when it's on the strut must be 340 pounds (3.4 inches x 100 pounds per inch rate). If there are 665 pounds acting through that corner of the car when the car is stationary, the spring will settle only 3.25 inches, because the first 340 pounds of the car's weight is needed just to overcome the pre-load. That leaves 325 pounds, which at a rate of 100 pounds/inch will cause the car to settle 3.25 inches. If there was no pre-load, the

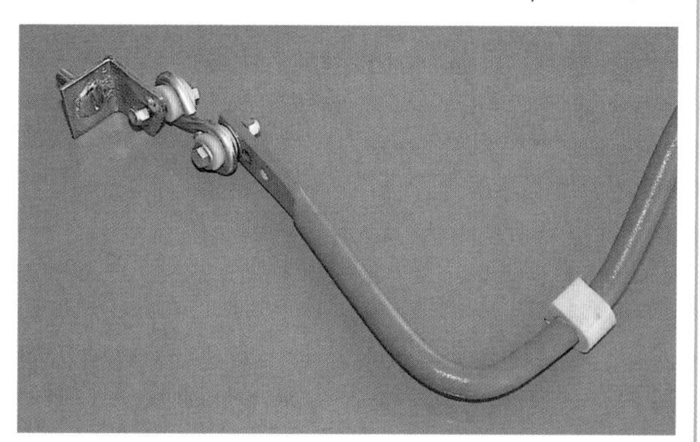

The use of adjustable swaybars allows the tweaking of the handling to suit the driver's preference and the tyres being used.

If you're on a very tight budget, standard swaybars can be twinned, as shown here.

car would settle by 6.65 inches (665 pounds divided by the 100 pounds/inch spring rate), so you can see the amount of pre-load is very important in determining ride height.

The stiffness of a coil spring is determined by its wire thickness, number of free coils and the diameter of the spring.

The higher the pre-load, the less the strut will compress when the car's weight is on it. In other words, this gives more bump and less rebound. Unless you are also moving the lower spring mount (or fitting an adjustable one) the greater the pre-load, the higher will be the ride height because the car's weight won't compress the spring as much. So when you select a spring that has a stiffer rate and a shorter free length, you cannot use those specifications to work out the ride height unless you know the spring pre-load when it is mounted on the car!

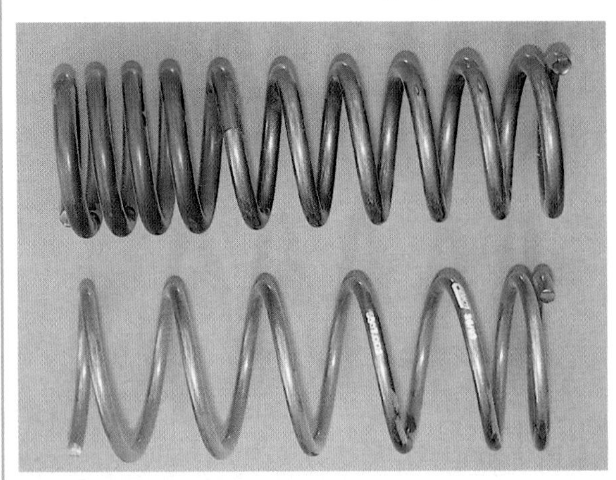

A variable-rate spring (top) gives initially soft deflections that harden as the spring is compressed to a greater extent. Such a spring can be used to give a ride that's softer over small bumps but firms with larger suspension movements.

For this and other reasons, it's best that you state to a spring manufacturer or suspension shop what you want the spring to achieve, rather than design the spring yourself. For example, you could say: "I want a spring that has a 30 per cent higher rate and gives a ride height that is 1.5 inches lower than standard." Where doing all the calculations yourself **can** be a real advantage is when you are on a very tight budget. In this situation, you can work out all the specs of a spring that you would like and then go along to a wrecking yard and find some standard springs from another car that already have the required specs! This can be a very cheap way indeed of getting good-quality, lower springs that have a higher rate.

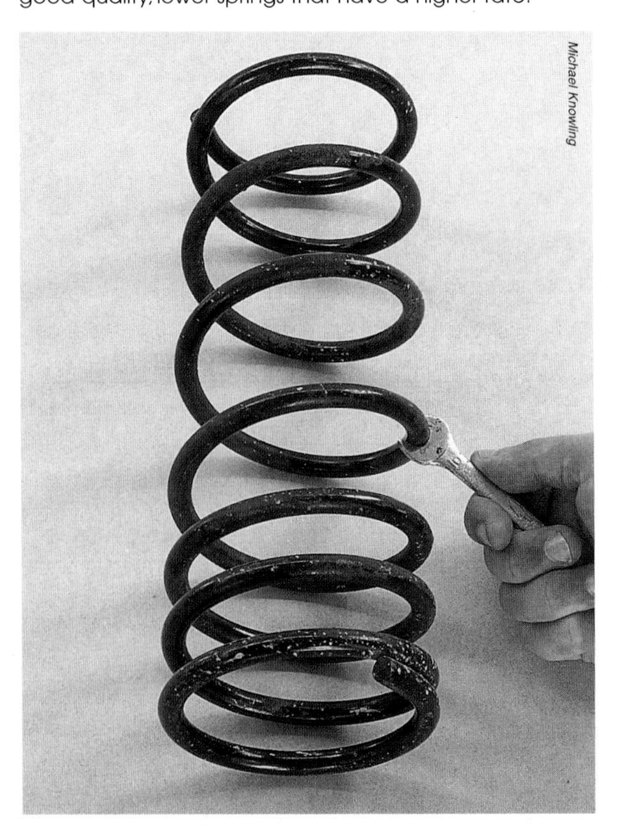

Michael Knowling

Finding a standard spring from another car can be a very cheap way of gaining a performance upgrade. A spanner can be used to quickly compare different wire thicknesses.

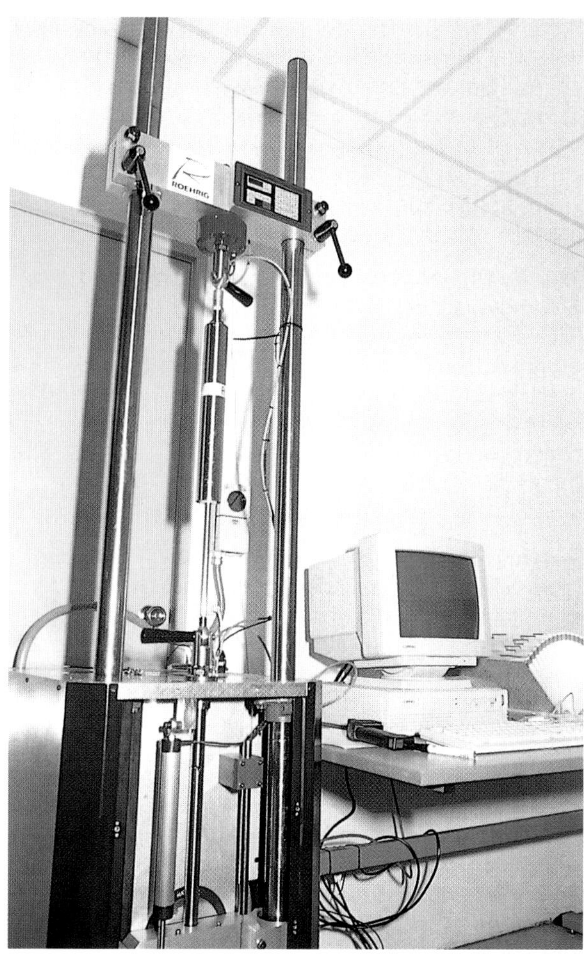

To accurately match a shock absorber to a spring, the shocker should be thoroughly tested on a shock absorber dyno. However, very few suspension workshops have such a piece of equipment...

If the spring is shorter and has a similar number of coils to the original spring, coil bind can occur. This happens when the spring is compressed to the extent that all the coils meet. When a spring coil binds, the effective rate suddenly jumps to infinity, which isn't good for handling or the integrity of the bodywork! To work out the greatest spring compression possible before this occurs, measure the gaps between all the coils and then measure the spring's length (either in or out of the car).

Minimum spring length = length spring - air gaps

This strut features two coils of different rates and height-adjustable spring platforms. The springs can be easily mixed and matched to provide good results, while screw adjustment can be used to change corner weights and ride heights.

Externally adjustable shock absorbers allow the fine-tuning of the suspension damping to suit driver preferences and the spring and tyre combination being used.

For example, a spring has the following air gaps: 1.5, 1.75, 1.75, 1.75, 1.75, 1.75, 1.5 inches. These add up to 11.75 inches. If the spring is 15 inches long, the most it can be compressed is 15 - 11.75 = 3.25 inches. The bump stop must be sufficiently long that this does not ever occur. One type of spring that does **not** follow this rule is the conical Minibloc spring used by GM in some of its suspension designs; this type of spring can compress almost flat without coil bind occurring.

Coil spring design can be summarised:

- The fewer the number of coils, the stiffer the spring.
- The thicker the wire, the stiffer the spring.
- A spring with more coils than standard needs to use either a lower spring seat (on a strut) or a longer bump stop to avoid coil binding.
- Ride height will be affected by three factors: the spring rate, the spring pre-load and (on a strut) the position of the lower spring mount

When specifying springs for a modified car, the original equipment springs should always be the starting point for the design. The input of a spring manufacturer in the area of spring life and strength in the given application should also be sought.

Most well set-up road cars use springs 20-30 per cent higher in rate and 1-1.5 inches (25-38mm) lower in ride height. Do not be tempted to massively lower your car if you want it to handle well — without sufficient suspension travel, it won't!

These very sophisticated rally car dampers use external fluid recirculation and custom everything! For most road car applications, off-the-shelf, good-quality, adjustable dampers are sufficient.

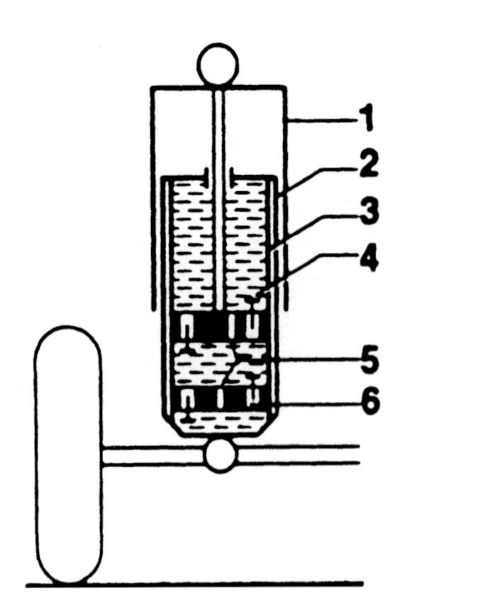

Figure 12.12
*A twin tube damper. **1** – protective tube, **2** – outer tube, **3** – inner tube, **4** – piston valve, **5** – restriction, **6** – bottom valve.*

SHOCK ABSORBERS

Shock absorbers — more properly called dampers — prevent the springs from oscillating freely up and down, as they would otherwise do after the car had passed over bumps. Each movement of the piston within the damper forces oil to flow through tiny spring-loaded valves that open and shut, depending on the direction and force of the oil flow. The speed and distance of the piston movement determines the damping firmness of a given shock absorber. Figures 12.12 and 12.13 show the internal construction of single-and twin-tube shock absorbers.

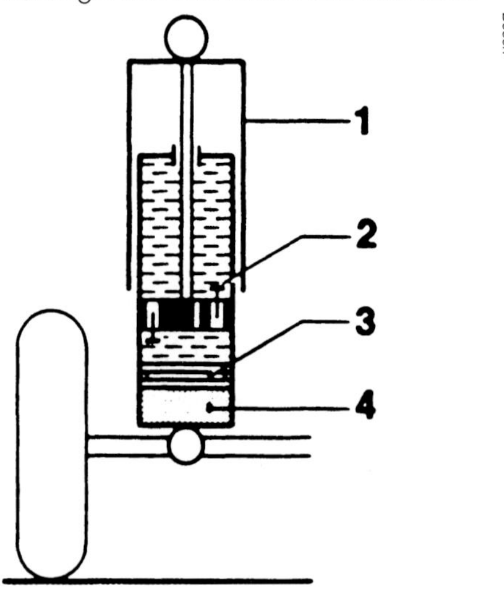

Figure 12.13
*A single tube damper. **1** – protective tube, **2** – piston valve, **3** – separating piston, **4** – gas under pressure.*

Every book on modifying car suspension will state something along the lines of "You must have a matched spring and damper combination" or "If you fit new springs, new dampers must be matched to them". However, not one in 100 suspension workshops has a damper dynamometer — even supposing they had the budget to buy dozens of dampers and then graph the characteristics of each! This means that, in practice, when you buy stiffer springs you should also simply buy stiffer, good-quality dampers. If you retain the standard dampers with stiffer springs, a jiggly, poor ride will be experienced. Handling will also be way below par. On some cars, fitting lower springs will require that the dampers have a shorter stroke, otherwise the new springs will lack pre-load, causing them to unseat themselves when the car is jacked up. A progressive-rate spring can also be used if spring unseating is a problem.

One way the dampers **can** be matched to your car (not only matched to the springs but also to your car's mass, tyres and your driving preferences) is to buy adjustable dampers. Dampers are available that are adjustable for bump and rebound, although on most only rebound can be changed. Figure 12.14 (opposite) is the printout of a dynamometer test undertaken on a Koni Sports damper. The range of rebound adjustment can be clearly seen.

Note that some models of adjustable dampers need to be removed from the car before they can be adjusted — making frequent fine-tuning adjustments somewhat time consuming and/or expensive! The best adjustable dampers have provision for an adjusting knob to be placed in a fitting on top or bottom of the damper. Other good dampers have an adjusting knob placed on the side of the damper body.

Dampers normally have softer bump than rebound characteristics. Figure 12.15 (opposite) shows the bump /rebound characteristics for the front damper in a standard Daewoo Lanos. As can be seen, rebound damping is approximately 6-7 times as firm as bump damping. This helps create a soft ride, because the initial vertical movement of the sprung mass is not transferred as strongly to the unsprung mass (that is, the car body). Sports dampers have stronger bump and rebound damping, as shown in Figure 12.16 (page 293).

In addition to fine-tuning of damper/spring matching, using adjustable dampers has the benefit of allowing adjustment to compensate for damper wear. It's not unknown for half of the total rebound adjustment to be used over 40,000km of road use, just to retain the car's original damping characteristics!

CORNER WEIGHTING

Cars that have suspension systems in which the spring preload is able to be easily adjusted can gain handling and braking benefits by being 'corner weighted'. This is the process by which the static mass acting through each wheel is equalised as much as possible front/rear and side/side. If the car's mass is spread evenly across the four tyres, each will be able to provide better cornering grip, while a car with equal side/side weight distribution will have better braking and acceleration performance through decreased tyre slippage.

While corner weight adjustment can be carried out by shimming spring seats, by far the most common approach is to adjust the height of the lower spring seats on struts. Aftermarket struts equipped with threaded collars allowing this adjustment are commonly available, with just a handful of production cars also equipped like this from the factory.

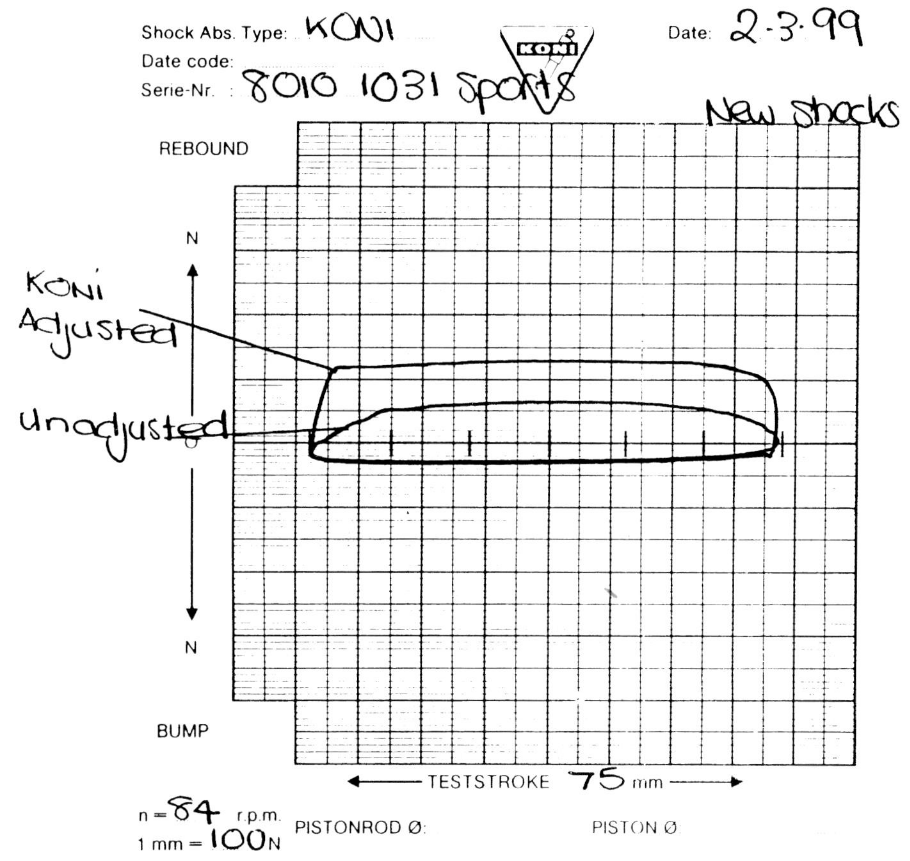

Figure 12.14
The range of rebound adjustment available on a Koni Sport damper. Most adjustable dampers can be altered only in rebound characteristics.

Fast Fours & Rotaries

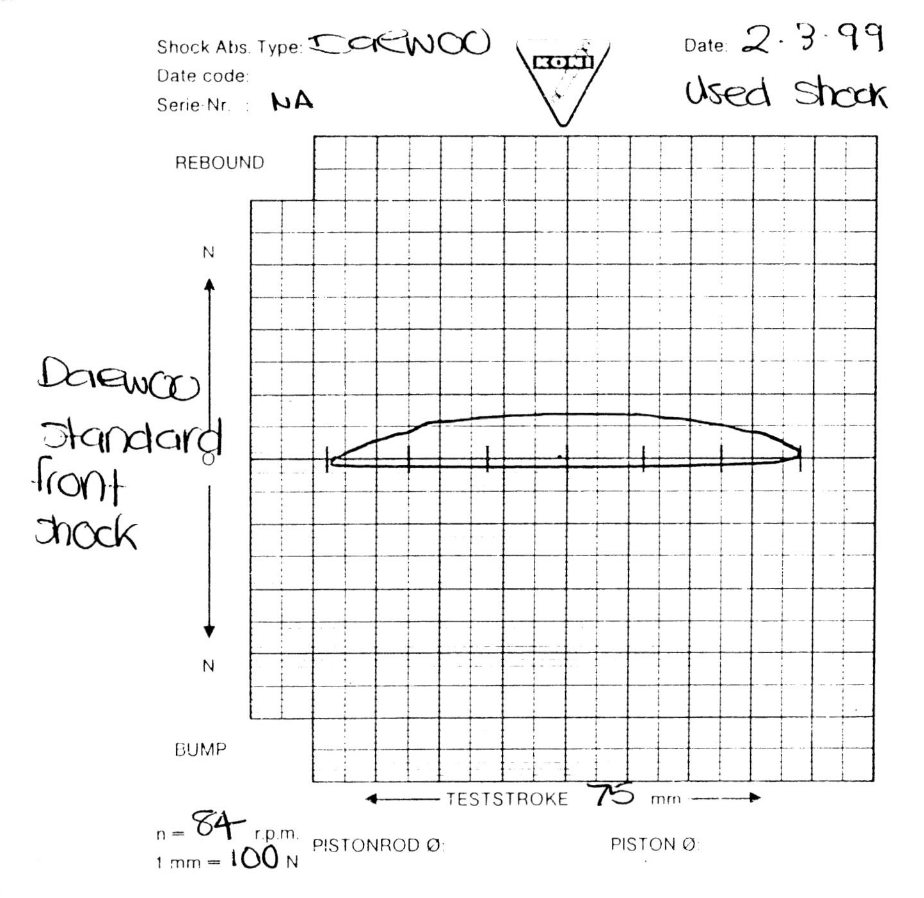

Figure 12.15
The bump/rebound characteristics of a standard Daewoo Lanos front damper. As can been seen, the bump damping is very soft when compared with the rebound damping.

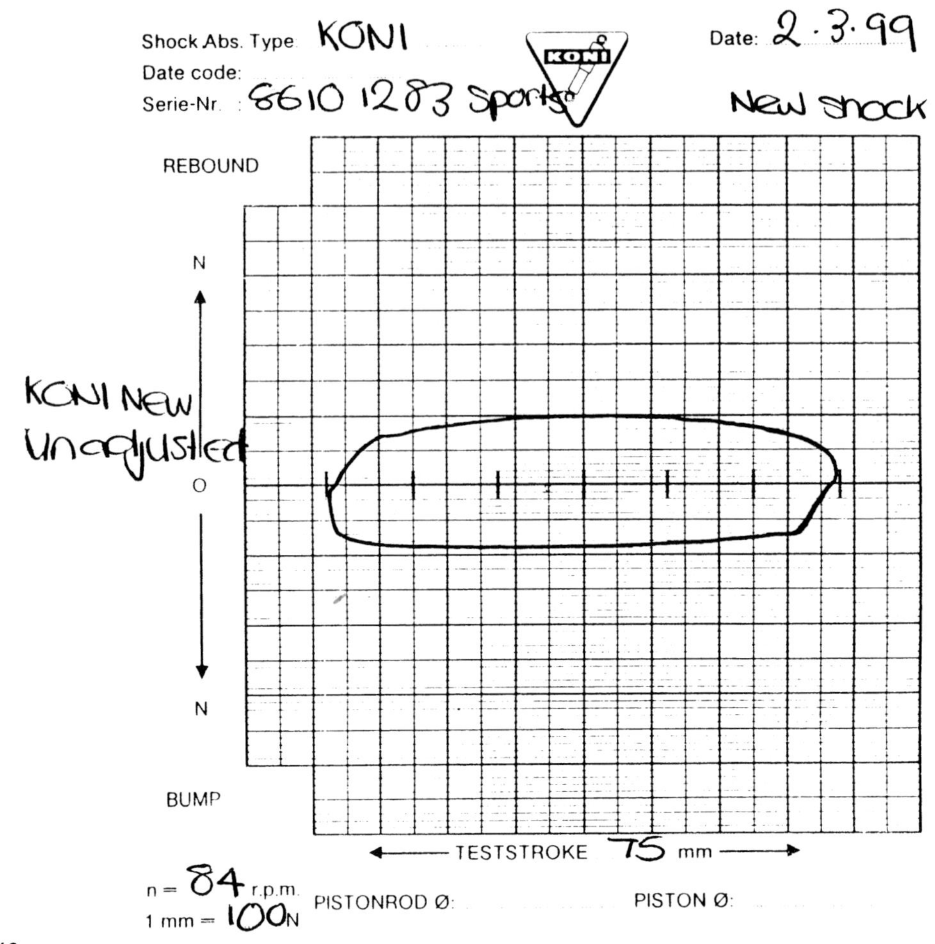

Figure 12.16
The damping characteristics of a Koni Sport damper in factory, unadjusted form. This damper was used to replace the Daewoo Lanos standard damper whose characteristics are shown in Figure 12.15.

Corner weighting requires a perfectly flat floor and an electronic corner weight scale system. Such a system uses four scales (one for each wheel) connected to a digital read-out panel that shows the individual masses, the total mass of the car and also the percentages of the total mass which is acting on the front, rear or diagonals. The car should be corner weighted with the driver in the vehicle (and the fuel tank half full), and with the sway-bars disconnected. The car should be bounced before each measurement to remove any sticking of the shock absorbers and to ensure that the springs are seated.

Corner weighting of a car by the use of electronic scales can be used to determine the mass of the car working through each tyre.

In a car with unequal front/rear mass distribution, heavy items such as the battery can be moved to the lighter end to help offset this difference.

If it's found that a wheel of the car has significantly lower measured mass than the other three corners, the spring seat can be wound up, causing that part of the car to be lifted slightly higher. That particular wheel then supports a greater portion of the car's mass. However, as indicated, the ride height will then also be higher — so there's a limit to how far you can go with this strategy! Many cars have an unequal front/rear weight distribution. This can be partially countered by moving portable, heavy items like the battery, sound system amplifiers and subwoofers from the heavy end to the light end. The lateral weight distribution can often be balanced by changing the spring pre-load.

BUSHES

Suspension systems have numerous pivot points that use rubber bushes. These bushes are designed to allow movement in certain planes and arcs, but when the suspension is loaded during hard cornering, braking or acceleration, they often deflect sufficiently to allow unwanted movements to also occur!

Before looking at replacement bushes, it's important to understand how a standard rubber bush works. Most bushes bond the rubber to both the inner and outer steel parts of the bush. The rubber consequently distorts in torsion as the suspension arm pivots. It's this internal shear characteristic of a rubber bush that allows it to be made accurately, means that it requires no lubrication and gives it excellent longevity.

The hardness of rubber bushes is often different in different planes, with voids and other manufacturing tricks incorporated to allow this to occur. Figure 12.17 shows an example of a tension rod bush that incorporates fluid-filed voids. This bush is designed to allow a measure of front toe-in under braking, improving vehicle stability. Figure 12.18 (overleaf) shows this process. Other suspension bushes in a car may allow some longitudinal deflection of the bush under low loads (absorbing small sharp bumps) while resisting much stronger vertical forces. Some rubber bushes are even designed so different frequencies of vibration are absorbed by differing amounts. In summary, most rubber bushes aren't quite the cheap and nasty design they can at first appear to be.

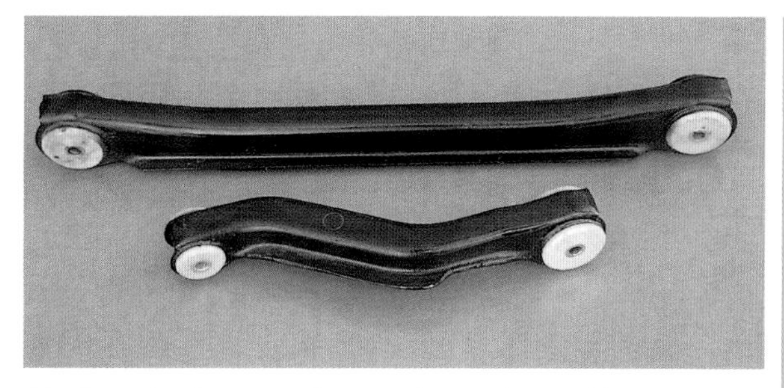

Urethane bushes are often used to replace rubber. Urethane bushes use a lubricated rotating steel sleeve to allow the suspension movement, whereas rubber bushes internally deflect to cater for this movement.

An offset plastic bush such as the one on the left can be used to give the suspension geometry further adjustment.

A common replacement for rubber is to use bushes made from urethane. The urethane deflects less under loads than most rubber bushes, maintaining better dynamic suspension geometry. However, urethane acts as an incompressible viscous liquid, rather than being able to change its volume under compression as rubber does. This means a captive urethane bush transfers greater forces directly to the bodywork. Also, when urethane bushes are used, there must be sliding movement between the pivot and the bush itself — the plastic bush will not internally deflect to allow for the suspension arm movement. To allow the rotational motion to occur,

Holden

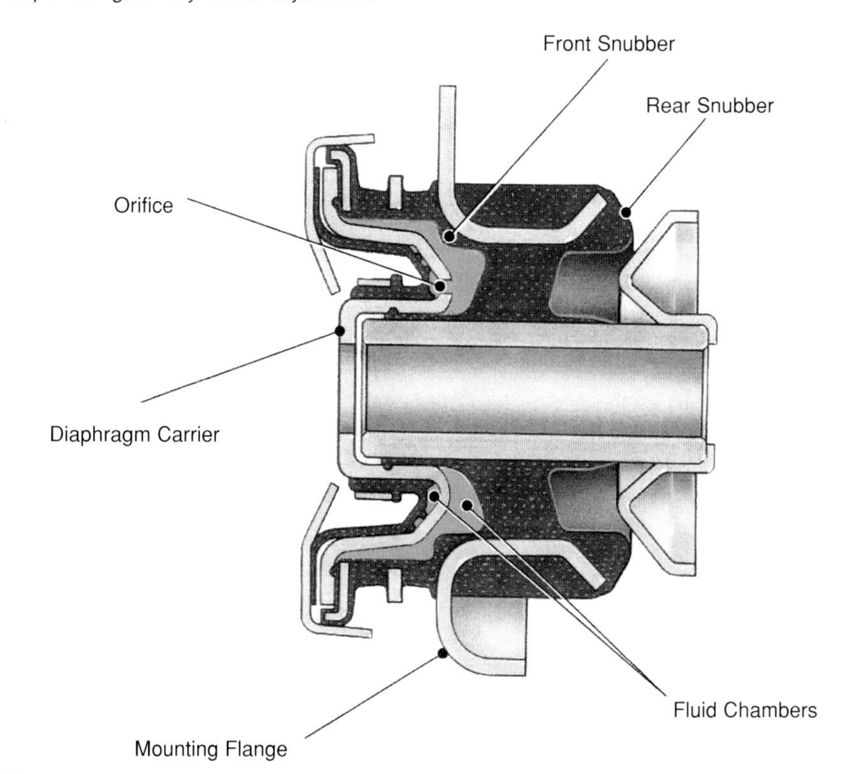

Figure 12.17
Many rubber suspension bushes are much more sophisticated than they at first appear. This tension rod bush contains fluid-filled chambers to allow deflection in certain directions while resisting movement in other directions.

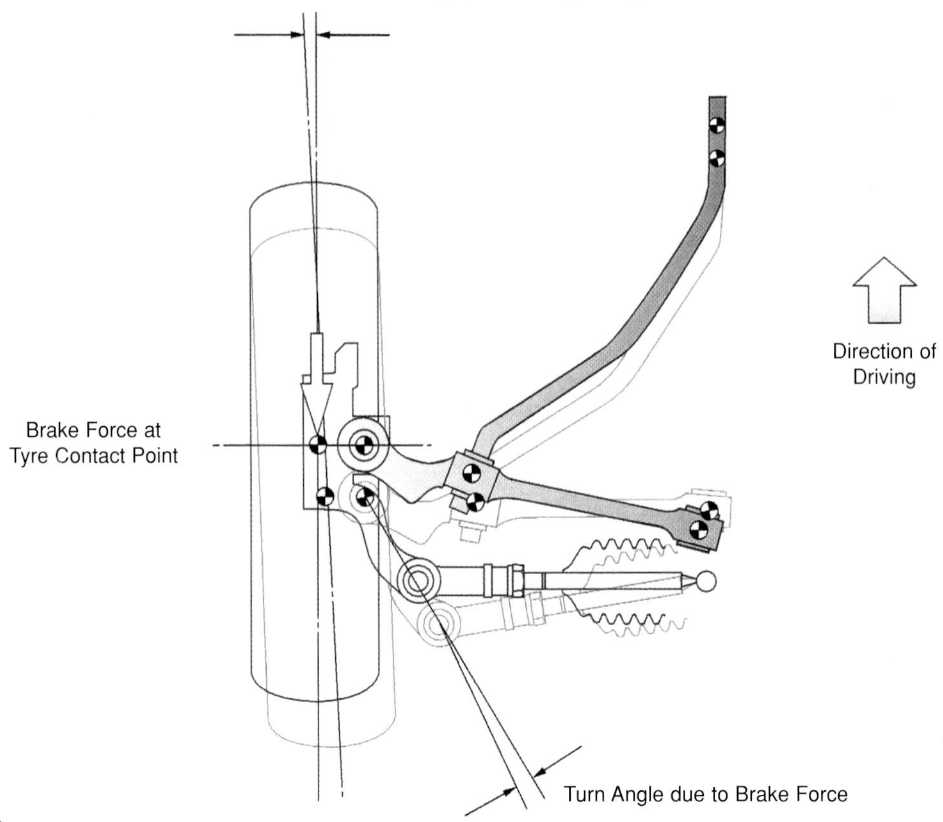

Figure 12.18
The front suspension on this car allows a small amount of toe-in to occur during braking, aiding stability. In this case, the bushes are designed to deflect so this occurs.

generally a steel sleeve is free to move within the urethane bush. A urethane bush therefore requires lubrication and accurate lateral location of the sleeve.

There is one situation where the new urethane bush does not require rotational lubrication, and that's when it replaces a rubber bush in a swaybar link. In this situation, the greater hardness of the urethane has no drawbacks and allows the swaybar to respond more quickly.

NVH is likely to be increased to a greater or lesser degree when urethane bushes are fitted, but far worse is the propensity for some urethane bushes to squeak. If the bush isn't equipped with grooves in which grease can be injected, or isn't made from the right material in the first place, it can be simply impossible to keep quiet. Squeaks, clicks and creaks rapidly ruin your appreciation of the more precise handling you might have gained with their fitment! I suggest that when buying any plastic suspension bush, you get a written, money-back guarantee that the new bushes will be quiet in normal operation. Note that the amount of noise generated by the bush is highly dependent on the fitting method — new surfaces must be clean and free of paint.

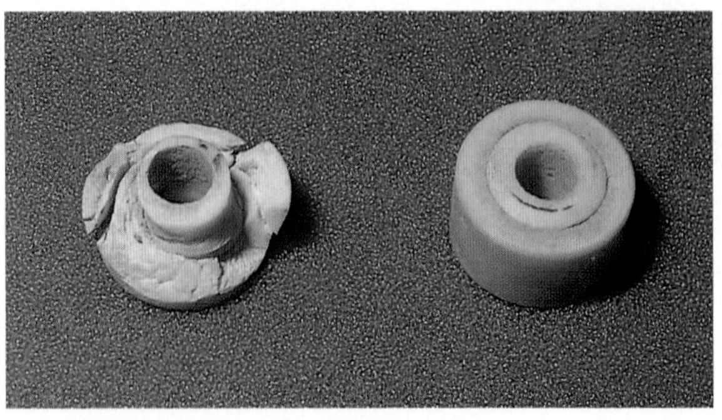

Not all urethane bushes are made equal. One of these bushes has become badly abraded, while other designs can constantly squeak, despite frequent lubrication.

It should be noted that changing rubber for urethane bushes in most cars does not give a major benefit on its own. However, if the cornering loads are far greater than standard through the implementation of other effective suspension mods (and you then drive the car hard!), urethane bushes will give some further improvement.

OVERSTEER AND UNDERSTEER

When a car is changing direction, the tyres must provide the required centripetal force. If the centripetal force developed before slip occurs is insufficient to provide the required cornering, the front wheels won't head in the direction in which they are pointed and the car will understeer. Understeering cars run wide in the corner, with the front tyres sliding first. A well set-up understeering car can be made to behave by momentarily reducing power, which transfers more weight to the front tyres and allows them to grip better. Manufacturers almost always design understeer on their road cars — it's safest for drivers who instinctively lift the throttle if the car starts to slide.

An oversteering car is one in which the rear tyres cannot generate the required centripetal force before slipping. The front wheels follow the line of the corner, but the rear

Front-wheel-drive cars tend towards understeer — the front of the car sliding first. This GTP Mitsubishi Magna shows major understeer and also wheelspin, despite having an LSD.

wheels aren't taking the same corner so sharply! The rear of the car slides and a spin will occur if no corrective action is taken. The recommended driver move is to unwind some steering lock (or even turn the wheel so the front wheels are pointing in the direction of the slide) to counteract the tail's behaviour.

Front-Wheel-Drive Cars

Front-wheel-drive cars understeer. This is the case because the front tyres are forced to perform two tasks: transferring torque to the road and turning the car. However, front-wheel-drive (FWD) cars have packaging advantages (the centre tunnel doesn't have to be large and the boot floor can be low without a rear differential getting in the way), so FWD cars are far more popular than rear-wheel-drive machines.

As indicated, in a FWD car the front tyres have to work much harder than the rear tyres. It is therefore logical to run wider tyres on the front than the back — in the same way that many powerful rear-wheel-drive cars run wider tyres at the back. However, fitting wider front tyres is rarely done, perhaps because the resulting appearance isn't considered fashionable! Even if not fitting wider front tyres, the front rubber can be of better quality to provide a similar advantage. Tyre pressures can also be adjusted, with the fronts running higher pressures. Some FWD performance cars have factory-recommended tyre pressures 3-4psi higher in the front than in the rear tyres.

More negative camber on the front wheels will allow the outside tyre to remain closer to vertical during cornering, so reducing understeer. Because in a FWD car it's the front outside tyre that's doing most of the work, the additional negative camber the inside front tyre will then have doesn't matter. Running lots of negative camber has a downside, though. If you don't frequently corner the car hard, the inside shoulders of the tyres will wear more quickly than the outside shoulders.

Braking performance will also be down a little. Increasing the amount of positive castor will increase the amount of dynamic camber that occurs, reducing understeer without the straight-line problems of running lots of static negative camber. If the front wheels have less cornering grip than the rear wheels, reducing the cornering grip of the rears will improve the overall handling of the car. Lower rear tyre pressures, less rear negative camber and even a little rear toe-out will each reduce the amount of understeer.

Many FWD cars respond to zero toe, a front negative camber of 1.5-2 degrees and a positive castor angle of 3 degrees (manual steering) or up to 5 degrees on power-steered cars. At the rear, 1.25-1.5 degrees of negative camber with a toe-out setting of 2-3mm is often effective.

Reducing bodyroll will result in the front camber angles being more consistently vertical. It will also decrease the effect of lurching weight transfers upsetting the car — through 'S' bends, for example. In a FWD car, roll is usually controlled by working at the **rear** of the car. The reason for this confuses many people, so I'll look at this more closely.

A swaybar (anti-roll bar) resists bodyroll by trying to lift the inside wheel when the car is cornering. If a thick swaybar is placed on the front of a FWD car, it will unload the inside wheel to a greater degree than occurred before its fitting. This means the outside front tyre is even more heavily loaded (doing its torque-and-turn duties), so it will slide earlier, not later. The result — more understeer, not less. Also, if the inside front wheel is being lifted, it will not be able to gain traction. Since most FWD cars don't have a limited-slip diff, the inside wheel will siphon away all the power, spinning freely just when you are trying to accelerate hard out of the corner! A thicker anti-roll bar fitted to the rear of the car (stiffness increases of 100 per cent are common) helps hold the car flat, counteracting understeer without causing the same problems. In the same way, higher-rate rear springs, rear dampers stiffer in bump and even extended rear bump rubbers will all help.

But can the rear roll stiffness be too great? Yes, it can. A FWD car with an over-stiff rear will have throttle-off oversteer. This can be a little tricky, especially in wet conditions or with an inexperienced driver. Someone who lifts off sharply when the car starts to understeer might be a little surprised when the back suddenly comes out! Very sporty FWD cars (and those set up for circuit use) frequently lift the inside rear wheel right off the ground because of the very stiff rear roll stiffness.

FWD cars that use extremely stiff front and rear bars often handle well in dry conditions, but on a wet road, can become very skatey. The car's handling then degenerates into plough understeer, where turning the wheel further makes no difference to the direction the car heads in. However, as with all anti-roll bar use, using adjustable bars allows the car's handling to be tuned to give the best results.

Changing the front/rear weight distribution can also be used to alter the car's handling characteristics. Heavy but movable items like the battery, sound system amplifiers and subwoofers should be located as far rearwards as possible. In some cars, you will even notice a handling difference when the fuel level is high or low.

Rear-Wheel-Drive Cars

Unlike FWD cars, those pushed from behind can be set up for near-neutral handling. With the torque load being borne by the back wheels and the front wheels responsible for just steering, the forces which the front and rear tyres are responsible for transmitting to the road can in many cars be near equal — giving handling dominated by neither understeer or oversteer. However, most RWD cars are set up for understeer when the throttle opening is held constant. It should be noted, though, that if the road is slippery or the engine powerful, even standard cars will power-oversteer when given a bootful of throttle.

The enormous advantage of a RWD car is that — if it has enough power — it can be balanced on the throttle. If a RWD car is understeering slightly through a corner, more throttle can be gently applied. This will cause the rear tyres to decrease in lateral grip, making understeer no longer the dominant trait. More or less throttle can be used to adjust the car's in-corner stance. However, if the car is over-tyred or under-powered, this level of throttle control will not be available. Also, if the engine's power delivery is non-linear (for example, in a highly boosted turbo car), the skill level of the driver will need to be high to allow this type of throttle control. This is one reason that, in turbo cars, the boost control system should be built with good throttle control in mind rather than having the boost arriving in a rush. Note also that if the car is set up so oversteer is excessive, the car will corner more slowly than it would with reduced oversteer. While it may look (and feel!) spectacular, when opposite lock steering is applied, the front wheels are pointing out of the corner rather than into it — that is, they're developing a cornering force in the wrong direction!

Because a RWD car transfers torque through the back wheels, it's necessary that they both stay on the ground. Lifting an inside rear wheel will give wheelspin (in cars without an LSD) and more power oversteer (in cars with an LSD). Note that the combination of an LSD and stiff rear bar can make the car quite twitchy, especially if the car uses a simple independent system like semi-trailing arms. RWD cars therefore tend to use greater front roll stiffness, gained by springs, dampers and anti-roll bars.

However, the approach taken will depend on the handling characteristics of the car. If understeer is strong, more front negative camber and a slightly stiffer rear bar can be used. Oversteer can be reduced by using wider and/or stickier rear tyres, fitting a larger front swaybar

This RWD BMW M3 is displaying oversteer — the rear of the car sliding under power as the corner is exited.

This Mazda MX5 (Miata) displays impeccable RWD handling — gentle understeer on entering the corner and gentle oversteer on the exit.

and softening rear spring and damper rates. Toe-in used at the back of a rear-wheel-drive car will also sometimes help to reduce oversteer, as will the use of increased rear negative camber.

Four-Wheel-Drive Cars

Four-wheel-drive performance cars can have an enormous advantage in cornering over their two-wheel-drive brethren. In fact, once you have owned a four-wheel-drive performance car, it can be difficult to return to a car that drives only half its wheels! The advantage comes from traction. Even a very powerful four-wheel-drive (4WD) car will be able to accept full throttle once you are past the apex, so that exit speeds can always be high — even with a safe, relatively slow entrance speed. However, the exact handling traits of a four-wheel-drive car are very dependent on the way the four-wheel-drive system works. Most 4WD performance cars use constant four-wheel drive; that is, all the wheels are driven all the time. However, this doesn't mean the front wheels and the rear wheels are always fed the same amount of torque. Many cars use a viscous coupling that initially splits the torque 50/50 front/rear and can then apportion torque either way up to about a 70/30 ratio. If the front wheels are spinning faster than the rears because they have lost traction, the rear wheels are fed more torque. The opposite occurs if the rear wheels are spinning faster than the front wheels. Rarer are cars that are predominantly front (or rear) drive, with four-wheel drive occurring only when slippage of the driven wheels is detected.

Those cars that are constant 4WD tend to understeer. This is because during cornering, the front tyres not only have to transmit a proportion of the torque to the pavement, they also have to turn the car. This understeer is most pronounced when a large amount of throttle is being used at the point of turn-in. Those 4WD cars that are normally two-wheel drive will have initial handling characteristics that depend on which end is being driven first.

Unlike a RWD or FWD car, all four-wheel-drive cars have the ability to change the wheels through which the majority of torque is being transmitted. Some cars can even change this torque split from side to side, not just end to end. The fact that this occurs automatically (not under direct driver control) needs to be considered when discussing the handling of these cars. Torque being transmitted through a wheel is normally reduced when the wheels at that end of the car are starting to rotate more quickly than the other wheels, **not** when that end of the car is starting to slide. This means that if a corner is entered too quickly and the accelerator is lifted at the apex, the torque split (or even the fact that it is a 4WD car) is largely irrelevant. Without power being applied, the car behaviour comes back to the basics of tyres, suspension geometry and weight distribution.

On the other hand, if a wet corner is being accelerated through and the front wheels start to rotate more quickly than the rears because they're losing traction, more torque will be fed to the rear wheels, so the front torque load will be reduced. If the car is power understeering at the time because of the large loads being applied through the tyres, the reduction in front axle power may well reduce the understeer. The same applies to the back wheels — if the car is power-oversteering with the rear wheels starting to spin, the automatic reduction in torque supplied to the back axle will decrease the oversteer. But these points apply only when the car is under power.

Because the handling behaviour depends on the tyres' longitudinal grip (not just lateral), tyre choice is critical on a 4WD car. A tyre with a soft sidewall but good 'acceleration' grip will affect the cornering torque split that occurs. In some cars fitted with tyres having these characteristics, the torque split will not alter or will be slow to change — the sidewall distortion will cause the tyre tread to lift (usually causing understeer) but the front wheels will not be spinning any faster than the rears as they slide across the road! Without the normal change in torque split, the car will feel very much like a FWD machine. If the torque split is able to be modified by mechanical or

Handling Solutions For Front-, Rear- and Four-Wheel Drives

Drivetrain Type	Excessive Oversteer	Excessive Understeer
Front Wheel Drive	• Change rear tyres – wider, softer, lower profile, higher tyre pressures • Increase rear toe-in • Increase rear negative camber • Stiffer front anti-roll bar • Stiffer front springs • Stiffer front shock absorbers • Soften rear anti-roll bar • Soften rear springs • Soften rear shock absorbers • Change driving style	• Change front tyres – wider, softer, lower profile, higher tyre pressures • Increase front negative camber • Increase front positive castor • Increase rear toe-out • Stiffer rear anti-roll bar • Stiffer rear springs • Stiffer rear shock absorbers • Soften front anti-roll bar • Soften front springs • Soften front shock absorbers • Fit front Limited Slip Differential • Fit anti-lift front suspension kit • Fit traction control system • Change driving style
Rear Wheel Drive	• Change rear tyres – wider, softer, lower profile, higher tyre pressures • If it is power oversteer, fit open rear differential • Increase rear negative camber • Increase rear toe-in • Stiffer front anti-roll bar • Stiffer front springs • Stiffer front shock absorbers • Soften rear anti-roll bar • Soften rear springs • Soften rear shock absorbers • Fit traction control system • Change driving style	• Change front tyres – wider, softer, lower profile, higher tyre pressures • Increase front negative camber • Increase front positive castor • Stiffer rear anti-roll bar • Stiffer rear springs • Stiffer rear shock absorbers • Soften front anti-roll bar • Soften front springs • Soften front shock absorbers • Change driving style

Drivetrain Type		Excessive Oversteer	Excessive Understeer
Four Wheel Drive	Constant four wheel drive	• Change rear tyres – wider, softer, lower profile, higher tyre pressures • Increase rear toe-in • Increase rear negative camber • Stiffer front anti-roll bar • Stiffer front springs • Stiffer front shock absorbers • Soften rear anti-roll bar • Soften rear springs • Soften rear shock absorbers • Alter torque split to give greater front bias • Change driving style	• Change front tyres – wider, softer, lower profile, higher tyre pressures • Increase negative camber • Increase positive castor • Increase rear toe-out • Stiffer rear anti-roll bar • Stiffer rear springs • Stiffer rear shock absorbers • Soften front anti-roll bar • Soften front springs • Soften front shock absorbers • Alter torque split to give greater rear bias • Fit anti-lift suspension kit to front • Change driving style
	Initially front or rear wheel drive, then four wheel drive	• Change rear tyres – wider, softer, lower profile, higher tyre pressures • Increase rear toe-in • Increase rear negative camber • Stiffer front anti-roll bar • Stiffer front springs • Stiffer front shock absorbers • Soften rear anti-roll bar • Soften rear springs • Soften rear shock absorbers • Alter torque split to give greater front bias • Change driving style	• Change front tyres – wider, softer, lower profile, higher tyre pressures • Increase negative camber • Increase positive castor • Rear toe-out • Stiffer rear anti-roll bar • Stiffer rear springs • Stiffer rear shock absorbers • Alter torque split to give greater rear bias • Fit anti-lift suspension kit to front • Change driving style

Note: Not all of the suggestions in these tables will be appropriate for every kind of situation.

For example, if the front and rear suspension of an understeering rear-wheel-drive car is simply much too soft, softening the front end further will probably make the understeering condition worse, not better! The easiest and cheapest initial fixes for handling problems are to alter tyre pressures and change (or fit) anti-roll bars.

Constant four-wheel-drive cars like this Pulsar GTiR tend towards understeer. The inside front tyre is doing very little work in this corner.

electronic means (see the next section), a major change can be made to the handling characteristics of a 4WD car in power-on situations.

With understeer the dominant characteristic of most 4WD cars, measures that reduce front-end push are often undertaken. This can include the use of a stiffer rear anti-roll bar and/or stiffer rear springs, increased front negative camber and/or less rear negative camber, slight rear toe-out and higher front tyre pressures. Increased castor (giving more dynamic camber) and measures that reduce front-end lift under power (such as suspension anti-lift kits) will also reduce understeer.

It should be remembered that a four-wheel-drive car still has only the four tyres through which all the cornering, acceleration and braking forces need to be transmitted. Therefore, the outright grip available (in all directions) to a four-wheel-drive car is no different from a two-wheel-drive car with the same-sized tyres! This is because a tyre only has so much grip potential — borrow some for traction and there will be less lateral grip available; borrow some to overcome the lateral forces and there will be reduced longitudinal grip! While four-wheel-drive cars flatter drivers, a loss of control in a four-wheel-drive car is just as deadly as in a two-wheel-drive machine...

FOUR-WHEEL STEERING

To a greater or lesser degree, many cars use four wheels to steer. At the 'lesser' end are those cars that use passive four-wheel steering, where the rear wheels are moved through small steering angles as the suspension bushes deflect under cornering or braking forces. The Porsche 928 was one of the first cars to use this technique. In the Porsche, the rear wheels toe-in under deceleration loads, reducing the likelihood of oversteer when braking or backing-off partway through a corner.

Most sporting cars these days use similar passive rear-wheel steering. For example, the Alfa Romeo Spider uses a rear steering suspension link on its double-wishbone suspension to momentarily steer the outside rear wheel out of phase with the front wheels to help turn-in, then in phase with the front wheels once the corner has been entered. (Out of phase means the front and rear wheels turn in opposite directions, while in phase means both pairs steer in the same direction.)

A more sophisticated approach is to use a hydraulic ram that actively moves the rear wheels, again often by deflecting the suspension bushes. Some Nissan Skylines take this approach, which is dubbed HICAS. In the HICAS system, the rear wheels can be steered through an angle

This 4WD Audi has very high rear roll stiffness so the work on the outside front tyre is reduced. As a result of this roll stiffness, the inside rear wheel is off the ground!

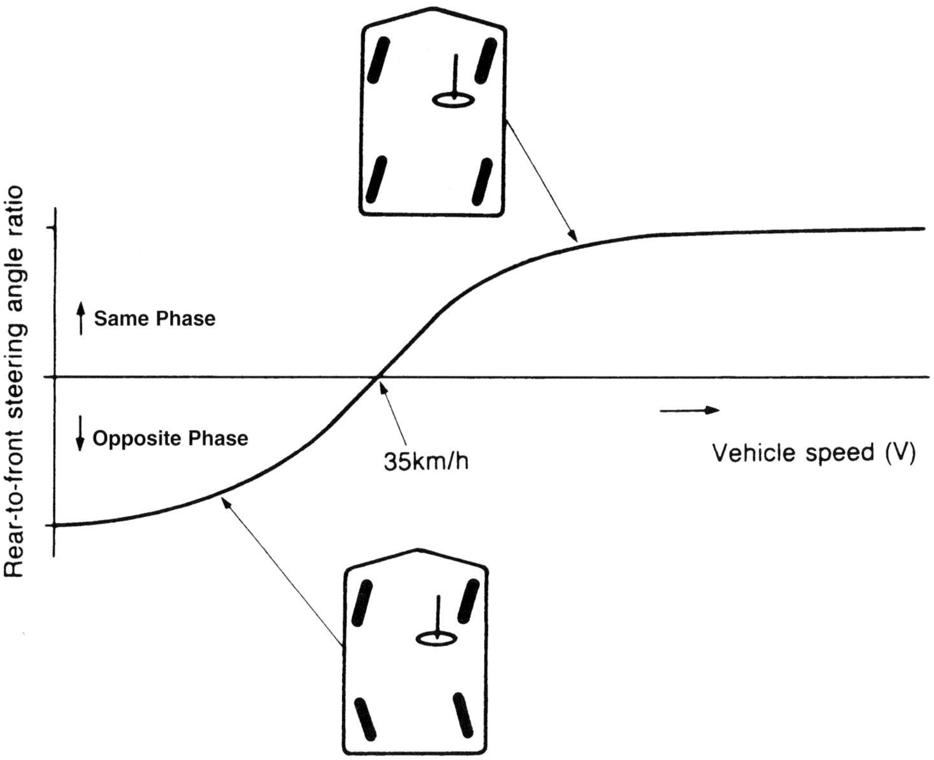

Figure 12.19
Cars with four-wheel steering usually turn the rear wheels out of phase with the fronts at low speeds and in phase at higher speeds.

of only 1 degree, versus, for example, the same car's front inner wheel steering angle of 38 degrees. Like the Alfa Romeo passive system mentioned above, HICAS momentarily steers the rear wheels out of phase to aid turn-in. The amount of rear steering angle is apparently dependent on the speed of the steering input. One HICAS system uses an electronic control unit with inputs from neutral, clutch and brake switches; a steering angle sensor; oil level switch; and high speed sensor. Its outputs are 'right' and 'left' solenoid valves, a power steering solenoid and a failsafe valve.

The most sophisticated four-wheel steering systems use a steering rack to move the rear wheels through larger angles — as much as 5 degrees in the case of one Mazda. In the Mazda system, the rear wheels are steered out of phase with the fronts at speeds of less than 35km/h, there is no rear wheel steering at 35km/h, and then the rear wheels are steered in phase at speeds of over 35km/h. Figure 12.19 shows this approach. The change in the phasing of the rear-wheel steering is achieved by the use of a stepper motor that controls components in the rear hydraulic system, with the steering angle itself transferred to the rear of the car via a long shaft. The rear steering system uses an electronic control unit, with the main inputs being two speed sensors and a front-to-rear steering ratio sensor. The outputs comprise the stepper motor and two solenoid valves.

While I have not modified a four-wheel steering system (and neither have I spoken to anyone who has), there's a few points to bear in mind. It's noticeable in at least one powerful four-wheel-steered FWD Mazda that at around the speed where the rear wheels do not steer (being midway through their 'out of phase' to 'in phase' transition), the car has much more understeer than at either lower or higher speeds. Since manufacturers use the low speed out-of-phase steering simply to give better manoeuvrability through the generation of a smaller turning circle, when a car is being driven hard it appears that it may respond better in pure handling terms if in-phase steering is used all the time. Altering the speed signal input to the steering electronic control unit may be able to be used to achieve this.

However, before modifying a four-wheel steering system, it's wise to temporarily disable it (usually easily achieved electronically by removing a fuse or using a similar technique) and drive the car in this two-wheel steering form. If the difference is minimal or not even noticeable (the case with some simple four-wheel steering systems), it's probably not worth making the effort to change the way it works. However, if there is a major improvement with the system operating, you may be able to improve car behaviour still further! Finally, in any modification of a four-wheel steering system, take care during testing to ensure that the steering system behaves in an expected manner. For example, don't jury-rig wiring that then falls off during high-speed driving — the results could be rather nasty!

ELECTRONIC STABILITY CONTROL SYSTEMS

Many of the world's car manufacturers are adopting electronically controlled systems that actively change the way a car handles. Systems such as Vehicle Stability Control, Active Yaw Control, Active Stability Control, ATTESA ET-S and Electronic Stability Program are all active systems that sense what the car is doing and take steps to change that behaviour. Unlike automatically adjusted dampers, for example, the effect on the car's handling can be quite dramatic. Modifying the handling of cars

Those cars with active four-wheel drive or Automatic Stability Control Systems may be able to be electronically modified with very good results. Here's the power oversteer that was present in this Skyline GT-R, in all-low gear, full-throttle applications before the torque split controller modifications...

equipped with any active electronic handling system should be carried out quite differently from those mods done on a conventional car's suspension.

The active systems work in two quite different ways. Most common is the braking of a single wheel, sometimes with a reduction in engine torque occurring at the same time. Understeer is controlled through the braking of the inside rear wheel. How this works can be understood if you imagine an understeering vehicle attempting to negotiate a righthand bend. The front of the car is sliding wide, so the system brakes the righthand rear wheel, while the other wheels continue to turn at their normal rate. This causes the car to attempt to pivot around the slowed wheel to the right, so reducing the understeer. When the vehicle is oversteering, the outside front wheel is braked to near lock-up. Again, if you picture this wheel nearly stopped but the others continuing at normal speed, it can be seen that the car will attempt to pivot around, reducing the amount of oversteer.

The other approach is to change the way torque is directed to the wheels. In an active four-wheel-drive car, the control strategy is based on reducing the amount of torque being transferred to the end of the car losing traction. This means if the rear wheels are spinning, torque is directed to the front. Alternatively, in some systems, torque can be directed from one side to the other, helping the vehicle turn the corner because the wheels on one side of the car are being driven with more torque than those on the other side.

Input sensors used in these systems include individual wheel speed, steering angle, yaw rate (how quickly the car is turning around a vertical axis) and acceleration

...and here's the stable and fast car, cornering hard after the electronic modifications had been performed!

Changes can be made to the output of the G-sensor so the driveline control computer believes the car is cornering with either greater or lesser lateral acceleration than it really is.

both laterally and longitudinally. The outputs can comprise (depending on the system) the ABS hydraulic control unit, wet multiplate clutches or active differentials.

It's very likely that enterprising companies will soon release 'hot chips' for these systems to dial up a little more oversteer than the standard systems will allow you, or simply prevent the systems cutting in until larger slip angles occur. However, it's still quite possible to make DIY changes. How? By changing the input signals the stability control ECU receives.

Electronically Modifying a Torque Split Control System

One of the reasons I'm covering electronic handling in this book is that my own car has an active handling control system I have changed substantially. The difference the modifications made to the car are so great that to equal the changes using the traditional approach of fitting new tyres, wheels, springs, dampers and swaybars would have cost many thousands of dollars. I know that, because other owners of the same type of car have taken just that expensive traditional route! Modifying the car electronically cost less than one-tenth of this and has other considerable advantages as well.

The way I modified my car can serve as an example of how the handling of current cars can be modified more cheaply and with better results than was once the case, even though at first such a task looks daunting, if not impossible!

This control module was built so the output of the lateral G-sensor in a Nissan Skyline could be varied.

The new control knob on the dashboard varies the output ratio of the lateral G-sensor, so changing the front-rear cornering torque split the car employs. Extreme changes to low-speed handling can be made by adjustment of this control.

The R32 Nissan Skyline GT-R is a twin turbo 2.6-litre six-cylinder high-performance car. Regarded by many as one of the best cars ever produced in Japan, it uses a four-wheel-drive system that sends all the torque to the rear wheels most of the time. I was unaware of this drivetrain trait when I bought the car and — having stepped out of a constant four-wheel-drive turbo car — was very disappointed with the amount of power oversteer present.

The car uses a wet multiplate clutch that can apportion torque to the front wheels on an infinitely variable basis, up to a maximum of 50 per cent. The clutch is controlled by a dedicated ECU, which directs a pump to vary hydraulic pressure to the clutch. The driver can see precisely what is occurring through the presence in the instrument cluster of a Torque Split gauge, which shows how much torque is being directed to the front wheels.

Unfortunately, from my point of view, the Torque Split gauge needle seemed stuck on zero whenever I cornered! The torque split computer worked well in a straight line, sending power to the front wheels whenever wheelspin was detected, but — somewhat disconcertingly — the same approach seemed to happen when cornering as well. This meant the front wheels only started pulling when the rear was sliding well sideways, back wheels literally spinning. However — and it's an important point — this was how Nissan had designed the car to operate. What I considered chronic power oversteer they saw as

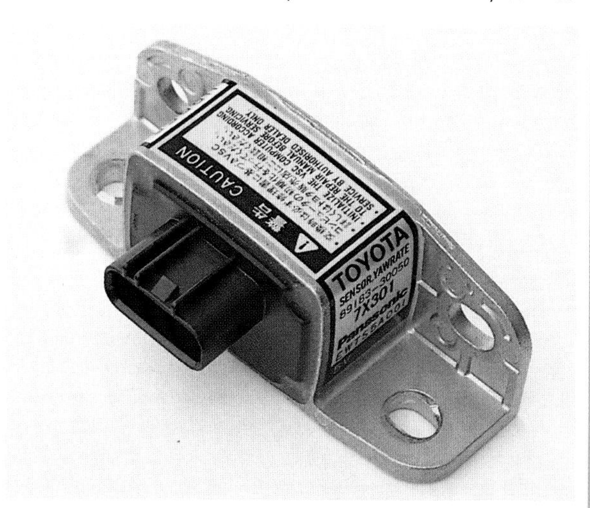

The yaw rate sensor is another sensor lending itself to the use of interceptor-style output changes.

	Longitudinal Accelerometer 1 (volts)	Longitudinal Accelerometer 2 (volts)	Lateral Accelerometer (volts)
Stationary	2.5	2.5	2.6
Moderate acceleration, 2nd gear	3.0	3.0	2.6
Strong acceleration, 1st gear	3.7	3.8	2.8
Left turn, high speed	2.6	2.7	4.0
Right turn, high speed	2.7	2.6	1.1

Figure 12.20
The measured voltage outputs of the G-sensor in a Nissan Skyline GT-R. These measurements were used to help determine what modifications needed to be made to modify the front/rear torque split of the four-wheel-drive system.

overcoming typical four-wheel-drive understeer. What I saw as a miserably handling beast liable to bite at any provocation, many motoring journalists saw as the best handling car ever!

My requirements for the car meant modification of the electronically controlled four-wheel-drive system was needed. More than anything else, it seemed to me that the four-wheel-drive system was simply too slow to come into action. My understanding through reading popular magazines was that four-wheel drive occurred only when there was rear wheelspin, a very large throttle opening was used or the G-sensor detected enough lateral acceleration.

Those cars with electronic damping control often have after-market systems available to allow in-cabin changes to be made.

The Japanese aftermarket manufacturers Blitz, Field and HKS all produced torque split controllers for the GT-R. I sourced a Japanese magazine review of one of the controllers and had the article translated. The translation showed that the device plugged into the centrally mounted G-sensor and that it certainly altered the handling characteristics! From this I decided that changing the output of the G-sensor would alter the torque split characteristics of the car.

The other avenue to pursue would have been to fool the torque split ECU into thinking there was wheelspin occurring, even when there wasn't. In fact, fitting different-sized tyres to either the front or back axle would achieve this (the tyres' rotational speed actually **would** be different), but I didn't want the car to be in four-wheel drive all the time if it was never intended by the designers that this be the case. Turning the car into a constant four-wheel drive would quickly wear out the small front differential, for example. Another way of making the ECU think there was wheelspin occurring would be to intercept the wheel-speed sensor outputs and change them. However, altering a frequency output signal like this (while possible) is much more difficult than changing a voltage signal.

So it was decided that the G-sensor (ie accelerometer) output be increased, so the four-wheel-drive ECU thought the car was cornering harder than it really was. I expected it to then direct more torque to the front wheels. The first step was to study the output characteristics of the G-sensor. At this stage no workshop manual was available (and when it was later purchased, it had little detail on the torque split system!), so, armed with a multimeter and an assistant, I measured the output of the G-sensor in all sorts of driving conditions. This testing proved there were actually three G-sensors within the one package. Figure 12.20 shows what was found.

It's important to realise that the sum knowledge of how the G-sensors behaved was found with a multimeter, careful road testing and a pen and paper. From these measurements it was possible to work out that two of the sensors measured longitudinal acceleration (ie acceleration and braking) and the other measured lateral acceleration (ie cornering). All three sensors had a 0-5 volt output signal. When there was no acceleration, the sensors all had about 2.5 volts output. The harder the car accelerated in a forward direction, the higher the voltage output was from the longitudinal G-sensors. When decelerating, the sensor voltage dropped below 2.5 volts. The lateral G-sensor had a voltage output that decreased below 2.5 on right turns and increased above 2.5 volts on left turns.

My first step was to increase the voltage going to the ECU from the lateral sensor. To do this, I experimented by wiring a 1.5-volt battery in series with the signal wire so its output was boosted by 1.5 volts. However, this caused the ECU to shut down the four-wheel-drive system completely (along with the ABS!) because the ECU sensed a fault condition with the engine idling but apparently the car cornering hard!

My next step was to approach some electronics experts with a clear requirement. I wanted an amplifier that increased the output swing of the voltage whenever it varied from 2.5 volts, while leaving a 2.5-volt input unchanged. That way, the ECU wouldn't know something was wrong. The amplifier had to have a gain (ie an amplification ratio) of up to two times and needed to work from a 12-volt automotive power supply. I might add that if you're getting lost in all of this, I do not have the electronic skills to design or build such an amplifier — but there are plenty of electronics technicians who do! All you must have is a quite precise idea of the requirements of the circuit.

The amplifier was built and wired into the lateral G-sensor output, so this signal was boosted whenever it swung away from 2.5 volts. In other words, as far as the ECU could detect, every corner I drove around gently was really being taken at an incredible pace! Sadly, there was absolutely no difference in the car's cornering behaviour and the dash-mounted torque split gauge also behaved as standard. Weeks of experimentation followed. The amp was working — when it was connected to the longitudinal sensors, I could boost straight-line acceleration torque going to the front wheels, but this gave only limited benefits when cornering.

I then thought carefully about the effect each of the G-sensors was having on the torque split. When I accelerated in a straight line, torque went to the front wheels — even when there was no wheelspin. If indeed the lateral accelerometer seemed to have little or no effect over cornering torque split, why not make the lateral sensor the longitudinal, and the 'powerful' longitudinal sensor the lateral? And how to easily do that? Simply by turning the whole G-sensor box through 90 degrees!

Doing this made a stunning difference to the handling. Instead of every slow corner exit requiring a gentle foot and/or opposite lock, full throttle could be used everywhere! Because it was a torque direction change rather than a handling change, the effect was uncanny. The immediate difference was far in excess of fitting, say, race rubber or new dampers. The whole car felt incredibly different — much less powerful, even. This was because the behaviour of the car was so much tamer.

However, this simple fix caused some problems of its own. First, in tight corners taken with a high entrance speed, the car could now understeer excessively. Also, the confused torque split computer allowed some wheelspin high in the rev range in first gear. With a wet road, the car also felt a little odd — something not quite right but hard to positively identify. I then happened to speak to a man who had been involved in the racing of Skyline GT-Rs. He suggested that, in fact, the lateral G-sensor output actually **decreased** the amount of torque going to the front wheels as you cornered harder! This meant that instead of trying to boost the lateral G-sensor signal, I should have been trying to **reduce** it.

I returned the G-sensor orientation to standard and organised the design and building of an adjustable amplifier that could give less than 1:1 gain (ie reduce the output swings instead of increasing them) while still leaving the 2.5 volt 'stationary' signal untouched. That way, with a calibrated knob on the dash, I could adjust the torque split system from standard right through to the lateral G-sensor having no influence at all.

And that's just what the final configuration is like. The knob is calibrated 0-10, with 0 being standard (no change to the G-sensor output) and 10 being for lots of front-wheel drive (the G-sensor output held at 2.5 volts all the time). In dry conditions I use 5, while 4 is used in really tight low-speed corners when I want the tail to come out a bit to turn the car in. When the road gets wet, 8 or 9 is selected, and when it's streaming with water I'm a 10 man! Set up in this way, the handling is absolutely fantastic. The car has immense stability, quite phenomenal wet weather handling and unchanged straight-line torque split control. The adjustability of the system means a turn of the knob can instantly suit the car to varying road surfaces, tyre wear and even driver mood. If I wish, I can even change the handling behaviour from corner to corner. And I'd like to stress that the difference in handling is extreme — far greater than I would have considered even possible.

While this example is obviously specific to the R32 Skyline GT-R, it's very likely that other cars with active torque split or stability control systems can be modified in a similar way.

BRAKES

A car modified for performance should have upgraded brakes. Not only can you safely slow from the higher speeds you will attain, you will also be able to use the brakes as an integral part of setting fast point-to-point times. Late braking in a powerful car with excellent retardation is nearly as much fun as exiting the corner fast!

Pads

Upgrading the brakepads can make a very significant improvement to braking performance. Reduction in fade, improvements in initial bite, and stopping distances that are dramatically decreased can all result from the fitting of very good brakepads. Unfortunately, selecting brakepads is a little like buying tyres — despite the specifications and brochures, you won't really know how good the pads are in your car and for your driving style until they're fitted! Often it's easiest to get the advice of

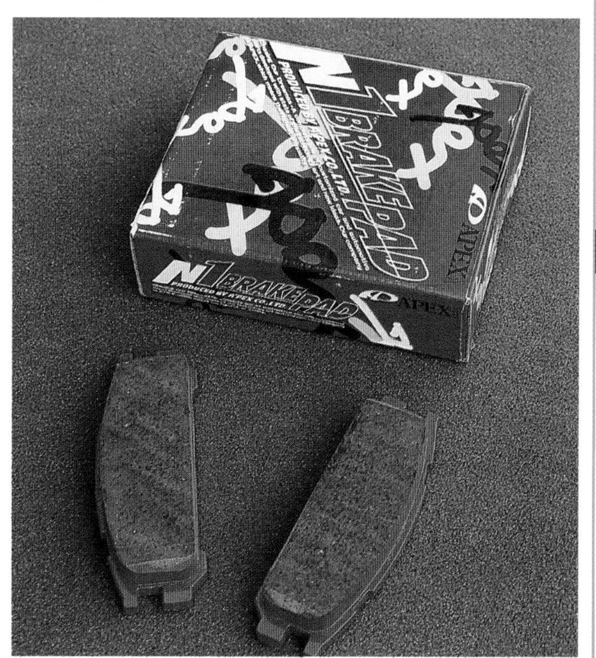

Good-quality brakepads can make a substantial difference to the braking performance of a car. The fitting of better-quality pads should be the first upgrade undertaken to a braking system.

The testing of brakepads and discs by manufacturers is often extreme!

someone else with a similar car who has made a successful pad upgrade. Members of car clubs who use their road cars to participate in club track sprints and the like are a good source of information on brakepads.

Brakepads designed to work in pure race applications are not usually suitable for the road, because they're effective only when they have reached high operating temperatures. If this type of pad is used in a road application, it will lack feel and require high pedal effort in normal intermittent applications.

The basic requirements for brake pad frictional material are that it must:

- Be able to cope with the thermal and mechanical stresses
- Have a stable coefficient of friction over the design temperatures, pressures, rubbing speeds and compressive loads
- Have an acceptable life with minimal disc wear

Inorganic material is normally used in brakepads in conjunction with organic friction modifiers. A single pad compound may have as many as 15 different reinforced fibres (inorganic material) and six organic friction modifiers. Steel and synthetic ceramic fibres are widely used. When steel fibres are used, the pads are referred to as being 'metallic'. Where some of the steel fibres are replaced with other types of fibre such as carbon, the pads are referred to as being of the 'semi-metallic' type.

The fitting of brake ducts is a cost-effective and simple way to reduce heat-generated fade. Here the disc backing plate has also been removed.

Asbestos fibre was previously widely used in brakepads but asbestos has now been banned in many countries because of the health risks. Semi-metallic pads have greater power dissipation per unit area than asbestos pads and are now widely used.

Most standard road brakepads work well in the range from ambient to about 300 degrees C, while better road pads will still be effective when the disc temperature reaches 600 degrees C. If data is available on the pads, the coefficient of friction of the pad should be as high as possible and it should be constant from ambient until very high temperatures are reached. A coefficient of friction of about 0.3 is typical of a conventional road pad, rising to 0.5 for very good-quality pads. If you're on a tight budget, selecting pads designed for towing or other heavy-duty work can result in a pad that works surprisingly well in a performance application. However, as with all high-efficiency pads, expect disc wear to be increased when using this type of brakepad.

In a MacPherson strut car an effective, cheap and easy brake upgrade can sometimes be made by swapping in a direct replacement strut already factory-fitted with larger brakes. However, the front/rear brake balance and pedal travel may both require further modifications.

Cooling

Improving the flow of cooling air to the front brakes (which do by far the bulk of the work) can make a dramatic reduction to fade. If the brakes go 'off' after 10 or 15 minutes of hard driving — but then recover relatively quickly — adding brake ducts will substantially increase the time before fade occurs.

Brake ducts can be easily made using a length of flexible high-temperature hose (1.5-3 inches in diameter, depending on the size of the car and the clearances available), some plastic plumbing fittings available from a hardware store and an assortment of clamps and brackets. To install the duct, a hole is drilled in the disc backing plate (or the backing plate is removed all together), with the hose either butting against the hole or aimed at the disc from a short distance. The other end of the flexible duct is then run to the front of the car, picking up air from a suitable high-pressure location such as a spoiler opening or a new forward-facing plastic bellmouth. Points to consider during the installation include provision of adequate clearance for the tyre at all steering locks and suspension deflections, and the choice of a high-temperature hose that will not catch fire or melt onto the brake caliper and disc when the going gets tough!

The drilling of normally undrilled discs (as has been done here) is not recommended. The discs will almost certainly subsequently crack.

Discs and Calipers

Replacement of the standard disc rotor with an upgraded disc that's slotted and/or cross-drilled is an easy step to improving braking performance. Taking this route also allows the retention of the standard calipers and pads, if desired. The term 'cross drilling' refers to the rows of holes drilled through the friction surfaces of the rotor, while 'slotting' describes the series of shallow grooves milled from near the centre of the disc towards the edge. These two procedures give the disc better ventilation, more pad bite, better out-gassing properties (ie the 'cushion' of gases being produced by the working pad is more easily removed) and a lower disc mass. The wiping motion of the slots and holes across the pad also help to deglaze them. Contrary to what might be expected, slotted and drilled discs do not increase pad wear in road cars. The reasons for this are the shaving effect of the slots (which ensure improved contact between pad and rotors) and the lower running temperatures.

Fitting grooved and drilled discs is a simple brake upgrade. The disc stays cooler and the pads work more effectively.

The slots milled in the rotors of this type are typically about 1.5mm deep and 3mm wide. Typically, about 36 holes per disc face are drilled, with each hole having a diameter of about 6.5mm and being chamfered around its opening to reduce pad abrasion and provide a smooth transition between hole and friction area, so combating surface cracking. To further counter the tendency for cracks to form between holes, only one hole per vane on ventilated discs is usually made. In road applications, the holes are not located near the edge of the disc — the part of the disc most subject to expansion and contraction through heating and cooling cycles. The removal of metal that occurs in the drilling and slotting processes makes this type of disc a little lighter than conventional discs. However, the weight saving is not great — about 3.8 per cent in one case. In addition, the pad swept area of drilled and slotted discs is decreased, being down by about 8 per cent over a conventional disc.

One of the problems in using cross-drilled discs in road applications is their propensity to crack around the holes. In fact, manufacturers of aftermarket discs featuring cross-drilling and slotting report a return rate more than four times higher than for the standard undrilled product. Returns were most often made because of major cracking. Note that very small cracks emanating from the holes are **not** usually a cause of concern.

A master cylinder brace that prevents the firewall flexing will improve braking feel.

Back-to-back testing of a car equipped with slotted and drilled discs versus the standard discs showed a significant reduction in stopping distance. From 100km/h, the average stopping distance with the upgraded discs was 38.6 metres, versus the 42.6 metres of the standard discs.

For many high-performance cars, major brake upgrade kits are also available off the shelf. These usually comprise new discs, calipers (often of racecar origin) and pads. Generally, the wheel diameter needs to be increased over standard to provide adequate caliper clearance. There are some superb kits available, but the quality of the components used are such that these kits usually cost a great deal. For this reason, check thoroughly the legality, warranty, service back-up and braking performance of the kits before you buy.

The fitting of a motorsport calipers and large discs can radically upgrade braking. However, check the legality, warranty and service backup of the kits before you buy.

Aerodynamics

Volkswagen

For many years, the aerodynamic behaviour of road cars was almost totally ignored. Manufacturers made sure the engine cooling and cabin ventilation systems worked effectively; then the car was taken out of the wind tunnel quick-smart! However, these days, all cars have undergone extensive aerodynamic development aimed at providing low drag and good high-speed stability. Unfortunately, though, many people modifying their car take no notice of aerodynamics and, while there are plenty of workshops specialising in suspension, engine work and transmissions, there are very few workshops specialising in the aerodynamics of road cars. Note that the vast majority of aftermarket bodykit manufacturers perform no aerodynamic research and read nothing on the subject, so know little about it...

The good news is aerodynamic testing is easily carried out. This allows you to see the pattern of the airflow over your car and makes it possible for you to build or select spoilers, skirts and the like that are at least likely to do more good than harm! The comic graphic in this chapter (page 316) provides a quick and easy overview of aerodynamics as it applies to cars, but for a more detailed look, read on!

AIRFLOW, DRAG AND LIFT

Imagine for a moment that air was as visible as water. As every car passed, you would be able to see the swirls and whirls of air disturbed by its passage. Some cars would drag behind them an enormous wake, larger even than the frontal area of the car. Others would have only a small area of disturbed air trailing them, these cars slipping easily through the air. If air could be seen, you can be certain that car aerodynamics would have never gone out of fashion — after all, boats with flat fronts are pretty rare!

There are two factors that decide how easily a car can pass through air. The most commonly quoted factor is Cd, or coefficient of drag. A flat dinner plate moved face-first through air has a Cd of about 1.1. More slippery shapes have a lower Cd, such as the 0.45 Cd of a sphere. The other important factor is size. The frontal cross-sectional area of a car is the height multiplied by the width, excluding open areas such as the space between the wheels and including additional areas such as those of the rear vision mirrors. (To gain an approximation of the frontal area, simply multiply height by width.)

Cross-sectional area depends on how big the car is, but the Cd is influenced by how the air flows over the car. The total drag for a given speed is proportional to Cd multiplied by the frontal cross-section, a figure termed CdA. Note that the larger the car, the easier it is for its designers to achieve a low Cd — but the CdA figure is the one that's more important! Incidentally, many manufacturers' Cd figures are quite rubbery — often, cars are retrospectively given a poorer Cd figure after the next model is released.

If you imagine air being a series of thin layers, when the airflow remains in layers (laminar) as it passes over the car, the drag acting on the body is low. Anything that causes the laminar flow of air to separate from the body — and so become turbulent — causes drag. The ultimate low-drag shape is a teardrop with its hemispherical front end and long, tapering tail. Theoretically, a teardrop shape has a Cd of only 0.05! Note that it's not the pointy end that faces forward, but the rounded end that goes first.

When the smoothly curved front end of the teardrop shape meets the air, the air is gently deflected around it, staying attached to the object in attached flow. The long tail of the drop allows the flows to rejoin with only very minor turbulence. If the air had not remained attached to the shape (because of a sudden step in the shape, for example), the flow would have become turbulent at that point. However, it's obviously impractical to have a car shaped like a teardrop! The closest road vehicles currently come to this are the solar race cars — vehicles that have incredibly low Cd values (eg 0.1) matched with very low cross-sectional areas. These cars can travel at 100km/h using only 1.5-2kW (2-2.7hp) of power! Even in normal road cars, the basic rules of having smooth surfaces with no abrupt transitions of shape, gentle front curves and long tapering tails continue to apply.

When a car moves, air is deflected above, below and around it. The point at which the air splits to pass above or below the car (termed the stagnation point) is important in deciding how slippery the car will be. The lower the stagnation point, the better, because then less air runs into the (usually) rough underside of the car. However, unless the car has a front spoiler extending almost to the ground, air **will** pass under the car. This has caused some manufacturers to start adopting low drag undersides for their cars. There is also another reason for entraining this air into a smooth flow — creating less lift, which I'll get to in a moment.

So if the bulk of air doesn't pass under the car, where does it go? Combustion air is drawn into the engine, usually from near the front of the car. This air movement normally doesn't create much drag but the flows of cooling air through the radiator **do** create a lot of drag. On a car with a front radiator, a huge amount of air enters through the cooling duct opening, is forced to flow through the

The radiator outlet ducts on the bonnet of this Lotus Elise direct the air over the windscreen. Venting the air in this way reduces the amount of air passing under the car, lowering drag and lift.

The airflow on this Saab 9-3 remains attached across the roof, around the curve of the rear window and onto the spoiler. The attached airflow across this curve would tend to generate lift — counteracted to some extent by the spoiler.

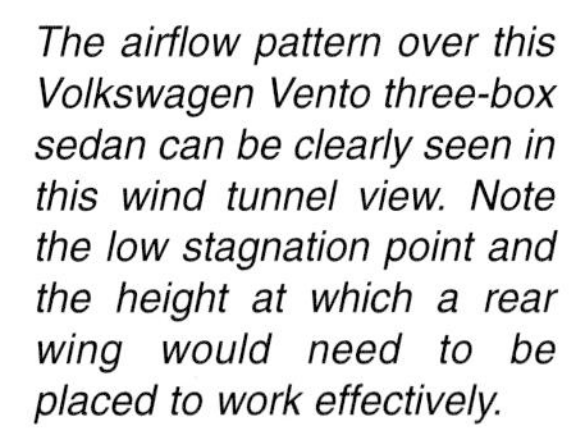

The airflow pattern over this Volkswagen Vento three-box sedan can be clearly seen in this wind tunnel view. Note the low stagnation point and the height at which a rear wing would need to be placed to work effectively.

radiator and then spills out untidily underneath the car. This turbulent movement of the radiator cooling air increases the Cd figure by as much as 10 per cent. For example, the radiator cooling airflow accounts for 8 per cent of the AU Ford Falcon's drag. This means that without this drag penalty, the car's Cd would drop from 0.295 to 0.271!

The use of a radiator intake duct that retains attached airflow for at least most of its length will give the best flow with the least drag. Controlling the flow of air after it leaves the radiator core is also important. Ducting air out through the wheel wells (a low pressure area) is efficient, but even better is the ducting of the air out through the top of the bonnet. Many sports homologation versions of road cars have taken the approach, fitting special vents

The flow transition from the front of the car to the bonnet needs to be gentle if the airflow is to remain attached. This Mercedes A Class has one long gentle curve from nose to tail.

to flow radiator air out through the bonnet. However, care needs to be taken that this flow does not disrupt the attached flow across the upper surface of the bonnet.

Air that's forced over the top of the car has to make its way from the relatively bluff nose area to the top of the bonnet, staying attached over the transition formed by the leading edge of the bonnet. This radius is critical. If this corner is too sharp, the air will separate from the car's surface. A separation bubble will form on the bonnet, leading to the presence of turbulence at the front of the car. It's for this reason that most modern cars have very gentle transitions of shape in this area. One of the all-time great aero specials — the 1970 Plymouth Roadrunner Superbird with its enormous rear wing — had a curved nosecone substituted for its normally bluff front. This extension prevented bonnet separation and also reduced the amount of air passing underneath the car.

The radius formed by the transition from windscreen to roof is also vital. If this angle is too sharp, the airflow may not remain attached to the roof, causing turbulence and further problems towards the back of the car — it's not much good having a rear wing placed in turbulent air!

What happens at the back of the car is **extremely** important in determining total drag, rear axle lift and, to a more limited extent, front axle lift. In many cases, the flow at the back of the car is more important than the flow behaviour at the front. The pattern of airflow at the rear of the car depends very much on the type of car being examined. A three-box sedan must have a very

The 1970 Plymouth Roadrunner Superbird (and its older sister the 1969 Dodge Charger Daytona), had curved nosecones substituted for the normally bluff fronts. This extension prevented bonnet separation and also reduced the amount of air passing underneath the car.

shallow-angled rear window if the airflow from the roof is to remain attached down onto the boot. It's very important that this flow **does** remain attached — the area of the wake will be reduced, dramatically lowering the car's overall Cd. When the profile of a three-box sedan is compared with an ultimate teardrop shape, it can be seen that separation at the roof/rear window transition can easily occur.

Cars with gently sloping rear windows — often hatchbacks or coupes — allow the airstream to remain attached right to the rear of the car, so producing only a small wake. The transitional curve between the roof and the hatch needs to be gentle if the airflow is to remain attached, and the angle of the hatch to the horizontal is also critical. It's important to note that while it may look 'obvious' to the eye that the airflow remains attached across a coupe or hatchback's rear, wool tuft testing (covered shortly) needs to be carried out to prove this.

Even quite minor changes in rear hatch angle can cause major changes in drag. Tests carried out by Volkswagen have shown that the Cd of the car can vary from 0.34 to 0.44 as a result of slight alterations to the rear hatch angle. At one angle (30 degrees to the horizontal in this case), the airflow separation point jumped back and forth from the end of the roof to the bottom of the hatch, depending on the curvature at the rear edge of the roof. It was this 30-degree rear hatch angle that produced the highest Cd value.

This sort of substantial change in the car's drag coefficient will have a large influence on the car's top speed and fuel consumption. For example, if the car were

For a bonnet-mounted intake duct to work effectively, the airflow across the leading surfaces of the bonnet must be attached — this Subaru Impreza WRX STi intercooler intake works well. Note also the height of the rear wing, designed to work in clean air.

On the Audi S3, the airflow separates at the end of the roof. The small roof extension spoiler makes sure this happens cleanly without a suction peak forming, as would occur if the flow wrapped partially around onto the rear window.

capable of 170km/h with a 0.44 Cd, with a drag coefficient of 0.34 this would jump to 185km/h! In some cars, even a 10 per cent reduction in drag will decrease open-road fuel consumption by 5 per cent.

Cars with a near-vertical rear hatch have airflow separation that occurs at the end of the roof. This means the wake is as large as the frontal cross-sectional area of the car. People who locate spoilers halfway down the rear

Most modern wing mirrors are designed to allow airflow onto the side glass and develop only low amounts of wind noise.

hatch should realise they're achieving nothing with this placement! A large wake is also present on most station wagon designs. For example, the Cd of the AU Ford Falcon wagon is sedan is 0.341, versus the sedan's 0.295.

Airflow along the sides of the car is also important. The use of flush-mounted side glass is one approach taken to reduce the surface roughness; however, this is implemented more for noise reduction than for lowering drag. Rear-vision mirrors have also changed in shape as their aerodynamic drag and the way they influence the behaviour of the airflow further down the side of the car is taken into account. For example, the Mitsubishi Magna mounts the rear-vision mirrors on the doors rather than the A-pillars, so there is sufficient air pressure against the side glass to keep it seated in the frameless door rubbers.

Also very important is the plan view shape of the A and C pillars. Very curved pillars are used to encourage the flow of air from the windscreen smoothly around onto the side glass. On three box sedans, the greater the flow of air that can be gained from the sides of the car onto the bootlid, the better. This air can be used with aerodynamic aids such as spoilers, and its presence also helps fill that 'hole' in the air created by the car's forward movement, thus lowering drag.

A classic case of modification of the shape of a car to achieve laminar flow down its sides occurred way back with the design of the very first Volkswagen Kombi. Initially, the vehicle had an almost flat front and very sharp corners — a bit like a moving shoebox! In this form, severe turbulence occurred down each side of the vehicle and it had a Cd of 0.76. Slight rounding of the nose was then carried out, with special attention paid to

The cabin ventilation air outlet is always located in a low pressure area. In this Toyota Corolla, the outlet is aerodynamically connected to the low-pressure wake by the bumper cover, which has been removed for this photo.

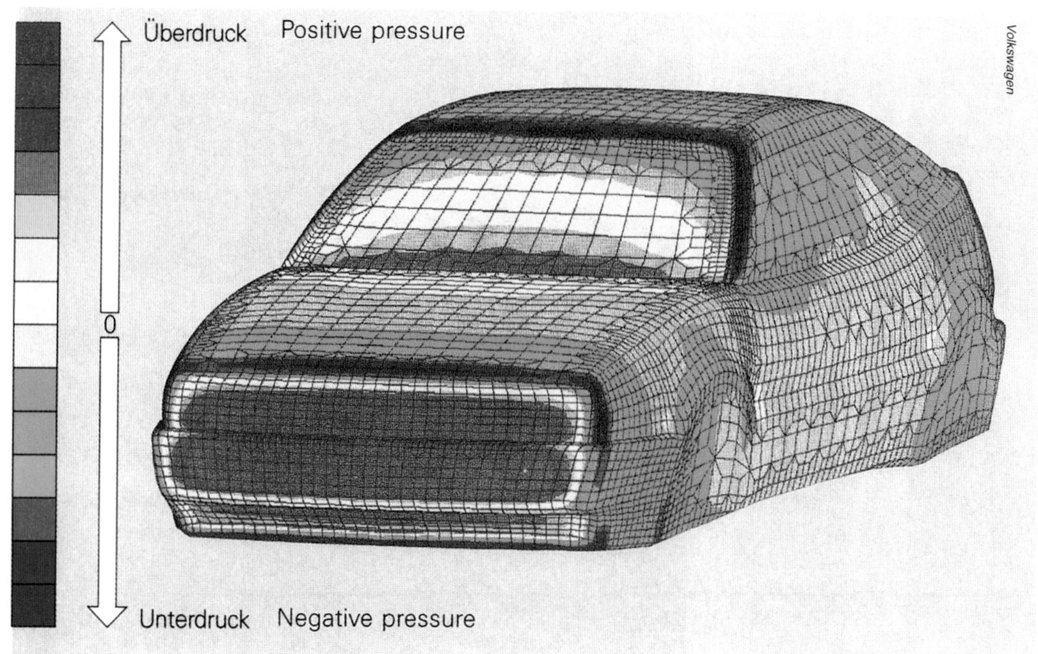

Figure 13.1
The modelled pressure distribution over a vehicle. Where the airflow wraps around curves on the upper body, the pressure is low (shown by blue and green). The pressure is high on the front of the car and at the base of the windscreen (shown by red and yellow).

smoothing the transition from the front to the sides of the vehicle. The Cd then dropped to just 0.42!

As indicated previously, a car draws along a wake of disturbed air. The smaller this disturbance, the lower will be the drag. A very long tail that tapers in both height and width will cause the least drag — that teardrop shape again! While long, tapered tails aren't practical in normal car design, a truncated version is often used. Cars using boat-tailing (a narrowing of the width of the rear when looked at in plan view) are common, with measured drag reductions of up to 13 per cent achieved with this approach. The Opel Calibra decreased its Cd by 0.01 with a total rear boat-tailing of 130mm. Cars with tails that taper in profile are also common, but often the taper is abruptly cut off. This is a called a Kamm tail.

The flow of air over the car will create high- and low-pressure areas. Low-pressure areas (creating lift and sometimes drag) occur most frequently where the airstream passes over an upper curved surface — the transition from grille to bonnet, windscreen to roof, and roof to rear window. The shape of the original Porsche 911 created very high lift coefficients; it's a good example of where laminar airflow wraps around a long, upper-body curve. High-pressure areas (creating downforce and sometimes drag) occur at the very front of the car and in the transition from bonnet to the base of the windscreen.

Figure 13.1 shows the typical pattern of pressures acting on the upper surface of a car. The way these pressures act will result in an overall lift coefficient, or Cl. The lift coefficient is normally expressed for both the front axle (Clf) and rear axle (Clr). A negative lift value (shown by a minus sign in front of the coefficient) indicates that downforce occurs — relatively rare in a road car.

The cabin ventilation inlets and outlets provide good clues as to the location of (respectively) high- and low-pressure areas. Intakes for the cabin ventilation system

This AU Ford Falcon XR6 has attached airflow across the roof and onto the boot, allowing the dual-plane rear wing to work very effectively. Older cars usually have flow separation occurring at the end of the roof.

Aerodynamic testing is vital if effective aerodynamic mods are to be made. Smoke testing can be carried out using a second vehicle, a mains-voltage portable generator, a stage-type smoke machine and a long wand made of plastic pipe.

David Bryant

are almost always at the base of the windscreen, while outlet vents are placed in a range of areas. On most recent cars, the outlets are hidden behind the bumper bar in the low-pressure wake, but in older cars, vents in the C-pillar or across the top of the rear window are used. While the location of these vents is not of direct use in modifying cars, looking at cars and thinking about the pressures present at these vent locations can be quite illuminating.

AERODYNAMIC TESTING

One of the most important aspects of aerodynamic modification is the testing of both the standard car and any modifications made to it. Very effective aero testing can be carried out away from the mega-dollar wind tunnels most people seem to think of as a prerequisite for any aerodynamic testing.

Wool Tufting

Wool tufting involves sticking small lengths of wool all over the area of the car being examined and then driving the vehicle. A speed of about 80km/h usually works very well, although both higher and lower speeds can also be used. The air flowing over that section of the bodywork blows the wool tufts around, allowing the tracking of the actual movement of air over the car. Where the airflow is attached, the wool tufts will lie parallel and flat with the body, their ends fluttering only slightly. In areas of turbulence, the tufts' behaviour will be very different — fluttering wildly and sometimes even spinning in circles.

The best way to wool tuft a car is to use a large number of brightly coloured tufts about 5-10cm long, each stuck at one end with masking tape to the bodywork. Place the tufts at least 10-20cm apart, otherwise they tend to touch one another and stick together as they move in the airflow.

Volkswagen

The cambered profile of some cars is very reminiscent of an aircraft wing, with the generated lift also being very strong!

THE ***AERODYNAMIC DESIGN*** OF CARS INFLUENCES HANDLING, PERFORMANCE AND ECONOMY. AERO MODS CAN BE USED TO IMPROVE ANY OF THESE FACTORS.

HOW EASILY A BODY ***SLIPS THROUGH AIR*** IS INDICATED BY ITS ***DRAG CO-EFFICIENT*** NUMBER, OR ***Cd***. THE Cd OF A PARACHUTE IS 1·4, WHILE THE Cd OF A GOOD CAR IS 0·30. THE ***LOWER*** THE CO-EFFICIENT NUMBER THE ***MORE SLIPPERY*** IS THE CAR.

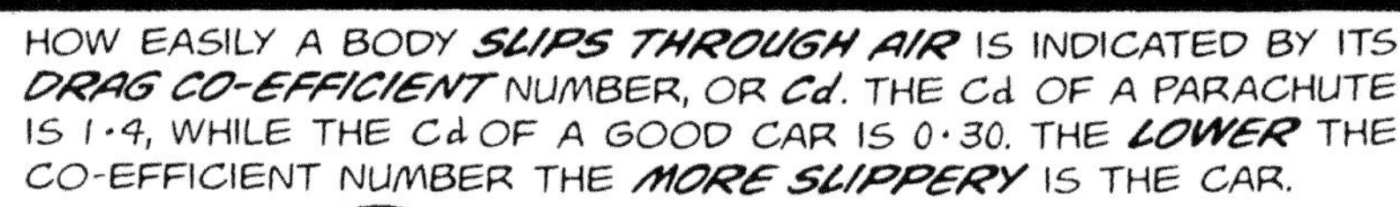

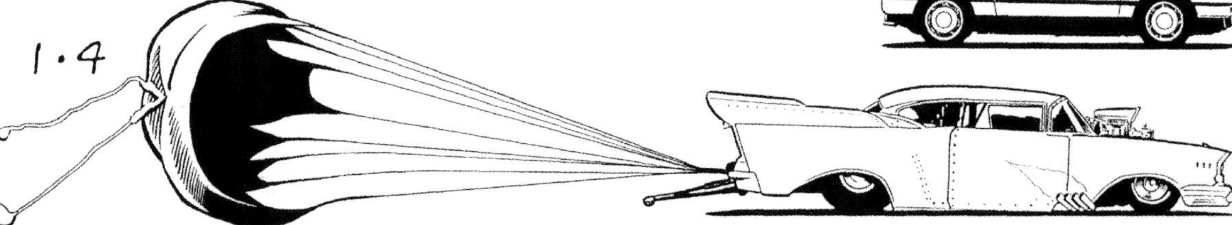

THE TOTAL DRAG OF A CAR IS ALSO AFFECTED BY ITS ***CROSS - SECTIONAL AREA***, WHICH IS THE HEIGHT MULTIPLIED BY THE WIDTH OF THE CAR. IT IS MEASURED IN SQUARE METRES. THE ***TOTAL DRAG*** IS FOUND BY MULTIPLYING THE Cd BY THE CROSS-SECTIONAL AREA. IT'S THIS TOTAL DRAG FIGURE WHICH IS MOST IMPORTANT.

Cd = 0.40
AREA = 1.61
TOTAL DRAG
= 0.64

PORSCHE 356B

Cd = 0.35
AREA = 1.82
TOTAL DRAG
= 0.64

PORSCHE 944

THERE ARE TWO TYPES OF AIRFLOW OVER A CAR - ***LAMINAR*** AND ***TURBULENT***. LAMINAR FLOW OCCURS WHEN THE AIR SLIDES OVER THE SURFACE IN SMOOTH LAYERS. WHEN YOU SEE A PICTURE OF A CAR IN A WIND TUNNEL, THE AIRFLOW SHOWN IS USUALLY LAMINAR.

LAMINAR FLOW CAN CREATE ***DRAG*** AND ***LIFT***. WHEN THE AIRFLOW WRAPS OVER CURVED SURFACES THE RESULT IS LIFT, WHICH IS HOW AN AEROPLANE WING WORKS. THE ARROWS SHOW THE AERODYNAMIC FORCES AT WORK ON THE CAR, WITH THE LONGER THE ARROW THE STRONGER THE FORCE.

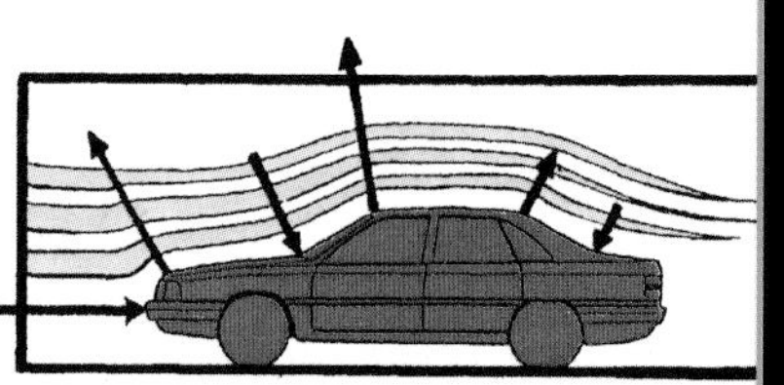

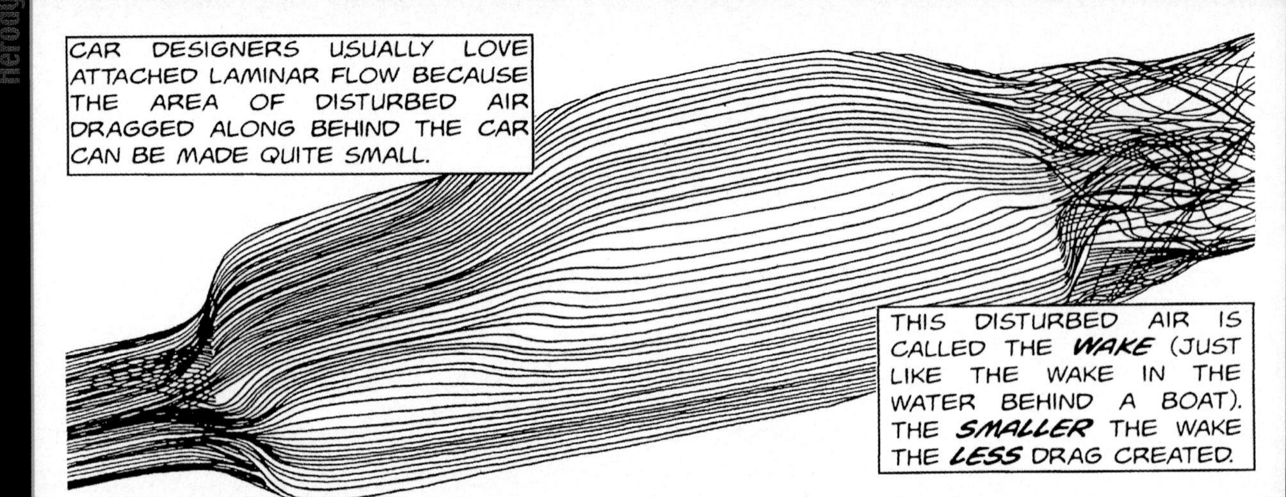

IF THE AIR MOVEMENT IS IN RANDOM SWIRLS AND EDDIES THEN THE AIR-FLOW IS ***TURBULENT***. TURBULENCE CAUSES LOTS OF DRAG BUT NOT MUCH LIFT. WHEN THE AIRFLOW BECOMES UNSTUCK FROM THE CAR'S SURFACE TURBULENCE OCCURS. IF THE AIRFLOW SEPARATES AT THE END OF A SEDAN'S ROOF THERE'S A HEAP OF TURBULENCE ON THE BOOT AND THE WAKE IS ALSO BIGGER. THE RESULT IS A POOR *Cd* FIGURE - THERE'S LOTS OF DRAG.

WHERE THE AIRFLOW DOES LEAVE THE CAR ***CLEAN SEPARATION*** IS REQUIRED. ON LOW-DRAG CARS THE AIRFLOW LEAVES AT THE END OF THE BOOT-LID AND IF THE FLOW CAN WRAP ITSELF AROUND THIS CORNER MORE DRAG IS CREATED. THESE CARS HAVE A SHARP CHANGE OF ANGLE AT THE TRAILING EDGE OF THE BOOT TO PREVENT THIS SUCTION PEAK FORMING.

THE MOST SLIPPERY SHAPE OF ALL IS A ***TEARDROP***. THE AIR FLOWS AROUND THE GENTLY CURVED FRONT WITHOUT SEPARATION AND TURBULENCE, AND THE WAKE IS ALMOST NON-EXISTENT.

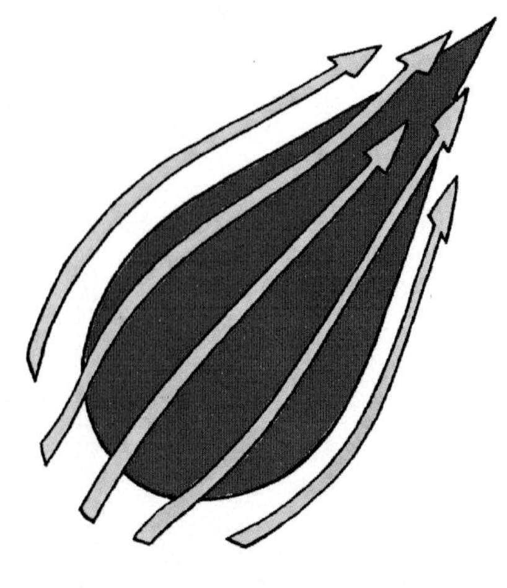

A *MITSUBISHI* MAGNA HAS THE VERY LOW Cd OF 0.28. IT'S EASY TO SEE WHY IT'S SO SLIPPERY.

BEFORE YOU MAKE AERO MODS IT'S BEST TO ***WOOL-TUFT*** THE CAR AND THEN DRIVE IT ON THE ROAD. OBSERVE FROM ANOTHER CAR WHAT HAPPENS TO THE WOOL. WHEN THE TUFTS LINE UP IN ROWS THE FLOW IS LAMINAR, AND WHEN THE TUFTS SPIN AROUND THE AIRFLOW IS TURBULENT.

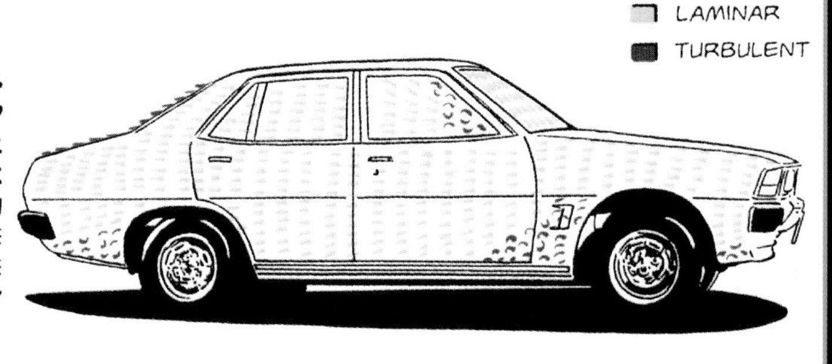

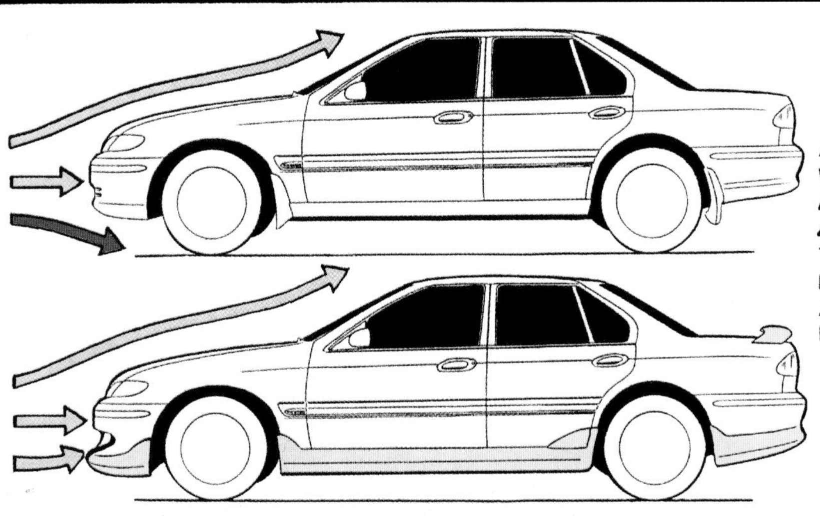

A LOW FRONT SPOILER WILL OFTEN ***REDUCE DRAG*** AND FRONT AXLE ***LIFT***. TURBULENCE UNDER THE CAR WILL BE REDUCED AND RADIATOR AIRFLOW WILL BE IMPROVED.

HOWEVER IT'S THE BACK OF THE CAR WHERE THERE'S MORE AERO ACTION. IF YOUR CAR'S A ***HATCH*** WITH A STEEP TAIL, AIM TO GET A CLEAN SEPARATION AT THE END OF THE ROOF. USING A ***ROOF EXTENSION*** SPOILER WILL DO THIS.

ON A ***LIFTBACK*** WHAT YOU DO WILL DEPEND ON WHERE THE AIRFLOW SEPARATES - AND DON'T ASSUME THAT THIS HAPPENS AT THE VERY BACK OF THE CAR. IF THE AIRFLOW LEAVES AT THE END OF THE ROOF USE A ***SMALL*** SPOILER TO GET CLEAN SEPARATION AT THIS POINT, WHILE IF SEPARATION OCCURS AT THE END OF THE LIFTBACK'S HATCH, USE A ***LARGE*** SPOILER TO CANCEL OUT SOME OF THE GENERATED LIFT, AS WELL AS TO PROVIDE CLEAN SEPARATION.

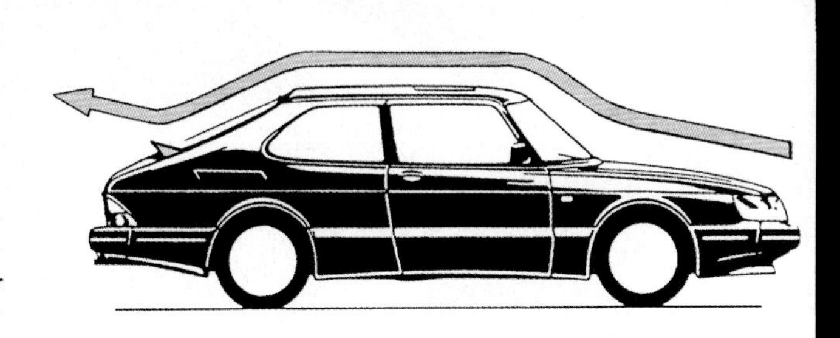

ON ***SEDANS*** WHICH ARE OLDER IN SHAPE THAN THE VN COMMODORE/EA FALCON, THE AIRFLOW SEPARATES AT THE END OF THE ROOF. USE A ***VERY LARGE*** SPOILER AND/OR INCREASE THE HEIGHT OF THE BOOTLID TO PROMOTE FLOW RE-ATTACHMENT ON THE BOOT. THE WALKINSHAW GROUP A WAS A CLASSIC EXAMPLE OF IMPROVING A CAR'S AERODYNAMICS.

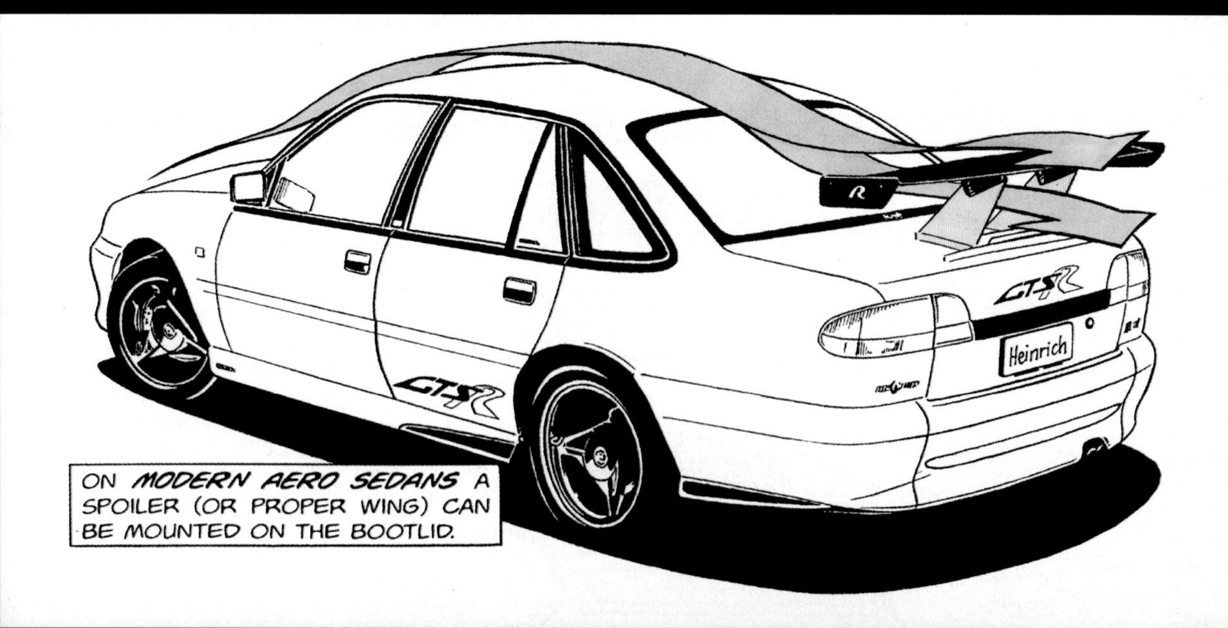

This is a good view of attached flow — everywhere you look on the front three-quarters of this Mitsubishi Magna, there is a lack of turbulence. Note there is still good flow onto the side window next to the mirror — as required to keep the glass pressed against the window rubber in this car that has frameless doors.

It's almost impossible to drive a tufted car and examine the behaviour of the tufts at the same time, so organise a video camera and a multi-laned road. Have someone travelling in another car to video the car being tested. Front and back views as well as side views should be taken. Use of a quality camera and VCR will allow the tape to be played back frame-by-frame to examine the airflow pattern very closely.

Alternatively, a normal camera can be used, as has been done to produce the photos shown here. Using a fast shutter speed will freeze the tufts, but sometimes it's better to allow some blur so that fluttering of the tufts is more visible.

Close inspection of the wool tufts on the videotape or photos will tell you where the turbulence is — and turbulence is drag! As an example, an early '80s sedan was tested in this way. The majority of turbulence on the car was in the 'step' between the roof and the bootlid. Air was actually heading **forwards** in the middle of the bootlid when the car was moving. In other words, swirls of turbulent flow were occurring on the bootlid at any speed over about 40km/h. There was some turbulence at the side of the rear-view mirrors, and also turbulence low on the doors, especially in front of the back wheels. One of the plastic headlight protectors was causing problems as air funnelled up between the plastic protector and the glass, before then spilling onto the bonnet. None of these characteristics was previously known.

After each wool tuft test is completed, you should examine the tufts carefully before removing them. Most types of wool will fray at the ends when the tufts have been fluttering violently — indicative of being positioned in areas of turbulence. Conversely, tufts that have been in areas of attached flow will remain unfrayed.

On-road wool-tuft testing should not be underestimated as a tool. If a still day is chosen, the camera car is not too

Smoke testing shows the general flow but does not show the separation that can be seen with the wool tufts. With smoke testing you can't tell whether the airstream remains glued to the bodywork or not.

Wool-tuft testing shows the airflow over the front half of a Volkswagen Golf cabriolet is very good. The airflow is generally attached, meaning there's little turbulence and drag is also low. If you look very closely, one area of turbulence can be seen adjacent to the righthand headlight. This tuft is whirling around in the air, showing there's some flow separation as the air passes from the headlight to the side of the car.

David Bryant

While the flow of air across the roof is attached, separation occurs at the trailing edge of the roof. This can be seen from the pattern of tufts on the rear window — they point in all directions! Note there is some flow around the gently contoured C pillar onto the bootlid, but not enough to achieve properly attached flow at the rear of the car.

David Bryant

Smoke testing can be used to show the main stream of air over the car. Here it can be clearly seen that the airstream doesn't wrap around onto the rear window —as the tufting shows, the separation point is at the trailing edge of the roof.

David Bryant

With the roof open and the windows down, there is massive air turbulence in the cabin. Note the behaviour of the tufts on the rollbar and the movement of the driver's hair!

The airflow over the front half of the car is identical whether the roof is up or down. This is as would be expected and shows that the wool-tuft testing is very repeatable. Note again the turbulence near the headlight and this time also around the exposed wipers.

Again, smoke testing shows what the wool tufting suggested — without the surface of the roof to follow, the air tumbles and eddies into the cabin.

The rear of the Magna is not quite as good as its front. Note the laminar flow from the roof onto the rear window and around the C pillars onto the boot. However, there is obvious turbulence around the centre part of the rear window and the leading edge of the central part of the bootlid.

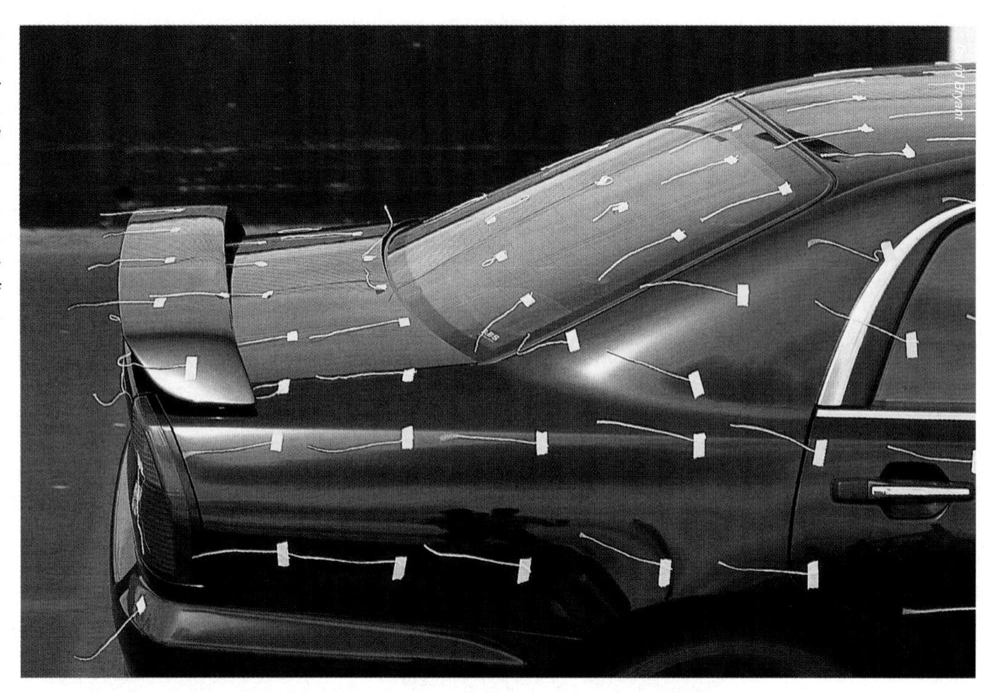

close and the results are carefully recorded and examined, a great deal of information about airflow over the car can be gathered at near zero cost.

Smoke Testing

Another form of testing that can be carried out uses smoke. Unlike wool-tuft testing, smoke testing needs a closed section of road or a racetrack. The test speeds are not high (only 10-20km/h) but the precision driving required and the presence of the smoke makes it all a little awkward on a public road! Smoke testing can be carried out using a mains-voltage portable generator, a stage-type smoke machine and a long wand made of plastic pipe. The plastic pipe used in the testing shown here had a piece of foam rubber wrapped around its end so that had it come into contact with paintwork, no scratches would be left. The vegetable-type oil heated by the smoke machine can at times drip from the end of the wand, but it has no apparent effect on paintwork.

The smoke production equipment can be mounted in the tray of a utility or light truck driven near the vehicle being tested. Purists will point out that the presence of a second vehicle in both this and wool-tuft testing will disrupt the flow of air over the test car, but in much testing I have carried out in this manner, such an effect has rarely been obvious. Anyway, cars **do** drive in close proximity to each other at highway speeds!

During the actual testing, the two cars are driven along at exactly the same speed. The operator places the smoke wand so the smoke stream flows over the areas of the test car that are of interest, and the pattern formed by the smoke is studied. Again, the movement of the smoke can be photographed or videotaped. However, while smoke testing looks much more sophisticated than the use of wool tufts, in practice wool tufting often gives as much (or even more) information and can be carried out much more simply. For this reason, you should carry out wool-tuft testing before you go to the trouble of smoke testing.

Smoke can be injected into the low-pressure area behind the car, allowing the size of the wake to be very clearly seen. Note how the pillars that support the spoiler on this car increase the size of the wake.

Figure 13.2
Before being fitted with a bodykit, the Holden VR Commodore had turbulent flow on the rear window and bootlid. Other areas of turbulence included the sill panels, behind the mirrors and, of course, in the wake.

Track Testing

The effectiveness of major aerodynamic changes made to a road car can be assessed by measuring its lap times at a racetrack. However, both the car and driver will need to be capable of completely consistent laps, otherwise the before-and-after testing will be affected by factors not relevant to aerodynamics! In other words, the brakes going 'off' will affect lap times to a greater extent than the development of 20kg less lift on the front axle!

I had track testing carried out on a VR Holden Commodore, with and without an extensive aftermarket bodykit. In standard form, the Commodore has a claimed Cd of around 0.34. The bodykit comprised a new bumper replacement front spoiler/grille, front fender 'eyebrows', sideskirts, a rear skirt, a large boot-mounted rear wing and bonnet vents. The rear wing was modelled on a factory wing from another car, but the rest of the additions were designed with appearance as the number one priority.

A professional test driver was employed to drive the car on the track, one he knew very well from doing many other back-to-back tests there. Before the hot laps, the car was wool tufted and the pattern of airflow studied in both standard and bodykitted forms. With the kit fitted, the turbulence of the standard car along the sill panels had been reduced but the kit's front wheel arch eyebrows added some turbulence that had not been there previously. Major turbulence across the rear window and boot of the standard car had been reduced, probably because better flow re-attachment now occurred. These changes in airflow over the surface of the car can be seen in Figures 13.2 and 13.3.

However, against the stopwatch there was only the tiniest of improvements — 1/10th of a second in lap times around 96 seconds. The professional test driver could feel **absolutely no difference** in the car's behaviour, even when passing through a long kink in the main straight at about 140km/h. From a performance point of view, it could only be concluded that the bodykit did little.

Figure 13.3
With the kit fitted, the turbulence of the standard car along the sill panels was reduced, but the kit's front wheel-arch eyebrows added some turbulence that had not been there previously. Major turbulence across the rear window and boot of the standard car had been reduced, probably because better flow re-attachment occurred.

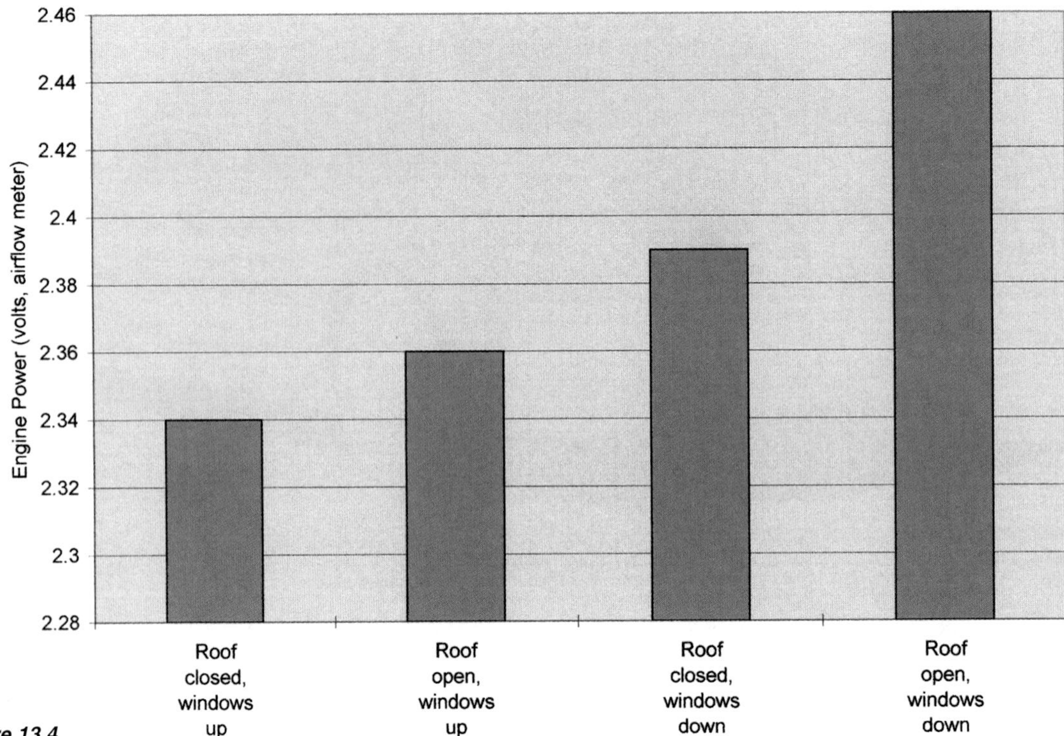

Figure 13.4
Measuring the airflow meter output voltage indicates how much power the engine is developing. At speed, most of this power is used to overcome aerodynamic drag. Here a Golf cabriolet was tested in different configurations, with the airflow meter voltage measured in fifth gear at a constant 100km/h. As can be seen, the drag is lowest with the roof and windows closed. Opening the windows actually creates more drag than having the roof open with the windows up, and, finally, the highest drag occurs with both the roof and windows open.

Other Testing Techniques

At higher speeds, the power developed by the engine is used almost entirely to overcome aerodynamic drag. If engine power is then measured in different aerodynamic configurations, changes to drag can be seen. An indication of the amount of engine power required to overcome drag can be found by measuring the engine's air consumption at a constant speed. This can most easily be carried out by measuring the airflow meter output voltage or frequency by means of a multimeter mounted in the cabin. (Note that while it may at first appear that a more accurate result would be gained from measuring injector duty cycle, in practice this figure jumps around a lot, even at a constant speed!)

A test was carried out on a Volkswagen Golf cabriolet, measuring the airflow meter voltage output. Soft-top cars have much higher drag when the top is down, because of the separation (and so turbulence) that then occurs. Figure 13.4 shows the results of the tests, which were all done in fifth gear at a constant 100km/h. As can be seen from the graph, the drag is lowest with the roof and windows closed. Opening the windows actually creates more drag than having the roof open with the windows up! Finally, the highest drag occurs with both the roof and windows open. This testing must obviously be done on a flat road on a windless day and with the speed held absolutely constant. Higher speeds will show aerodynamic drag variations more clearly.

The following technique can be used to measure the lift or downforce acting on the car. A linear potentiometer ('pot') is attached to the car such that the wiper arm of the pot is moved with the suspension travel. This can be done using a lever pivoting from a front or rear swaybar, or with the pot mounted parallel with a strut. A multimeter is placed inside the cabin of the car, reading off the pot's resistance. As the car passes over bumps, the value being read on the multimeter will constantly change but on a smooth road driven at speed, a change in **average** ride height can be seen. For example, at high speed, the average resistance of the pot might change from 500 to 600 ohms. With the car again stationary, an assistant can then be used to press down the front or rear of the car's body (or lift it) until a similar reading is gained inside the car. The force applied by the assistant will then show the magnitude of the lift (or downforce) occurring on that axle. This type of test will be most effective on a car with soft suspension.

If a wing is placed at the rear end of a boot or fastback car's hatch, the downforce developed by it can be approximated by the following test. For the sake of clarity, the test procedure is described only for a three-box sedan with a wing mounted on the trailing edge of the boot.

First, a spring is temporarily attached to the boot hinge assembly so the boot is held slightly open. This means that closing the boot the last few centimetres extends the stiff spring. An assistant is placed inside the car. He or she accurately describes the height of the wing by lining it up with a mark made on the rear window. Obviously, the passenger's eye level must be held constant throughout the testing. Before testing commences, the boot lid is closed against the spring tension, being shut as far as it will go **before** it starts to compress the boot's rubber seal. The assistant notes this position, and a piece of masking tape is applied to the rear window to show this 'closed' level.

The car is then driven with the boot latched shut. At about 50km/h, the boot is opened using the remote release, allowing it to spring up a little. As the car is then driven faster, the wing pushes down on the bootlid, closing it against the spring tension. The minimum speed that holds the wing in its previously marked 'boot closed' position is then noted. For example, with a certain spring preload, the car might need to travel at 110km/h before the bootlid is pushed down to its 'closed' position. With the car then stopped, the wing is loaded with weights until the bootlid reaches this same closed position. The minimum weight placed on the wing that closes the bootlid to this position is the downforce being exerted by the wing at that speed.

A test was carried out in this way on an R32 Nissan Skyline GT-R. At 100km/h the standard wing exerted a measured downwards force of just under 6kg.

The calculation of a car's approximate drag coefficient can also be carried out. If you know the amount of power the car has available at the wheels in watts (kilowatts x 1000), the car's frontal area in square metres (~ height x width) and the car's top speed in metres per second (km/h x 0.278),

The equation used is: $$\mathbf{Cd} = \frac{\text{Power at wheels}}{0.6 \times \text{Area} \times \text{Speed}^3}$$

For example, for an R32 Skyline GT-R with a top speed of 255km/h that becomes:

$$\mathbf{Cd} = \frac{200{,}000}{0.6 \times 2.35 \times (70.9 \times 70.9 \times 70.9)} = 0.40$$

While this equation is good fun to play with, it has several limitations. It does not take into account rolling resistance (although this is only a small proportion of the total resistance at full speed) and you also soon discover that altering a variable like engine power or speed makes a dramatic difference to the calculated Cd value! So you cannot use this equation to give a definitive figure — but it can give some interesting results.

Finally, a manometer can be used to measure the pressures present on the surface of the car. The type of manometer described in Chapter 5 can be used for this purpose. However, to be really accurate, the manometer probe opening needs to be flush with the surface of the car, requiring that holes be drilled in the bodywork wherever you want to take measurements! This doesn't usually meet with the approval of the average car nut. However, in some circumstances, manometer testing can give interesting results. The pressure under a wing, in a brake cooling or engine intake duct, or in the wheel well can all be sensed using a manometer tube. With this type of testing, high speeds are sometimes required to get a sufficient pressure variation to allow easy measurement.

AERODYNAMIC PERFORMANCE MODS

The fitting of a front spoiler is a simple and usually effective modification. A front spoiler prevents air flowing under the car, so creating beneath the car a wake-like effect with its associated low pressure. Both Cd and Clf (front axle lift) are reduced with the fitting of a front

This Ford Escort shows the aerodynamic behaviour of a car built without the benefit of visiting a wind tunnel! Note the separation that occurs near the headlight — one tuft is pointing sideways, while another has become shredded because of its turbulent fluttering!

Separation that occurs on the roof/rear window/boot transitions is common on older cars. Note the tuft standing vertically on the boot lid, and the differing behaviours of the tufts across the rear window.

The attached flow across the rear hatch of this Mazda Astina shows the wake of this car is quite small, with the rear spoiler promoting clean separation. However, the attached flow across the roof/rear window would create substantial lift.

Pop-up headlights can do awful things to airflow! Here the airflow is almost all attached, the area directly behind the mudflap excepted.

But with the headights up, there is massive turbulence across about one-third of the bonnet and a substantial part of the guard! The top speed of this car fell by 10km/h with the headlights raised...

The R32 Nissan Skyline GT-R uses a true wing (its section is shown in Figure 13.11) located in the airstream at the trailing edge of the boot. As can be seen from this smoke testing, airflow occurs on both sides of the wing — a requirement for its effective operation.

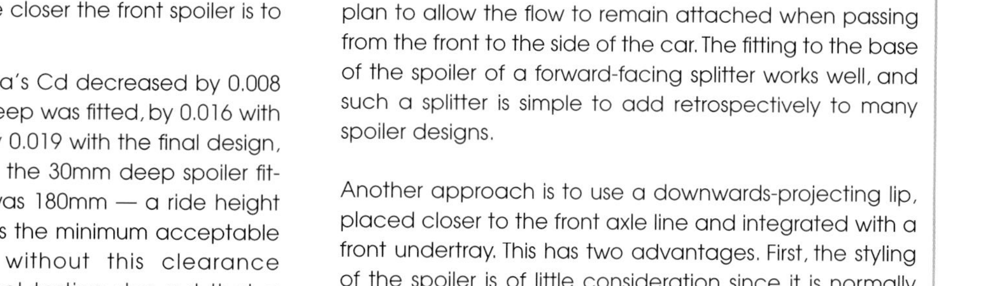
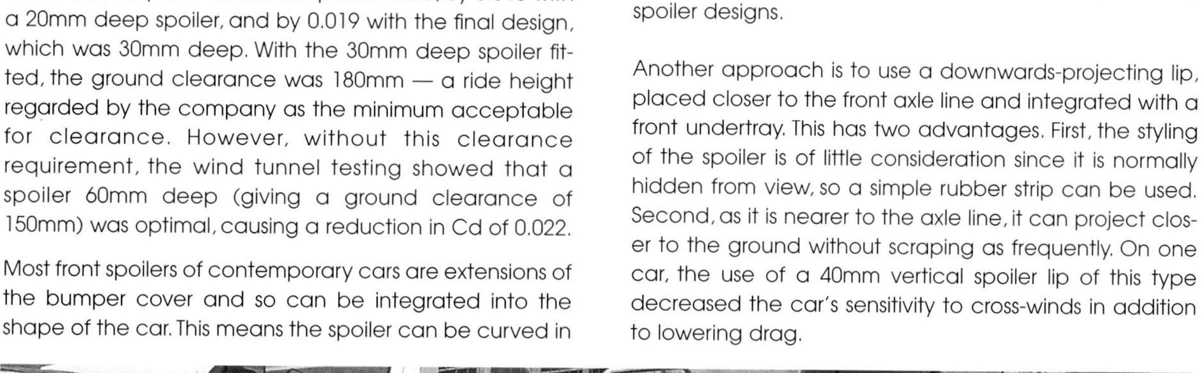

spoiler, and on most cars, the closer the front spoiler is to the ground, the better.

For example, the Opel Calibra's Cd decreased by 0.008 when a front spoiler 10mm deep was fitted, by 0.016 with a 20mm deep spoiler, and by 0.019 with the final design, which was 30mm deep. With the 30mm deep spoiler fitted, the ground clearance was 180mm — a ride height regarded by the company as the minimum acceptable for clearance. However, without this clearance requirement, the wind tunnel testing showed that a spoiler 60mm deep (giving a ground clearance of 150mm) was optimal, causing a reduction in Cd of 0.022.

Most front spoilers of contemporary cars are extensions of the bumper cover and so can be integrated into the shape of the car. This means the spoiler can be curved in plan to allow the flow to remain attached when passing from the front to the side of the car. The fitting to the base of the spoiler of a forward-facing splitter works well, and such a splitter is simple to add retrospectively to many spoiler designs.

Another approach is to use a downwards-projecting lip, placed closer to the front axle line and integrated with a front undertray. This has two advantages. First, the styling of the spoiler is of little consideration since it is normally hidden from view, so a simple rubber strip can be used. Second, as it is nearer to the axle line, it can project closer to the ground without scraping as frequently. On one car, the use of a 40mm vertical spoiler lip of this type decreased the car's sensitivity to cross-winds in addition to lowering drag.

Examination of the wake area again shows there is some airflow occurring under (as well as over) the wing.

Lexus

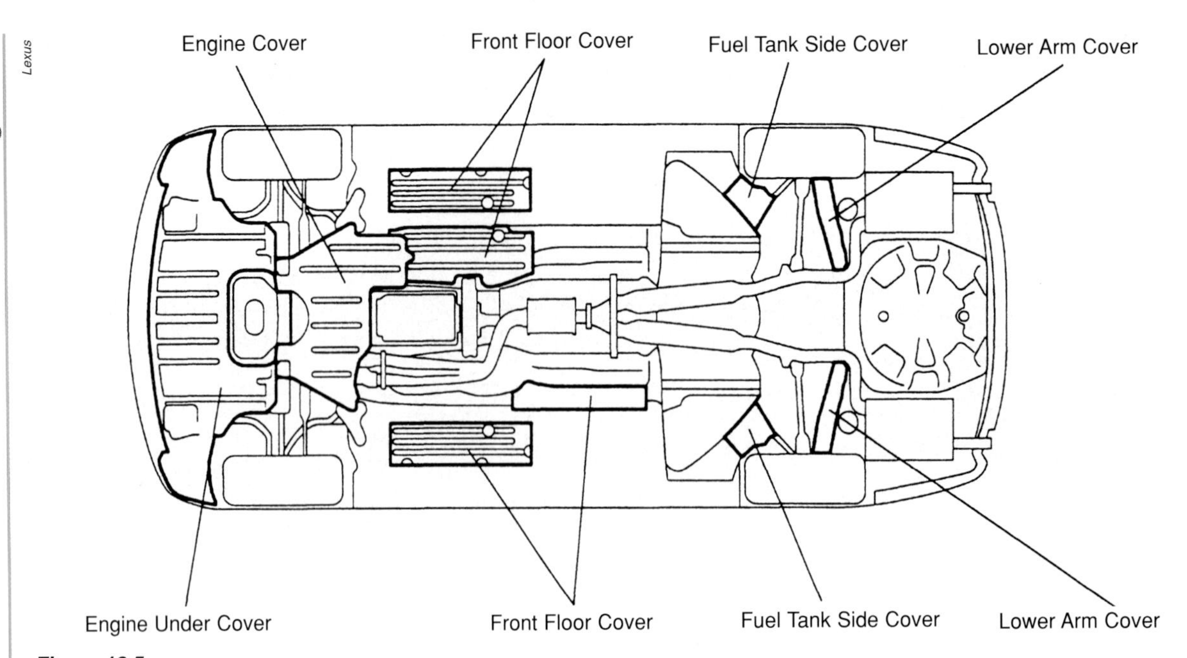

Figure 13.5
The Lexus GS300 uses underbody additions to promote attached flow, resulting in reduced wind noise and drag.

Both aero drag and lift will be reduced if the car is lowered at the front to give a slight rake. As an example, one tested car dropped in drag by 9 per cent and lift decreased by 25 per cent when given a rake of only 2 degrees downwards. The AU Ford Falcon decreased in drag by 2.9 per cent when the front of the car was lowered by 20mm, but dropped in drag by only another 0.1 per cent when the rear of the car was also lowered by 20mm.

An exception to the 'low spoiler is good' school of thought occurs if the car's undersides have been designed to entrain the air to good aerodynamic effect. For example, if a flat undertray and rear diffuser have been fitted, it would be wise to leave the front spoiler alone! The type of front spoiler fitted to such a car is likely to have a raised central opening, allowing the required amount of air to flow under the car. However, the vast majority of cars do not (yet) have sophisticated underfloor aerodynamics. One car that has had attention paid to the underbody aerodynamics is the Lexus GS300. Figure 13.5 shows the GS300 underbody additions, which are designed to promote attached flow to reduce wind noise and drag. Figure 13.6 shows the wheel fairings placed ahead of the front and rear wheels. The Porsche

This splitter fitted to the front of this BMW 2-litre race car lowers the stagnation point, with the result that more air goes over the top of the car rather than passes the rough underside. Lift and drag are both reduced.

Boxster also uses a smooth undertray, with Porsche quoting a Cd reduction of 6 per cent and a very substantial Clf reduction of 36 per cent from its fitment.

If your wool-tuft testing reveals flow separation soon after the leading edge of the bonnet, the angle between the

Lexus

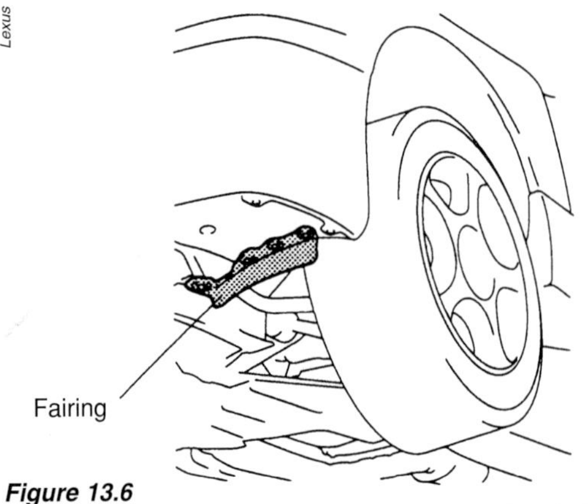

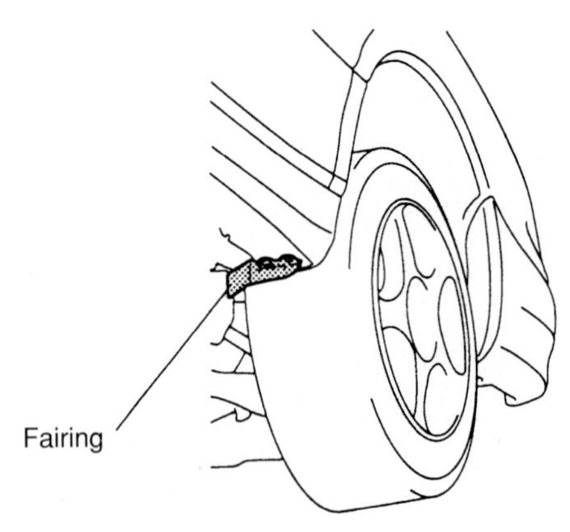

Figure 13.6
The Lexus GS300 also uses fairings placed ahead of the front wheels (left) and rear wheels (right), again for reduced wind noise and drag.

The lowering of a car will decrease both drag and lift. However, both adequate ground and tyre clearance need to be kept if the car is to remain useful on the road!

grille and the bonnet needs to be reduced. New noses aren't available over the counter like front spoilers, but for some cars they do exist. What you are after is a gen - tle radius from the grille around onto the bonnet, so that air flows smoothly across the surface. Separation that occurs around headlights can sometimes be reduced by the use of slightly curved headlight covers, which can be formed from clear plastic.

Make sure your radiator intake duct leads only to the radiator — then at least the drag induced by this air is being put to good use! Many cars have gaping holes in and under the radiator support panel through which air can pass, creating turbulence while also not cooling the radiator core. Gaps between the bonnet and its locking platform, around the headlights and between the bumper and its surrounding sheet metalwork should also be sealed.

When a car can get by with less cooling air, some car manufacturers have in the past blocked off part of the

This Lotus Elise uses a flat undertray for much of the floor area, smoothing the flow and so decreasing turbulence. Flat floors and rear diffusers are likely to become much more widespread on road cars as designers chase lower drag and less lift.

A front spoiler that lowers itself only at high speed allows good ground clearance when entering and leaving driveways, while still being aerodynamically beneficial when reduced lift and drag are required. This is the mechanism that allows the front spoiler of a R31 Nissan Skyline to pivot up and down.

The front diffusers on this racing sedan accelerate the air passing around the leading edge of the spoiler, generating downforce.

grille to improve the Cd — the Ford Escort, VW Golf, Walkinshaw Commodore and Opel Calibra are all examples. If the radiator takes up only a proportion of the grille width, blank the rest off from inside. Painting the panel black will make it invisible. Placing outlet grilles on the bonnet directly behind the radiator will help extract cooling air, reduce drag and reduce lift. Try to direct the air out along the bonnet, otherwise turbulence will be created.

When you get to the back of the car, things become much more complex. Starting at the base of the car's rear, a diffuser has aero benefit. A diffuser consists of a panel that slopes upward towards the back of the car. A properly shaped diffuser can both reduce drag and reduce lift. A diffuser works in a number of ways, but it is simplest to think of it as having the shape of an upside-down fastback car. Just as the airflow wraps around the upper surfaces of a fastback car — creating lift and reducing the size of the wake — so the airflow around the curved throat of the diffuser creates downforce, with the smaller wake reducing drag.

It's important to note that it's the leading edge of the diffuser that develops the low pressure — the increasing volume of the diffuser following the throat just acts as the 'pump', encouraging the flow of air under the car. Many cars with petrol tanks mounted under their rear have de facto diffusers — the tank floor is often angled upwards towards the rear of the car to aid departure angle ground clearance! The flow of air under the car must be attached if a diffuser is to work, suggesting that a full undertray and a front spoiler with a central raised section are additional requirements.

On most cars, the lower the front spoiler, the better. This is because a low-pressure area is created under the car and air is prevented from running into the obstructions that are normally present.

What happens in the aerodynamics of rear spoilers and/or wings depends very much on the shape of the rear of the car. If the devices are carefully designed and wind tunnel proven, quite small aero devices can be effective. Volkswagen found in testing that a 40mm high rubber lip on the back of an early Scirocco lowered the Cd from 0.41 to 0.38 and reduced lift by 20 per cent. BMW, on their M5 V8, used a tiny lip — only 10mm or 15mm high — to develop 40kg downforce at 250km/h. However, these sorts of results are achieved with only **very** extensive testing.

Ford

An alternative to a bumper extension spoiler is the use of a downwards-projecting lip integrated with a front undertray. On this AU Ford Falcon, the use of a 40mm vertical lip decreased the car's sensitivity to crosswinds in addition to lowering drag.

The Opel Calibra uses a number of aerodynamic devices to achieve both low lift and drag coefficients. These include a front spoiler with a raised centre section, rear diffuser, rear boat-tailing and deep sideskirts.

Venting the flow of radiator cooling air through the bonnet, as has been done on this Mazda RX-7 SP, reduces drag.

Any airflow allowed to wrap around curves at the rear of the car is likely to cause both drag and lift. Unless that air is then to be made further use of by another aerodynamic device, this should be prevented. For example, on some older three-box sedans, the airflow wraps around the roof/rear window transition before separating part-way down the rear glass. Flow re-attachment does not subsequently occur on the bootlid, so the wake is already large. In this case, it's better that the separation point occurs cleanly at the end of the roof, preventing the development of a suction peak across the top of the rear window. In the same way, attached airflow that's allowed to partially wrap around the trailing edge of the bootlid will create more drag and lift than when a boot extension spoiler is fitted to provide clean separation.

Front spoilers should have a gentle transitional curve from the front to side of the car, so the airflow remains attached around this curve.

Wool-tuft testing of the Skyline shows a separation bubble across the middle of the rear window, but attached flow occurs across the rear half of the bootlid and on both sides of the wing.

This close-up view of the rear window, bootlid and spoiler on a Subaru Liberty (Legacy) RS clearly shows the separation on the central part of the rear window and leading edge of the boot, with the flow then reattaching in front of the rear spoiler.

Another view of the Skyline wing shows the attached flow across both surfaces. The behaviour of the top tufts across the Gurney Flap (raised upper section) of the wing is interesting.

Rover

Figure 13.7
Rover suggestions for rear spoiler placements.

Most cars — especially those with long, sloping tails — have lift forces acting on them at speed. Rear spoilers can be placed so the laminar flow over the curved surfaces is 'spoiled', causing a higher pressure to occur over the rear of the car. The best location for a rear spoiler will depend on where separation actually occurs — something easily seen through wool-tuft testing. Figure 13.7 shows the suggested location of rear spoilers on a variety of car shapes, based on testing carried out by Rover.

To work, true wings need to be in areas of non-turbulent 'clean air'. This is because a wing requires attached flow over both its upper and lower surfaces. The air flowing

This BMW uses tiny rubber lips on both elements of the spoiler. When a full-sized wind tunnel is available to easily measure changes in lift and drag, these subtle additions can be positioned for maximum effectiveness.

This Mazda RX-7 SP uses a high-mounted wing with a thin section. Its height ensures it will be in 'clean' air, making it effective as a genuine wing rather than a spoiler.

over the lower surface of the wing takes a longer path around the greater curve than the air flowing across the upper surface. This means the air flowing across the lower surface has to flow more quickly, creating a lower pressure.

The sections of the wings (that is, their cross-sectional shapes) used on cars fall into two quite different design types. These can be loosely termed 'banana wings' and 'traditional wings'. In sedan racing cars, banana-type

The roof extension spoiler of the Pulsar GTiR prevents the airflow wrapping around the roof/hatch transition, so reduces drag and lift.

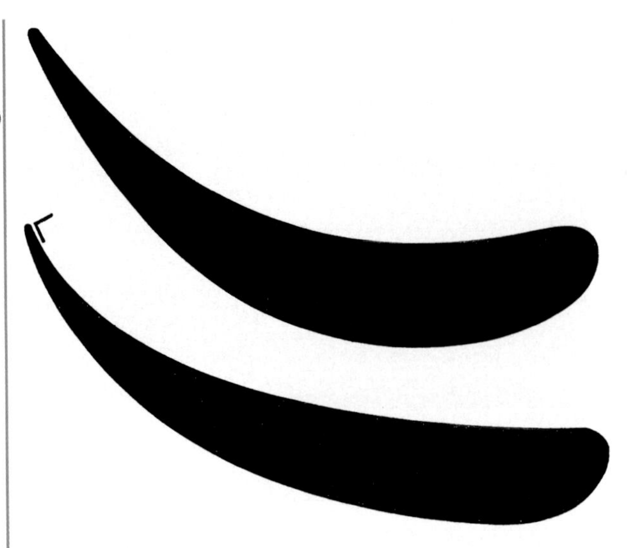

Figure 13.8
'Banana' wing sections are used in motorsport applications, where increased downforce always gives faster lap times, irrespective of the increased drag.

wings are almost always used. The sections of two of this type of wing are shown in Figure 13.8. While providing a large amount of downforce, banana wings are seldom used on road cars. This is the case for two reasons. First, such a wing has a high amount of drag. In a road car, an increase in open-road fuel consumption resulting from the fitting of a banana wing is not normally a viable trade-off for the decrease in rear-end lift. Second, a large amount of rear downforce will cause an increase in lift at the front of the car, unless the front is also equipped with a device giving downforce! Such a device is normally a front spoiler/splitter with very little ground clearance, which makes it impracticable on a road car. For this reason, traditional, thin section wings are more frequently used in road applications.

The sections used for traditional wings can be most easily derived from model aircraft. These have airspeeds similar to that of a fast road car, develop real lift (downforce when inverted) and have very low drag coefficients. Most model aircraft manufacturers have wing sections available, together with complete data on lift, drag and the best attack angle. Some model aircraft manufacturers use software that allows the printing out of full-size wing sections. From these templates, the internal ribs of the wing can be cut, or the model aircraft manufacturer can make a foam core with this cross-sectional shape. The core can then be fibreglassed or covered in carbon fibre. Making an accurate, strong wing in this manner can be quite cheap — much cheaper than approaching a motorsport manufacturer!

Figure 13.9
Thinner section wings such as this Clark Y aerofoil are effective if the wing is mounted high in 'clean' air. This type of wing has very low drag.

One wing section that's available is the Clark Y aerofoil. This was the wing section mounted high across the rear of the classic aero car — the Plymouth Superbird. In NASCAR racing at speeds of up to 320km/h, the car was very successful. With this aerofoil, the peak Cl of 1.17 (downforce when turned upside down!) occurs at an attack angle of 10 degrees. The Clark Y aerofoil section is shown in Figure 13.9 and another model aircraft section — the FX 63-137 — in Figure 13.10. The latter aerofoil has a maximum Cl of 1.6 but still has very low drag. Sketches of the sections of some standard wings used on a variety of road cars are shown in Figure 13.11.

Figure 13.10
The FX 63-137 is a wing section normally used in model aircraft. It has a maximum Cl of 1.6 at an attack angle of 11 degrees.

It should be noted that the aerodynamic behaviour of a wing when tested in isolation is likely to be quite different from when it's fitted to a car. For example, when does a wing cease acting as a wing and start acting as a spoiler? In other words, when does it use laminar flow over both upper and lower surfaces to generate a pressure differential, and when does it affect the global airflow over other parts of the car? Wings — even those with 'proper' sections — mounted relatively close to the bodywork often act as spoilers for at least part of the time. Once again, the result of the modification can really only be assessed through testing.

Note also that while people casually talk about 'downforce' in road cars, the vast majority of cars develop positive lift, even with aero addenda attached. For example, the Porsche Boxster has a Clf of 0.13 and a Clr of 0.10. So

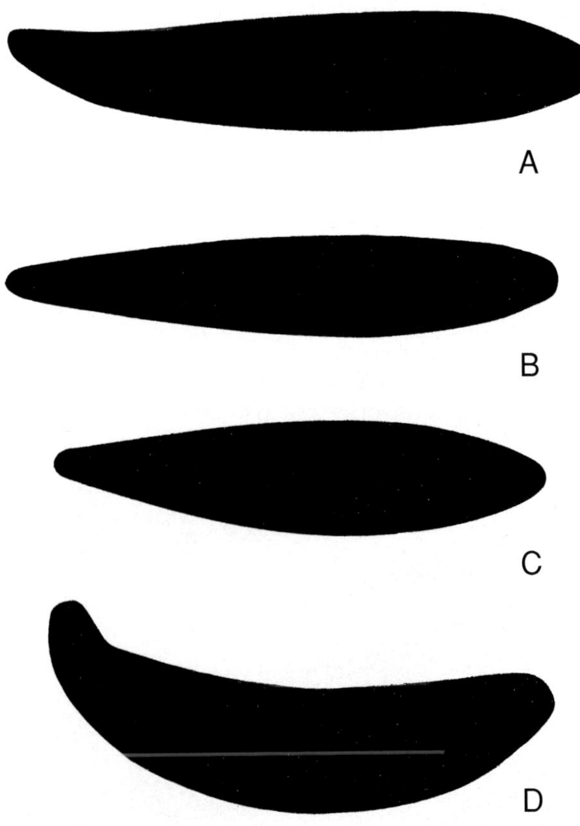

Figure 13.11
*Sketches (**not** dimensionally accurate drawings!) of the wing sections used on a variety of road cars. (a) Nissan R33 Skyline GT-R; (b) Toyota Supra RZ; (c) Mitsubishi Lancer Evolution IV; (d) Nissan R32 Skyline GT-R.*

The Toyota Supra uses heavily sculptured sideskirts to smooth the airflow past both the front and rear wheels.

The Holden Commodore GTS-R uses a very high-mounted rear wing, complete with endplates. However, with the wing set at an attack angle that provides rear downforce, excessive front axle lift occurs. Holden chose to angle the wing to provide no aerodynamic gains — it's there for looks alone.

The rear lip spoiler on the Mitsubishi Cordia Turbo will be effective if there is attached flow right to the trailing edge of the hatch. Wool-tuft testing can easily show if this is the case.

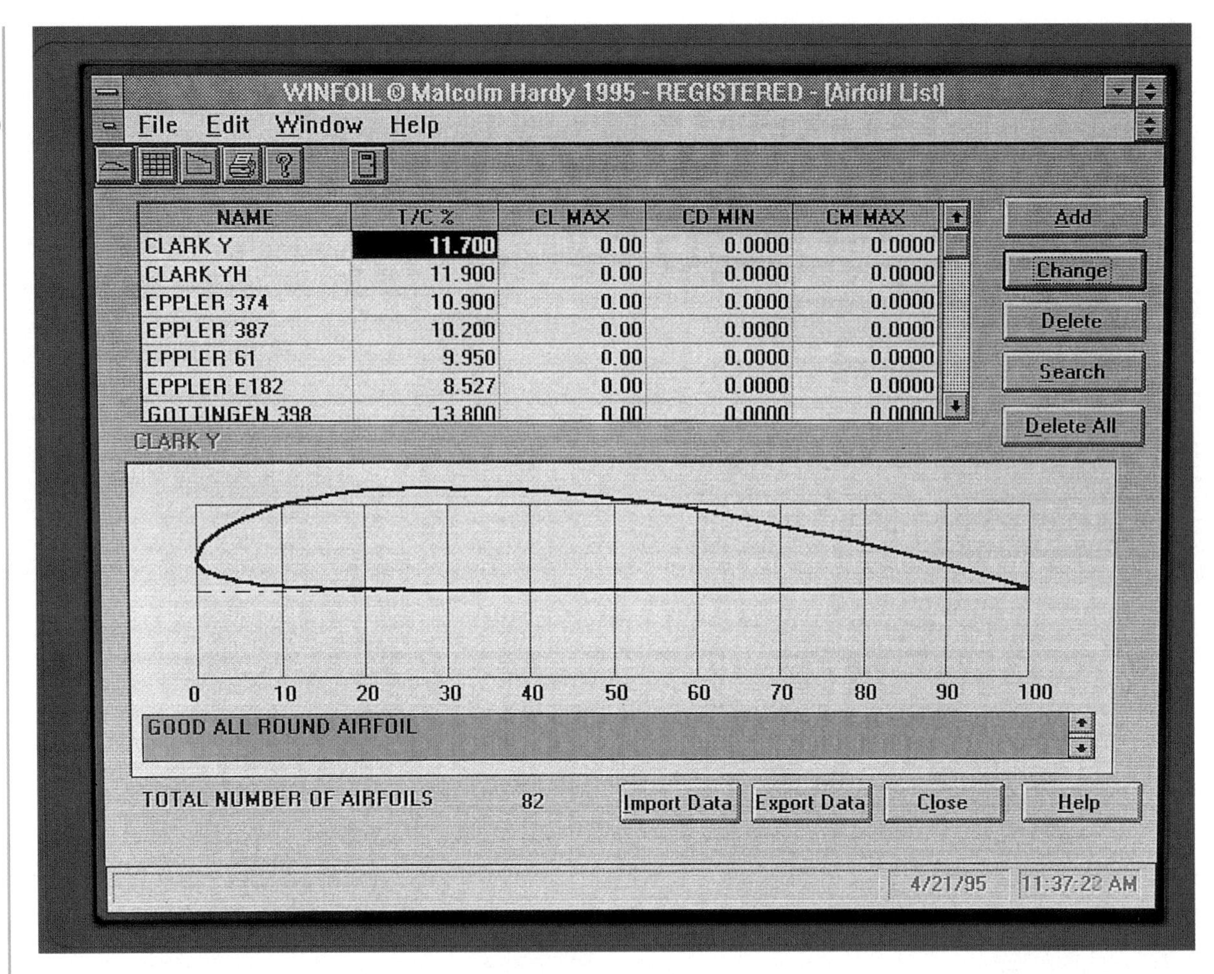

Software designed for plotting model aircraft wing profiles can be used to generate automotive wing sections. This program can be used to print out full-size wing sections.

what **really** should be discussed when describing spoilers and wings is the reduction in lift that has occurred. The AU Falcon is one car for which the change in rear lift with different spoilers fitted is known. In standard form, the Clr = 0.125. The fitting of a single-plane rear spoiler decreased this to a Clr = 0.05 (a reduction of 60 per cent!), with an increase in drag of 1.9 per cent. A dual-plane spoiler was also available. Fitting this gave a Clr = —0.025 (ie in this car, positive downforce was developed), with drag increased by another 5.8 per cent.

A full-size paper template of the Clark Y wing section, produced by a printer using a model aircraft wing database software program. A foam-cored car wing of this profile could be made in fibreglass or carbon fibre.

The actual lift forces occurring on the front and rear axles can be quantified if the frontal area of the car and the coefficients of lift for the front and rear axles are known.

$$\text{Lift force} = A \times V^2 \times Cl \times 0.6$$

Where lift is in Newtons, A is the frontal area in square metres, V is velocity in metres per second, and Cl is the coefficient of lift.

To convert the lift force to kilograms, divide by 9.81.

For example, a car with a frontal area of 1.7m^2, travelling at a speed of 120km/h (120km/h x 0.278 = 33.4m/s), and with a coefficient of lift at the front axle of 0.118, will give:

$$\text{Lift force} = 1.7 \times 33.4^2 \times 0.118 \times 0.6$$
$$= 134 \text{ Newtons, or } 13.7\text{kg.}$$

AERODYNAMIC BALANCE

Reducing the amount of lift that occurs at one end of the car is likely to have the opposite effect on the other end. For example, the fitting of an effective rear spoiler that actually generates downforce will cause front lift to increase, simply because a push down on the back will cause a seesawing lift at the front! The coefficient of lift of both the front and back axles will also depend heavily on the presence or otherwise of a crosswind. When the airflow has a yaw angle (ie a crosswind component is present), the coefficient of lift often increases dramatically. This means effective aerodynamic modification needs to take into account the relationship between the lift (or downforce) occurring on both axles, and also how it alters with yaw airflow.

Fitting just a front spoiler (or just a rear spoiler) may not give an overall benefit. BMW on the installation instructions for the E30 3-series M Technic aero kit states: "Important safety information: In view of the effect on the car's driving characteristics, authorisation is given only for installation of the entire package (front apron, side skirts, side panel trims, rear apron and rear spoiler). No responsibility will be accepted by BMW AG for damage arising if this urgent recommendation is not complied with." Perhaps BMW just wants to sell all the bits of the kit, but it's an interesting statement nonetheless!

Another example of a sideskirt designed to facilitate the passage of air past the rear wheel.

Smoke testing of a different kind! Where the airflow separates at the end of the roof, a roof extension spoiler will decrease lift. However, the size of the wake will be increased, resulting in increased drag. The factory-fitted 'tea tray' spoiler can be seen above the rear window of this Toyota Supra.

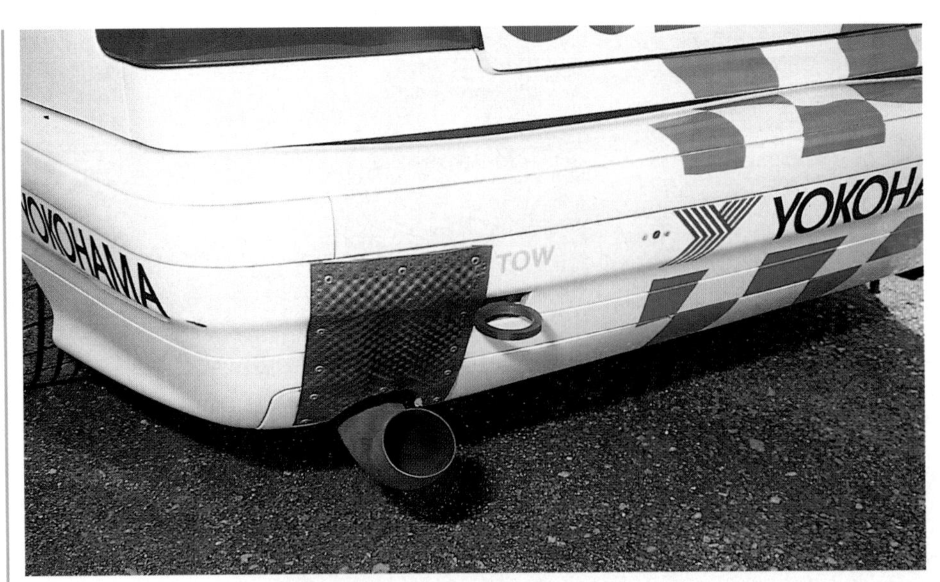

This BMW 2-litre racing car has the exhaust pipe aimed into the wake — presumably to aid in pressure recovery behind the vehicle. If this actually occurred, drag would be reduced.

A wing/spoiler will cause a change of airflow pattern over the rear part of the car. It may reduce lift by upsetting the normally attached flow of air over the upper curves, or it may develop an aerodynamic force through the differing flows on each side of the wing. On some cars both effects occur.

The rear diffuser tunnels on the Jaguar XJ220 show what can be done when the designer starts with a clean sheet of paper!

The effectiveness of most aerodynamic additions can only be determined through testing. What looks good may or may not work!

Although an old example, the mid-'70s Porsche 911 is another interesting car to examine in terms of its aerodynamic development. Its aerodynamic lift problems (especially with yaw angles) were so bad that very major changes needed to be made to the body, and care was needed to ensure that the front/rear lift balance remained under control.

The original 911 was designed to have a relatively low drag coefficient, and in fact Cd = 0.381 was achieved for the first cars. However, the similarity of the shape of the classic Porsche to an aircraft wing was overlooked! Testing was carried out on a later 1969 'C' series that indicated that it had a Cd of 0.408, a Clf of 0.190 and a Clr of 0.264. With a cross-sectional area of $1.75m^2$, the lift at

The large rear wing of the Evolution III Mitsubishi Lancer GSR has an upper element substantially above boot level. The driver of this rally car spoke of the dust-laden wake rising to the height of the upper element, suggesting there's not a lot of airflow on the underside of the upper element.

This Holden VR Commodore was tested both before and after the fitting of a bodykit. Both track and wool-tuft tests were carried out.

150km/h was 35kg on the front and 49kg on the rear axle. An even worse situation existed when the airflow was passing over the car at an angle. With an airflow yaw angle of 22.5 degrees, the Clr and Clf both increased to 0.43. This means that at 150km/h, lift on each axle was 80kg! At the car's top speed of 230km/h, there was a lift of nearly 190kg occurring on **each** axle. Scary stuff .... especially considering the car weighed only about 1100kg!

A number of iterations of front and rear spoiler design occurred, including the use of a 'ducktail' rear spoiler and a front spoiler. With the front spoiler extended downwards until ground clearance was reduced to 190mm the following drag and lift figures were realised: Cd = 0.375, Clf = – 0.055 (ie positive downforce) and Clr = 0.113. However, the very effective front spoiler meant that a pitching (nose down) effect was introduced: the front spoiler needed to be matched with a more

With the car standard, there was turbulence occurring on the rear window and bootlid. Other turbulent areas included the sill panels, behind the rear vision mirrors and in the wake.

After the kit was fitted, the airflow was subtly changed. Note the airflow down each side of the rear window is now laminar, and the flow on the bootlid is improved.

The front wheel arch 'eyebrows' of the kit have created turbulence, and there is also turbulence on the leading edge of the sideskirt. Note the wool tufts being sucked into the wheel arch — this area is normally one of low pressure.

effective rear design. The famous Turbo-type 911 rear spoiler was then introduced. With this and the front spoiler fitted, Cd = 0.414, Clf = 0.010 and Clr = 0.025.

A more recent example of front/rear lift balance can be seen in the testing carried out by Rover on a number of small and medium sized cars. The cars — identified only as A, B, C, etc — were assessed for their high speed straight-line and lane change stability. They were also aerodynamically tested with and without a variety of rear spoilers. Importantly, the changes in high-speed handling were very pronounced, even with relatively minor changes in aero lift.

Both drag and lift coefficients were assessed at zero yaw (ie no side wind) in the wind tunnel. The wind tunnel

The Australian Holden VL Group A Commodore (the "Walkinshaw") is one of the world's best examples of a major manufacturer retrospectively making an un-aerodynamic car very aerodynamic! In this view, the blocked-off grille (previously between the headlights) and the very deep front spoiler are visible. The central inlet is for the radiator with the other two inlets for front brake ducts.

Initially in the car's development, the window in the C-pillar was filled in, probably to promote flow onto the bootlid and rear spoiler. However, the governing bodies disallowed this in the homologation car.

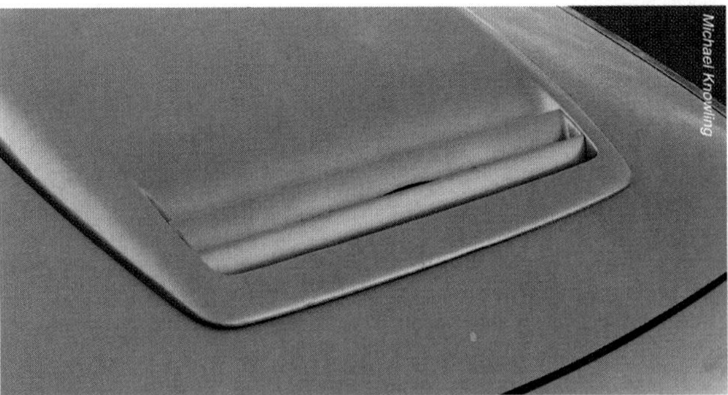

The bonnet vent contains louvres designed to direct the air that has passed through the radiator onto the upper surface of the bonnet, without disrupting the attached flow occurring in this region.

The bootlid of the Walkinshaw was raised considerably to promote flow reattachment, diminishing the size of the wake and allowing the large (and rearwards-positioned) spoiler to work effectively. The car had considerably reduced drag and lift over standard models.

Full race cars don't make particularly good role models for aerodynamic mods to road cars — the race car has almost no ground clearance and a very high drag coefficient.

results for one coupe are interesting. In standard form it had a Clf of 0.161 and a Clr of 0.044. A front spoiler was then fitted, along with a rear boot-mounted wing that was able to be adjusted for its attack angle. With the spoilers fitted, the front lift dropped to Clf = 0.096, but the rear axle lift actually increased to Clr = 0.112. Adjusting the angle of the rear spoiler only made the rear axle lift worse, until in fact it reached Clr = 0.206. With the increase in rear lift, the front lift decreased (as expected), dropping to Clf = 0.078. So in this case, the rear spoiler actually **increased** rear lift, rather than reducing it.

The effect of increased rear lift on the car's high-speed lane change handling was dramatic. Rover drivers (testing the car at no less than 180km/h!) rated the standard car as 'good' to 'very good' in its lane change stability. However, as rear lift increased, the rating dropped, falling to 'barely acceptable' when the Clr = 0.206.

Fitting a rear spoiler was more successful on another car, a three-door hatchback. In standard form Clf = 0.037 and Clr = 0.147. Four spoilers were tried, with the best giving Clf = 0.022 and Clr = 0.021. In testing, the Rover drivers' rating of the standard car's stability decreased quickly as speeds went up, while with the best spoiler package, the car was rated nearly as well for stability at high speeds as for low speeds.

In other words, at high speed, changing the lift coefficients acting on the front and rear axles of a car has a real, measurable impact on stability. This is the case even in violent swerve-and-recover manoeuvres carried out at high speed, where you might expect other factors to overwhelm the aerodynamic considerations.

So is it preferable to have less lift on the front axle or less lift on the rear axle? (Or, on which end should there be more downforce?) It appears that it's preferable to have less lift at the rear of the car, so it is at the back where the greatest efforts should be expended in placing spoilers and/or wings. This is the opposite of what might be first thought, since it seems logical that a large front spoiler would be more likely to guarantee high-speed stability. In addition to stability, it has also been shown that reducing rear axle lift helps high-speed braking.

CONVERSION TABLES

Length

To convert:	to:	multiply by:
Inches	Millimetres	25.4
Centimetres	Inches	0.394
Thousandths of an inch	Millimetres	0.0254
Millimetres	Thousandths of an inch	39.4
Feet	Centimetres	30.5

Mass

To convert:	to:	multiply by:
Pounds	Kilograms	0.454
Kilograms	Pounds	2.20

Area

To convert:	to:	multiply by:
Square inches	Square centimetres	6.45
Square centimetres	Square inches	0.155
Square feet	Square metres	0.093
Square metres	Square feet	10.8

Miscellaneous

To convert:	to:	multiply by:
Pounds per inch	Newtons per millimetre	0.175
Newtons per millimetre	Pounds per inch	5.71
Kilograms per millimetre	Pound per inch	56.0
Pounds per inch	Kilograms per millimetre	0.0179
Metres per second, per second	g	0.102
g	Metres per second, per second	9.81

Volume and Flow

To convert:	to:	multiply by:
English gallons	Litres	4.55
Litres	English gallons	0.220
US gallons	Litres	3.79
Litres	US gallons	0.264
US gallons	English gallons	0.833
English gallons	US gallons	1.20
Cubic inches	Cubic centimetres	16.4
Cubic centimetres	Cubic inches	0.061
Cubic feet	Cubic metres	0.0283
Cubic metres	Cubic feet	35.3
Cubic feet per minute	Cubic metres per hour	1.70
Cubic metres per hour	Cubic feet per minute	0.589
Litres per hour	English gallons per hour	0.220
English gallons per hour	Litres per hour	4.55
Litres per hour	US Gallons per hour	0.264
US gallons per hour	Litres per hour	3.79
Litres per second	Cubic metres per hour	3.60
Cubic metres per hour	Litres per second	0.278
Litres per second	Cubic feet per minute	2.12
Cubic feet per minute	Litres per second	0.472
Litres per second	Cubic feet per hour	127
Cubic feet per hour	Litres per second	0.00786
Kilograms of air per second	Pounds of air per second	2.20
Pounds of air per second	Kilograms of air per second	0.0454
Pounds of air per second	Kilograms of air per hour	1630
Kilograms of air per hour	Pounds of air per second	0.000612
Pounds of air per second	Pounds of air per hour	3600
Pounds of air per hour	Pounds of air per second	0.000278
Kilograms of air per second	Pounds of air per hour	7940
Pounds of air per hour	Kilograms of air per second	0.000126

Power

To convert:	to:	multiply by:
Horsepower	Kilowatts	0.746
Kilowatts	Horsepower	1.34
PS (Pferdestarke)	Horsepower	0.986
Horsepower	PS (Pferdestarke)	1.01
PS (Pferdestarke)	Kilowatts	0.735
Kilowatts	PS (Pferdestarke)	1.36

Torque

To convert:	to:	multiply by:
Foot-pounds	Newton metres	1.36
Newton metres	Foot-pounds	0.738
Kilogram metres	Foot pounds	7.23
Foot pounds	Kilogram metres	0.138
Kilogram metres	Newton metres	9.81
Newton metres	Kilogram metres	0.102

Temperature

To convert:	to:	
Degrees Fahrenheit	Degrees Celsius	Subtract 32 and then multiply by 0.556
Degrees Celsius	Degrees Fahrenheit	Multiply by 1.8 and then add 32
Kelvin	Degrees Celsius	Subtract 273.15
Degrees Celsius	Kelvin	Add 273.15

Fuel Consumption

	Divided by:	Equals:
282.48	Miles per English gallon	Litres per 100 kilometres
282.48	Litres per 100 kilometres	Miles per English gallon
235.21	Miles per US gallon	Litres per 100 kilometres
235.21	Litres per 100 kilometres	Miles per US gallon

Velocity

To convert:	to:	multiply by:
Miles per hour	Kilometres per hour	1.61
Kilometres per hour	miles per hour	0.622
Metres per second	Kilometres per hour	3.60
Kilometres per hour	Metres per second	0.278
Metres per second	Miles per hour	2.24
Miles per hour	Metres per second	0.447

Pressure

To convert:	to:	multiply by:
Pounds per square inch	Bar	0.069
Bar	Pounds per square inch	14.50
Pounds per square inch	Kilopascals	6.90
Kilopascals	Pounds per square inch	0.145
Inches of mercury	Kilopascals	3.39
Kilopascals	Inches of mercury	0.295
Millimetres of mercury	Kilopascals	0.133
Kilopascals	Millimetres of mercury	7.50
Inches of water	Kilopascals	0.248
Kilopascals	Inches of water	4.03
Inches of water	Pounds per square inch	0.0361
Pounds per square inch	Inches of water	27.69
Centimetres of water	Kilopascals	0.0976
Kilopascals	Centimetres of water	10.2

Index

Note: the page numbers refer to the beginning of the sections that deal with the topic.